A Systematic Approach
to Learning
Robot Programming
with ROS

A Systematic Approach to Learning Robot Programming with ROS

Wyatt S. Newman

CRC Press
Taylor & Francis Group
Boca Raton   London   New York

CRC Press is an imprint of the
Taylor & Francis Group, an **informa** business

A CHAPMAN & HALL BOOK

CRC Press
Taylor & Francis Group
6000 Broken Sound Parkway NW, Suite 300
Boca Raton, FL 33487-2742

International Standard Book Number-13: 978-1-4987-7782-7 (Paperback)
International Standard Book Number-13: 978-1-138-09630-1 (Hardback)

Library of Congress Cataloging-in-Publication Data

Names: Newman, Wyatt S., author.
Title: A systematic approach to learning robot programming with ROS / Wyatt S. Newman.
Description: Boca Raton : CRC Press, [2017] | Includes index.
Identifiers: LCCN 2017013872| ISBN 9781138096301 (hardback : alk.paper) | ISBN 9781498777827 (pbk. : alk. paper)
Subjects: LCSH: Robots--Programming. | Robots--Control systems. | Operating systems (Computers)
Classification: LCC TJ211.45 .N49 2017 | DDC 629.8/925133--dc23
LC record available at https://lccn.loc.gov/2017013872

Visit the Taylor & Francis Web site at
http://www.taylorandfrancis.com

and the CRC Press Web site at
http://www.crcpress.com

To my wife, Peggy, with thanks for her constant encouragement and support in this effort.

Contents

Section II Simulation and Visualization in ROS

Chapter   3 ▪ Simulation in ROS   95

Chapter   4 ▪ Coordinate Transforms in ROS   153

List of Figures

Preface

The Robot Operating System (ROS) is becoming the *de facto* standard programming approach for modern robotics. The ROS wiki (http://www.ros.org/history/) claims:

> The ROS ecosystem now consists of tens of thousands of users worldwide, working in domains ranging from tabletop hobby projects to large industrial automation systems.

Why ROS? In 1956, Joseph Engelberger founded Unimation, Inc., the world's first robotics company[7]. Over the last half century, however, advances in robotics have been disappointing. Robotics research around the world has largely produced demonstrations and curiosities that died in the lab. New researchers in the field typically started from scratch, building new robots from the ground up, tackling the problems of actuator and sensor interfacing, building low-level servo controls, and typically running out of resources before achieving higher levels of robot competence. These custom machines and custom software were seldom re-used for follow-on work.

It was recognized that this pattern of building Towers of Babel was unproductive, and that the task of building more intelligent robots would require a sustained, collaborative effort that could build on foundations to reach higher levels of competence. In 1984, Vincent Hayward and Richard Paul introduced RCCL (Robot Control C Library) [15] as an approach to address this long-standing problem. Unfortunately, RCCL did not gain sufficient acceptance among robotics researchers. Both National Instruments [24] and Microsoft [39], [40] introduced products to attempt to bring standardization to robot programming. However, researchers found these approaches burdensome and/or expensive.

The origins of the ROS were initiated at the Stanford Artificial Intelligence Lab in 2007 [26]. This attempt to unify fragmented approaches to robotics was adopted by Google and supported via Willow Garage from 2008 to 2013 [12], and subsequently by Google's Open Source Robotics Foundation [10] (OSRF) from 2013 to present. The ROS approach followed the modern trend of open-source software and distributed collaboration. Further, it bridged to and leveraged parallel open-source efforts, including OpenCV [28], PointCloudLibrary [21], Open Dynamics Engine [8], Gazebo [19], and Eigen [20]. While ROS may be similar to RCCL in its openness and accessibility to researchers, Google's 7 years of ongoing support may be credited with ROS surviving a crucial incubation period.

ROS is now used throughout the world in academia, industry and research institutions. Developers have contributed thousands of packages, including solutions from some of the world's leading experts in targeted areas. New robot companies are offering ROS interfaces with their products, and established industrial robot companies are introducing ROS interfaces as well. With widespread adoption of ROS as the *de facto* standard approach to robot programming, there is new hope for accelerating the capabilities of robots. In the recent DARPA Robotics Challenge, most of the teams that qualified used ROS. Developers of new self-driving cars are exploiting ROS. New robot companies are springing up with leaps in capability, in part driven by ROS resources. Given the momentum and track record of ROS, it is clear that today's robotics engineer must be skilled in ROS programming.

What is ROS? The name "robot operating system" is arguably a misnomer. Defining ROS succinctly is difficult, since it encompasses myriad aspects, including style of programming (notably, relying on loosely-coupled, distributed nodes); interface definitions and paradigms for communications among nodes; interface definitions for incorporation of libraries and packages; a collection of tools for visualization, debugging, data logging and system diagnostics; a repository of shared source code; and bridges to multiple useful, independent open-source libraries. ROS is thus more of a way of life for robot programmers than simply an operating system. Definitions of ROS are drawn from the following sources.

From the ROS wiki (`http://wiki.ros.org/ROS/Introduction`):

> ROS is an open-source, meta-operating system for your robot. It provides the services you would expect from an operating system, including hardware abstraction, low-level device control, implementation of commonly-used functionality, message-passing between processes, and package management. It also provides tools and libraries for obtaining, building, writing, and running code across multiple computers.
>
> The primary goal of ROS is to support code reuse in robotics research and development. ROS is a distributed framework of processes (aka **nodes**) that enables executables to be individually designed and loosely coupled at runtime. These processes can be grouped into Packages, which can be easily shared and distributed. ROS also supports a federated system of code Repositories that enable collaboration to be distributed as well. This design, from the filesystem level to the community level, enables independent decisions about development and implementation, but all can be brought together with ROS infrastructure tools.

Online comment by Brian Gerkey:[1]

> I usually explain ROS in the following way:
>
> 1. plumbing: ROS provides publish-subscribe messaging infrastructure designed to support the quick and easy construction of distributed computing systems.
> 2. tools: ROS provides an extensive set of tools for configuring, starting, introspecting, debugging, visualizing, logging, testing, and stopping distributed computing systems.
> 3. capabilities: ROS provides a broad collection of libraries that implement useful robot functionality, with a focus on mobility, manipulation, and perception.
> 4. ecosystem: ROS is supported and improved by a large community, with a strong focus on integration and documentation. ros.org is a one-stop-shop for finding and learning about the thousands of ROS packages that are available from developers around the world.

From the text *ROS by Example* [13]:

> The primary goal of ROS (pronounced "Ross") is to provide a unified and open source programming framework for controlling robots in a variety of real world and simulated environments.

---

[1]`http://answers.ros.org/question/12230/what-is-ros-exactly-middleware-framework-operating-system/`

Regarding ROS plumbing (from *ROS by Example* [13]):

> The core entity in ROS is called a node. A node is generally a small program written in Python or C++ that executes some relatively simple task or process. Nodes can be started and stopped independently of one another and they communicate by passing messages. A node can publish messages on certain topics or provide services to other nodes.
>
> For example, a publisher node might report data from sensors attached to your robot's microcontroller. A message on the `/head_sonar` topic with a value of 0.5 would mean that the sensor is currently detecting an object 0.5 meters away. Any node that wants to know the reading from this sensor need only subscribe to the `/head_sonar` topic. To make use of these values, the subscriber node defines a callback function that gets executed whenever a new message arrives on the subscribed topic. How often this happens depends on the rate at which the publisher node updates its messages.
>
> A node can also define one or more services. A ROS service produces some behavior or sends back a reply when sent a request from another node. A simple example would be a service that turns an LED on or off. A more complex example would be a service that returns a navigation plan for a mobile robot when given a goal location and the starting pose of the robot.

Approach to this Presentation: ROS has many features, tools, style expectations and quirks. ROS has a steep learning curve, since there is much detail to master before one can be productive. The ROS wiki has links to documentation and a sequence of tutorials. However, these can be hard to follow for one new to ROS, as the definitions are scattered and the level of detail presented varies broadly from unexplained examples to explanations oriented toward sophisticated users. The intent of this book is to introduce the reader to essential components of ROS with detailed explanations of simple code examples along with the corresponding theory of operation. This introduction only scratches the surface, but it should get the reader started on building useful ROS nodes and bring the reader to a state where the tutorials become more readable.

ROS code can be written in either C++ or Python. This text uses C++ exclusively. For Python, the reader is referred to *Programming Robots with ROS: A Practical Introduction to the Robot Operating System* [34].

Accompanying code examples for this text assume use of Linux Ubuntu 14.04 and ROS Indigo. If one is using a PC running Windows or a Mac running OSX, an option is to install `VirtualBox` to set up a virtual Linux computer that can be run without disrupting the native operating system. Instructions for installing `VirtualBox`, `Ubuntu`, `ROS` and the accompanying code examples and tools are included at: `https://github.com/wsnewman/learning_ros` within the subdirectory `additional_documents`. (This directory also includes an introductory guide to using `git`.)

Getting one's computer configured to use ROS can be a challenge. It may be helpful to refer to *A Gentle Introduction to ROS* [27] to further illuminate and assist with ROS installation and obtain greater detail and behind-the-scenes explanations of ROS organization and communications. (Additional books on ROS are listed at: `http://wiki.ros.org/Books`.)

Installation of ROS is described on-line at: `http://wiki.ros.org/indigo/Installation/Ubuntu`.

When installing ROS, one has the option to name a ROS `workspace`. In this text, it will be assumed that this directory resides under the user's home directory and is named `ros_ws`. If you choose another name for your ROS workspace, substitute that name for `ros_ws` in all places referred to herein by `ros_ws`.

Code examples referred to herein may be found at: `https://github.com/wsnewman/learning_ros`. Some additional packages used with this code are contained in `https://github.com/wsnewman/learning_ros_external_packages`. Both repositories should be cloned into the user's ROS workspace in the subdirectory `~/ros_ws/src` to be able to compile the examples. To install these manually, after setting up your ROS environment, `cd` to `~/ros_ws/src` and enter:

```
git clone https://github.com/wsnewman/learning_ros.git
```

and

```
git clone https://github.com/wsnewman/learning_ros_external_packages.git
```

This will cause all of the example code referred to herein to appear within these two subdirectories.

Alternatively (and recommended), a repository `https://github.com/wsnewman/learning_ros_setup_scripts` contains scripts that automate installing ROS, installing the example code for for this text, and installing additional useful tools, and setting up your ROS workspace. Instructions for obtaining and running these scripts are provided on-line at that site. These instructions are applicable to either a native Ubuntu 14.04 installation, or a VirtualBox installation of Ubuntu 14.04. (Note, though, that VirtualBox can bog down when running computationally intensive or graphically intensive code. A native Linux installation and a compatible GPU processor are preferable.)

This text is not intended to be exhaustive. The interested student, researcher, automation engineer or robot enthusiast will find thousands of ROS packages to explore. Further, on-line tutorials describe details and extensions of the basics presented here. Rather, the intent of this text is to provide an introduction that will enable readers to understand the organization of ROS, ROS packages, ROS tools, incorporation of existing ROS packages into new applications, and developing new packages for robotics and automation. It is also the intent to facilitate continuing education by preparing the reader to better understand the existing on-line documentation.

This text is organized in six sections:

- ROS Foundations

- Simulation and Visualization with ROS

- Perceptual Processing in ROS

- Mobile Robots in ROS

- Robot Arms in ROS

- System Integration and Higher-Level Control

Each of these topics lies within broad fields with considerable specialization and evolving research developments. This text cannot attempt to teach the reader these fields. However, a robot system requires integration of elements spanning hardware integration, human/machine interfacing, control theory, kinematics and dynamics, manipulation planning, motion planning, image processing, scene interpretation, and an array of topics in artificial intelligence. The robot engineer must be a generalist and thus needs an understanding of at least rudimentary practice in each of these areas. An intent of the ROS ecosystem is to import existing packages that contribute in each of these areas and integrate them into a custom system without having to be an expert in every field. Thus understanding ROS interfacing and ROS approaches to each of these areas offers value to the system integrator who can ideally leverage the expertise contributed by robotics researchers, computer scientists, and software engineers around the world.

Acknowledgements

Part of the joy of working in academia is regular contact with bright and enthusiastic students. I want to thank my former advisees Chad Rockey and Eric Perko for dragging me into ROS back in 2010. I've since converted from a ROS skeptic to a ROS evangelist. I am grateful for many students since, who have helped me in my conversion and in learning new ROS tricks, including Tony Yanick, Bill Kulp, Ed Venator, Der-Lin Chow, Devin Schwab, Neal Aungst, Tom Shkurti, TJ Pech, and Luc Bettaieb.

My thanks, also, to Prof. Sethu Vijayakumar and the Scottish Informatics and Computer Science Alliance for their support while I initiated lectures in ROS and the groundwork for this text at University of Edinburgh. Thanks as well to associates from U. Edinburgh, including Chris Swetenham, Vladimir Ivan and Michael Camilleri, who contributed ROS programming as teammates in the DARPA Robotics Challenge, teaching me additional ROS tricks in the process.

I am grateful for the support of the Hung Hing Ying family, whose foundation helped make possible my stay at the Unversity of Hong Kong as a Hung Hing Ying Distinguised Visiting Professor while organizing and working with the HKU DARPA Robotics Challenge team. This experience was a valuable immersion in ROS. My thanks as well to Kei Okada and his students from U. Tokyo for their contributions to our HKU team, including valuable insights and techniques in ROS.

My thanks to Randi Cohen, senior acquistions editor at Taylor and Francis, for encouraging and guiding the publication of this book. I also want to thank the reviewers, Dr. Craig Carignan of NASA Goddard Space Flight Center and U. Maryland and Prof. Juan Rojas at Guangdong University of Technology, for their valuable feedback.

Finally, I am grateful for the support of my wife, Peggy Gallagher, and daughters Clea and Alair Newman, for constant encouragement and assistance.

The success of ROS would not have been possible without the support of Google and the Open Source Robotics Foundation, as well as all of the many online contributors who created valuable ROS packages and online tutorials, as well as responded with answers to many posted ROS questions.

Author

Wyatt S. Newman is a professor in the department of Electrical Engineering and Computer Science at Case Western Reserve University, where he has taught since 1988. His research is in the areas of mechatronics, robotics and computational intelligence, in which he has 12 patents and over 150 technical publications. He received an S.B. degree from Harvard College in Engineering Science, an S.M. degree in Mechanical Engineering from M.I.T. in thermal and fluid sciences, an M.S.E.E. degree from Columbia University in control theory and network theory, and a Ph.D. in Mechanical Engineering from M.I.T. in design and control. A former NSF Young Investigator in robotics, Prof. Newman has also held appointments as: a senior member of research staff, Philips Laboratories; visiting scientist at Philips Natuurkundig Laboratorium; visiting faculty at Sandia National Laboratories, Intelligent Systems and Robotics Center; NASA summer faculty fellow at NASA Glenn Research Center; visiting fellow in neuroscience at Princeton University; distinguished visiting fellow at Edinburgh University, School of Informatics; and the Hung Hing Ying Distinguished Visiting Professor at the University of Hong Kong. Prof. Newman led robotics teams competing in the 2007 DARPA Urban Challenge and the 2015 DARPA Robotics Challenge, and he continues to be interested in wide-ranging aspects and applications of robotics.

I

ROS Foundations

INTRODUCTION
This section introduces the foundations of the robot operating system (ROS), beginning with essential concepts, tools and constructs. Creation and use of ROS packages, nodes and tools are demonstrated. Basics of ROS communications covering publishers and subscribers, services and clients, action servers and action clients, and parameter servers are introduced. These elements introduce the essence of ROS as a precursor to discussions of robotic specifics.

## CHAPTER 1

Introduction to ROS: ROS tools and nodes

CONTENTS

INTRODUCTION

This introductory chapter will focus on the concept of **nodes** in ROS, starting with minimal examples. A few ROS tools will be introduced as well to help illuminate the behavior of ROS nodes. The simplest means of ROS communications—publish and subscribe—will be used here. Alternative communications means (**services** and **actionservers**) will be deferred until Chapter 2.

1.1 SOME ROS CONCEPTS

Communications among nodes are at the heart of ROS. A **node** is a ROS program that uses ROS's middleware for communications. A node can be launched independently of other nodes and in any order among launches of other nodes. Many nodes can run on the same computer, or nodes may be distributed across a network of computers. A node is useful only if it can communicate with other nodes and ultimately with sensors and actuators.

Communication among nodes uses the concepts of `messages`, `topics`, `roscore`, `publishers` and `subscribers`. (`Services` are also useful, and these are closely related to publishers and subscribers). All communications among nodes is serialized network communications. A `publisher` publishes a `message`, which is a packet of data that is interpretable using an associated key. Since each message is received as a stream of bits, it is necessary to consult the key (the message type description) to know how to parse the bits and reconstruct the corresponding data structure. A simple example of a message is `Float64`, which is defined in the package `std_msgs`, which comes with ROS. The message type helps the publisher pack a floating-point number into the defined stream of 64 bits, and it also helps the subscriber interpret how to unpack the bitstream as a representation of a floating-point number.

A more complex example of a message is a `twist`, which consists of multiple fields describing three-dimensional translational and rotational velocities. Some messages also accommodate optional extras, such as time stamps and message identifier codes.

When data is published by a publisher node, it is made available to any interested subscriber nodes. Subscriber nodes must be able to make connections to the published data. Often, the published data originates from different nodes. This can happen because these publishers have changed due to software evolution or because some publisher nodes are relevant in some contexts and other nodes in other contexts. For example, a publisher node responsible for commanding joint velocities may be a stiff position controller, but in other scenarios a compliant-motion controller may be needed. This hand-off can occur by changing the node publishing the velocity commands. This presents the problem that the subscriber does not know who is publishing its input. In fact, the need to know what node is publishing complicates construction of large systems. This problem is addressed by the concept of a `topic`.

A topic may be introduced and various publishers may take turns publishing to that topic. Thus a subscriber only needs to know the name of a topic and does not need to know what node or nodes publish to that topic. For example, the topic for commanding velocities may be `vel_cmd`, and the robot's low-level controller should subscribe to this named topic to receive velocity commands. Different publishers may be responsible for publishing velocity-command messages on this topic, whether these are nodes under experimentation or trusted nodes that are swapped in to address specific task needs.

Although creating the abstraction of a `topic` helps some, a publisher and a subscriber both need to know how to communicate via a topic. This is accomplished through communications middleware in ROS via the provided executable node `roscore`. The `roscore` node is responsible for coordinating communications, like an operator. Although there can be many ROS nodes distributed across multiple networked computers, there can be only one instance of `roscore` running, and the machine on which `roscore` runs establishes the `master` computer of the system.

A publisher node initiates a topic by informing `roscore` of the topic (and the corresponding message type). This is called `advertising` the topic. To accomplish this, the publisher instantiates an object of the class `ros::Publisher`. This class definition is part of the ROS distribution, and using publisher objects allows the designer to avoid having to write communications code. After instantiating a publisher object, the user code invokes the member function `advertise` and specifies the message type and declares the desired topic name. At this point, the user code can start sending messages to the named topic using the publisher member function `publish`, which takes as an argument the message to be published.

Since a publisher node communicates with `roscore`, `roscore` must be running before any ROS node is launched. To run `roscore`, open a terminal in Linux and enter `roscore`. The response to this command will be a confirmation "started core services." It will also print the ROS_MASTER_URI, which is useful for informing nodes running on non-master computers how to reach roscore. The terminal running roscore will be dedicated to roscore, and it will be unavailable for other tasks. The roscore node should continue running as long as the robot is actively controlled (or as long as desired to access the robot's sensors).

After roscore has been launched, a publisher node may be launched. The publisher node will advertise its topic and may start sending messages (at any rate convenient to the node, though at a frequency limited by system capacity). Publishing messages at 1 kHz rate is normal for low-level controls and sensor data.

Introducing a sensor to a ROS system requires specialized code (and possibly specialized hardware) that can communicate with the sensor. For example, a LIDAR sensor may require RS488 communications, accept commands in a specific format, and start streaming data in a predefined format. A dedicated microcontroller (or a node within the main computer) must communicate with the LIDAR, receive the data, then publish the data with a ROS message type on a ROS topic. Such specialized nodes convert the specifics of individual sensors into the generic communications format of ROS.

When a publisher begins publishing, it is not necessary that any nodes are listening to the messages. Alternatively, there may be many subscribers to the same topic. The publisher does not need to be aware of whether it has any subscribers nor how many subscribers there may be. This is handled by the ROS middleware. A subscriber may be receiving messages from a publisher, and the publisher node may be halted and possibly replaced with an alternative publisher of the same topic, and the subscriber will resume receiving messages with no need to restart the subscriber.

A ROS subscriber also communicates with roscore. To do so, it uses an object of class ros::Subscriber. This class has a member function called subscribe that requires an argument of the named topic. The programmer must be aware that a topic of interest exists and know the name of the topic. Additionally, the subscriber function requires the name of a callback function. This provides the necessary hook to the ROS middleware, such that the callback function will start receiving messages. The callback function suspends until a new message has been published, and the designer may include code to operate on the newly received message.

Subscriber functions can be launched before the corresponding publisher functions. ROS will allow the subscriber to register its desire to receive messages from a named topic, even though that topic does not exist. At some point, if or when a publisher informs roscore that it will publish to that named topic, the subscriber's request will be honored, and the subscriber will start receiving the published messages.

A node can be both a subscriber and a publisher. For example, a control node would need to receive sensor signals as a subscriber and send out control commands as a publisher. This only requires instantiating both a subscriber object and a publisher object within the node. It is also useful to pipeline messages for sequential processing. For example, a low-level image processing routine (*e.g.* for edge finding) could subscribe to raw camera data and publish low-level processed images. A higher-level node might subscribe to the edge-processed images, look for specific shapes within those images, and publish its results (*e.g.* identified shapes) for further use by still higher-level processes. A sequence of nodes performing successive levels of processing can be modified incrementally by replacing one node at a time. To replace one such link in a chain, the new node needs only to continue to use the same input and output topic names and message types. Although the implementation of algorithms within the modified node may be dramatically different, the other nodes within the chain will be unaffected.

The flexibility to launch publisher and subscriber nodes in any order eases system design. Additionally, individual nodes may be halted at any time and additional nodes may be hot-swapped into the running system. This can be exploited, for example, to launch some specialized code when it is needed and then halt the (potentially expensive) computations when no longer needed. Additionally, diagnostic nodes (*e.g.* interpreting and reporting on published messages) may be run and terminated *ad hoc*. This can be useful to examine the output of selected sensors to confirm proper functioning.

It should be appreciated that the flexibility of launching and halting publishers and subscribers at any time within a running system can also be a liability. For time-critical code—particularly control code that depends on sensor values to generate commands to actuators—an interruption of the control code or of the critical sensor publishers could result in physical damage to the robot or its surroundings. It is up to the programmer to make sure that time-critical nodes remain viable. Disruptions of critical nodes should be tested and responded to appropriately (*e.g.* with halting all actuators).

From the system architecture point of view, ROS helps implement a desired software architecture and supports teamwork in building large systems. Starting from a predetermined software architecture, one can construct a skeleton of a large system constructed as a collection of nodes. Initially, each of the nodes might be dummy nodes, capable of sending and receiving messages via predefined topics (the software interfaces). Each module in the architecture could then be upgraded incrementally by swapping out an older (or dummy) node for a newer replacement and no changes would be required elsewhere throughout the system. This supports distributed software development and incremental testing, which are essential for building large, complex systems.

1.2 WRITING ROS NODES

In this section, design of a minimal publisher node and a minimal subscriber node will be detailed. The concept of a ROS `package` is introduced, along with instructions on how to compile and link the code via the associated files `package.xml` and `CMakeLists`. Several ROS tools and commands are introduced, including `rosrun`, `rostopic`, `rosnode`, and `rqt_graph`|. Specific C++ code examples for a publisher and a subscriber are detailed, and results of running the compiled nodes are shown.

The code examples used in this section are contained in the accompanying code repository, within the directory package `minimal_nodes` under `learning_ros/Part_1/minimal_nodes`. This introduction begins with instructions on how the example code was created. In following these instructions it is important to avoid ambiguity from naming conflicts. In this section, the names used will be altered (as necessary) from the provided code examples for the purpose of illustrating to how the example code was created. In subsequent sections, the example code may be used verbatim.

Before creating new ROS code, one must establish a directory (a ROS workspace) where ROS code will reside. One creates this directory somewhere convenient in the system (for example, directly under `home`). A subdirectory called `src` must exist, and this is where source code (packages) will reside. The operating system must be informed of the location of your ROS workspace (typically done automatically through edits to the start-up script `.bashrc`). Setting up a ROS workspace (and automating defining ROS paths) needs to be done only once. The process is described at: `http://wiki.ros.org/ROS/Tutorials/InstallingandConfiguringROSEnvironment`. It is important that the necessary environment variables be set in Linux, or the OS will not know where to find your code for compilation and execution. Formerly (ROS Fuerte and older), ROS used its own build system called `rosbuild` that was replaced by the `path{catkin}indexcatkin` build system, which is faster, but can be more complex. A useful simplification is `catkin_simple`, which reduces the detail required of the programmer to specify how to build a project.

For the following, it is assumed that you already have ROS Indigo installed, that you have a ROS workspace defined (called `ros_ws` in the examples to follow), that it has a `src` subdirectory, and that the OS has been informed of the path to this workspace (via environment variables). These requirements will satisfied if your setup uses the `learning_ros_setup_scripts` recommended in the preface. We proceed with creating new code within this workspace.

1.2.1 Creating ROS packages

The first thing to do when starting to design new ROS code is to create a `package`. A `package` is a ROS concept of bundling multiple, necessary components together to ease building ROS code and coupling it with other ROS code. Packages should be logical groups of code, *e.g.* separate packages for low-level joint control, for high-level planning, for navigation, for mapping, for vision, etc. Although these packages normally would be developed separately, nodes from separate packages would ultimately be run together simultaneously on a robot (or on a network of computers collectively controlling the robot).

One creates a new `package` using the `catkin_create_pkg` command (or the alternative `cs_create_pkg`, which will be the preferred approach in this presentation). The `catkin_create_pkg` command is part of the ROS installation. For a given package, package-creation needs to be done only once. One can go back to this package and add more code incrementally (without needing to create the package again). However, the code added to a package should logically belong to that package (*e.g.* avoid putting low-level joint-control code in the same package as mapping code).

New packages should reside under the `src` directory of your catkin workspace (*e.g.* `ros_ws/src`). As a specific example, consider creation of a new package called `my_minimal_nodes`, which will contain source code in C++ and depend on the basic, pre-defined message types contained in `std_msgs`. To do so, open a terminal and navigate (cd) to the `ros_ws` directory. A shortcut for this is `roscd`, which will bring you to `~/ros_ws`. From here, move to the subdirectory `/src` and enter the following command:

```
catkin_create_pkg    my_minimal_nodes    roscpp    std_msgs
```

The effect of this command is to create and populate a new directory: `~/ros_ws/src/my_minimal_nodes`.

Every package name in ROS must be unique. By convention, package names should follow common C variable naming conventions: lower case, start with a letter, use underscore separators, *e.g.* `grasp_planner`. (See `http://wiki.ros.org/ROS/Patterns/Conventions`.) Every package used within your system must be uniquely named. As noted at `http://wiki.ros.org/ROS/Patterns/Conventions`, you can check whether a name is already taken via `http://www.ros.org/browse/list.php`. For the present illustration, the name `my_minimal_nodes` was chosen so as not to conflict with the package `minimal_nodes`, which resides in the example code repository (under `~/ros_ws/src/learning_ros/Part_1/minimal_nodes`).

Moving to the newly created package directory, `~/ros_ws/src/my_minimal_nodes`, we see that it is already populated with `package.xml`, `CMakeLists` and the subdirectories `src` and `include`. The `catkin_create_pkg` command just created a new `package` by the name of `my_minimal_nodes`, which will reside in a directory of this name.

As we create new code, we will depend on some ROS tools and definitions. Two dependencies were listed during the `catkin_create_pkg` command: `roscpp` and `std_msgs`. The `roscpp` dependency establishes that we will be using a C++ compiler to create our ROS code, and we will need C++ compatible interfaces (such as the classes `ros::Publisher` and `ros::Subscriber`, referred to earlier). The `std_msgs` dependency says that we will need to rely on some datatype definitions (standard messages) that have been predefined in ROS. (an example is `std_msgs::Float64`).

The package.xml file: A ROS package is recognized by the build system by virtue of the fact that it has a `package.xml` file. A compatible `package.xml` file has a specific structure that names the package and lists its dependencies. For the new package `my_minimal_nodes`, a `package.xml` file was auto-generated, and its contents are shown in Listing 1.1.

Listing 1.1: Contents of `package.xml` for minimal nodes package

```xml
1   <?xml version="1.0"?>
2   <package>
3     <name>my_minimal_nodes</name>
4     <version>0.0.0</version>
5     <description>The my_minimal_nodes package</description>
6
7     <!-- One maintainer tag required, multiple allowed, one person per tag -->
8     <!-- Example:  -->
9     <!-- <maintainer email="jane.doe@example.com">Jane Doe</maintainer> -->
10    <maintainer email="wyatt@todo.todo">wyatt</maintainer>
11
12
13    <!-- One license tag required, multiple allowed, one license per tag -->
14    <!-- Commonly used license strings: -->
15    <!--   BSD, MIT, Boost Software License, GPLv2, GPLv3, LGPLv2.1, LGPLv3 -->
16    <license>TODO</license>
17
18
19    <!-- Url tags are optional, but multiple are allowed, one per tag -->
20    <!-- Optional attribute type can be: website, bugtracker, or repository -->
21    <!-- Example: -->
22    <!-- <url type="website">http://wiki.ros.org/my_miminal_nodes</url> -->
23
24
25    <!-- Author tags are optional, multiple are allowed, one per tag -->
26    <!-- Authors do not have to be maintainers, but could be -->
27    <!-- Example: -->
28    <!-- <author email="jane.doe@example.com">Jane Doe</author> -->
29
30
31    <!-- The *_depend tags are used to specify dependencies -->
32    <!-- Dependencies can be catkin packages or system dependencies -->
33    <!-- Examples: -->
34    <!-- Use build_depend for packages you need at compile time: -->
35    <!--   <build_depend>message_generation</build_depend> -->
36    <!-- Use buildtool_depend for build tool packages: -->
37    <!--   <buildtool_depend>catkin</buildtool_depend> -->
38    <!-- Use run_depend for packages you need at runtime: -->
39    <!--   <run_depend>message_runtime</run_depend> -->
40    <!-- Use test_depend for packages you need only for testing: -->
41    <!--   <test_depend>gtest</test_depend> -->
42    <buildtool_depend>catkin</buildtool_depend>
43    <build_depend>roscpp</build_depend>
44    <build_depend>std_msgs</build_depend>
45    <run_depend>roscpp</run_depend>
46    <run_depend>std_msgs</run_depend>
47
48
49    <!-- The export tag contains other, unspecified, tags -->
50    <export>
51      <!-- Other tools can request additional information be placed here -->
52
53    </export>
54  </package>
```

The `package.xml` file is merely ASCII text using XML formatting, and thus you can open it with any editor (`gedit` will do). In Listing 1.1, most of the lines are merely comments, such as `<!-- Example:  -->`, where each comment is delimited by an opening of   `<!--`

and closing of `-->`. The comments instruct you how to edit this file appropriately. It is recommended that you edit the values to enter your name and e-mail address as author of the code, particularly if you intend to share your contribution publicly.

The line `<name>my_minimal_nodes</name>` corresponds to the name of the new package. It is important that this name correspond to the name of your package. You cannot merely create a new directory (with a new name) and copy over the contents of another package. Because of the mismatch between your directory name and the package name in the `package.xml` file, ROS will be confused.

Within the `package.xml` file, the lines:

```
<build_depend>roscpp</build_depend>
<build_depend>std_msgs</build_depend>
<run_depend>roscpp</run_depend>
<run_depend>std_msgs</run_depend>
```

explicitly declare dependency on the package `roscpp` and on the package `std_msgs`, both of which were explicitly listed as dependencies upon creation of this package. Eventually, we will want to bring in large bodies of third-party code (other packages). In order to integrate with these packages (*e.g.* utilize objects and datatypes defined in these packages), we will want to add them to the `package.xml` file. This can be done by editing our package's `package.xml` file and adding `build_depend` and `run_depend` lines naming the new packages to be utilized, emulating the existing lines that declare dependence on `roscpp` and `std_msgs`.

The `src` directory is where we will put our user-written C++ code. We will write illustrative nodes `minimal_publisher.cpp` and `minimal_subscriber.cpp` as examples. It will be necessary to edit the `CMakeLists.txt` file to inform the compiler that we have new nodes to be compiled, which will be described further later.

1.2.2 Writing a minimal ROS publisher

In a terminal window, move to the `src` directory within the `my_minimal_nodes` package that has been created. Open an editor, create a file called `minimal_publisher.cpp` and enter the following code. (Note: if you attempt to copy/paste from an electronic version of this text, you likely will get copying errors, including undesired newline symbols, which will confuse the C++ compiler. Rather, refer to the corresponding example code on the associated github repository at `https://github.com/wsnewman/learning_ros`, under package `~/ros_ws/src/learning_ros/Part_1/minimal_nodes`.)

Listing 1.2: Minimal Publisher

```
1   #include <ros/ros.h>
2   #include <std_msgs/Float64.h>
3
4   int main(int argc, char **argv) {
5       ros::init(argc, argv, "minimal_publisher"); // name of this node will be "↵
            minimal_publisher"
6       ros::NodeHandle n; // two lines to create a publisher object that can talk to ROS
7       ros::Publisher my_publisher_object = n.advertise<std_msgs::Float64>("topic1", 1);
8       //"topic1" is the name of the topic to which we will publish
9       // the "1" argument says to use a buffer size of 1; could make larger, if expect ↵
            network backups
10
11      std_msgs::Float64 input_float; //create a variable of type "Float64",
12      // as defined in: /opt/ros/indigo/share/std_msgs
13      // any message published on a ROS topic must have a pre-defined format,
14      // so subscribers know how to interpret the serialized data transmission
15
16      input_float.data = 0.0;
```

```
17
18       // do work here in infinite loop (desired for this example), but terminate if ←
             detect ROS has faulted
19       while (ros::ok())
20       {
21           // this loop has no sleep timer, and thus it will consume excessive CPU time
22           // expect one core to be 100% dedicated (wastefully) to this small task
23           input_float.data = input_float.data + 0.001; //increment by 0.001 each ←
                 iteration
24           my_publisher_object.publish(input_float); // publish the value--of type ←
                 Float64--
25           //to the topic "topic1"
26       }
27   }
```

The above code is dissected here. On line 1,

```
#include <ros/ros.h>
```

is needed to bring in the header files for the core ROS libraries. This should be the first line of any ROS source code written in C++.

Line 2,

```
#include <std_msgs/Float64.h>
```

brings in a header file that describes objects of type: `std_msgs::Float64`, which is a message type we will use in this example code.

As you incorporate use of more ROS message types or ROS libraries, you will need to include their associated header files in your code, just as we have done with **std_msgs**.

Line 4,

```
int main ( int argc , char ** argv )
```

declares a **main** function. For all ROS nodes in C++, there must be one and only one **main()** function per node. Our **minimal_publisher.cpp** file has **main()** declared in the standard "C" fashion with generic **argc**, **argv** arguments. This gives the node the opportunity to use command-line options, which are used by ROS functions (and thus these arguments should always be included in **main()**).

The code lines 5 through 7 all refer to functions or objects defined in the core ROS library.

Line 5:

```
ros::init(argc, argv, "minimal_publisher");
```

is needed in every ROS node. The argument **minimal_publisher** will be the name that the new node will use to register itself with the ROS system upon start-up. (This name can be overridden, or **remapped**, upon launch, but this detail is deferred for now.) The node name is required, and every node in the system must have a unique name. ROS tools take advantage of the node names, *e.g.* to monitor which nodes are active and which nodes are publishing or subscribing to which topics.

Line 6 instantiates a ROS **NodeHandle** object with the declaration:

```
ros::NodeHandle n;
```

A **NodeHandle** is required for establishing network communications among nodes. The

`NodeHandle` name `n` is arbitrary. It will be used infrequently (typically, for initializing communications objects). This line can simply be included in every node's source code (and no harm is done in the rare instances in which it is not needed).

Line 7:

```
ros::Publisher my_publisher_object = n.advertise<std_msgs::Float64>("topic1", 1);
```

instantiates an object to be called `my_publisher_object` (the name is the programmer's choice). In instantiating this object, the ROS system is informed that the current node (here called `minimal_publisher`) intends to publish messages of type `std_msgs::Float64` on a topic named `topic1`. In practice, one should choose topic names that are helpful and descriptive of the type of information carried via that topic.

On line 11:

```
std_msgs::Float64 input_float;
```

the program creates an object of type `std_msgs:Float64` and calls it `input_float`. One must consult the message definition in `std_msgs` to understand how to use this object. The object is defined as having a member called `data`. Details of this message type can be found by looking in the corresponding directory with: `roscdstd_msgs`. The subdirectory `msg` contains various files defining the structures of numerous standard messages, including `Float64.msg`. Alternatively, one can examine the details of any message type with the command `rosmsg show ...`, *e.g.* from a terminal, entering the command:

```
rosmsg show std_msgs/Float64
```

will display the fields of this message, which results in the response:

```
float64 data
```

In this case, there is only a single field, named `data`, which holds a value of type `float64` (a ROS primitive).

On line 16,

```
input_float.data = 0.0;
```

the program initializes the `data` member of `input_float` to the value 0.0. An infinite loop is then entered, that will self terminate upon detecting that the ROS system has terminated, which is accomplished using the function `ros::ok()` in line 19:

```
while (ros::ok())
```

This approach can be convenient for shutting down a collection of nodes by merely halting the ROS system (*i.e.*, by killing `roscore`).

Inside the `while` loop, the value of `input_float.data` is incremented by 0.001 per iteration. This value is then published (line 24) using:

```
my_publisher_object.publish(input_float);
```

It was previously established (upon instantiation of the object `my_publisher_object` from the class `ros::Publisher`) that the object `my_publisher_object` would publish messages of type `std_msg::Float64` to the topic called `topic1`. The publisher object, `my_publisher_object` has a member function `publish` to invoke publications. The publisher expects an argument of compatible type. Since the object `input_float` is of type `std_msgs::Float64`, and since the publisher object was instantiated to expect arguments of this type, the `publish` call is consistent.

The example ROS node has only 14 active lines of code. Much of this code is ROS-specific and may seem cryptic. However, most of the lines are common boilerplate, and becoming familiar with these common lines will make other ROS code easier to read.

1.2.3 Compiling ROS nodes

ROS nodes are compiled by running `catkin_make`. This command must be executed from a specific directory. In a terminal, navigate to your ROS workspace (`~/ros_ws`). Then enter `catkin_make`.

This will compile all packages in your workspace. Compiling large collections of code can be time consuming, but compilation will be faster on subsequent edit, compile and test iterations. Although compilation is sometimes slow, the compiled code can be very efficient. Particularly for CPU-intensive operations (*e.g.* for point-cloud processing, image processing or intensive planning computations), compiled C++ code typically runs faster than Python code.

After building a catkin package, the executables will reside in a folder in `ros_ws/devel/lib` named according to the source package.

Before we can compile, however, we have to inform `catkin_make` of the existence of our new source code, `minimal_publisher.cpp`. To do so, edit the file `CMakeLists.txt`, which was created for us in the package `my_minimal_nodes` when we ran `catkin_create_pkg`. This file is quite long (187 lines for our example), but it consists almost entirely of comments.

The comments describe how to modify `CmakeLists.txt` for numerous variations. For the present, we only need to make sure we have our package dependencies declared, inform the compiler to compile our new source code, and link our compiled code with any necessary libraries.

`catkin_package_create` already fills in the fields:

Listing 1.3: Snippet from `CMakeLists.txt`

```
find_package(catkin REQUIRED COMPONENTS
  roscpp
  std_msgs
)

include_directories(
  ${catkin_INCLUDE_DIRS}
)
```

However, we need to make two modifications, as follows:

Listing 1.4: Adding the new node and linking it with libraries

```
## Declare a cpp executable
add_executable(my_minimal_publisher src/minimal_publisher.cpp)

## Specify libraries to link a library or executable target against
target_link_libraries(my_minimal_publisher ${catkin_LIBRARIES} )
```

These modifications inform the compiler of our new source code, as well as which libraries with which to link.

In the above, the first argument to `add_executable` is a chosen name for the executable to be created, here chosen to be `my_minimal_publisher`, which happens to be the same root name as the source code.

The second argument is where to find the source code relative to the package directory. The source code is in the `src` subdirectory and it is called `minimal_publisher.cpp`. (It is typical for the source code to reside in the `src` sub-directory of a package.)

There are a few idiosyncrasies regarding the node name. In general, one cannot run two nodes with the same name. The ROS system will complain, kill the currently running node, and start the new (identically named) node. ROS does allow for different packages to re-use node names, although only one of these at a time may be run. Although (executable) node names are allowed to be duplicated across packages, the `catkin_make` build system gets confused by such duplication (though the build can be forced by compiling packages one at a time). For simplicity, it is best to avoid replicating node names.

Having edited the `CMakeLists` file, we can compile our new code. To do so, from a terminal, navigate to the `ros_ws` directory and enter:

```
catkin_make
```

This will invoke the C++ compiler to build all packages, including our new `my_minimal_nodes` package. If the compiler output complains, find and fix your bugs.

Assuming compilation success, if you look in the directory `ros_ws/devel/lib/my_minimal_nodes`, you will see a new, executable file there named `my_minimal_publisher`. This is the name that was chosen for the output file (executable node) with the addition of `add_executable(my_minimal_publisher src/minimal_publisher.cpp)` in `CMakeLists`.

1.2.4   Running ROS nodes

As noted in Section 1.1, there must be one and only one instance of `roscore` running before any nodes can be started. In a terminal, enter:

```
roscore
```

This should respond with a page of text concluding with "started core services." You can then shrink this window and leave it alone. ROS nodes can be started and stopped at random without needing to start a new `roscore`. (If you kill `roscore` or the window running `roscore`, however, all nodes will stop running.)

Next, start the new publisher node by entering (from a new terminal):

```
rosrun    my_minimal_nodes    my_minimal_publisher
```

The arguments to the command `rosrun` are the package name (`my_minimal_nodes`) and the executable name (`my_minimal_publisher`).

The `rosrun` command can seem confusing at times due to re-use of names. For example, if we wanted to make a LIDAR publisher node, we might have a package called `lidar_publisher`, a source file called `lidar_publisher.cpp`, an executable called `lidar_publisher`, and a node name (declared within the source code) of `lidar_publisher`. To run this node, we would type:

```
rosrun     lidar_publisher      lidar_publisher
```

This may seem redundant, but it still follows the format:

```
rosrun     package_name     executable_name
```

Once the command has been entered, the ROS system will recognize a new node by the name of `lidar_publisher`. This name re-use may seem to lead to confusion, but in many instances, there is no need to invent new names for the package, the source code, the executable name and the node name. In fact, this can help simplify recognizing named entities–as long as the context is clear (package, source code, executable, node name).

1.2.5 Examining running minimal publisher node

After entering: `rosrun my_minimal_nodes my_minimal_publisher`, the result may seem disappointing. The window that launched this node seems to hang and provides no feedback to the user. Still, `minimal_publisher` is running. To see this, we can invoke some ROS tools.

Open a new terminal and enter: `rostopic`. You will get the following response:

```
rostopic is a command-line tool for printing information about ROS Topics.

Commands:
rostopic bw display bandwidth used by topic
rostopic echo print messages to screen
rostopic find find topics by type
rostopic hz display publishing rate of topic
rostopic info print information about active topic
rostopic list list active topics
rostopic pub publish data to topic
rostopic type print topic type

Type rostopic <command> -h for more detailed usage, e.g. 'rostopic echo -h'
```

This shows that the command `rostopic` has eight options. If we type:

```
rostopic list
```

the result is:

```
/rosout
/rosout_agg
/topic1
```

We see that there are three active topics—two that ROS created on its own and the topic created by our publisher, `topic1`.

Entering:

```
rostopic   hz   topic1
```

results in the following output:

```
average rate: 38299.882
min: 0.000s max: 0.021s std dev: 0.00015s window: 50000
average rate: 38104.090
min: 0.000s max: 0.024s std dev: 0.00016s window: 50000
```

This output shows that our minimal publisher (on this particular computer) is publishing its data at roughly 38 kHz (with some jitter). Viewing the system monitor would show that one CPU core is fully saturated just running the minimal publisher. This is because the `while` loop within our ROS node has no pauses. It is publishing as fast as it can.

Entering:

```
rostopic   bw   topic1
```

yields the following output:

```
average: 833.24KB/s
mean: 0.01KB min: 0.01KB max: 0.01KB window: 100
average: 1.21MB/s
mean: 0.00MB min: 0.00MB max: 0.00MB window: 100
average: 746.32KB/s
mean: 0.01KB min: 0.01KB max: 0.01KB window: 100
```

This display shows how much of our available communications bandwidth is consumed by our minimal publisher (nominally 1 MB/s). This rostopic option can be useful for identifying nodes that are over-consuming communications resources.

Entering:

```
rostopic info topic1
```

yields:

```
Type: std_msgs/Float64

Publishers:
 * /minimal_publisher (http://Wall-E:56763/)

Subscribers: None
```

This tells us that `topic1` involves messages of type `std_msgs/Float64`. At present, there is a single publisher to this topic (which is the norm), and that publisher has a node name of `minimal_publisher`. As noted above, this is the name we assigned to the node within the source code on line 5:

```
ros::init(argc,argv,"minimal_publisher");
```

Entering:

```
rostopic echo topic1
```

causes `rostopic` to try to print out everything published on `topic1`. A sample of the output is:

```
data: 860619.606909
---
data: 860619.608909
---
data: 860619.609909
---
data: 860619.612909
---
```

In this case, the display has no hope of keeping up with the publishing rate, and most of the messages are dropped between lines of display. If the echo could keep up, we would expect to see values that increase by increments of 0.001, which is the increment used in the `while` loop of our source code.

The `rostopic` command tells us much about the status of our running system, even though there are no nodes receiving the messages sent by our minimal publisher. Additional handy ROS commands are summarized in the "ROS cheat sheet" (see `http://www.ros.org/news/2015/05/ros-cheatsheet-updated-for-indigo-igloo.html`).

For example, entering:

```
rosnode list
```

results in the following output:

```
/minimal_publisher
/rosout
```

We see that there are two nodes running: `rosout` (a generic process used for nodes to display text to a terminal, launched by default by `roscore`) and `minimal_publisher` (our node).

Although `minimal_publisher` does not take advantage of the capability of displaying output to its terminal, the link is nonetheless available through the topic `rosout`, which would get processed by the display node `rosout`. Using `rosout` can be helpful, since one's code does not get slowed down by output (*e.g.* `cout`) operations. Rather, messages get sent rapidly by publishing the output to the `rosout` topic, and a separate node (`rosout`) is responsible for user display. This can be important, *e.g.* in time-critical code where some monitoring is desired, but not at the expense of slowing the time-critical node.

1.2.6 Scheduling node timing

We have seen that our example publisher is abusive of both CPU capacity and communications bandwidth. In fact, it would be unusual for a node within a robotic system to require updates at 30 kHz. A more reasonable update rate for even time-critical, low-level nodes is 1 kHz. In the present example, we will slow our publisher to 1 Hz using a ROS timer.

A modified version of source code for `minimal_publisher.cpp`, called `sleepy_minimal_publisher.cpp`, is shown below:

Listing 1.5: Minimal Publisher with Timing: `sleepy_minimal_publisher.cpp`

```cpp
#include <ros/ros.h>
#include <std_msgs/Float64.h>

int main(int argc, char **argv) {
    ros::init(argc, argv, "minimal_publisher2"); // name of this node will be "↩
        minimal_publisher2"
    ros::NodeHandle n; // two lines to create a publisher object that can talk to ROS
    ros::Publisher my_publisher_object = n.advertise<std_msgs::Float64>("topic1", 1);
    //"topic1" is the name of the topic to which we will publish
    // the "1" argument says to use a buffer size of 1; could make larger, if expect ↩
        network backups

    std_msgs::Float64 input_float; //create a variable of type "Float64",
    // as defined in: /opt/ros/indigo/share/std_msgs
    // any message published on a ROS topic must have a pre-defined format,
    // so subscribers know how to interpret the serialized data transmission

    ros::Rate naptime(1.0); //create a ros object from the ros Rate class;
    //set the sleep timer for 1Hz repetition rate (arg is in units of Hz)
```

```
18
19      input_float.data = 0.0;
20
21      // do work here in infinite loop (desired for this example), but terminate if ↩
            detect ROS has faulted
22      while (ros::ok())
23      {
24          // this loop has no sleep timer, and thus it will consume excessive CPU time
25          // expect one core to be 100% dedicated (wastefully) to this small task
26          input_float.data = input_float.data + 0.001; //increment by 0.001 each ↩
                iteration
27          my_publisher_object.publish(input_float); // publish the value--of type ↩
                Float64--
28          //to the topic "topic1"
29      //the next line will cause the loop to sleep for the balance of the desired period
30          // to achieve the specified loop frequency
31      naptime.sleep();
32      }
33  }
```

There are only two new lines in the above program: line 16

```
ros::Rate naptime(1); //set the sleep timer for 1Hz repetition rate
```

and line 31:

```
naptime.sleep();
```

The ROS class **Rate** is invoked to create a **Rate** object that was named "naptime". In doing so, **naptime** is initialized with the value "1", which is a specification of the desired frequency (1 Hz). After creating this object, it is used within the **while** loop, invoking the member function **sleep()**. This causes the node to suspend (thus ceasing to consume CPU time) until the balance of the desired period (1 second) has expired.

After re-compiling the modified code (with **catkin_make**), we can run it (with **rosrun**) and examine its behavior by entering (from a new terminal):

```
rostopic hz topic1
```

which produces the display below:

```
average rate: 1.000
min: 1.000s max: 1.000s std dev: 0.00000s window: 2
average rate: 1.000
min: 1.000s max: 1.000s std dev: 0.00006s window: 3
average rate: 1.000
min: 1.000s max: 1.000s std dev: 0.00005s window: 4
average rate: 1.000
```

This output indicates that **topic1** is being updated at 1 Hz with excellent precision and very low jitter. Further, an inspection of the system monitor shows negligible CPU time consumed by our modified publisher node.

If we enter

```
rostopic echo topic1
```

from a terminal, example output looks like the following:

```
data: 0.153
---
```

```
data: 0.154
---
data: 0.155
---
```

Each message sent by the publisher is displayed by `rostopic echo`, as evidenced by the increments of 0.001 between messages. This display is updated once per second, since that is the rate at which new data is now published.

1.2.7 Writing a minimal ROS subscriber

The complement to the publisher is a subscriber (a listener node). We will create this node in the same package, `my_minimal_nodes`. The source code will go in the subdirectory `src`.

Open an editor and create the file `minimal_subscriber.cpp` in the directory `~/ros_ws/src/my_minimal_nodes/src`. Enter the following code (which may be found in `~/ros_ws/src/learning_ros/Part_1/minimal_nodes/src/minimal_subscriber.cpp`):

Listing 1.6: Minimal Subscriber

```cpp
#include<ros/ros.h>
#include<std_msgs/Float64.h>
void myCallback(const std_msgs::Float64& message_holder)
{
   // the real work is done in this callback function
   // it wakes up every time a new message is published on "topic1"
   // Since this function is prompted by a message event,
   //it does not consume CPU time polling for new data
   // the ROS_INFO() function is like a printf() function, except
   // it publishes its output to the default rosout topic, which prevents
   // slowing down this function for display calls, and it makes the
   // data available for viewing and logging purposes
   ROS_INFO("received value is: %f",message_holder.data);
   //really could do something interesting here with the received data...but all we do ↵
       is print it

}

int main(int argc, char **argv)
{
   ros::init(argc,argv,"minimal_subscriber"); //name this node
   // when this compiled code is run, ROS will recognize it as a node called "↵
       minimal_subscriber"
   ros::NodeHandle n; // need this to establish communications with our new node
   //create a Subscriber object and have it subscribe to the topic "topic1"
   // the function "myCallback" will wake up whenever a new message is published to ↵
       topic1
   // the real work is done inside the callback function

   ros::Subscriber my_subscriber_object= n.subscribe("topic1",1,myCallback);

   ros::spin(); //this is essentially a "while(1)" statement, except it
   // forces refreshing wakeups upon new data arrival
   // main program essentially hangs here, but it must stay alive to keep the callback ↵
       function alive
   return 0; // should never get here, unless roscore dies
}
}
```

Most of the code within the minimal subscriber is identical to that of the minimal publisher (though the node name in line 19 has been changed to "`minimal_subscriber`"). There are four important new lines to examine.

Most notably, the subscriber is more complex than the publisher, since it requires a `callback`, which is declared in line 3,

```
void myCallback(const std_msgs::Float64& message_holder)
```

This function has an argument of a reference pointer (indicated by the & sign) to an object of type `std_msgs::Float64`. This is the message type associated with `topic1`, as published by our minimal publisher.

The importance of the callback function is that it is awakened when new data is available on its associated topic (which is set to `topic1` in this example). When new data is published to the associated topic, the callback function runs, and the published data appears in the argument `message_holder`. (This message holder must be of a type compatible with the message type published on the topic of interest, in this case `std_msgs::Float64`).

Within the callback function, the only action taken is to display the received data, implemented on line 13.

```
ROS_INFO("received value is: %f",message_holder.data);
```

Display is performed using `ROS_INFO()` instead of `cout` or `printf`. `ROS_INFO()` uses message publishing, which avoids slowing time-critical code for display driving. Also, using `ROS_INFO()` makes the data available for logging or monitoring. However, as viewed from the terminal from which this node is run, the output is displayed equivalently to using `cout` or `printf`. The argument to `ROS_INFO()` is the same as `printf` in C.

An alternative to using `ROS_INFO()` is `ROS_INFO_STREAM()`. Line 13 could be replaced with:

```
ROS_INFO_STREAM("received value is: "<<message_holder.data<<std::endl);
```

which produces the same output, but uses syntax of `cout`.

Once the callback function executes, it goes back to sleep, ready to be re-awakened by arrival of a new message on `topic1`.

In the main program, a key new concept is on line 26:

```
ros::Subscriber my_subscriber_object= n.subscribe("topic1",1,myCallback);
```

The use of `ros::Subscriber` is similar to the use of `ros::Publisher` earlier. An object of type `Subscriber` is substantiated; `Subscriber` is a class that exists within the ROS distribution. There are three arguments used in instantiating the subscriber object. The first argument is the topic name; `topic1` is chosen as the topic to which our minimal publisher publishes. (For this example, we want our subscriber node to listen to the output of our example publisher node.)

The second argument is the queue size. If the callback function has trouble keeping up with published data, the data may be queued. In the present case, the queue size is set to one. If the callback function cannot keep up with publications, messages will be lost by being overwritten by newer messages. (Recall that in the first example, `rostopic echo topic1` could not keep up with the 30 kHz rate of the original minimal publisher. Values displayed skipped many intermediate messages.) For control purposes, typically only the most recent sensor value published is of interest. If a sensor publishes faster than the callback function can respond, there is no harm done in dropping messages, as only the most recent message would be needed. In this (and many cases) a queue size of one message is all that is needed.

The third argument for instantiating the `Subscriber` object is the name of the callback function that is to receive data from `topic1`. This argument has been set to `myCallback`, which is the name of our callback function, described earlier. Through this line of code, we associate our callback function with messages published on `topic1`.

Finally, line 28:

```
ros::spin();
```

introduces a key ROS concept that is non-obvious but essential. The callback function should awaken whenever a new message appears on `topic1`. However, the main program must yield some time for the callback function to respond. This is accomplished with a `spin()` command. In the present case, a `spin` causes the main program to suspend, but keeps the callback function alive. If the main program were to run to conclusion, the callback function would no longer be poised to react to new messages. The `spin()` command keeps `main()` alive without consuming significant CPU time. As a result, the minimal subscriber is quite efficient.

1.2.8 Compiling and running minimal subscriber

For our new node to get compiled, we must include reference to it in `CMakeLists`. This requires adding two lines, very similar to what we did to enable compiling the minimal publisher. The first new line is simply:

```
add_executable(my_minimal_subscriber src/minimal_subscriber.cpp)
```

The arguments are the desired executable name (chosen to be `my_minimal_subscriber`), and the relative path to the source code (`src/minimal_subscriber.cpp`).

The second line added is:

```
target_link_libraries(my_minimal_subscriber  ${catkin_LIBRARIES} )
```

which informs the compiler to link our new executable with the declared libraries.

After updating `CMakeLists`, the code is newly compiled with the command `catkin_make` (which must be run from the `ros_ws` directory).

The code example should compile without error, after which a new executable, `my_minimal_subscriber`, will appear in the directory: `~/ros_ws/devel/lib/my_minimal_nodes`. Note: it is not necessary to recall this lengthy path to the executable file. The executable will be found when running `rosrun` with the package name and node name as arguments.

After recompiling, we now have two nodes to run. In one terminal, enter:

```
rosrun my_minimal_nodes sleepy_minimal_publisher
```

and in a second terminal enter:

```
rosrun my_minimal_nodes my_minimal_subscriber
```

It does not matter which is run first. An example of the display in the terminal of the `my_minimal_subscriber` node is shown below:

```
[ INFO] [1435555572.972403158]: received value is: 0.909000
[ INFO] [1435555573.972261535]: received value is: 0.910000
[ INFO] [1435555574.972258968]: received value is: 0.911000
```

This display was updated once per second, since the publisher published to `topic1` at

1 Hz. The messages received differ by increments of 0.001, as programmed in the publisher code.

In another terminal, entering:

```
rosnode list
```

results in the output below.

```
/minimal_publisher2
/minimal_subscriber
/rosout
```

This shows that we now have three nodes running: the default `rosout`, our minimal publisher (which we named node `minimal_publisher2` for the timed version) and our minimal subscriber, `minimal_subscriber`.

In an available terminal, entering:

```
rqt_graph
```

produces a graphical display of the running system, which is shown in Fig 1.1.

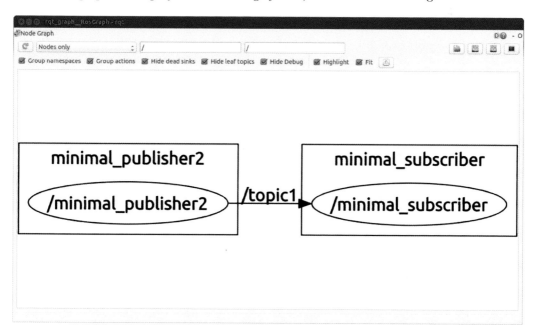

Figure 1.1: Node topology as illustrated by rqt_graph

The graphical display shows our two nodes. Our minimal publisher node is shown publishing to `topic1` and our minimal subscriber is shown subscribing to this topic.

1.2.9 Minimal subscriber and publisher node summary

We have seen the basics for how to create our own publisher and subscriber nodes. A single node can subscribe to multiple topics by replicating the corresponding lines of code to create additional subscriber objects and callback functions. Similarly, a node can publish to multiple topics by instantiating multiple publisher objects. A node can also be both a subscriber and a publisher.

In a ROS-controlled robot system, custom device-driver nodes must be designed that can read and publish sensor information and one or more nodes that subscribe to actuator (or controller setpoint) commands and impose these on actuators. Fortunately, there is already a large body of existing ROS drivers for common sensors, including LIDARs, cameras, the Kinect sensor, inertial measurement units, encoders, etc. These may be imported as packages and used as-is in your system (though perhaps requiring some tweaking to reference specifics of your system). There are also packages for driving some servos (*e.g.* hobby-servo style RCs and Dynamixel motors). There are also some ROS-Industrial interfaces to industrial robots, which only require publishing and subscribing to and from robot topics. In some cases, the user may need to design device driver nodes to interface to custom actuators. Further, hard real-time, high-speed servo loops may require a non-ROS dedicated controller (although this might be as simple as an Arduino microcontroller). The user then assumes the responsibility for designing the hard-real-time controller and writing a ROS-compatible subscriber interface to run on the control computer.

1.3   MORE ROS TOOLS: CATKIN_SIMPLE, ROSLAUNCH, RQT_CONSOLE, AND ROSBAG

Having introduced minimal ROS talkers (publishers) and listeners (subscribers), one can already begin to appreciate the value of some additional ROS tools. The tools and facilities introduced here are `catkin_simple`, `roslaunch`, `rqt_console` and `rosbag`.

1.3.1   Simplifying `CMakeLists.txt` with `catkin_simple`

As seen in Section 1.2.3, the `CMakeLists.txt` file generated with a new package creation is quite long. While the required edits to this file were relatively simple, introducing additional features can require tedious, non-obvious changes. A package called `catkin_simple` helps to simplify the `CMakeLists.txt` file. This package can be found at `https://github.com/catkin/catkin_simple.git`. A copy has been cloned into our external-packages repository at `https://github.com/wsnewman/learning_ros_external_packages.git`, which should already be cloned into your `~/ros_ws/src` directory.

Additionally, the external-packages repository has a Python script that assists with creating new packages that use `catkin_simple`. To run this script, it is convenient to define an alias to point to this script as a command. Within a terminal, enter:

```
alias cs_create_pkg='~/ros_ws/src/learning_ros_external_packages/cs_create_pkg.py'
```

Subsequently, from this same terminal, you can create a package that uses `catkin_simple` with the command `cs_create_pkg`. For example, navigate to `~/ros_ws/src` and create a new package called `my_minimal_nodes2` by entering:

```
cs_create_pkg my_minimal_nodes2 roscpp std_msgs
```

Note that this command will be recognized only in the terminal from which the alias `cs_create_pkg` was defined. More conveniently, the alias definition should be including in your `.bashrc` file. To do so, navigate to your home directory and edit the (hidden) file `.bashrc`. Append the line

```
alias cs_create_pkg='~/ros_ws/src/learning_ros_external_packages/cs_create_pkg.py'
```

to this file and save it. Subsequently, this alias will be recognized by all new terminals that are opened. If you have used the recommended setup scripts, this .bashrc file edit will already be performed for you.

After having invoked our cs_create_pkg command, the new package my_minimal_nodes2 contains the expected structure, including subdirectories src and include, the package.xml file, a CMakeLists.txt file and a new file, README.md. The README file should be edited to describe the purpose of the new package and how to run examples within the package. The README file uses markdown formatting (see https://guides.github.com/features/mastering-markdown/ for a description of markdown). Such formatting allows your README file to be displayed with attractive formatting when viewed from your repository via a browser.

The package.xml file is similar to that created by catkin_create_pkg, except it includes the additional dependency:

```
<buildtool_depend>catkin_simple</buildtool_depend>
```

The CMakeLists.txt file is considerably simplified. A copy of the default generated file is:

Listing 1.7: CMakeLists.txt with catkin_simple

```
cmake_minimum_required(VERSION 2.8.3)
project(my_minimal_nodes2)

find_package(catkin_simple REQUIRED)

#uncomment next line to use OpenCV library
#find_package(OpenCV REQUIRED)

#uncomment the next line to use the point-cloud library
#find_package(PCL 1.7 REQUIRED)

#uncomment the following 4 lines to use the Eigen library
#find_package(cmake_modules REQUIRED)
#find_package(Eigen3 REQUIRED)
#include_directories(${EIGEN3_INCLUDE_DIR})
#add_definitions(${EIGEN_DEFINITIONS})

catkin_simple()

# example boost usage
# find_package(Boost REQUIRED COMPONENTS system thread)

# C++0x support - not quite the same as final C++11!
# use carefully;  can interfere with point-cloud library
# SET(CMAKE_CXX_FLAGS "${CMAKE_CXX_FLAGS} -std=c++0x")

# Libraries: uncomment the following and edit arguments to create a new library
# cs_add_library(my_lib src/my_lib.cpp)

# Executables: uncomment the following and edit arguments to compile new nodes
# may add more of these lines for more nodes from the same package
# cs_add_executable(example src/example.cpp)

#the following is required, if desire to link a node in this package with a library ↩
    created in this same package
# edit the arguments to reference the named node and named library within this package
# target_link_libraries(example my_lib)

cs_install()
cs_export()
```

The line catkin_simple() invokes actions to automatically perform much of the tedious editing of CMakeLists.txt. The commented lines are reminders of how to exploit CMake variations, including linking with popular libraries, such as Eigen, PCL and OpenCV. To modify CMakeLists.txt to compile our desired code, uncomment and edit the line:

```
cs_add_executable(example src/example.cpp)
```

By copying `minimal_publisher.cpp` to the new package, `my_minimal_nodes2`, we can specify that this node should be compiled by modifying the above line in `CMakeLists.txt` to:

```
cs_add_executable(minimal_publisher3 src/minimal_publisher.cpp)
```

This indicates that we wish to compile `minimal_publisher.cpp` and call the executable `minimal_publisher3` (a name that will not conflict with other instances of minimal publisher nodes). We can add commands to compile additional nodes by inserting more `cs_add_executable` lines of the same form. It is not necessary to specify linking with libraries. This may not seem to be much of a simplification at this point, but `catkin_simple` will become much more valuable when we start to link with more libraries, create libraries, and create custom messages.

1.3.2 Automating starting multiple nodes

In our minimal example, we ran two nodes: one publisher and one subscriber. To do so, we opened two separate terminals and typed in two **rosrun** commands. Since a complex system may have hundreds of nodes running, we need a more convenient way to bring up a system. This can be done using `launch` files and the command **roslaunch**. (See `http://ros.org/wiki/roslaunch` for more details and additional capabilities, such as setting parameters.)

A launch file has the suffix `.launch`. It is conventionally named the same as the package name (though this is not required). It is also conventionally located in a subdirectory of the package by the name of `launch` (also not required). A launch file can also invoke other launch files to bring up multiple nodes from multiple packages. However, we will start with a minimal launch file. In our package `my_minimal_nodes`, we may create a subdirectory `launch` and within this directory create a `my_minimal_nodes.launch` file containing the following lines:

```
<launch>
<node name="publisher" pkg="my_minimal_nodes" type="sleepy_minimal_publisher"/>
<node name="subscriber" pkg="my_minimal_nodes" type="my_minimal_subscriber"/>
</launch>
```

In the above, using XML syntax, we use the keyword "node" to tell ROS that we want to launch a ROS node (an executable program compiled by **catkin_make**). We must specify three key/value pairs to launch a node: the package name of the node (value of "pkg"), the binary executable name of the node (value of "type"), and the name by which the node will be recognized by ROS when launched (value of "name"). In fact, we already specified node names within the source code (e.g. `ros::init(argc,argv, "minimal_publisher2")` within `sleepy_minimal_publisher.cpp`). The launch file gives you the opportunity to rename the node when it is launched. For example, by setting: `name="publisher"` in the launch file, we would still start running an instance of the executable called `sleepy_minimal_publisher` within the package `my_minimal_nodes`, but it will be known to `roscore` by the name `publisher`. Similarly, by assigning `name="subscriber"` to the executable `my_minimal_subscriber`, this node will be known to `roscore` as `subscriber`.

We can execute the launch file by typing:

```
roslaunch my_minimal_nodes my_minimal_nodes.launch
```

Recall that for using **rosrun** to start nodes, it was necessary that **roscore** be running first. When using **roslaunch**, it is not necessary to start **roscore** running. If **roscore** is already running, the **roslaunch** will launch the specified nodes. If **roscore** is not already running, **roslaunch** will detect this and will start **roscore** before launching the specified nodes.

The terminal in which we executed **roslaunch** displays the following:

```
SUMMARY
========

PARAMETERS
 * /rosdistro: indigo
 * /rosversion: 1.11.13

NODES
  /
    publisher (my_minimal_nodes/sleepy_minimal_publisher)
    subscriber (my_minimal_nodes/my_minimal_subscriber)

ROS_MASTER_URI=http://localhost:11311

core service [/rosout] found
process[publisher-1]: started with pid [18793]
process[subscriber-2]: started with pid [18804]
```

From another terminal, entering:

```
rosnode list
```

produces the output:

```
/publisher
/rosout
/subscriber
```

which shows that nodes known as "publisher" and "subscriber" are running (using the new names assigned to these nodes by the launch file).

Most often, we will not want to change the name of the node from its original specification, but ROS launch files nonetheless require that this option be used. To decline changing the node name, the default name (embedded in the source code) may be used as the desired node name.

After launching our nodes, **rostopic list** shows that **topic1** is alive, and the command **rostopic info topic1** shows that the node **publisher** is publishing to this topic and the node **subscriber** has subscribed to this topic. Clearly, this will be useful when we have many nodes to launch.

One side effect, though, is that we no longer see the output from our subscriber, which formerly appeared in the terminal from which the subscriber was launched. However, since we used **ROS_INFO()** instead of **printf** or **cout**, we can still observe this output using the **rqt_console** tool.

1.3.3 Viewing output in a ROS console

A convenient tool to monitor ROS messages is **rqt_console**, which can be invoked from a terminal by entering: **rqt_console**. With our two nodes running, an example of using this tool is shown in Fig 1.2.

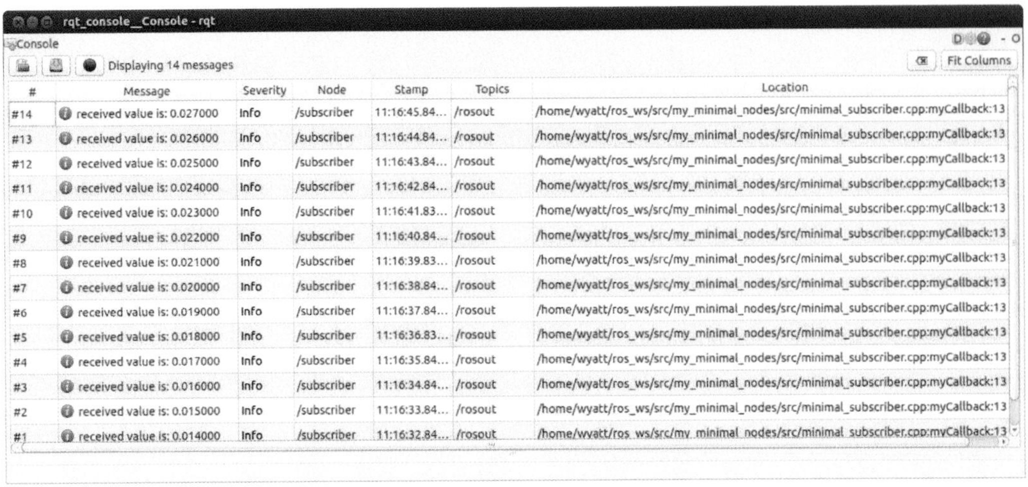

Figure 1.2: Output of `rqt_console` with minimal nodes launched

In this instance, `rqt_console` shows values output by the minimal subscriber from the time `rqt_console` was started and until `rqt_console` was paused (using the `rqt_console` "pause" button). The lines displayed show that the messages are merely informational and not warnings or errors. The console also shows that the node responsible for posting the information is our minimal subscriber (known by the node name "subscriber"). `rqt_console` also notes the time stamp at which the message was sent.

Multiple nodes using `ROS_INFO()` may be run simultaneously, and their messages may be viewed with `rqt_console`, which will also note new events, such as starting and stopping new nodes. Another advantage of using `ROS_INFO()` instead of `printf()` or `cout` is that the messages can be logged and run in playback. A facility for doing this is `rosbag`.

1.3.4 Recording and playing back data with `rosbag`

The `rosbag` command is extremely useful for debugging complex systems. One can specify a list of topics to record while the system is running, and `rosbag` will subscribe to these topics and record the messages published, along with time stamps, in a "bag" file. `rosbag` can also be used to play back bag files, thus recreating the circumstances of the recorded system. (This playback occurs at the same clock rate at which the original data was published, thus emulating the real-time system.)

When running `rosbag`, the resulting log (bag) files will be saved in the same directory from which `rosbag` was launched. For our example, move to the `my_minimal_nodes` directory and create a new subdirectory `bagfiles`. With our nodes still running (which is optional; nodes can be started later), navigate to the `bagfile` directory (wherever you chose to store your bags) and enter:

```
rosbag record topic1
```

With this command, we have asked to record all messages published on `topic1`. Running `rqt_console` will display data from `topic1`, as reported by our subscriber node using `ROS_INFO()`. In the screenshot of `rqt_console` shown in Fig 1.3, the `rosbag` recording startup is noted at line number 34; at this instant, the value output from our subscriber node is 0.236 (i.e., 236 seconds after the nodes were launched).

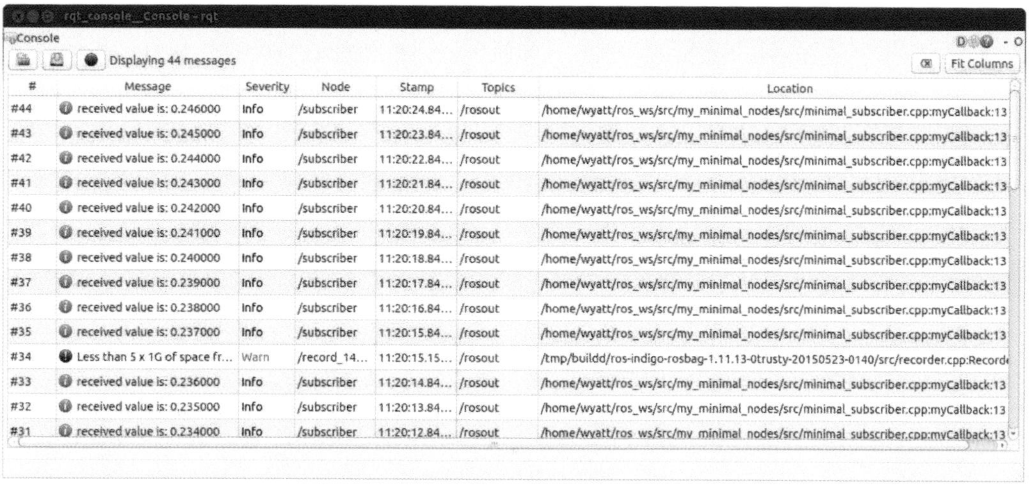

Figure 1.3: Output of `rqt_console` with minimal nodes running and `rosbag` running

The `rosbag` node was subsequently halted with a control-C in the terminal from which it was launched. Looking in the `bagfiles` directory (from which we launched `rosbag`), we see there is a new file, named according to the date and time of the recording, and with the suffix `.bag`.

We can play back the recording using `rosbag` as well. To do so, first kill the running nodes, then type:

```
rosbag   play   fname.bag
```

where "fname" is the file name of the recording. The `rosbag` terminal shows a playback time incrementing, but there is otherwise no noticeable effect. The screen output is:

```
rosbag play 2016-01-07-11-20-15.bag
[ INFO] [1452184943.589921994]: Opening 2016-01-07-11-20-15.bag

Waiting 0.2 seconds after advertising topics... done.

Hit space to toggle paused, or 's' to step.
wyatt@Wall-E:~/ros_ws$ 452183621.541102    Duration: 5.701016 / 750.000001
```

The `rqt_console` display indicates that the bagfile has been opened, but no other information is displayed. What is happening at this point is that `rosbag` is publishing the recorded data (recorded from `topic1`) to `topic1` at the same rate as the data was recorded.

To see that this is taking place, halt all nodes (including `rosbag`). Run `rqt_console`. In a terminal window, start the subscriber, but not the publisher, using:

```
rosrun   my_minimal_nodes   my_minimal_subscriber
```

At this point, this terminal is suspended, because `topic1` is not yet active.

From another terminal, navigate to the `bagfiles` directory and enter:

```
rosbag  play fname.bag
```

where "fname" is (again) the name of the recording that was bagged previously. `rosbag` now assumes the role formerly taken by `sleepy_minimal_publisher`, recreating the messages that were formerly published. The `my_minimal_subscriber` window reports the recorded data, updating once per second. `rqt_console` also shows the data posted by the minimal subscriber. As can be seen from the screenshot in Fig 1.4, the playback corresponds to the original recording, with the first output (console line 2) displaying a value of 0.238 from the subscriber node.

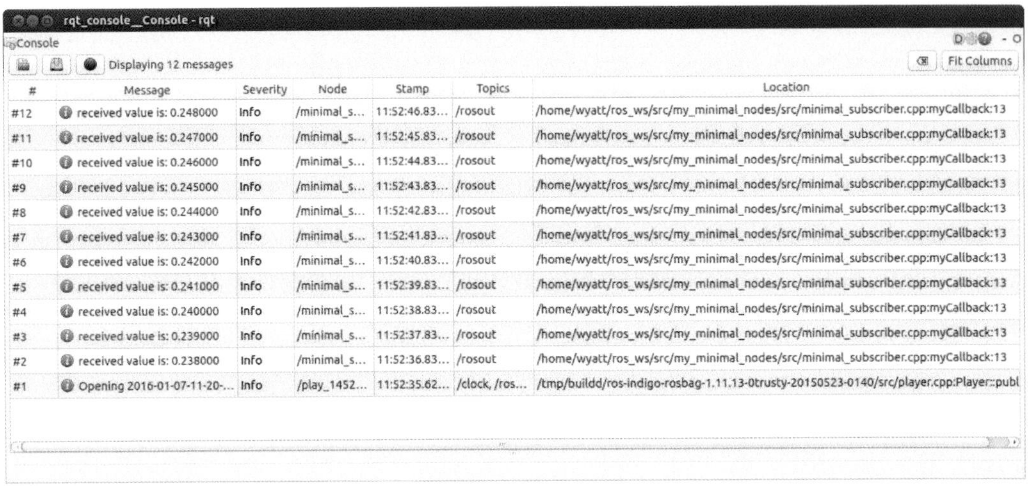

#	Message	Severity	Node	Stamp	Topics	Location
#12	received value is: 0.248000	Info	/minimal_s...	11:52:46.83...	/rosout	/home/wyatt/ros_ws/src/my_minimal_nodes/src/minimal_subscriber.cpp:myCallback:13
#11	received value is: 0.247000	Info	/minimal_s...	11:52:45.83...	/rosout	/home/wyatt/ros_ws/src/my_minimal_nodes/src/minimal_subscriber.cpp:myCallback:13
#10	received value is: 0.246000	Info	/minimal_s...	11:52:44.83...	/rosout	/home/wyatt/ros_ws/src/my_minimal_nodes/src/minimal_subscriber.cpp:myCallback:13
#9	received value is: 0.245000	Info	/minimal_s...	11:52:43.83...	/rosout	/home/wyatt/ros_ws/src/my_minimal_nodes/src/minimal_subscriber.cpp:myCallback:13
#8	received value is: 0.244000	Info	/minimal_s...	11:52:42.83...	/rosout	/home/wyatt/ros_ws/src/my_minimal_nodes/src/minimal_subscriber.cpp:myCallback:13
#7	received value is: 0.243000	Info	/minimal_s...	11:52:41.83...	/rosout	/home/wyatt/ros_ws/src/my_minimal_nodes/src/minimal_subscriber.cpp:myCallback:13
#6	received value is: 0.242000	Info	/minimal_s...	11:52:40.83...	/rosout	/home/wyatt/ros_ws/src/my_minimal_nodes/src/minimal_subscriber.cpp:myCallback:13
#5	received value is: 0.241000	Info	/minimal_s...	11:52:39.83...	/rosout	/home/wyatt/ros_ws/src/my_minimal_nodes/src/minimal_subscriber.cpp:myCallback:13
#4	received value is: 0.240000	Info	/minimal_s...	11:52:38.83...	/rosout	/home/wyatt/ros_ws/src/my_minimal_nodes/src/minimal_subscriber.cpp:myCallback:13
#3	received value is: 0.239000	Info	/minimal_s...	11:52:37.83...	/rosout	/home/wyatt/ros_ws/src/my_minimal_nodes/src/minimal_subscriber.cpp:myCallback:13
#2	received value is: 0.238000	Info	/minimal_s...	11:52:36.83...	/rosout	/home/wyatt/ros_ws/src/my_minimal_nodes/src/minimal_subscriber.cpp:myCallback:13
#1	Opening 2016-01-07-11-20-...	Info	/play_1452...	11:52:35.62...	/clock, /ros...	/tmp/buildd/ros-indigo-rosbag-1.11.13-0trusty-20150523-0140/src/player.cpp:Player::publ

Figure 1.4: Output of `rqt_console` with `minimal_subscriber` and `rosbag play` of recorded (bagged) data

Dynamically, these values are posted at the original 1 Hz rate `sleepy_minimal_publisher` published them. The `rosbag` player terminates when it gets to the end of the recorded data.

Note that our subscriber is oblivious to what entity is publishing to `topic1`. Consequently, playback of previously recorded data is indistinguishable from receiving live data. This is very useful for development. For example, a robot may be teleoperated through an environment of interest while it publishes sensor data from cameras, LIDAR, etc. Using `rosbag`, this data may be recorded verbatim. Subsequently, sensor processing may be performed on the recorded data to test machine vision algorithms, for example. Once a sensory-interpretation node is shown to be effective on the recorded data, the same node may be tried verbatim on the robot system. Note that no changes to the developed node are needed. In live experiments, this node would merely receive messages published by the real-time system instead of by `rosbag playback`.

1.4   MINIMAL SIMULATOR AND CONTROLLER EXAMPLE

Concluding this introduction, we consider a pair of nodes that both publish and subscribe. One of these nodes is a minimal simulator, and the other is a minimal controller. The minimal simulator simulates $F = ma$ by integrating acceleration to update velocities. The input force is published to the topic `force_cmd` by some entity (eventually the controller). The resulting system state (velocity) is published to the topic `velocity` by the minimal simulator.

The simulator code is shown in Listing 1.8, and the source code is in the accompanying repository, in `https://github.com/wsnewman/learning_ros/tree/master/Part_1/minimal_nodes/src`.

The `main()` function initializes both a publisher and a subscriber. Formerly, we saw nodes as dedicated publishers or dedicated subscribers, but a node can (and often does) perform both actions. Equivalently, the simulator node behaves like a link in a chain, processing incoming data and providing timely output publications.

Another difference from our previous publisher and subscriber nodes is that both the callback function and the `main()` routine perform useful actions. The minimal simulator has a callback routine that checks for new data on the topic `force_cmd`. When the callback receives new data, it copies it to a global variable, `g_force`, so that the `main()` program has access to it. The `main()` function iterates at a fixed rate (set to 100 Hz). In order for the callback function to respond to incoming data, the `main()` function must provide `spin` opportunities. Formerly, our minimal subscriber used the `spin()` function, but this resulted in the `main()` function ceasing to contribute new computations.

An important new feature in the minimal simulator example is the use of the ROS function: `ros::spinOnce()`. This function, executed within the 100 Hz loop of the simulator, allows the callback function to process incoming data at 10 ms intervals. If a new input (a force stimulus) is received, it is stored in `g_force` by the callback function. In the complementary minimal controller node, values of force stimulus are published at only 10 Hz. Consequently, 9 out of 10 times there will be no new messages on the `force_command` topic. The callback function will not block, but neither will it update the value of `g_force` when there is no new transmission available. The main loop will still repeat its iterations at 100 Hz, albeit re-using stale data in `g_force`. This behavior is realistic, since a simulator should emulate realistic physics. For an actuated joint that is digitally controlled, controller effort (force or torque) commands typically behave with a sample-and-hold output between controller updates (sample periods).

Listing 1.8: Minimal Simulator

```
1   // minimal_simulator node:
2   // wsn example node that both subscribes and publishes
3   // does trivial system simulation, F=ma, to update velocity given F specified on topic↵
        "force_cmd"
4   // publishes velocity on topic "velocity"
5   #include<ros/ros.h>
6   #include<std_msgs/Float64.h>
7   std_msgs::Float64 g_velocity;
8   std_msgs::Float64 g_force;
9   void myCallback(const std_msgs::Float64& message_holder)
10  {
11  // checks for messages on topic "force_cmd"
12  ROS_INFO("received force value is: %f",message_holder.data);
13  g_force.data = message_holder.data; // post the received data in a global var for ↵
        access by
14  // main prog.
15  }
16  int main(int argc, char **argv)
17  {
18  ros::init(argc,argv,"minimal_simulator"); //name this node
19  // when this compiled code is run, ROS will recognize it as a node called "↵
        minimal_simulator"
20  ros::NodeHandle nh; // node handle
21  //create a Subscriber object and have it subscribe to the topic "force_cmd"
22  ros::Subscriber my_subscriber_object= nh.subscribe("force_cmd",1,myCallback);
23  //simulate accelerations and publish the resulting velocity;
24  ros::Publisher my_publisher_object = nh.advertise<std_msgs::Float64>("velocity",1);
25  double mass=1.0;
26  double dt = 0.01; //10ms integration time step
27  double sample_rate = 1.0/dt; // compute the corresponding update frequency
```

```
28  ros::Rate naptime(sample_rate);
29  g_velocity.data=0.0; //initialize velocity to zero
30  g_force.data=0.0; // initialize force to 0; will get updated by callback
31  while(ros::ok())
32  {
33  g_velocity.data = g_velocity.data + (g_force.data/mass)*dt; // Euler integration of
34  //acceleration
35  my_publisher_object.publish(g_velocity); // publish the system state (trivial--1-D)
36  ROS_INFO("velocity = %f",g_velocity.data);
37  ros::spinOnce(); //allow data update from callback
38  naptime.sleep(); // wait for remainder of specified period; this loop rate is faster ←
        than
39  // the update rate of the 10Hz controller that specifies force_cmd
40  // however, simulator must advance each 10ms regardless
41  }
42  return 0; // should never get here, unless roscore dies
43  }
```

The minimal simulator may be compiled and run. Running `rqt_console` shows that the velocity has a persistent value of 0.

The result can be visualized graphically with the ROS tool `rqt_plot`. To do so, use command-line arguments for the values to be plotted, *e.g.*:

```
rqt_plot velocity/data
```

will plot the velocity command against time. This output will be boring, at present, since the velocity is always zero.

One can manually publish values to a topic from a command line. For example, enter the following command in a terminal:

```
rostopic pub -r 10 force_cmd std_msgs/Float64 0.1
```

This will cause the value 0.1 to be published repeatedly on the topic `force_cmd` at a rate of 10 Hz using the consistent message type `std_msgs/Float64`. This can be confirmed (from another terminal) with:

```
rostopic echo force_cmd
```

which will show that the `force_cmd` topic is receiving the prescribed value.

Additionally, invoking:

```
rqt_plot velocity/data
```

will show that the velocity is increasing linearly, and `rqt_console` will print out the corresponding values (for both force and velocity).

Instead of publishing force-command values manually, these can be computed and published by a controller. Listing 1.9 displays a compatible minimal controller node. (This code is also contained in the accompanying examples repository under `https://github.com/wsnewman/learning_ros/tree/master/Part_1/minimal_nodes/src`.)

The minimal controller subscribes to two topics, (`velocity` and `vel_cmd`), and publishes to the topic `force_cmd`. At each control cycle (set to 10 Hz), the controller checks for the latest system state (velocity), checks for any updates to the commanded velocity, and computes a proportional error feedback to derive (and publish) a force command. This simple controller attempts to drive the simulated system to the user-commanded velocity setpoint.

Again, the `ros::spinOnce()` function is used to prevent blocking in the timed main loop. Callback functions put received message data in the global variables `g_velocity` and `g_vel_cmd`.

Listing 1.9: Minimal Controller

```
1   // minimal_controller node:
2   // wsn example node that both subscribes and publishes--counterpart to ↵
        minimal_simulator
3   // subscribes to "velocity" and publishes "force_cmd"
4   // subscribes to "vel_cmd"
5   #include<ros/ros.h>
6   #include<std_msgs/Float64.h>
7   //global variables for callback functions to populate for use in main program
8   std_msgs::Float64 g_velocity;
9   std_msgs::Float64 g_vel_cmd;
10  std_msgs::Float64 g_force; // this one does not need to be global...
11  void myCallbackVelocity(const std_msgs::Float64& message_holder)
12  {
13  // check for data on topic "velocity"
14  ROS_INFO("received velocity value is: %f",message_holder.data);
15  g_velocity.data = message_holder.data; // post the received data in a global var for ↵
        access by
16  //main prog.
17  }
18  void myCallbackVelCmd(const std_msgs::Float64& message_holder)
19  {
20  // check for data on topic "vel_cmd"
21  ROS_INFO("received velocity command value is: %f",message_holder.data);
22  g_vel_cmd.data = message_holder.data; // post the received data in a global var for ↵
        access by
23  //main prog.
24
25   }
26  int main(int argc, char **argv)
27  {
28  ros::init(argc,argv,"minimal_controller"); //name this node
29  // when this compiled code is run, ROS will recognize it as a node called "↵
        minimal_controller"
30  ros::NodeHandle nh; // node handle
31  //create 2 subscribers: one for state sensing (velocity) and one for velocity commands
32  ros::Subscriber my_subscriber_object1= nh.subscribe("velocity",1,myCallbackVelocity);
33  ros::Subscriber my_subscriber_object2= nh.subscribe("vel_cmd",1,myCallbackVelCmd);
34  //publish a force command computed by this controller;
35  ros::Publisher my_publisher_object = nh.advertise<std_msgs::Float64>("force_cmd",1);
36  double Kv=1.0; // velocity feedback gain
37  double dt_controller = 0.1; //specify 10Hz controller sample rate (pretty slow, but
38  //illustrative)
39  double sample_rate = 1.0/dt_controller; // compute the corresponding update frequency
40  ros::Rate naptime(sample_rate); // use to regulate loop rate
41  g_velocity.data=0.0; //initialize velocity to zero
42  g_force.data=0.0; // initialize force to 0; will get updated by callback
43  g_vel_cmd.data=0.0; // init velocity command to zero
44  double vel_err=0.0; // velocity error
45  // enter the main loop: get velocity state and velocity commands
46  // compute command force to get system velocity to match velocity command
47  // publish this force for use by the complementary simulator
48  while(ros::ok())
49  {
50  vel_err = g_vel_cmd.data - g_velocity.data; // compute error btwn desired and actual
51  //velocities
52  g_force.data = Kv*vel_err; //proportional-only velocity-error feedback defines ↵
        commanded
53  //force
54  my_publisher_object.publish(g_force); // publish the control effort computed by this
55  //controller
56  ROS_INFO("force command = %f",g_force.data);
57  ros::spinOnce(); //allow data update from callback;
58  naptime.sleep(); // wait for remainder of specified period;
59  }
60  return 0; // should never get here, unless roscore dies
61  }
```

Once the two nodes are compiled with `catkin_make` (which requires editing `CMakeLists.txt` to add these executables to the package), they can be run (with `rosrun`) from separate terminal windows (assuming `roscore` is running). Running `rqt_console` reveals that the force command is updated once for every 10 updates of velocity (as expected for the simulator at 100 Hz and the controller at 10 Hz).

The velocity command may be input from another terminal using a command line, *e.g.*:

```
rostopic pub r 10 vel_cmd  std_msgs/Float64 1.0
```

publishes the value 1.0 to the topic `vel_cmd` repeatedly at a rate of 10 Hz. Watching the output on `rqt_console` shows the velocity converge exponentially on the desired value of `vel_cmd`.

The result can be visualized graphically with the ROS tool `rqt_plot`. To do so, use command line arguments for the values to be plotted, for example:

```
rqt_plot vel_cmd/data,velocity/data,force_cmd/data
```

will plot the velocity command, the actual velocity and the force commands on the same plot versus time. For the minimal simulator and minimal controller, the velocity command was initially set to 0.0 via `rostopic pub`. Subsequently, the command was set to 2.0. A screenshot of the resulting `rqt_plot` is shown in Fig 1.5. The control effort (in red) reacts to accelerate the velocity closer to the goal of 2.0, then the control effort decreases. Ultimately, the system velocity converges on the goal value and the required control effort decreases to zero.

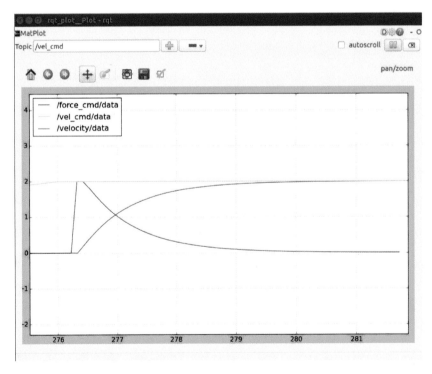

Figure 1.5: Output of `rqt_plot` with `minimal_simulator`, `minimal_controller` and step velocity-command input via console

1.5   WRAP-UP

This chapter has introduced the reader to some basics of ROS. It has been shown that multiple asynchronous nodes can communicate with each other through the mechanisms of publish and subscribe of messages on topics, as coordinated by `roscore`. Some ROS tools were introduced for the purpose of compiling nodes (`catkin_create_pkg`, `cs_create_pkg` and `catkin_make`), for logging and replaying published information (`rosbag`) and for visualizing ROS output (`rqt_console` and `rqt_plot`).

In the next chapter, additional communications topics are introduced, including defining custom message types, using client–server interactions, the common design paradigm of action servers, and the ROS parameter server.

Messages, Classes and Servers

CONTENTS

INTRODUCTION

The previous chapter explained how ROS nodes can communicate via publishing and subscribing. Example (trivial) message types were introduced. To make publishing and subscribing more general, it is helpful to be able to define custom messages. Also, using publish and subscribe is not always appropriate. Sometimes a peer-to-peer communication is necessary. This capability is realized in ROS through a client–service interaction. Client–service interactions address concerns of publish and subscribe in terms of knowledge of the source of communications and guaranteed receipt of messages. A limitation of client–service communications is that these transactions are blocking, and thus the client node is suspended until the service node responds. Often, the service to be performed can require significant time to complete, in which cases it is desirable that the interaction be non-blocking. For this purpose, a third interaction mechanism, action clients and action servers, is available. These three options will be covered in this chapter.

2.1 DEFINING CUSTOM MESSAGES

ROS message types are based on 14 primitives (built-in types), plus fixed or variable-length arrays of these. (See `http://wiki.ros.org/msg`.) Using these built-in types, more complex message types can be constructed.

Our minimal nodes illustrated use of standard messages (`std_msgs`) for communicating via publish and subscribe. The `std_msgs` package defines 32 message types, most of which correspond to a single built-in field (data) type. A notable exception is `Header.msg`, which is composed of three fields, each of which corresponds to a built-in type.

More sophisticated message types can be defined by including other defined message types (not necessarily primitive, built-in types). For example, it is common to include the header message type within higher level message definitions, since including a time stamp is a common need. The ability to include defined message types within new message definitions is recursively extensible (with messages that include messages that include messages ...); ultimately, though, the higher level message type is entirely definable in terms of the built-in primitives.

Some useful (more complex) messages are defined in additional packages, such as `geometry_msgs`, `sensor_msgs`, `nav_msgs`, `pcl_msgs`, `visualization_msgs`, `trajectory_msgs`, and `actionlib_msgs`.

Defined messages can be examined interactively using `rosmsg show`, followed by the package and message name of the message of interest. For example, entering:

```
rosmsg show std_msgs/Header
```

outputs:

```
uint32 seq
time stamp
string frame_id
```

which shows that `Header` is comprised of three fields: `seq`, `stamp` and `frame_id`. These message names store data of primitive types `uint32`, `time` and `string`, respectively.

If a message type already exists in the standard distribution of ROS, you should use that message. However, it is sometimes necessary to define one's own message. Defining custom messages is described at `http://wiki.ros.org/ROS/Tutorials/DefiningCustomMessages`, which can be consulted for more details. The next section introduces the basics of defining custom messages.

2.1.1 Defining a custom message

The following description refers to corresponding code in the accompanying repository within the package `example_ros_msg`.

The package `example_ros_msg` was created using:

```
cs_create_pkg example_ros_msg roscpp std_msgs
```

This creates a directory structure under `example_ros_msg`. By using `cs_create_pkg`, we will be able to use the the abbreviated `Cmakelists.txt`.

To define a new message type, we create a subdirectory in this package called `msg`. Within this `msg` directory, we create a new text file by the name of `ExampleMessage.msg`. The example message file contains only three relevant lines:

```
Header header
int32 demo_int
float64 demo_double
```

This message type will have three fields, which may be referred to as **header**, **demo_int** and **demo_double**. Their types are **Header**, **int32** and **float64**, respectively, which are all message types defined in the package **std_msgs**.

To inform the compiler that we need to generate new message headers, the "package.xml" file must be edited. Insert (or uncomment) the following lines:

```
<build_depend>message_generation</build_depend>
```

and

```
<run_depend>message_runtime</run_depend>
```

The abbreviated **CMakeLists.txt** file (with unnecessary comments removed) is shown in Listing 2.1.

Listing 2.1: CMakeLists.txt using catkin_simple

```
1   cmake_minimum_required(VERSION 2.8.3)
2   project(example_ros_msg)
3
4   find_package(catkin_simple REQUIRED)
5
6   catkin_simple()
7
8   # Executables
9   #cs_add_executable(example_ros_message_publisher src/example_ros_message_publisher.cpp↩
    )
10
11  cs_install()
12  cs_export()
```

Note the **cs_add_executable** is commented out. We will enable this once we have our anticipated source code for a test node, **example_ros_message_publisher.cpp**.

Having defined a message type, we can generate corresponding header files suitable for C++ file inclusion. Compiling the code with **catkin_make** produces a header file, which it installs in the directory: **~/ros_ws/devel/include/example_ros_msg/ExampleMessage.h**. (Reminder: here and throughout this text, it will be assumed that the ROS workspace is called **ros_ws**.)

Source code for nodes that want to use this new message type should depend on the package **example_ros_msg** (in the corresponding **package.xml** file) and should include the new header with the line:

```
#include <example_ros_msg/ExampleMessage.h>
```

in the C++ source code of the node using this message type. An illustrative example follows.

The accompanying code repository (in **https://github.com/wsnewman/learning_ros/tree/master/Part_1/minimal_nodes**) includes a source file under **example_ros_msg/src/example_ros_message_publisher.cpp**. The source code is shown in Listing 2.2.

Listing 2.2: `example_ros_message_publisher`: example node using custom message type

```
1   #include <ros/ros.h>
2   #include <example_ros_msg/ExampleMessage.h>
3   #include <math.h>
4
5   int main(int argc, char **argv) {
6       ros::init(argc, argv, "example_ros_message_publisher"); // name of this node
7       ros::NodeHandle n; // two lines to create a publisher object that can talk to ROS
8       ros::Publisher my_publisher_object = n.advertise<example_ros_msg::ExampleMessage>(←
            "example_topic", 1);
9       //"example_topic" is the name of the topic to which we will publish
10      // the "1" argument says to use a buffer size of 1; could make larger, if expect ←
            network backups
11
12      example_ros_msg::ExampleMessage  my_new_message;
13      //create a variable of type "example_msg",
14      // as defined in this package
15
16      ros::Rate naptime(1.0); //create a ros object from the ros Rate class;
17      //set the sleep timer for 1Hz repetition rate (arg is in units of Hz)
18
19      // put some data in the header.  Do: rosmsg show std_msgs/Header
20      //  to see the definition of "Header" in std_msgs
21      my_new_message.header.stamp = ros::Time::now(); //set the time stamp in the header←
            ;
22      my_new_message.header.seq=0; // call this sequence number zero
23      my_new_message.header.frame_id = "base_frame"; // would want to put true reference←
            frame name here, if needed for coord transforms
24      my_new_message.demo_int= 1;
25      my_new_message.demo_double=100.0;
26
27      double sqrt_arg;
28      // do work here in infinite loop (desired for this example), but terminate if ←
            detect ROS has faulted
29      while (ros::ok())
30      {
31          my_new_message.header.seq++; //increment the sequence counter
32          my_new_message.header.stamp = ros::Time::now(); //update the time stamp
33          my_new_message.demo_int*=2.0; //double the integer in this field
34          sqrt_arg = my_new_message.demo_double;
35          my_new_message.demo_double = sqrt(sqrt_arg);
36
37          my_publisher_object.publish(my_new_message); // publish the data in new ←
                message format on topic "example_topic"
38          //the next line will cause the loop to sleep for the balance of the desired period
39              // to achieve the specified loop frequency
40          naptime.sleep();
41      }
42  }
```

This node uses the new message type as follows. It defines a publisher object as:

```
ros::Publisher my_publisher_object = n.advertise<example_ros_msg::←
    ExampleMessage>("example_topic", 1);
```

which says that `example_topic` will carry messages of type `example_ros_msg::ExampleMessage`. (The message type is identified by referring to the package that contains it, `example_ros_msg`, followed by the preamble of the file name that details the format of the message, *i.e.* `ExampleMessage` taken from filename `ExampleMessage.msg`.)

We also instantiate an object of type `example_ros_msg::ExampleMessage` with the line:

```
example_ros_msg::ExampleMessage  my_new_message;
```

Note that when referring to the header file, we use the notation `example_ros_msg/ExampleMessage.h` (path to the header file), but when instantiating an object based on this definition (or referring to the datatype for publication), we use the class notation `example_ros_msg::ExampleMessage`.

Within the source code of `example_ros_message_publisher.cpp`, the various fields of the new message object, `my_new_message`, are populated, and then this message is published.

Populating fields of the new message type is simple, *e.g.*:

```
my_new_message.demo_int= 1;
```

Accessing elements of hierarchical fields requires drilling down deeper, as in:

```
my_new_message.header.stamp = ros::Time::now(); //set the time stamp in the header;
```

Here, `stamp` is a field within the `Header` message type for the field `header`. Additionally, this line of code illustrates another useful ROS function: `ros::Time::now()`. This looks up the current time and returns it in a form compatible with `header` (consisting of separate fields for seconds and nanoseconds). Note: the absolute time is essentially meaningless. However, differences in time can be used as valid time increments.

By uncommenting the line in `CMakeLists.txt`,

```
cs_add_executable(example_ros_message_publisher src/example_ros_message_publisher.cpp)
```

and re-running `catkin_make`, a new node is created, with the name `example_ros_message_publisher`.

Running this node (assuming `roscore` is running):

```
rosrun example_ros_msg example_ros_message_publisher
```

produces no output. However, (from a separate terminal), running:

```
rostopic list
```

reveals that there is a new topic, `example_topic`. We can examine the output of this topic with:

```
rostopic echo example_topic
```

which produces the following output:

```
header:
  seq: 1
  stamp:
    secs: 1452225386
    nsecs: 619262393
  frame_id: base_frame
demo_int: 4
demo_double: 3.16227766017
---
header:
  seq: 2
  stamp:
    secs: 1452225387
    nsecs: 619259445
  frame_id: base_frame
demo_int: 8
demo_double: 1.77827941004
```

```
---
header:
  seq: 3
  stamp:
    secs: 1452225388
    nsecs: 619234854
  frame_id: base_frame
demo_int: 16
demo_double: 1.33352143216
---
```

We see that our new node successfully uses the new message type. Sequence numbers increase monotonically. The `demo_int` field is doubled at each iteration (per the logic of the source code). The `demo_double` field displays sequential square-roots (starting from 100). The `secs` field of the header increments by 1 second each iteration (since the iteration rate timer was set to 1 Hz). The string `base_frame` appears in the `frame_id` field.

Following the same process, one can create more customized message types. After defining a new message type, nodes within the same package or nodes in other packages can use the new message type, provided the external packages list `example_ros_msg` as a dependency (in the corresponding `package.xml` file).

2.1.2 Defining a variable-length message

One very useful extension of the ROS message primitives is the ability to send and receive arbitrary-length vectors of message types. To illustrate how to do this, refer to the package `custom_msgs` within the `Part_1` folder of the accompanying code repository. The package `custom_msgs` contains no source code, but it does have files `CMakeLists.txt` and `package.xml`, as well as a `msg` folder that contains a message file, `VecOfDoubles.msg` with the following contents (a single line):

```
float64[] dbl_vec
```

Although our new package has no source files to compile, we still need to execute the process that generates header files from `*.msg` files. To invoke this, uncomment two lines in the `package.xml` file:

```
<build_depend>message_generation</build_depend>
<run_depend>message_runtime</run_depend>
```

With these edits, the new package can by built by running `catkin_make`. This yields a new header file, `VecOfDoubles.h`, appearing in the folder `:~/ros_ws/devel/include/custom_msgs`. One can confirm that the new message has been installed by entering:

```
rosmsg show custom_msgs/VecOfDoubles
```

which yields the response:

```
float64[] dbl_vec
```

When this message is received by a subscriber, it can be interpreted as a C++ `vector` of doubles.

An example publisher using the new message type is `vector_publisher.cpp`, which resides in the `src` folder of the `example_ros_msg` package. The code listing of the `vector_publisher.cpp` appears in Listing 2.3.

Listing 2.3: `vector_publisher.cpp`: publisher node using custom message `VecOfDoubles`

```
1   #include <ros/ros.h>
2   //next line requires a dependency on custom_msgs within package.xml
3   #include <custom_msgs/VecOfDoubles.h> //this is the message type we are testing
4
5   int main(int argc, char **argv) {
6       ros::init(argc, argv, "vector_publisher"); // name of this node
7       ros::NodeHandle n; // two lines to create a publisher object that can talk to ROS
8       ros::Publisher my_publisher_object = n.advertise<custom_msgs::VecOfDoubles>("↩
            vec_topic", 1);
9
10      custom_msgs::VecOfDoubles vec_msg; //create an instance of this message type
11      double counter=0;
12      ros::Rate naptime(1.0); //create a ros object from the ros Rate class; set 1Hz ↩
            rate
13
14      vec_msg.dbl_vec.resize(3); //manually resize it to hold 3 doubles
15      //After setting the size, one can access elements of this array conventionally, e.↩
            g.
16      vec_msg.dbl_vec[0]=1.414;
17      vec_msg.dbl_vec[1]=2.71828;
18      vec_msg.dbl_vec[2]=3.1416;
19
20      //Alternatively, one can use the vector member function push_back() to append data↩
            to an existing array, e.g.:
21      vec_msg.dbl_vec.push_back(counter); // this makes the vector longer, to hold ↩
            additional data
22      while(ros::ok()) {
23      counter+=1.0;
24      vec_msg.dbl_vec.push_back(counter);
25      my_publisher_object.publish(vec_msg);
26      naptime.sleep();
27      }
28  }
```

On line 2 of Listing 2.3, the message header for our new vector message is:

```
#include <custom_msgs/VecOfDoubles.h> //this is the message type we are testing
```

On line 8, a publisher is instantiated, as before, but this time using our new message type.

```
ros::Publisher my_publisher_object = n.advertise<custom_msgs::VecOfDoubles>("vec_topic↩
    ", 1);
```

The topic to which it will publish is `vec_topic`.

An instance of our new message type, named **vec_msg** is declared on line 10:

```
custom_msgs::VecOfDoubles vec_msg; //create an instance of this message type
```

The variable-length message can be resized, if desired, as on line 14:

```
vec_msg.dbl_vec.resize(3); //manually resize it to hold 3 doubles
```

Note that one must refer to the field name **dbl_vec** within the message **vec_msg**, from which one can invoke member functions of the vector object including **size()**, **resize()**, and **push_back()**. After resizing the vector, individual elements may be accessed by indexing it. Three values are stored in the first three locations with lines 16 through 18: .

```
vec_msg.dbl_vec[0]=1.414;
vec_msg.dbl_vec[1]=2.71828;
vec_msg.dbl_vec[2]=3.1416;
```

Alternatively, one may append data to the vector, which is automatically resized to accommodate the additional data. This is done with the **push_back** member function, as in line 21:

```
vec_msg.dbl_vec.push_back(counter); // this makes the vector longer, to hold ↩
    additional data
```

The example program then goes into a timed loop, incrementing a counter by 1 per iteration, appending this value to the variable-length vector, and publishing the result via line 25:

```
my_publisher_object.publish(vec_msg);
```

In order to compile this new node in our package **example_ros_msg**, both the **package.xml** and the **CMakeLists.txt** within this package must be edited. The **package.xml** must be informed of the new dependency on our new message. This is done by inserting the lines:

```
<build_depend>custom_msgs</build_depend>
<run_depend>custom_msgs</run_depend>
```

Logically, these can be inserted following the same tags for **std_msgs**, although the order is flexible.

The second required change is to edit **CMakeLists.txt** to invoke compiling of our new node. This is done by inserting the line:

```
cs_add_executable(vector_publisher src/vector_publisher.cpp)
```

With these changes, the new node is compiled with **catkin_make**. The new node can be run with:

```
rosrun example_ros_msg vector_publisher
```

This results in publishing a variable-length vector to the topic **vec_topic** once per second. This effect can be observed by running:

```
rostopic echo vec_topic
```

which produces the following output:

```
dbl_vec: [1.414, 2.71828, 3.1416, 0.0, 1.0, 2.0]
---
dbl_vec: [1.414, 2.71828, 3.1416, 0.0, 1.0, 2.0, 3.0]
---
dbl_vec: [1.414, 2.71828, 3.1416, 0.0, 1.0, 2.0, 3.0, 4.0]
---
dbl_vec: [1.414, 2.71828, 3.1416, 0.0, 1.0, 2.0, 3.0, 4.0, 5.0]
---
dbl_vec: [1.414, 2.71828, 3.1416, 0.0, 1.0, 2.0, 3.0, 4.0, 5.0, 6.0]
---
dbl_vec: [1.414, 2.71828, 3.1416, 0.0, 1.0, 2.0, 3.0, 4.0, 5.0, 6.0, 7.0]
```

This shows that the variable-length vector message is increasing in length at each iteration, due to the **push_back** operations.

A complementary subscriber node is also included in the src folder of the example_ros_msg package: vector_subscriber.cpp. It largely follows the example minimal_subscriber.cpp. The contents of vector_subscriber.cpp are shown in Listing 2.4.

Listing 2.4: vector_subsriber.cpp: subscriber node using custom message VecOfDoubles

```
1  #include<ros/ros.h>
2  #include <custom_msgs/VecOfDoubles.h> //this is the message type we are testing
3  void myCallback(const custom_msgs::VecOfDoubles& message_holder)
4  {
5      std::vector <double> vec_of_doubles = message_holder.dbl_vec; //can copy contents of
          message to a C++ vector like this
6      int nvals = vec_of_doubles.size(); //ask the vector how long it is
7      for (int i=0;i<nvals;i++) {
8          ROS_INFO("vec[%d] = %f",i,vec_of_doubles[i]); //print out all the values
9      }
10     ROS_INFO("\n");
11 }
12
13 int main(int argc, char **argv)
14 {
15     ros::init(argc,argv,"vector_subscriber"); //default name of this node
16     ros::NodeHandle n; // need this to establish communications with our new node
17
18     ros::Subscriber my_subscriber_object= n.subscribe("vec_topic",1,myCallback);
19
20     ros::spin();
21     return 0; // should never get here, unless roscore dies
22 }
```

In Listing 2.4, line 2 brings in the header file of our new message, as was done with the publisher as well:

```
#include <custom_msgs/VecOfDoubles.h> //this is the message type we are testing
```

A subscriber is instantiated in the main() function, line 18:

```
ros::Subscriber my_subscriber_object= n.subscribe("vec_topic",1,myCallback);
```

This is identical to the earlier minimal_subscriber.cpp example, except that the topic name has been changed to vec_topic, to be consistent with our vector publisher node.

The callback function, starting on line 3:

```
void myCallback(const custom_msgs::VecOfDoubles& message_holder)
```

is defined to accept an argument based on our new variable-length vector message.

The callback function is awakened by the receipt of a new transmission on the vec_topic topic. The received data is contained within message_holder, the contents of which can be copied to a C++ vector, as shown in line 5:

```
std::vector <double> vec_of_doubles = message_holder.dbl_vec;
```

Subsequently, this vector object can be used with all of its member functions. For example, the length of the vector can be found, as in line 6:

```
int nvals = vec_of_doubles.size(); //ask the vector how long it is
```

The callback function then iterates through all components of the vector, displaying them with `ROS_INFO()`.

To compile the subscriber node, add the following line to the **CMakeLists.txt** file:

```
cs_add_executable(vector_subscriber src/vector_subscriber.cpp)
```

then compile with **catkin_make**.

To test the operation of the complementary nodes, (with roscore running) in one terminal, enter:

```
rosrun example_ros_msg vector_publisher
```

In a second terminal, start the subscriber with:

```
rosrun example_ros_msg vector_subscriber
```

As a result, the subscriber displays the following example output:

```
[ INFO] [1452450667.220847072]: vec[0] = 1.414000
[ INFO] [1452450667.220985967]: vec[1] = 2.718280
[ INFO] [1452450667.221041396]: vec[2] = 3.141600
[ INFO] [1452450667.221097413]: vec[3] = 0.000000
[ INFO] [1452450667.221151168]: vec[4] = 1.000000
[ INFO] [1452450667.221208351]: vec[5] = 2.000000
[ INFO] [1452450667.221265818]:

[ INFO] [1452450668.220694702]: vec[0] = 1.414000
[ INFO] [1452450668.220802727]: vec[1] = 2.718280
[ INFO] [1452450668.220898220]: vec[2] = 3.141600
[ INFO] [1452450668.220989838]: vec[3] = 0.000000
[ INFO] [1452450668.221095457]: vec[4] = 1.000000
[ INFO] [1452450668.221200559]: vec[5] = 2.000000
[ INFO] [1452450668.221308814]: vec[6] = 3.000000
[ INFO] [1452450668.221393760]:

[ INFO] [1452450669.220683996]: vec[0] = 1.414000
[ INFO] [1452450669.220777258]: vec[1] = 2.718280
[ INFO] [1452450669.220820689]: vec[2] = 3.141600
[ INFO] [1452450669.220884177]: vec[3] = 0.000000
[ INFO] [1452450669.220937618]: vec[4] = 1.000000
[ INFO] [1452450669.220999832]: vec[5] = 2.000000
[ INFO] [1452450669.221057884]: vec[6] = 3.000000
[ INFO] [1452450669.221115040]: vec[7] = 4.000000
[ INFO] [1452450669.221172102]:
```

This shows that our publisher and subscriber are communicating with each other successfully using the new variable-length message type.

In practice, when using variable-length messages, one must be careful of two issues:

1. Make sure your variable-length vectors do not get overly large. It is all too easy to forget that the vector continues to grow, and thus you could consume all of your memory and all of your communications bandwidth as the message grows too large.

2. As with conventional arrays, if you try to access memory that has not been allocated, you will get errors (typically a segmentation fault). For example, if your received message copied to `vec_of_doubles` is three elements long, trying to access `vec_of_doubles[3]` (the fourth element) will result in run-time errors (but no compiler warnings).

This concludes the introduction to defining message types. As will be shown, the same process is used for defining additional message types used for ROS services and ROS action servers.

2.2 INTRODUCTION TO ROS SERVICES

So far, our primary means of communications among nodes consisted of publish and subscribe. In this mode of communications, the publisher is unaware of its subscribers, and a subscriber knows only of a topic, not which node might publish to that topic. Messages are sent at unknown intervals, and subscribers may miss messages. This style of communications is appropriate for repetitive messages, such as publication of sensor values. For such cases, the sensor publisher should not need to know which or how many nodes subscribe to its output, nor will the publisher change its message in response to any requests from consumers. This form of communication is simple and flexible. Provided the preceding restrictions are not of concern, publish and subscribe is preferred.

Alternatively, it is sometimes necessary to establish bi-directional, one-to-one, reliable communication. In this case, the client sends a request to a service, and the service sends a response to the client. The question and answer interaction is on demand, and the client is aware of the name of the service provider.

It should be noted that ROS services are intended to be fast responses. When a client sends a request, the client is suspended until an answer returns. ROS code should be tolerant of such delays. If the request involves extensive calculations or delays to respond, then you probably should use yet another alternative, the action server and action client mechanism (to be introduced in Section 2.5).

The example below is contained in the package `example_ros_service` in the accompanying repository. It is introduced by first describing a service message, then constructing a service provider node, and finally constructing a service client node.

2.2.1 Service messages

Defining a custom service message requires describing the data types and field names for both a request and a response. This is done by a ROS template, called a `*.srv` file. Contents of this file auto-generate C++ headers that can be included in C++ files.

To create a new service message within the desired package (in this example, `example_ros_service`), create a sub-directory called `srv`. (This directory is at the same level as `src`.) Inside this directory, create a text file named `*.srv`. In the current example, this text file has been named `ExampleServiceMsg.srv`. The contents of the example service message are given below:

```
string name
---
bool on_the_list
bool good_guy
int32 age
string nickname
```

In the above, the request structure is defined by the lines above "`---`", and the response structure is defined by the lines below "---". The request (for this simple example) consists of a single component, referenced as **name**, and the datatype contained in this field is a (ROS) string.

For the response part of this example, there are four fields, named **on_the_list**, **good_guy**, **age** and **nickname**. These have respective datatypes of **bool**, **bool**, **int32** and **string** (all of which are defined as built-in ROS message types).

Although the service message is described very simply in a text file, the compiler will be instructed to parse this file and build a C++ header file that can be included in a C++ program. Just as we did with defining custom message types for publish and subscribe, we inform the compiler that we need to generate new message headers via the **package.xml** file. This file must be edited to insert (or uncomment) the following lines:

```
<build_depend>message_generation</build_depend>
```

and

```
<run_depend>message_runtime</run_depend>
```

The package **example_ros_service** was created with the help of **cs_create_pkg**, which uses **catkin_simple** to simplify the **CMakeLists.txt** file. Anticipating that we will want to compile two test nodes, **example_ros_service.cpp** and **example_ros_client.cpp**, corresponding lines in the **CMakeLists.txt** file specify that these source files should be compiled. The **CMakeLists.txt** file (with instructional comments removed, and with the **cs_add_executable** lines temporarily commented out) appears in Listing 2.5 below.

Listing 2.5: CMakeLists.txt using catkin_simple for example client and service nodes

```
1   cmake_minimum_required(VERSION 2.8.3)
2   project(example_ros_service)
3
4   find_package(catkin_simple REQUIRED)
5
6   catkin_simple()
7
8   #cs_add_executable(example_ros_service src/example_ros_service.cpp)
9   #cs_add_executable(example_ros_client src/example_ros_client.cpp)
10
11  cs_install()
12  cs_export()
```

Having defined the service message, one can test the package compilation even before any C++ source code has been written. The new package can be compiled by running **catkin_make** (as always, from the **ros_ws** directory).

Although the **CMakeLists.txt** file has no active **cs_add_executable** instructions, the catkin build system will recognize that there is a new **srv** file, and it will auto-generate C++ compatible header files. In fact, it is common to have ROS packages that contain message definitions, but no C++ code. The pre-defined messages in **std_msgs** are contained within a package (named **std_msgs**), although this package contains no C++ source code. This is a common technique for defining messages that may be useful among multiple packages.

The header files that have been auto-generated reside within the directory **catkin/devel/include/example_ros_service**. It is not necessary to remember this directory path, because **catkin_make** will know where to look for these headers if you list a dependency on the package **example_ros_service**. As with the creation of new message types for publish and subscribe, the auto-generated header file shares the same base name as the user-define *.srv file. For our example, after compilation the directory catkin/devel/include/example_ros_service contains a header file called ExampleServiceMsg.h. What is slightly more complex than a publish and subscribe message type is that two more service header files were created in the catkin/devel/include/example_ros_service directory: ExampleServiceMsgRequest.h and ExampleServiceMsgResponse.h. Both of these header files are included within the ExampleServiceMsg.h header file. Note that the names of these header files contain the base name, ExampleServiceMsg, but they also have an appended component to their names, either Request or Service.

To use the new service message type in a node, include its associated header file in the C++ source code with the line:

```
#include <example_ros_service/ExampleServiceMsg.h>
```

Note that this header inclusion references the name of the package where the message is defined (example_ros_service), along with the base name of the message (ExampleServiceMsg), with the suffix .h appended. The delimiters <...> tell the compiler to look for this file in the expected locations (in this case, catkin/devel/include).

In the present example, the source code for our illustrative nodes is part of the same example_ros_service package. If we wanted to use our new service message within nodes of a separate package, we would add: #include<example_ros_service/ExampleServiceMsg.h> in our C++ code identically. However, there will be two differences in the new package.xml file. For one, it will not be necessary to depend on message_generation or message_runtime, since the message type needs to be generated only once (and this is performed with the package example_ros_service is compiled). The second difference is that the new package's package.xml file must list a dependency on package example_ros_service by including the line: <build_depend>example_ros_service</build_depend>.

2.2.2 ROS service nodes

An example ROS service node is defined in the package example_ros_service, in the subdirectory src, named example_ros_service.cpp. The CMakeLists.txt file is edited to instruct the compiler to compile this code by adding (or uncommenting) the line:

```
cs_add_executable(example_ros_service src/example_ros_service.cpp)
```

The source code of example_ros_service.cpp is given in Listing 2.6.

Listing 2.6: example_ros_service.cpp: example service node

```
1  //example ROS service:
2  // run this as: rosrun example_ROS_service example_ROS_service
3  // in another window, tickle it manually with (e.g.):
4  //    rosservice call lookup_by_name "Ted"
5
6
7  #include <ros/ros.h>
8  #include <example_ros_service/ExampleServiceMsg.h>
9  #include <iostream>
10 #include <string>
11 using namespace std;
12
13 bool callback(example_ros_service::ExampleServiceMsgRequest& request, ↵
        example_ros_service::ExampleServiceMsgResponse& response)
```

```
14  {
15      ROS_INFO("callback activated");
16      string in_name(request.name); // convert this to a C++-class string, so can use ↩
            member funcs
17      //cout<<"in_name:"<<in_name<<endl;
18      response.on_the_list=false;
19
20      // here is a dumb way to access a stupid database...
21      // hey: this example is about services, not databases!
22      if (in_name.compare("Bob")==0)
23      {
24          ROS_INFO("asked about Bob");
25          response.age = 32;
26          response.good_guy=false;
27          response.on_the_list=true;
28          response.nickname="BobTheTerrible";
29      }
30        if (in_name.compare("Ted")==0)
31      {
32          ROS_INFO("asked about Ted");
33          response.age = 21;
34          response.good_guy=true;
35          response.on_the_list=true;
36          response.nickname="Ted the Benevolent";
37      }
38
39    return true;
40  }
41
42  int main(int argc, char **argv)
43  {
44    ros::init(argc, argv, "example_ros_service");
45    ros::NodeHandle n;
46
47    ros::ServiceServer service = n.advertiseService("lookup_by_name", callback);
48    ROS_INFO("Ready to look up names.");
49    ros::spin();
50
51    return 0;
52  }
```

In the example service code, note the inclusion of the new header file:

```
#include <example_ros_service/ExampleServiceMsg.h>
```

Within the body of **main()**, the line:

```
ros::ServiceServer service = n.advertiseService("lookup_by_name", callback);
```

is similar to creating a publisher in ROS. In this case, a service is created, and it will be known as **lookup_by_name**. When a request to this service arrives, the named callback function (in this case, simply named **callback**) will be invoked. The service does not have a timed loop. Rather (using **ros::spin();** in the main function), it sleeps until a request comes in, and incoming requests are serviced by the callback function.

Less obvious attributes of the service node construction are the type declarations of the arguments of the service callback.

```
bool callback(example_ros_service::ExampleServiceMsgRequest& request, ↩
    example_ros_service::ExampleServiceMsgResponse& response)
```

The argument:

```
example_ros_service::ExampleServiceMsgRequest& request
```

declares that the argument `request` is a reference pointer of type `example_ros_service::ExampleServiceMsgRequest`. Similarly, the second argument of the callback function is:

```
example_ros_service::ExampleServiceMsgResponse& response
```

which declares that the argument named `response` is a reference pointer to an object of type `example_ros_service::ExampleServiceMsgResponse`.

This may seem strange, since we did not define a datatype called `ExampleServiceMsgRequest` within our package `example_ros_service`. The build system created this datatype as part of auto-generating the message header file. The name `ExampleServiceMsgRequest` is created by appending `Request` to our service message name, `ExampleServiceMsg` (and similarly for `Response`). When you define a new service message, you can assume that the system will create these two new datatypes for you.

When the service callback function is invoked, the callback can examine the contents of the incoming request. In the present example, the request has only one field, `name` in its definition. The line:

```
string in_name(request.name);
```

creates a C++ style `string` object from the characters contained in `request.name`. With this string object, we can invoke the member function `compare()`, *e.g.* as in line 22 of the code to test whether the name is identical to "Bob."

```
if (in_name.compare("Bob")==0)
```

Within the callback routine, fields are filled in for the response. When the callback returns, this response message is transmitted back to the client that invoked the request. The mechanism for performing this communication is hidden from the programmer; it is performed as part of the ROS service paradigm (*e.g.*, you do not have to invoke an action similar to `publish(response)`).

Once the ROS service example is compiled, it can be run (assuming a `roscore` is running) with:

```
rosrun example_ros_service example_ros_service
```

2.2.3   Manual interaction with ROS services

Once the node `example_ros_service` is running, we can see that a new service is available. From a command prompt, enter:

```
rosservice list
```

The response will show a service named `/lookup_by_name` (which we declared to be the service name of our node).

We can interact manually with the service from the command line, *e.g.* by typing:

```
rosservice call lookup_by_name 'Ted'
```

The response to this is:

```
on_the_list: True
good_guy: True
age: 21
nickname: Ted the Benevolent
```

We see that the service reacted appropriately to our request. More generally, service requests would be invoked from other ROS nodes.

2.2.4 Example ROS service client

To interact with a ROS service programmatically, one composes a ROS `client`. An example is given in Listing 2.7 (`example_ros_client.cpp` in the package `example_ros_service`).

Listing 2.7: `example_ros_client.cpp`: example client node

```cpp
1   //example ROS client:
2   // first run: rosrun example_ROS_service example_ROS_service
3   // then start this node:  rosrun example_ROS_service example_ROS_client
4
5
6
7   #include <ros/ros.h>
8   #include <example_ros_service/ExampleServiceMsg.h> // this message type is defined in ↩
        the current package
9   #include <iostream>
10  #include <string>
11  using namespace std;
12
13  int main(int argc, char **argv) {
14      ros::init(argc, argv, "example_ros_client");
15      ros::NodeHandle n;
16      ros::ServiceClient client = n.serviceClient<example_ros_service::ExampleServiceMsg↩
            >("lookup_by_name");
17      example_ros_service::ExampleServiceMsg srv;
18      bool found_on_list = false;
19      string in_name;
20      while (ros::ok()) {
21          cout<<endl;
22          cout << "enter a name (x to quit): ";
23          cin>>in_name;
24          if (in_name.compare("x")==0)
25              return 0;
26          //cout<<"you entered "<<in_name<<endl;
27          srv.request.name = in_name; //"Ted";
28          if (client.call(srv)) {
29              if (srv.response.on_the_list) {
30                  cout << srv.request.name << " is known as " << srv.response.nickname ↩
                        << endl;
31                  cout << "He is " << srv.response.age << " years old" << endl;
32                  if (srv.response.good_guy)
33                      cout << "He is reported to be a good guy" << endl;
34                  else
35                      cout << "Avoid him; he is not a good guy" << endl;
36              } else {
37                  cout << srv.request.name << " is not in my database" << endl;
38              }
39
40          } else {
41              ROS_ERROR("Failed to call service lookup_by_name");
42              return 1;
43          }
44      }
45      return 0;
46  }
47  %
```

In this program, we include the same message header as in the service node:

```
#include <example_ROS_service/ExampleServiceMsg.h>
```

Note that if this node were defined within another package, we would need to list the package dependency `example_ros_service` within the `package.xml` file.

There are two key lines in the client program. First, line 16:

```
ros::ServiceClient client =
    n.serviceClient<example_ROS_service::ExampleServiceMsg>("lookup_by_name");
```

creates a ROS `ServiceClient` that expects to communicate requests and responses as defined in `example_ros_service::ExampleServerMsg`. Also, this service client expects to communicate with a named service called `lookup_by_name`. (This is the service name we defined inside of our example service node.)

Second, on line 17 we instantiate an object of a consistent type for requests and responses with:

```
example_ros_service::ExampleServiceMsg srv;
```

The above type specifies the package name in which the service message is defined (`example_ros_service`), and the name of the service message (`ExampleServiceMsg`). In this case `srv` contains a field for request and a field for response. To send a service request, we first populate the fields of the request message (in this case, the request message has only a single component), as per line 27:

```
srv.request.name = in_name; //e.g., manually test with contents: "Ted";
```

We then perform the transaction with the named service through the following call (line 28):

```
client.call(srv)
```

This call will return a Boolean to let us know whether the call was successful. If the call is successful, we may expect that the components of `srv.response` will be filled in (as provided by the ROS service). These components are examined and displayed by the example code. In this client example, the fields `on_the_list`, `name`, `age` and `good_guy` are evaluated, as shown in lines 31 through 33. For example, line 31:

```
cout << "He is " << srv.response.age << " years old" << endl;
```

looks up and reports the `age` field of the service response message.

2.2.5 Running example service and client

With `roscore` running, start the service with:

```
rosrun example_ros_service example_ros_service
```

The service displays "Ready to look up names" then suspends action until a service request comes in. Although the service is active, it consumes negligible resources (CPU cycles and bandwidth) while waiting for service requests.

In another terminal, start up the example client with:

```
rosrun example_ros_service example_ros_client
```

This results in prompting the user with `enter a name (x to quit):`. The following output resulted from responding with Ted, Bob, Amy, then x.

```
enter a name (x to quit): Ted
Ted is known as Ted the Benevolent
He is 21 years old
He is reported to be a good guy

enter a name (x to quit): Bob
Bob is known as BobTheTerrible
He is 32 years old
Avoid him; he is not a good guy

enter a name (x to quit): Amy
Amy is not in my database

enter a name (x to quit): x
```

While the client node is active, it can send repeated requests to the service, as shown above. The `ros::ServiceClient` only needs to be instantiated once, and `client.call(srv)` can be re-invoked with updated values in the `srv` message. Although the client may run to conclusion, the service would remain active (unless it is deliberately killed). Other client nodes may use the service as well, and such additional clients may run concurrently. The peer-to-peer communication scheme used in the service–client construction assures that the service will return its response only to the unique client that issued the corresponding request.

In summary, using services is appropriate for one-to-one, guaranteed communications. Although a sensor (*e.g.* a joint encoder) can simply keep publishing its current value, this would not be appropriate for a command such as `close_gripper`. In the latter case, we would want to send the command once and be assured that the command was received and acted on. A client–server interaction would be appropriate for such needs.

Some extra work is required to define the service message and to make sure both the server and the client include the corresponding message headers. ROS does include a package `std_srvs` with pre-defined service messages. However, unlike the fairly rich set of ROS messages for publish and subscribe use, there are only two service messages defined in this package, and they are of limited value. Typically, every time one composes a new ROS service node, it is also typically required that a corresponding service message be defined.

Services should be used only as quick request and response interactions. For interactions that can require long durations until a response is ready (*e.g.* `plan_path`), a more appropriate interface is action-servers and action-clients. Before introducing action servers, it is important to see how we can use C++ classes in constructing our nodes.

2.3   USING C++ CLASSES IN ROS

ROS code can quickly become overly long. To improve productivity and code re-use, it is desirable to use classes.

Use of classes in C++ is detailed in any text on C++. In general, it is desirable to:

- Define a class in a header file

     – Define prototypes for all member functions of the class

     – Define private and public data members

     – Define a prototype for a class constructor function

- Write a separate implementation file that:

     – Includes the above header file

     – Contains the working code for the declared member functions

     – Contains code to encapsulate necessary initializations in the constructor

An example is provided in the example code repository within the package `example_ros_class`. The header file of this example is given in Listing 2.8.

Listing 2.8: Header file of `example_ros_class`

```
1   // example_ros_class.h header file //
2   // wsn; Feb, 2015
3   // include this file in "example_ros_class.cpp"
4
5   // here's a good trick--should always do this with header files:
6   // create a unique mnemonic for this header file, so it will get included if needed,
7   // but will not get included multiple times
8   #ifndef EXAMPLE_ROS_CLASS_H_
9   #define EXAMPLE_ROS_CLASS_H_
10
11  //some generically useful stuff to include...
12  #include <math.h>
13  #include <stdlib.h>
14  #include <string>
15  #include <vector>
16
17  #include <ros/ros.h> //ALWAYS need to include this
18
19  //message types used in this example code;  include more message types, as needed
20  #include <std_msgs/Bool.h>
21  #include <std_msgs/Float32.h>
22  #include <std_srvs/Trigger.h> // uses the "Trigger.srv" message defined in ROS
23
24  // define a class, including a constructor, member variables and member functions
25  class ExampleRosClass
26  {
27  public:
28      ExampleRosClass(ros::NodeHandle* nodehandle); //"main" will need to instantiate a ↵
              ROS nodehandle, then pass it to the constructor
29      // may choose to define public methods or public variables, if desired
30  private:
31      // put private member data here;  "private" data will only be available to member ↵
              functions of this class;
32      ros::NodeHandle nh_; // we will need this, to pass between "main" and constructor
33      // some objects to support subscriber, service, and publisher
34      ros::Subscriber minimal_subscriber_; //these will be set up within the class ↵
              constructor, hiding these ugly details
35      ros::ServiceServer minimal_service_;
36      ros::Publisher   minimal_publisher_;
37
38      double val_from_subscriber_; //example member variable: better than using globals;↵
              convenient way to pass data from a subscriber to other member functions
39      double val_to_remember_; // member variables will retain their values even as ↵
              callbacks come and go
40
41      // member methods as well:
42      void initializeSubscribers(); // we will define some helper methods to encapsulate↵
              the gory details of initializing subscribers, publishers and services
43      void initializePublishers();
44      void initializeServices();
45
```

```
46      void subscriberCallback(const std_msgs::Float32& message_holder); //prototype for ↵
            callback of example subscriber
47      //prototype for callback for example service
48      bool serviceCallback(std_srvs::TriggerRequest& request, std_srvs::TriggerResponse&↵
            response);
49  }; // note: a class definition requires a semicolon at the end of the definition
50
51  #endif   // this closes the header-include trick...ALWAYS need one of these to match #↵
            ifndef
```

The header file in Listing 2.8 defines the structure of a new class, `ExampleRosClass`. This class defines prototypes for a constructor (declared on line 28: `ExampleRosClass(ros::NodeHandle*nodehandle);`), and a variety of private member objects and functions. The header file also declares member variables. These entities are all accessible by member functions of the class.

Defining objects for **publisher**, **subscriber** and **service** allows setup for these to be performed by the constructor, thus simplifying the main program.

Prototypes for the member functions are defined there, declaring the names, return types and argument types. The executable code constituting the implementation of these functions is contained in one or more separate `*.cpp` files. One should avoid putting implementation code in a header (except for very short implementations, such as `get()` and `set()` functions).

As a matter of style, private member objects and variables, *e.g.* `minimal_publisher_`, are named with a trailing underscore to provide a reminder to the user that these variables are accessible only to member functions within `ExampleRosClass`.

Note that the entire header file is contained within a compiler macro starting with `#ifndef EXAMPLE_ROS_CLASS_H_` and ending with `#endif`. This trick should always be used in writing header files. It helps the compiler avoid redundant copies of headers.

The implementation code corresponding to the header file of Listing 2.8 is shown in Listings 2.9 through 2.12, which breaks down the file `example_ros_class.cpp` separately in terms of the constructor, helper methods, callback functions, and the main program.

Listing 2.9: Implementation of `example_ros_class`: preamble and constructor

```
1   //example_ros_class.cpp:
2   //wsn, Jan 2016
3   //illustrates how to use classes to make ROS nodes
4   // constructor can do the initialization work, including setting up subscribers, ↵
            publishers and services
5   // can use member variables to pass data from subscribers to other member functions
6
7   // can test this function manually with terminal commands, e.g. (in separate terminals↵
            ):
8   // rosrun example_ros_class example_ros_class
9   // rostopic echo exampleMinimalPubTopic
10  // rostopic pub -r 4 exampleMinimalSubTopic std_msgs/Float32 2.0
11  // rosservice call exampleMinimalService 1
12
13
14  // this header incorporates all the necessary #include files and defines the class "↵
            ExampleRosClass"
15  #include "example_ros_class.h"
16
17  //CONSTRUCTOR:  this will get called whenever an instance of this class is created
18  // want to put all dirty work of initializations here
19  // odd syntax: have to pass nodehandle pointer into constructor for constructor to ↵
            build subscribers, etc
20  ExampleRosClass::ExampleRosClass(ros::NodeHandle* nodehandle):nh_(*nodehandle)
21  { // constructor
22      ROS_INFO("in class constructor of ExampleRosClass");
23      initializeSubscribers(); // package up the messy work of creating subscribers; do ↵
            this overhead in constructor
```

```
24      initializePublishers();
25      initializeServices();
26
27      //initialize variables here, as needed
28      val_to_remember_=0.0;
29
30      // can also do tests/waits to make sure all required services, topics, etc are ↵
            alive
31  }
```

Listing 2.10: Implementation of **example_ros_class**: helper member methods

```
33  //member helper function to set up subscribers;
34  // note odd syntax: &ExampleRosClass::subscriberCallback is a pointer to a member ↵
        function of ExampleRosClass
35  // "this" keyword is required, to refer to the current instance of ExampleRosClass
36  void ExampleRosClass::initializeSubscribers()
37  {
38      ROS_INFO("Initializing Subscribers");
39      minimal_subscriber_ = nh_.subscribe("example_class_input_topic", 1, &↵
            ExampleRosClass::subscriberCallback,this);
40      // add more subscribers here, as needed
41  }
42
43  //member helper function to set up services:
44  // similar syntax to subscriber, required for setting up services outside of "main()"
45  void ExampleRosClass::initializeServices()
46  {
47      ROS_INFO("Initializing Services");
48      minimal_service_ = nh_.advertiseService("example_minimal_service",
49                                              &ExampleRosClass::serviceCallback,
50                                              this);
51      // add more services here, as needed
52  }
53
54  //member helper function to set up publishers;
55  void ExampleRosClass::initializePublishers()
56  {
57      ROS_INFO("Initializing Publishers");
58      minimal_publisher_ = nh_.advertise<std_msgs::Float32>("example_class_output_topic"↵
            , 1, true);
59      //add more publishers, as needed
60      // note: COULD make minimal_publisher_ a public member function, if want to use it↵
            within "main()"
61  }
```

Listing 2.11: Implementation of **example_ros_class**: callback functions

```
65  // a simple callback function, used by the example subscriber.
66  // note, though, use of member variables and access to minimal_publisher_ (which is a ↵
        member method)
67  void ExampleRosClass::subscriberCallback(const std_msgs::Float32& message_holder) {
68      // the real work is done in this callback function
69      // it wakes up every time a new message is published on "exampleMinimalSubTopic"
70
71      val_from_subscriber_ = message_holder.data; // copy the received data into member ↵
            variable, so ALL member funcs of ExampleRosClass can access it
72      ROS_INFO("myCallback activated: received value %f",val_from_subscriber_);
73      std_msgs::Float32 output_msg;
74      val_to_remember_ += val_from_subscriber_; //can use a member variable to store ↵
            values between calls; add incoming value each callback
75      output_msg.data= val_to_remember_;
76      // demo use of publisher--since publisher object is a member function
77      minimal_publisher_.publish(output_msg); //output the square of the received value;
78  }
79
80
```

```
81  //member function implementation for a service callback function
82  bool ExampleRosClass::serviceCallback(std_srvs::TriggerRequest& request, std_srvs::↩
        TriggerResponse& response) {
83      ROS_INFO("service callback activated");
84      response.success = true; // boring, but valid response info
85      response.message = "here is a response string";
86      return true;
87  }
```

Listing 2.12: Implementation of <code>example_ros_class</code>: main program

```
91   int main(int argc, char** argv)
92   {
93       // ROS set-ups:
94       ros::init(argc, argv, "exampleRosClass"); //node name
95
96       ros::NodeHandle nh; // create a node handle; need to pass this to the class ↩
             constructor
97
98       ROS_INFO("main: instantiating an object of type ExampleRosClass");
99       ExampleRosClass exampleRosClass(&nh);  //instantiate an ExampleRosClass object and↩
             pass in pointer to nodehandle for constructor to use
100
101      ROS_INFO("main: going into spin; let the callbacks do all the work");
102      ros::spin();
103      return 0;
104  }
```

In `example_ros_class.cpp`, the ROS service definition and the ROS publisher are similar to the examples in our `minimal_nodes` package introduced in Chapter 1. Note, though, that the subscriber is able to invoke the publisher, since the publisher is a member object that is accessible to all member functions. Further, the subscriber can copy its received data to member variables (`val_from_subscriber_`, in this example), making this data available to all member functions. In addition, the subscriber callback function illustrates using a member variable (`val_to_remember_`) to store results between calls, since this member variable persistently holds its data between calls to the subscriber. The member variables thus behave similarly to global variables, but these variables are available only to class member methods, and thus this construction is preferred.

The constructor is responsible for setting up the example publisher, example subscriber and example service. A somewhat odd notation is required, though. Main() must instantiate a node handle for the constructor to use, and this value must be passed to the constructor. But the constructor is responsible for initializing class variables, which creates a chicken-and-egg problem. This is resolved with the somewhat odd notation using constructor initializer lists (line 20):

```
ExampleRosClass::ExampleRosClass(ros::NodeHandle* nodehandle):nh_(*nodehandle)
```

This initialization technique allows the main program to create an instance of the class `ExampleRosClass`, passing into the constructor (a pointer to) the nodehandle created by `main()` (line 98):

```
    ExampleRosClass exampleRosClass(&nh);
```

Note also the somewhat odd set-up of the subscriber and the service, which is performed within the constructor:

```
minimal_subscriber_ = nh_.subscribe("example_class_input_topic", 1, &↵
    ExampleRosClass::subscriberCallback,this);
```

and

```
minimal_service_ = nh_.advertiseService("example_minimal_service",&ExampleRosClass::↵
    serviceCallback,this);
```

The notation: &ExampleRosClass::subscriberCallback provides a pointer to the callback function to be used with the subscriber. This callback function is defined as a member function of the current class. Additionally, the keyword this tells the compiler that we are referring to the current instance of this class.

Note that the callback function associated with the service uses a pre-defined service message (line 82):

```
bool ExampleRosClass::serviceCallback(std_srvs::TriggerRequest& request, std_srvs::↵
    TriggerResponse& response)
```

The service message Trigger is defined in std_srvs. To compile the example ROS class package, a dependency on std_srvs is listed in package.xml. For custom service messages (as with custom ROS messages defined generally), it can be convenient to put these into a separate package, as has been done with the ROS-provided std_srvs package.

Although the example class notation is somewhat cumbersome, it is convenient to design our ROS nodes in this fashion. The constructor is thus able to encapsulate the details of setting up publishers, subscribers and services, as well as initialize all important variables. The constructor may also test that required topics and services from companion nodes are active and healthy before releasing control to the "main" program.

In the present example, the main program is very short. It merely creates an instance of the new class, and then it goes into a spin. All of the program work is then performed by callbacks of the new class object.

The example code can be tested (with roscore running) with the command:

```
rosrun example_ros_class example_ros_class
```

Upon start-up, this node displays the following output:

```
[ INFO] [1452372936.675150947]: main: instantiating an object of type ExampleRosClass
[ INFO] [1452372936.675272621]: in class constructor of ExampleRosClass
[ INFO] [1452372936.675323574]: Initializing Subscribers
[ INFO] [1452372936.682326226]: Initializing Publishers
[ INFO] [1452372936.683708338]: Initializing Services
[ INFO] [1452372936.685131573]: main: going into spin; let the callbacks do all the work
```

With this node running, one can peek and poke at the various I/O options from a command line. From a separate terminal, enter:

```
rosservice call example_minimal_service
```

This terminal responds with:

```
success: True
message: here is a response string
```

And the terminal running example_ros_class displays:

```
service callback activated
```

This shows the expected behavior. When a client (in this case, invoked manually from the command line) sends a request to the service named `example_minimal_service`, the corresponding callback function within the node `example_ros_class` is awakened. The callback function (line 83) outputs the text "service callback activated." It then populates the fields of the response message (lines 84 and 85):

```
response.success = true; // boring, but valid response info
response.message = "here is a response string";
```

and these values are returned to the service client via the service response message. The received values are displayed by the (manually invoked) service client, appearing as:

```
success: True
message: here is a response string
```

The subscriber callback function within the `example_ros_class` node can also be tested manually. From a separate terminal, enter:

```
rostopic pub -r 2 example_class_input_topic  std_msgs/Float32 2.0
```

The `rostopic` command accepts arguments in YAML syntax. (See `http://wiki.ros.org/ROS/YAMLCommandLine` for more detail.) The option `-r 2` means that the publication will be repeated with a rate of 2 Hz. After running this command, the output in terminal of node `example_ros_class` appears as:

```
[ INFO] [1452374662.370650465]: myCallback activated: received value 2.000000
[ INFO] [1452374662.870560823]: myCallback activated: received value 2.000000
[ INFO] [1452374663.370577645]: myCallback activated: received value 2.000000
```

With this minimal example, we see how to incorporate ROS publishers, subscribers and services within a C++ class. This example ROS class can be used as a starting point to create new ROS classes. Starting with this example, we rename the class and its services and topics. More services, topics, subscribers and publishers may be added, as desired. Additional member methods can be declared in the header and implementations provided in the main cpp file. Using classes helps to encapsulate some of the tedious boilerplate of ROS, allowing the programmer to focus on algorithms. Additionally, by using classes, one can create libraries that promote ease of code re-use, which is introduced next.

2.4   CREATING LIBRARY MODULES IN ROS

So far, we have created independent packages that take advantage of existing libraries. However, as source code gets lengthier, it is desirable to break it up into smaller modules. If work may be re-used by future modules, then it is best to create a new library. This section describes the steps to creating a library. It will refer to the examples in the accompanying code repository within two separate packages: `creating_a_ros_library` and `using_a_ros_library`.

The package `creating_a_ros_library` was created using:

```
cs_create_pkg creating_a_ros_library roscpp std_msgs std_srvs
```

This sets up the expected package structure, including folders for `src` and `include` and `CMakeLists.txt`, `package.xml` and `README.md` files.

From the previous section, we borrow code from the package `example_ros_class`. The implementation source file `example_ros_class.cpp` is copied to the `src` subdirectory of the new package `creating_a_ros_library`. This code is edited to remove the main() program. Also, the header file inclusion is modified to refer to:

```
#include <creating_a_ros_library/example_ros_class.h>
```

The source code file is otherwise identical to `example_ros_class`. For completeness, the edited file is shown in Listings 2.13 through 2.15, breaking this file into the preamble and constructor, helper member methods, and callback functions.

Listing 2.13: Implementation of library `example_ros_class`: preamble and constructor

```
 1
 2  //example_ros_class.cpp:
 3  //wsn, Jan 2016
 4  //illustrates how to use classes to make ROS nodes
 5  // constructor can do the initialization work, including setting up subscribers, ←
        publishers and services
 6  // can use member variables to pass data from subscribers to other member functions
 7
 8  // can test this function manually with terminal commands, e.g. (in separate terminals←
        ):
 9  // rosrun example_ros_class example_ros_class
10  // rostopic echo exampleMinimalPubTopic
11  // rostopic pub -r 4 exampleMinimalSubTopic std_msgs/Float32 2.0
12  // rosservice call exampleMinimalService 1
13
14
15  // this header incorporates all the necessary #include files and defines the class "←
        ExampleRosClass"
16  #include <creating_a_ros_library/example_ros_class.h>
17
18  //CONSTRUCTOR:  this will get called whenever an instance of this class is created
19  // want to put all dirty work of initializations here
20  // odd syntax: have to pass nodehandle pointer into constructor for constructor to ←
        build subscribers, etc
21  ExampleRosClass::ExampleRosClass(ros::NodeHandle* nodehandle):nh_(*nodehandle)
22  { // constructor
23      ROS_INFO("in class constructor of ExampleRosClass");
24      initializeSubscribers(); // package up the messy work of creating subscribers; do ←
            this overhead in constructor
25      initializePublishers();
26      initializeServices();
27
28      //initialize variables here, as needed
29      val_to_remember_=0.0;
30
31      // can also do tests/waits to make sure all required services, topics, etc are ←
            alive
32  }
```

Listing 2.14: Implementation of library `example_ros_class`: helper member methods

```
34  //member helper function to set up subscribers;
35  // note odd syntax: &ExampleRosClass::subscriberCallback is a pointer to a member ←
        function of ExampleRosClass
36  // "this" keyword is required, to refer to the current instance of ExampleRosClass
37  void ExampleRosClass::initializeSubscribers()
38  {
39      ROS_INFO("Initializing Subscribers");
40      minimal_subscriber_ = nh_.subscribe("example_class_input_topic", 1, &←
            ExampleRosClass::subscriberCallback,this);
41      // add more subscribers here, as needed
```

```
42  }
43
44  //member helper function to set up services:
45  // similar syntax to subscriber, required for setting up services outside of "main()"
46  void ExampleRosClass::initializeServices()
47  {
48      ROS_INFO("Initializing Services");
49      minimal_service_ = nh_.advertiseService("example_minimal_service",
50                                          &ExampleRosClass::serviceCallback,
51                                          this);
52      // add more services here, as needed
53  }
54
55  //member helper function to set up publishers;
56  void ExampleRosClass::initializePublishers()
57  {
58      ROS_INFO("Initializing Publishers");
59      minimal_publisher_ = nh_.advertise<std_msgs::Float32>("example_class_output_topic"←
            , 1, true);
60      //add more publishers, as needed
61      // note: COULD make minimal_publisher_ a public member function, if want to use it←
            within "main()"
62  }
```

Listing 2.15: Implementation of library `example_ros_class`: callback functions

```
66  // a simple callback function, used by the example subscriber.
67  // note, though, use of member variables and access to minimal_publisher_ (which is a ←
        member method)
68  void ExampleRosClass::subscriberCallback(const std_msgs::Float32& message_holder) {
69      // the real work is done in this callback function
70      // it wakes up every time a new message is published on "exampleMinimalSubTopic"
71
72      val_from_subscriber_ = message_holder.data; // copy the received data into member ←
            variable, so ALL member funcs of ExampleRosClass can access it
73      ROS_INFO("myCallback activated: received value %f",val_from_subscriber_);
74      std_msgs::Float32 output_msg;
75      val_to_remember_ += val_from_subscriber_; //can use a member variable to store ←
            values between calls; add incoming value each callback
76      output_msg.data= val_to_remember_;
77      // demo use of publisher--since publisher object is a member function
78      minimal_publisher_.publish(output_msg); //output the square of the received value;
79  }
80
81
82  //member function implementation for a service callback function
83  bool ExampleRosClass::serviceCallback(std_srvs::TriggerRequest& request, std_srvs::←
        TriggerResponse& response) {
84      ROS_INFO("service callback activated");
85      response.success = true; // boring, but valid response info
86      response.message = "here is a response string";
87      return true;
88  }
```

The header file with the class prototype needs to reside in a location where `catkin_make` can find it when it is desired to link our library to nodes in new packages. By convention, the header file is located under the `/include` directory–but not directly. Instead, a subdirectory of the `/include` directory is created (with the same name as the package) and the library's header file is placed here, *i.e.* in `.../creating_a_ros_library/include/creating_a_ros_library`. The original header file, `example_ros_class.h`, is copied to this directory. Subsequently, nodes may import this header file by including the line:

```
#include <creating_a_ros_library/example_ros_class.h>
```

(the same as done with the library).

The `CMakeLists.txt` file is edited to include (or uncomment) the line:

```
cs_add_library(example_ros_library src/example_ros_class.cpp)
```

This informs **catkin_make** that a new library is to be created, that this new library will be named **example_ros_library**, and the source code for this library resides in src/example_ros_class.cpp.

At this point, the library can be compiled by running **catkin_make** (from the **ros_ws** directory). After compilation, a new file appears in the directory ~/ros_ws/devel/lib, called **libexample_ros_library.so**. The base name, **example_ros_library.so**, follows from the name assignment in **cs_add_library**, and the build system prepends the name "lib." However, specifying linking to this library does not require knowledge of this new name. It is only necessary to note the package dependency in **package.xml** and to include the associated header file with **#include <creating_a_ros_library/example_ros_class.h>**.

To see whether our new library is working, we can create a test **main()** program in the same package. The file **example_ros_class_test_main.cpp** is shown in Listing 2.16.

Listing 2.16: **example_ros_class_test_main.cpp**: example of using a class

```
1   #include <creating_a_ros_library/example_ros_class.h>
2
3   int main(int argc, char** argv)
4   {
5       // ROS set-ups:
6       ros::init(argc, argv, "example_lib_test_main"); //node name
7
8       ros::NodeHandle nh; // create a node handle; need to pass this to the class ↵
            constructor
9
10      ROS_INFO("main: instantiating an object of type ExampleRosClass");
11      ExampleRosClass exampleRosClass(&nh);  //instantiate an ExampleRosClass object and↵
            pass in pointer to nodehandle for constructor to use
12
13      ROS_INFO("main: going into spin; let the callbacks do all the work");
14      ros::spin();
15      return 0;
16  }
```

This test program is identical to the corresponding **main()** function in **example_ros_class/src/example_ros_class.cpp**, except that the header file is specified as **#include <creating_a_ros_library/example_ros_class.h>** (line 1).

To compile the test main, two changes to **CMakeLists.txt** are required. The first is to specify compilation of a new executable, using the line:

```
cs_add_executable(ros_library_test_main src/example_ros_class_test_main.cpp)
```

The second change specifies that our new node should be linked with our new library, which requires the line:

```
target_link_libraries(ros_library_test_main example_ros_library)
```

This linker command references our executable-file name, **ros_library_test_main**, as the first argument and our new library name, **example_ros_library**, as the second argument.

We can run our new test node with the command:

```
rosrun creating_a_ros_library example_ros_class_test_main
```

which, as usual, references our package name (**creating_a_ros_library**) and the name of an executable in that package (`example_ros_class_test_main`). The result is an output identical (except for the time stamps) to that obtained in the previous section:

```
[ INFO] [1452395548.594901501]: main: instantiating an object of type ExampleRosClass
[ INFO] [1452395548.595099162]: in class constructor of ExampleRosClass
[ INFO] [1452395548.595126811]: Initializing Subscribers
[ INFO] [1452395548.599454035]: Initializing Publishers
[ INFO] [1452395548.600454593]: Initializing Services
[ INFO] [1452395548.601531005]: main: going into spin; let the callbacks do all the work
```

Our example `main()`, `example_ros_class_test_main`, shows how to use our new library within a node. Typically, however, new code will need to reference a library that exists in a separate package. Doing so is quite simple, as illustrated with the separate package `using_a_ros_library`. This package (included in the accompanying code repository) was created with:

```
cs_create_pkg using_a_ros_library roscpp std_msgs std_srvs creating_a_ros_library
```

Note that the above use of `cs_create_pkg` declares a dependency on the package `creating_a_ros_library`. This was not done when creating the package `creating_a_ros_library` (since this would have been circular, attempting to depend on the package being created). Because of this, the `CMakeLists.txt` file required the linker line `target_link_libraries(ros_library_test_main example_ros_library)` to note that the test main should be linked with the library created in the same package.

The test main program `example_ros_class_test_main.cpp` was copied from the `src` folder of the package `creating_a_ros_library` to the `src` folder of the new package `using_a_ros_library`. No changes were made to the test main source file. To compile this new node, the `CMakeLists.txt` file was edited to include the line:

```
cs_add_executable(ros_library_external_test_main src/example_ros_class_test_main.cpp)
```

However, using `catkin_simple`, it is not necessary to include a command to link with the new library. Since the `package.xml` file already notes a dependency on the package `creating_a_ros_library`, and since the source code includes the header file `#include <creating_a_ros_library/example_ros_class.h>`, catkin_simple recognizes that the executable should be linked with the library `example_ros_library` from package `creating_a_ros_library`. Using `catkin_make` without `catkin_simple` would require editing several additional lines of `CMakeLists.txt` to declare header locations, libraries to link in, and (in more general cases) compilation of custom messages.

Our new test node can be run with:

```
rosrun using_a_ros_library ros_library_external_test_main
```

The output and the behavior are identical to the corresponding cases in packages `example_ros_class` and `creating_a_ros_library`.

The above explanation illustrates how to create a ROS library. As code becomes more complex, using libraries is increasingly valuable for encapsulating detail and re-using software.

2.5 INTRODUCTION TO ACTION SERVERS AND ACTION CLIENTS

We have seen ROS communications mechanisms of publish–subscribe and service–client. A third important communications paradigm in ROS is the action server–action client pattern.

The action server–action client approach is similar to service–client communications, in that there are peer-to-peer communications between the service and the client. The client always knows the recipient (service) of the client's requests, and the service always knows the client to which it should respond. A limitation of the service–client approach is that the client "blocks" waiting on the service to respond. If the desired behavior requires a long time (*e.g.*, `clear_the_table`), it is desirable that the client continue to run other important tasks (*e.g.* `check_the_battery_voltage`) while the requested behavior is performed. Action servers–action clients fill this role.

There are many options and variations in construction of action servers and action clients; only simple examples are treated here. Further details can be found on-line at `http://wiki.ros.org/actionlib`. Example code corresponding to the present introduction can be found in the accompanying repository within the package `example_action_server`.

From `http://wiki.ros.org/actionlib`:

> In any large ROS-based system, there are cases when someone would like to send a request to a node to perform some task, and also receive a reply to the request. This can currently be achieved via ROS services.

> In some cases, however, if the service takes a long time to execute, the user might want the ability to cancel the request during execution or get periodic feedback about how the request is progressing. The actionlib package provides tools to create servers that execute long-running goals that can be preempted. It also provides a client interface in order to send requests to the server.

A common use of an action server is for execution of a pre-planned trajectory. If one has computed a sequence of joint-space poses to be realized at specified time steps, this entire message can be delivered to an action server for execution. (This is the approach used in ROS-Industrial, to transmit desired trajectories to industrial robots, where the trajectory is subsequently executed using the native controller.) Designing with action servers allows one to exploit `SMACH` (see `http://wiki.ros.org/smach`) for higher-level state-machine programming, as well as alternative decision-making coordination packages.

2.5.1   Creating an action server package

A new action server package may be created by navigating to `ros_ws/src` (or some subdirectory from here) and (using the `catkin_simple` option) invoking:

```
cs_create_pkg example_action_server roscpp actionlib
```

This has already been done in the accompanying example-code repository (see `https://github.com/wsnewman/learning_ros/tree/master/Part_1/example_action_server`). The package `example_action_server` uses the library `actionlib`.

Invoking `cs_create_pkg` (or the long version of `CMakeLists.txt`, via `catkin_create_pkg`) does much of the preparatory work, including establishing a `package.xml` file, a `CmakeLists.txt` file and subdirectories `include` and `src`. However, we will need an additional subdirectory to define a custom action message. Creation of an action message is similar to creation of a service message. Navigate to within the new package directory and invoke:

```
mkdir action
```

We will use this directory to define our communications message format between our new server and its future clients. To have the new action message pre-processed by `catkin_make`, it is also necessary to edit the `package.xml` file (as was done previously for `.msg` and `.srv` message creation) and uncomment the lines:

```
<build_depend>message_generation</build_depend>
```

and

```
<run_depend>message_runtime</run_depend>
```

This will result in auto-generating various header files to define message types for action client–action server communications.

2.5.2 Defining custom action-server messages

We have seen previously how to define custom messages for publish–subscribe communications, as well as custom service messages for client–service communications. The required steps included creating a corresponding folder in the package directory (`msg` or `srv` for messages and service messages, respectively) and composing a simple text file describing the format of the desired messages (`*.msg` or `*.srv` files, respectively). Service messages are slightly more complex than simple messages, since service messages require specification of both a request and a response.

Action server–action client messages are created similarly. Within the `action` subdirectory of a package, create a new file with the suffix `.action`. Like service messages, action messages have multiple regions, with three regions for action messages (versus two regions for service messages). These regions are the `goal`, `result` and `feedback` components.

In our example, the `action` folder contains the file `demo.action` with the following text:

```
#goal definition
#the lines with the hash signs are merely comments
#goal, result and feedback are defined by this fixed order, and separated by 3 hyphens
int32 input
---
#result definition
int32 output
int32 goal_stamp
---
#feedback
int32 fdbk
```

In the above, we prescribed three fields: `goal`, `result` and `feedback`. The goal definition in this simple case contains only a single component, called `input`, which is of type `int32`. Note that the # sign is a comment delimiter. The labels for `goal`, `result` and `feedback` are just reminders. Message generation will ignore these comments and assume that there are three fields in fixed order (goal, result, feedback), separated by three dashes.

Following three dashes, the `result` message is defined. In this example, the `result` definition consists of two components: `output` and `goal_stamp`, both of which are of type `int32`. The final definition, `feedback`, is also separated by three dashes, and the example contains only one field, called `fdbk`, also of type `int32`.

An action message must be written in the above format. The dashes and the order are important; the comments are optional, but helpful.

The components defined in these fields can contain any existing message definitions, provided the corresponding message packages have been named in the `package.xml` file and the corresponding headers are included in the source code file (to be composed, as described below).

Because our `package.xml` file has `message_generation` enabled, once we perform `catkin_make` on our new package, the build system will create multiple new *.h header files, although this is not obvious. These files will be located in `~/ros_ws/devel/include` within a subdirectory corresponding to our package name, `example_action_server`. Within this directory, there will be seven *.h files created, each with a name that starts with `demo` (the name we chose for our action message specification). Six of these are included within the seventh header file, `demoAction.h`. By including this composite header in action node code, we may then refer to message types such as `demoGoal` and `demoResult`.

The example action server source code, `example_action_server.cpp`, is shown in Listings 2.17 through 2.20, separated by class definition, constructor, callback function, and main program, respectively.

Listing 2.17: `example_action_server`: class definition

```
1   // example_action_server: a simple action server
2   // this version does not depend on actionlib_servers hku package
3   // wsn, October, 2014
4
5   #include<ros/ros.h>
6   #include <actionlib/server/simple_action_server.h>
7   //the following #include refers to the "action" message defined for this package
8   // The action message can be found in: .../example_action_server/action/demo.action
9   // Automated header generation creates multiple headers for message I/O
10  // These are referred to by the root name (demo) and appended name (Action)
11  #include<example_action_server/demoAction.h>
12
13  int g_count = 0;
14  bool g_count_failure = false;
15
16  class ExampleActionServer {
17  private:
18
19      ros::NodeHandle nh_;  // we'll need a node handle; get one upon instantiation
20
21      // this class will own a "SimpleActionServer" called "as_".
22      // it will communicate using messages defined in example_action_server/action/demo↩
                .action
23      // the type "demoAction" is auto-generated from our name "demo" and generic name "↩
                Action"
24      actionlib::SimpleActionServer<example_action_server::demoAction> as_;
25
26      // here are some message types to communicate with our client(s)
27      example_action_server::demoGoal goal_; // goal message, received from client
28      example_action_server::demoResult result_; // put results here, to be sent back to↩
                the client when done w/ goal
29      example_action_server::demoFeedback feedback_; // not used in this example;
30      // would need to use: as_.publishFeedback(feedback_); to send incremental feedback↩
                to the client
31
32
33
34  public:
35      ExampleActionServer(); //define the body of the constructor outside of class ↩
                definition
36
37      ~ExampleActionServer(void) {
38      }
39      // Action Interface
40      void executeCB(const actionlib::SimpleActionServer<example_action_server::↩
                demoAction>::GoalConstPtr& goal);
41  };
```

Listing 2.18: `example_action_server`: constructor

```
53  //implementation of the constructor:
54  // member initialization list describes how to initialize member as_
55  // member as_ will get instantiated with specified node-handle, name by which this ←
        server will be known,
56  //  a pointer to the function to be executed upon receipt of a goal.
57  //
58  // Syntax of naming the function to be invoked: get a pointer to the function, called ←
        executeCB, which is a member method
59  // of our class exampleActionServer.  Since this is a class method, we need to tell ←
        boost::bind that it is a class member,
60  // using the "this" keyword.  the _1 argument says that our executeCB takes one ←
        argument
61  // the final argument  "false" says don't start the server yet.  (We'll do this in the←
        constructor)
62
63  ExampleActionServer::ExampleActionServer() :
64      as_(nh_, "example_action", boost::bind(&ExampleActionServer::executeCB, this, _1),←
        false)
65  // in the above initialization, we name the server "example_action"
66  //  clients will need to refer to this name to connect with this server
67  {
68      ROS_INFO("in constructor of exampleActionServer...");
69      // do any other desired initializations here...specific to your implementation
70
71      as_.start(); //start the server running
72  }
```

Listing 2.19: `example_action_server`: callback function

```
74  //executeCB implementation: this is a member method that will get registered with the ←
        action server
75  // argument type is very long.  Meaning:
76  // actionlib is the package for action servers
77  // SimpleActionServer is a templated class in this package (defined in the "actionlib"←
        ROS package)
78  // <example_action_server::demoAction> customizes the simple action server to use our ←
        own "action" message
79  // defined in our package, "example_action_server", in the subdirectory "action", ←
        called "demo.action"
80  // The name "demo" is prepended to other message types created automatically during ←
        compilation.
81  // e.g.,  "demoAction" is auto-generated from (our) base name "demo" and generic name ←
        "Action"
82  void ExampleActionServer::executeCB(const actionlib::SimpleActionServer<←
        example_action_server::demoAction>::GoalConstPtr& goal) {
83      //ROS_INFO("in executeCB");
84      //ROS_INFO("goal input is: %d", goal->input);
85      //do work here: this is where your interesting code goes
86
87      //....
88
89      // for illustration, populate the "result" message with two numbers:
90      // the "input" is the message count, copied from goal->input (as sent by the ←
            client)
91      // the "goal_stamp" is the server's count of how many goals it has serviced so far
92      // if there is only one client, and if it is never restarted, then these two ←
            numbers SHOULD be identical...
93      // unless some communication got dropped, indicating an error
94      // send the result message back with the status of "success"
95
96      g_count++; // keep track of total number of goals serviced since this server was ←
            started
97      result_.output = g_count; // we'll use the member variable result_, defined in our←
            class
98      result_.goal_stamp = goal->input;
99
100     // the class owns the action server, so we can use its member methods here
101
```

```
102        // DEBUG: if client and server remain in sync, all is well--else whine and ←
               complain and quit
103        // NOTE: this is NOT generically useful code; server should be happy to accept new←
               clients at any time, and
104        // no client should need to know how many goals the server has serviced to date
105        if (g_count != goal->input) {
106            ROS_WARN("hey--mismatch!");
107            ROS_INFO("g_count = %d; goal_stamp = %d", g_count, result_.goal_stamp);
108            g_count_failure = true; //set a flag to commit suicide
109            ROS_WARN("informing client of aborted goal");
110            as_.setAborted(result_); // tell the client we have given up on this goal; ←
                   send the result message as well
111        }
112        else {
113            as_.setSucceeded(result_); // tell the client that we were successful acting ←
                   on the request, and return the "result" message
114        }
115  }
```

Listing 2.20: `example_action_server`: main program

```
107  int main(int argc, char** argv) {
108      ros::init(argc, argv, "demo_action_server_node"); // name this node
109
110      ROS_INFO("instantiating the demo action server: ");
111
112      ExampleActionServer as_object; // create an instance of the class "←
               ExampleActionServer"
113
114      ROS_INFO("going into spin");
115      // from here, all the work is done in the action server, with the interesting ←
               stuff done within "executeCB()"
116      // you will see 5 new topics under example_action: cancel, feedback, goal, result,←
               status
117      while (!g_count_failure) {
118          ros::spinOnce(); //normally, can simply do: ros::spin();
119          // for debug, induce a halt if we ever get our client/server communications ←
                   out of sync
120      }
121
122      return 0;
123  }
```

In `example_action_server.cpp`, line 6:

```
#include <actionlib/server/simple_action_server.h>
```

brings in the header file associated with the `simple_action_server` package, which is necessary to use this library. Line 11:

```
#include <example_action_server/demoAction.h>
```

brings in the action message descriptions, which are auto-generated header files created by `catkin_make`, as prescribed in our `action` subdirectory of our new package, `example_action_server`.

Line 16,

```
class ExampleActionServer {
```

begins the definition of class `ExampleActionServer`. This class includes an object from the `SimpleActionServer` class defined in the `actionlib` library. This object, given the name `as_`, is declared on line 24:

```
actionlib::SimpleActionServer<example_action_server::demoAction> as_;
```

The prototype for class `ExampleActionServer` also incorporates three message objects: `goal_`, `result_` and `feedback_`, declared in lines 27 through 29:

```
example_action_server::demoGoal goal_; // goal message, received from client
example_action_server::demoResult result_; // put results here, to be sent back to↩
    the client when done w/ goal
example_action_server::demoFeedback feedback_; // not used in this example;
```

Both client and server code will need to refer to these new messages to communicate with each other. To further illuminate, line 28 of the example C++ code instantiates a variable called `result_` that is of type `demoResult` as defined in the package `example_action_server`. Subsequently, we may refer to components of `result_`, such as in line 87:

```
result_.output = g_count;
```

When the server returns the goal message to its client, the value assigned to `result_.output` will be received by the client (as well as other fields of the result that are populated by the server).

Our class `ExampleActionServer` has a constructor, declared on line 35, the body of which is implemented in lines 53 through 62.

The most important component of the class `ExampleActionServer` is the callback function, the prototype for which appears on line 40.

```
void executeCB(const actionlib::SimpleActionServer<example_action_server::demoAction↩
    >::GoalConstPtr& goal);
```

This callback function will be associated with the `SimpleActionServer` object `as_`. The implementation of the callback function (lines 72 through 105, in this example) contains the heart of the code that will get executed to perform requested services. The prototype for this callback function is fairly long-winded. The method name `executeCB` is our arbitrary choice. The argument of this function is a pointer to a goal message. Declaration of the goal message refers to the `actionlib` library, the templated class `SimpleActionServer`, and our own action message, referred to by `example_action_server::demoAction`, which further has an auto-generated type `GoalConstPtr`. This syntax is tedious. However, the example may be copied verbatim and the programmer may change the specific function, package and message names as desired.

The constructor implementation, which starts at line 53, is also fairly cryptic:

```
ExampleActionServer::ExampleActionServer() :
    as_(nh_, "example_action", boost::bind(&ExampleActionServer::executeCB, this, _1),↩
        false)
```

The purpose of this line is to initialize the object `as_`. The initializer arguments specify that the new action server will be known to the ROS system by the name `example_action`.

Additionally, we wish to affiliate a callback function with the action server. Our class `ExampleActionServer` has a member method `executeCB` that we wish to use. This is accomplished in initialization of `as_` using `boost::bind`. In the arguments of `boost::bind` we specify that the callback function `executeCB` defined within the namespace of class `ExampleActionServer` is to be used. The keyword `this` indicates that `executeCB` is a member of the current object. The argument **_1** is used by **boost::bind** to tell the simple action server object that our defined callback function takes one argument.

The final initialization argument, **false**, tells the constructor that we do not yet wish to start the new action server running. Instead, our action server is started running by line 61 in the constructor:

```
as_.start(); //start the server running
```

Action server start-up is deferred to this point to avoid a race condition, assuring that initialization of the server is completed before it starts running.

The **main()** function is composed in lines 107 through 123. Line 112:

```
ExampleActionServer as_object; // create an instance of the class "↵
    ExampleActionServer"
```

instantiates an object of the new class **ExampleActionServer**. As is the case with simpler ROS services, the main program can now simply go into **spin()**, letting the callback function of the new action server do all the work. Also like ROS services, this action server node consumes negligible CPU or bandwidth resources while waiting for action client goal requests.

In the callback function **executeCB()**, the code may refer to the information contained within the goal message. Contents of the goal message are transmitted by an action client to the action server. The callback function may have information to be returned to the client; such information should be returned to the client by populating the expected fields of **result_**.

The simple example here does not illustrate how to send feedback messages to the client. (For further details, consult the on-line ROS tutorials.) The purpose of the feedback message is to provide the client with status reports while goal execution is in progress. Feedback messages are not required, but they are often helpful.

Once the important work within **executeCB** has been performed, this function must conclude by invoking either **as_.setAborted(result_)** or **as_.setSucceeded(result_)**. In either case, providing the argument **result_** causes these member methods to transmit the message **result_** back to the client. Additionally, the client is informed whether the server concluded successfully or aborted.

In the minimal example provided, the action server merely performs a simple communications diagnostic (as implemented in the callback function). Within **executeCB()** (in this example), the server keeps track of how many times it has serviced goals, and it copies this value into the **output** field of the result message (lines 86 and 87):

```
g_count++;
result_.output = g_count;
```

Also, on line 88, the **input** field of the goal message is copied to the **goal_stamp** field of the **result_** message.

```
result_.goal_stamp = goal->input;
```

The client will receive back the values in these fields of the **result_** message.

For this simple diagnostic server, if the server's record of number of goals achieved differs from the client's input, the server outputs an error message and shuts down (lines 95 through 101). Otherwise, the server reports success and returns the result message to the client (line 103):

```
as_.setSucceeded(result_);
```

In designing a new action server, the preceding complexity largely can be ignored by copying and pasting the provided example code. The heart of the server is contained within the implementation of **executeCB**, and this is where new code should be composed for new server behaviors.

Aside from the callback function implementation, designing of a new action server can re-use the example code here, provided necessary name changes are performed consistently. The new action server would reside in a new package, and this package name should replace each instance of **example_action_server** in the example code. Also, a more mnemonic class name should be chosen, and this new name should be substituted for each occurrence of **ExampleActionServer**.

A mnemonic action message appropriate for the new action server should be created, and the base name of this action message should be substituted everywhere for the example name "demo" (*e.g.* change ::demoAction, *etc.* to use the new action message base name).

The variable names **goal_**, **result_** and **feedback_** can be changed, but these names should be valid to stay as-is. The callback function name, **executeCB**, may be changed if desired, but the example name may be retained without confusion.

Importantly, the server name (currently **example_action**) should be changed to a name that is meaningful for the new server. Finally, the node name in **main()** (**demo_action_server_node**) should also be changed to something relevant and mnemonic.

2.5.3 Designing an action client

A compatible client of our new action server also resides in the same example package, **example_action_server**, with source file called **example_action_client.cpp**. The code is shown in Listing 2.21.

Listing 2.21: **example_action_client**

```
1   // example_action_client:
2   // wsn, October, 2014
3
4   #include <ros/ros.h>
5   #include <actionlib/client/simple_action_client.h>
6
7   //this #include refers to the new "action" message defined for this package
8   // the action message can be found in: .../example_action_server/action/demo.action
9   // automated header generation creates multiple headers for message I/O
10  // these are referred to by the root name (demo) and appended name (Action)
11  // If you write a new client of the server in this package, you will need to include ↩
            example_action_server in your package.xml,
12  // and include the header file below
13  #include <example_action_server/demoAction.h>
14
15
16  // This function will be called once when the goal completes
17  // this is optional, but it is a convenient way to get access to the "result" message ↩
            sent by the server
18  void doneCb(const actionlib::SimpleClientGoalState& state,
19          const example_action_server::demoResultConstPtr& result) {
20      ROS_INFO(" doneCb: server responded with state [%s]", state.toString().c_str());
21      int diff = result->output - result->goal_stamp;
22      ROS_INFO("got result output = %d; goal_stamp = %d; diff = %d",result->output,↩
            result->goal_stamp,diff);
23  }
24
25  int main(int argc, char** argv) {
26      ros::init(argc, argv, "demo_action_client_node"); // name this node
27      int g_count = 0;
```

```
28        // here is a "goal" object compatible with the server, as defined in ←
              example_action_server/action
29        example_action_server::demoGoal goal;
30
31        // use the name of our server, which is: example_action (named in ←
              example_action_server.cpp)
32        // the "true" argument says that we want our new client to run as a separate ←
              thread (a good idea)
33        actionlib::SimpleActionClient<example_action_server::demoAction> action_client←
              ("example_action", true);
34
35        // attempt to connect to the server:
36        ROS_INFO("waiting for server: ");
37        bool server_exists = action_client.waitForServer(ros::Duration(5.0)); // wait ←
              for up to 5 seconds
38        // something odd in above: does not seem to wait for 5 seconds, but returns ←
              rapidly if server not running
39        //bool server_exists = action_client.waitForServer(); //wait forever
40
41        if (!server_exists) {
42            ROS_WARN("could not connect to server; halting");
43            return 0; // bail out; optionally, could print a warning message and retry
44        }
45
46
47        ROS_INFO("connected to action server");  // if here, then we connected to the ←
              server;
48
49        while(true) {
50        // stuff a goal message:
51        g_count++;
52        goal.input = g_count; // this merely sequentially numbers the goals sent
53        //action_client.sendGoal(goal); // simple example--send goal, but do not ←
              specify callbacks
54        action_client.sendGoal(goal,&doneCb); // we could also name additional ←
              callback functions here, if desired
55        //    action_client.sendGoal(goal, &doneCb, &activeCb, &feedbackCb); //e.g., ←
              like this
56
57        bool finished_before_timeout = action_client.waitForResult(ros::Duration(5.0))←
              ;
58        //bool finished_before_timeout = action_client.waitForResult(); // wait ←
              forever...
59        if (!finished_before_timeout) {
60            ROS_WARN("giving up waiting on result for goal number %d",g_count);
61            return 0;
62        }
63        else {
64            //if here, then server returned a result to us
65        }
66
67        }
68
69    return 0;
70 }
```

Like the action server, the action client must bring in a header file from the actionlib library, specifically for the `simple_action_client` (line 5)

```
#include <actionlib/client/simple_action_client.h>
```

The action client (like the action server) must also refer to the custom action message, defined in the current package (code line 13):

```
#include<example_action_server/demoAction.h>
```

Often, different client programs may use the same server, and these clients may be defined in separate packages. In that case, it is necessary to specify in `package.xml` a dependency on the package that contains the action file that defines the action message to be used (and the corresponding header file must be included in the client's source code).

The main program of our action client (starting on line 25) instantiates an object of type `example_action_server::demoGoal` on line 29:

```
example_action_server::demoGoal goal;
```

This object is used to communicate goals to the action server.

An object called `action_client` is instantiated from the (templated) class `actionlib::SimpleActionClient`. We specialize this object to use our defined action messages with the template specification (line 33):

```
actionlib::SimpleActionClient<example_action_server::demoAction> action_client("↩
    example_action", true);
```

The new action client is constructed with the argument `example_action`. This is the name chosen in the design of our action server, as specified in the class constructor:

```
exampleActionServer::exampleActionServer() :
as_(nh_, "example_action", boost::bind(&exampleActionServer::executeCB, this, _1),↩
    false)
```

Action client nodes (like service clients) need to know the names of their respective servers in order to connect.

One of the member functions of an action client is `waitForServer()`. Line 37:

```
bool server_exists = action_client.waitForServer(ros::Duration(5.0)); // wait up to 5 ↩
    sec
```

causes the action client to attempt to connect to the named server. The format allows for waiting only up to some time limit, or waiting indefinitely (if the duration argument is omitted). This method returns `true` if successful connection is made.

In the example program, the main program keeps a count of the number of goals sent to the action server, and the goal message has its input field filled with the current iteration count (lines 51 and 52):

```
    g_count++;
    goal.input = g_count;
```

The server is requested to perform its service, as specified in the goal message, by invoking the member function (line 54):

```
action_client.sendGoal(goal,&doneCb);
```

This form of `sendGoal` includes reference to a callback function (arbitrarily) named `doneCb`.

The client then invokes line 57:

```
bool finished_before_timeout =
action_client.waitForResult(ros::Duration(5.0));
```

which causes the client to suspend, waiting on the server, but with a specified time-out limit.

If the server returns within the specified time limit, the goalCB function will be triggered. This function receives the result message provided by the server.

In the example code, the callback function (lines 18 through 23) compares the number of goals serviced by the server so far to the number of goals this client requested. It prints out the difference between the two. If only this client requests goals from the server and the client and server are both started in the same session, we should expect that the value of diff is always zero, since the number of goals served by the server and the number of goals requested by this client should be identical. However, if either the client or server is stopped and restarted (or if a second client is started), there will be a count mismatch.

This client example may be re-used for designing new action clients. For a new client package, the client code must refer to the respective server's package name, its respective server name, and its corresponding action file. In place of all instances of example_action_server, substitute the new client's package name. For all occurrences of action message name demo, substitute the new action message name. The new client's node name (within ros::init() on line 26) should be changed to something appropriate, mnemonic, and unique. In instantiating the action client (line 33), the name of the desired action server should be substituted for example_action. Most importantly, the doneCb() function (which may or may not be renamed) should contain some useful application code that accomplishes the desired objectives.

2.5.4 Running the example code

To run the example code, first make sure a roscore is running. The server and client can then be started in either order with the following commands in separate terminals:

```
rosrun example_action_server example_action_server
```

and

```
rosrun example_action_server example_action_client
```

The output of the action client looks like:

```
[ INFO] [1452388037.084421813]: got result output = 21756; goal_stamp = 21756; diff = 0
[ INFO] [1452388037.086694318]:  doneCb: server responded with state [SUCCEEDED]
[ INFO] [1452388037.086733133]: got result output = 21757; goal_stamp = 21757; diff = 0
[ INFO] [1452388037.088981490]:  doneCb: server responded with state [SUCCEEDED]
[ INFO] [1452388037.089020387]: got result output = 21758; goal_stamp = 21758; diff = 0
[ INFO] [1452388037.091277514]:  doneCb: server responded with state [SUCCEEDED]
[ INFO] [1452388037.091318054]: got result output = 21759; goal_stamp = 21759; diff = 0
```

With these nodes running, there will be five new topics under /example_action: cancel, feedback, goal, result, and status. You can watch the nodes communicate by running:

```
rostopic echo example_action/goal
```

which produces an output like:

```
---
header:
  seq: 21758
  stamp:
    secs: 1452388037
```

```
      nsecs: 89057580
    frame_id: ''
goal_id:
  stamp:
    secs: 1452388037
    nsecs: 89057840
  id: /demo_action_client_node-21759-1452388037.89057840
goal:
  input: 21759
---
header:
  seq: 21759
  stamp:
    secs: 1452388037
    nsecs: 91357355
  frame_id: ''
goal_id:
  stamp:
    secs: 1452388037
    nsecs: 91357624
  id: /demo_action_client_node-21760-1452388037.91357624
goal:
  input: 21760
---
```

or examine the result message with:

```
rostopic echo example_action/result
```

which produces an output like:

```
---
header:
  seq: 21759
  stamp:
    secs: 1452388037
    nsecs: 92533965
  frame_id: ''
status:
  goal_id:
    stamp:
      secs: 1452388037
      nsecs: 91357624
    id: /demo_action_client_node-21760-1452388037.91357624
  status: 3
  text: ''
result:
  output: 21760
  goal_stamp: 21760
---
```

You can try killing and restarting these nodes to observe the behaviors to timeouts or goal-count mismatches.

An idiosyncrasy of action servers in ROS is that reconnection (or new connection) from client to server appears to be somewhat unreliable. It can be helpful to include automatic connection re-tries to achieve more reliable connection. For example, the following code snippet persists in re-trying to connect a client `cart_move_action_client_` to its associated action server.

```
// attempt to connect to the server:
ROS_INFO("waiting for server: ");
bool server_exists = false;
while ((!server_exists)&&(ros::ok())) {
    server_exists = cart_move_action_client_.waitForServer(ros::Duration(0.5)); //
    ros::spinOnce();
    ros::Duration(0.5).sleep();
    ROS_INFO("retrying...");
}
ROS_INFO("connected to action server"); // if here, then we connected to the ←
    server;
```

A second action server–action client illustrative example is given in `example_action_server_w_fdbk.cpp` and `timer_client.cpp`. The contents of the action server code appear in Listings 2.22 through 2.25.

Listing 2.22: `example_action_server_w_fdbk.cpp`: class definition

```
1  // example_action_server: 2nd version, includes "cancel" and "feedback"
2  // expects client to give an integer corresponding to a timer count, in seconds
3  // server counts up to this value, provides feedback, and can be cancelled any time
4  // re-use the existing action message, although not all fields are needed
5  // use request "input" field for timer setting input,
6  // value of "fdbk" will be set to the current time (count-down value)
7  // "output" field will contain the final value when the server completes the goal ←
       request
8
9  #include<ros/ros.h>
10 #include <actionlib/server/simple_action_server.h>
11 //the following #include refers to the "action" message defined for this package
12 // The action message can be found in: .../example_action_server/action/demo.action
13 // Automated header generation creates multiple headers for message I/O
14 // These are referred to by the root name (demo) and appended name (Action)
15 #include<example_action_server/demoAction.h>
16
17 int g_count = 0;
18 bool g_count_failure = false;
19
20 class ExampleActionServer {
21 private:
22
23     ros::NodeHandle nh_;  // we'll need a node handle; get one upon instantiation
24
25     // this class will own a "SimpleActionServer" called "as_".
26     // it will communicate using messages defined in example_action_server/action/demo←
           .action
27     // the type "demoAction" is auto-generated from our name "demo" and generic name "←
           Action"
28     actionlib::SimpleActionServer<example_action_server::demoAction> as_;
29
30     // here are some message types to communicate with our client(s)
31     example_action_server::demoGoal goal_; // goal message, received from client
32     example_action_server::demoResult result_; // put results here, to be sent back to←
           the client when done w/ goal
33     example_action_server::demoFeedback feedback_; // for feedback
34     //  use: as_.publishFeedback(feedback_); to send incremental feedback to the ←
           client
35     int countdown_val_;
36
37
38 public:
39     ExampleActionServer(); //define the body of the constructor outside of class ←
           definition
40
41     ~ExampleActionServer(void) {
```

```
42      }
43          // Action Interface
44          void executeCB(const actionlib::SimpleActionServer<example_action_server::↩
                demoAction>::GoalConstPtr& goal);
45  };
```

Listing 2.23: `example_action_server_w_fdbk.cpp`: constructor

```
47  //implementation of the constructor:
48  // member initialization list describes how to initialize member as_
49  // member as_ will get instantiated with specified node-handle, name by which this ↩
        server will be known,
50  //   a pointer to the function to be executed upon receipt of a goal.
51  //
52  // Syntax of naming the function to be invoked: get a pointer to the function, called ↩
        executeCB,
53  // which is a member method of our class exampleActionServer.
54  // Since this is a class method, we need to tell boost::bind that it is a class member↩
        ,
55  // using the "this" keyword.  the _1 argument says that our executeCB function takes ↩
        one argument
56  // The final argument, "false", says don't start the server yet. (We'll do this in ↩
        the constructor)
57
58  ExampleActionServer::ExampleActionServer() :
59      as_(nh_, "timer_action", boost::bind(&ExampleActionServer::executeCB, this, _1),↩
            false)
60  // in the above initialization, we name the server "example_action"
61  //   clients will need to refer to this name to connect with this server
62  {
63      ROS_INFO("in constructor of exampleActionServer...");
64      // do any other desired initializations here...specific to your implementation
65
66      as_.start(); //start the server running
67  }
```

Listing 2.24: `example_action_server_w_fdbk.cpp`: callback function

```
69  //executeCB implementation: this is a member method that will get registered with the ↩
        action server
70  // argument type is very long.  Meaning:
71  // actionlib is the package for action servers
72  // SimpleActionServer is a templated class in this package (defined in the "actionlib"↩
        ROS package)
73  // <example_action_server::demoAction> customizes the simple action server to use our ↩
        own "action" message
74  // defined in our package, "example_action_server", in the subdirectory "action", ↩
        called "demo.action"
75  // The name "demo" is prepended to other message types created automatically during ↩
        compilation.
76  // e.g., "demoAction" is auto-generated from (our) base name "demo" and generic name ↩
        "Action"
77  void ExampleActionServer::executeCB(const actionlib::SimpleActionServer<↩
        example_action_server::demoAction>::GoalConstPtr& goal) {
78      ROS_INFO("in executeCB");
79      ROS_INFO("goal input is: %d", goal->input);
80      //do work here: this is where your interesting code goes
81      ros::Rate timer(1.0); // 1Hz timer
82      countdown_val_ = goal->input;
83      //implement a simple timer, which counts down from provided countdown_val to 0, in↩
            seconds
84      while (countdown_val_>0) {
85          ROS_INFO("countdown = %d",countdown_val_);
86
87          // each iteration, check if cancellation has been ordered
88          if (as_.isPreemptRequested()){
89              ROS_WARN("goal cancelled!");
```

```
90          result_.output = countdown_val_;
91          as_.setAborted(result_); // tell the client we have given up on this goal; ←
                    send the result message as well
92          return; // done with callback
93        }
94
95        //if here, then goal is still valid; provide some feedback
96        feedback_.fdbk = countdown_val_; // populate feedback message with current ←
                    countdown value
97        as_.publishFeedback(feedback_); // send feedback to the action client that ←
                    requested this goal
98        countdown_val_--; //decrement the timer countdown
99        timer.sleep(); //wait 1 sec between loop iterations of this timer
100       }
101       //if we survive to here, then the goal was successfully accomplished; inform the ←
                    client
102       result_.output = countdown_val_; //value should be zero, if completed countdown
103       as_.setSucceeded(result_); // return the "result" message to client, along with "←
                    success" status
104  }
```

Listing 2.25: `example_action_server_w_fdbk.cpp`: main program

```
106  int main(int argc, char** argv) {
107      ros::init(argc, argv, "timer_action_server_node"); // name this node
108
109      ROS_INFO("instantiating the timer_action_server: ");
110
111      ExampleActionServer as_object; // create an instance of the class "←
                ExampleActionServer"
112
113      ROS_INFO("going into spin");
114      // from here, all the work is done in the action server, with the interesting ←
                stuff done within "executeCB()"
115      // you will see 5 new topics under example_action: cancel, feedback, goal, result,←
                status
116      ros::spin();
117
118      return 0;
119  }
```

The program `example_action_server_w_fdbk.cpp` is mostly identical to `example_action_server.cpp`, except for the name by which ROS will know the action server, `timer_action` (line 58), and the body of the callback function (lines 77 through 104). In the callback function, the goal is interpreted as a timer value, and the job of the callback function is to count down by the specific number of seconds, then return.

The callback function has two important new elements. First, lines 88 through 93 test whether the client has requested that the current goal be preempted. If so, the current goal is aborted, the client is informed of this status, and the callback function concludes. It is the designer's responsibility to test for this status within the action server at whatever frequency is appropriate.

Checking for goal cancellation is important. For example, a trajectory command might be sent to a robot that is subsequently found to be a bad idea (*e.g.* a person walks into the robot's workspace). It may be necessary to stop the commanded motion suddenly, based on an alarm computed via a separate node. Alternatively, the commanded motion might be stalled, *e.g.* by a barrier blocking a specified navigation path. This could result in a deadlock, with the action server never returning a result to the action client. The client should be able to evaluate progress of a commanded goal and cancel it if necessary.

The second important feature of this action server appears in lines 96 and 97. Here, the feedback message is updated and published. Such feedback can be used by the action client to evaluate progress, *e.g.* to detect and resolve deadlocks. It is up to the designer to provide a corresponding client subscription to receive and evaluate such feedback.

One could design an equivalent node that subscribed to a goal topic and published to feedback and result topics. However, this could create confusion when multiple client nodes attempt to communicate with this action node. Feedback and result messages that are published would be received by all clients, not just the client for which the current goal is being processed. Using the action server–action client means of communication avoids this confusion. Only the client of the current goal receives feedback and result information. When a goal is completed, any other client can request a new goal.

This example uses the "simple" action server library. The simple version does not attempt to handle multiple goals in parallel, nor does it attempt to queue requests. Any goal requested while a goal is still in process will cause pre-emption of the goal in process. This is true whether the conflicting goals come from a single client or from multiple clients running in parallel. The designer must consider implications of this behavior when using simple action servers.

A corresponding action client that interacts with `example_action_server_w_fdbk.cpp` is `timer_client.cpp`, shown in Listings 2.26 and 2.27.

Listing 2.26: `timer_client.cpp`: preamble and callback functions

```
1   // timer_client: works together with action server called "timer_action"
2   // in source: example_action_server_w_fdbk.cpp
3   // this code could be written using classes instead (e.g. like the corresponding ←↩
        server)
4   //  see: http://wiki.ros.org/actionlib_tutorials/Tutorials/Writing%20a%20Callback%20←↩
        Based%20Simple%20Action%20Client
5
6   #include<ros/ros.h>
7   #include <actionlib/client/simple_action_client.h>
8   #include<example_action_server/demoAction.h> //reference action message in this ←↩
        package
9
10  using namespace std;
11
12  bool g_goal_active = false; //some global vars for communication with callbacks
13  int g_result_output = -1;
14  int g_fdbk = -1;
15
16  // This function will be called once when the goal completes
17  // this is optional, but it is a convenient way to get access to the "result" message ←↩
        sent by the server
18  void doneCb(const actionlib::SimpleClientGoalState& state,
19          const example_action_server::demoResultConstPtr& result) {
20      ROS_INFO(" doneCb: server responded with state [%s]", state.toString().c_str());
21      ROS_INFO("got result output = %d",result->output);
22      g_result_output= result->output;
23      g_goal_active=false;
24  }
25
26  //this function wakes up every time the action server has feedback updates for this ←↩
        client
27  // only the client that sent the current goal will get feedback info from the action ←↩
        server
28  void feedbackCb(const example_action_server::demoFeedbackConstPtr& fdbk_msg) {
29      ROS_INFO("feedback status = %d",fdbk_msg->fdbk);
30      g_fdbk = fdbk_msg->fdbk; //make status available to "main()"
31  }
32
33  // Called once when the goal becomes active; not necessary, but could be useful ←↩
        diagnostic
34  void activeCb()
35  {
36      ROS_INFO("Goal just went active");
37      g_goal_active=true; //let main() know that the server responded that this goal is in←↩
        process
38  }
```

Listing 2.27: `timer_client.cpp`: main program

```
40
41  int main(int argc, char** argv) {
42      ros::init(argc, argv, "timer_client_node"); // name this node
43      ros::NodeHandle n;
44      ros::Rate main_timer(1.0);
45      // here is a "goal" object compatible with the server, as defined in ←
            example_action_server/action
46      example_action_server::demoGoal goal;
47
48      // use the name of our server, which is: timer_action (named in ←
            example_action_server_w_fdbk.cpp)
49      // the "true" argument says that we want our new client to run as a separate ←
            thread (a good idea)
50      actionlib::SimpleActionClient<example_action_server::demoAction> action_client←
            ("timer_action", true);
51
52      // attempt to connect to the server: need to put a test here, since client ←
            might launch before server
53      ROS_INFO("attempting to connect to server: ");
54      bool server_exists = action_client.waitForServer(ros::Duration(1.0)); // wait ←
            for up to 1 second
55      // something odd in above: sometimes does not wait for specified seconds,
56      //  but returns rapidly if server not running; so we'll do our own version
57      while (!server_exists) { // keep trying until connected
58          ROS_WARN("could not connect to server; retrying...");
59          server_exists = action_client.waitForServer(ros::Duration(1.0)); // retry ←
                every 1 second
60      }
61      ROS_INFO("connected to action server");  // if here, then we connected to the ←
            server;
62
63      int countdown_goal = 1; //user will specify a timer value
64      while(countdown_goal>=0) {
65          cout<<"enter a desired timer value, in seconds (0 to abort, <0 to quit): ";
66          cin>>countdown_goal;
67          if (countdown_goal==0) { //see if user wants to cancel current goal
68              ROS_INFO("cancelling goal");
69              action_client.cancelGoal(); //this is how one can cancel a goal in ←
                    process
70          }
71          if (countdown_goal<0) { //option for user to shut down this client
72              ROS_INFO("this client is quitting");
73              return 0;
74          }
75          //if here, then we want to send a new timer goal to the action server
76          ROS_INFO("sending timer goal= %d seconds to timer action server",←
                countdown_goal);
77          goal.input = countdown_goal; //populate a goal message
78          //here are some options:
79          //action_client.sendGoal(goal); // simple example--send goal, but do not ←
                specify callbacks
80          //action_client.sendGoal(goal,&doneCb); // send goal and specify a callback←
                function
81          //or, send goal and specify callbacks for "done", "active" and "feedback"
82          action_client.sendGoal(goal, &doneCb, &activeCb, &feedbackCb);
83
84          //this example will loop back to the the prompt for user input. The main ←
                function will be
85          // suspended while waiting on user input, but the callbacks will still be ←
                alive
86          //if user enters a new goal value before the prior request is completed, ←
                the prior goal will
87          // be aborted and the new goal will be installed
88
89      }
90      return 0;
91  }
```

This client illustrates several additional features. Instead of one callback function, there are three. The callback function `activeCb()` (lines 34 through 38) is invoked when the action server accepts the client's goal request. This callback function is not required, but it can be useful for diagnostics. In the example, the callback function sets a flag to inform the `main()` function that the new goal is acknowledged and in process.

The callback function `feedbackCb()` (lines 28 through 31) receives feedback messages from the action server. It copies the results to global variables to make them accessible to `main()`. From such updates, the main function can confirm that the action server is still alive and progressing. A useful example of action-server feedback would be to report back on accomplishment of subgoals, *e.g.* from a goal request to navigate to a sequence of intermediate points in space or a goal request to perform a sequence of manipulations.

In instantiating the action client, the corresponding action server, `timer_action`, is named (line 49).

Lines 53 through 60 implement attempting to connect to the named action server, with re-tries once per second until the connection is successful. This is a useful construction, since launching code may result in action clients starting up before their respective action servers. Re-trying connections will allow this node to wait patiently until a successful connection is achieved. An alternative is for the client to wait forever with the function:

```
action_client.waitForServer();
```

A limitation of this, however, is that there will be no output from the client that may be suspended indefinitely without warning to the operator.

To associate the feedback functions with the goal request, the feedback functions are named in the `sendGoal` command (line 81):

```
action_client.sendGoal(goal, &doneCb, &activeCb, &feedbackCb);
```

With this construction, the action server will communicate its results back to this action client node.

Another important feature of this action client is the absence of the call `action_client.waitForResult()`. Our client can proceed with other business while its requested goal is carried out by the action server. For example, the client might be a co-ordinator responsible for checking evolving status, such as low battery voltage, detection of an anomaly that is important to investigate, failure of a navigation goal to progress, or detection of an error in the status of an on-going assembly. If the coordinator is not blocked waiting on the action server, it can evaluate such conditions and, as desired, abort the current goal and re-plan to generate an alternative goal.

In the simple timer client example, the client prompts an operator for a keyboard input corresponding to a timer duration (an integer number of seconds). If the goal is 0, this is interpreted (in this example) to mean that the operator wishes to cancel the current goal. This is done with the call (line 68):

```
action_client.cancelGoal();
```

If the requested goal is positive, then the client sends this value as a new goal. The default behavior of the simple action server is to accept the new goal, pre-empting the previous goal if it is not concluded.

As with service servers and service clients, the action server can remain alive indefinitely, serving various clients that may come and go.

Figure 2.1 is a screenshot displaying the output from the client and server interaction for the timer example, along with a display of `rqt_console`. In this example, the client was started first. The console display shows that the node `timer_client_node` was attempting to connect to the server. Messages 2 through 6 report unsuccessful retries to connect.

Message 7 is reported by the timer action server, which was started after the client. At this point, the server is waiting for goals and the client is waiting for user input.

After about about 8 seconds (as shown by the time stamps on messages 8 and 9), the user enters 100, and the client echoes that the value 100 is being sent as a goal to the action server (console message 9). The action server reports that its callback function is activated (message 10). With acceptance of the goal, the client node's corresponding callback function is activated (message 13). Both the server and client display the countdown status (messages 12 through 18). This shows that the client is successfully receiving feedback messages from the server (although the client's main function is ignoring this information in this example).

Note that during this countdown, the client's main program is suspended, waiting on new input from the user. However, the client's callback functions continue to respond to messages from the action server in spite of the fact that there are no `ros::spin()` nor `ros::spinOnce()` calls. This is because, when the action client was instantiated (`timer_client.cpp` line 49):

```
actionlib::SimpleActionClient<example_action_server::demoAction> action_client("↩
    timer_action", true);
```

the argument `true` specified that the action client should be run as a separate thread. As a result, the callback functions continue to respond to messages even when the main program is blocked, and no `spin` calls are required.

While the client is suspended, the action server continues to count down, and the client's callback functions continue to receive feedback. After a few seconds of countdown, the user enters "0" to the client timer prompt, and the console messages 19 and 20 show that the client interprets this as a request to cancel the current goal. Message 22 shows that the action server acknowledges the cancellation request. The client and server are both ready to accept new goals.

A second client may be started while the first client is still connected, with or without a goal completion pending. However, if the second client attempts to send a goal while the first client's goal is still in process, the server cancels the first goal, and the first client's `doneCB()` callback function receives notification and prints out:

```
doneCb: server responded with state [ABORTED]
```

The designer should be aware of this behavior. It can be useful to cancel an ongoing goal initiated by one client then cancelled by another client. This can help dislodge a deadlock if the first client misbehaves.

An action client that sends a goal can conclude (or be killed) without stopping the action server from continuing to process the goal. Even though the action server will attempt to report back to the (deceased) action client, this communication is merely a publication, and failure to receive this publication does not impede the action server from continuing along normally.

This concludes the introduction to action servers. A fourth means of ROS communications known as the parameter server is introduced next.

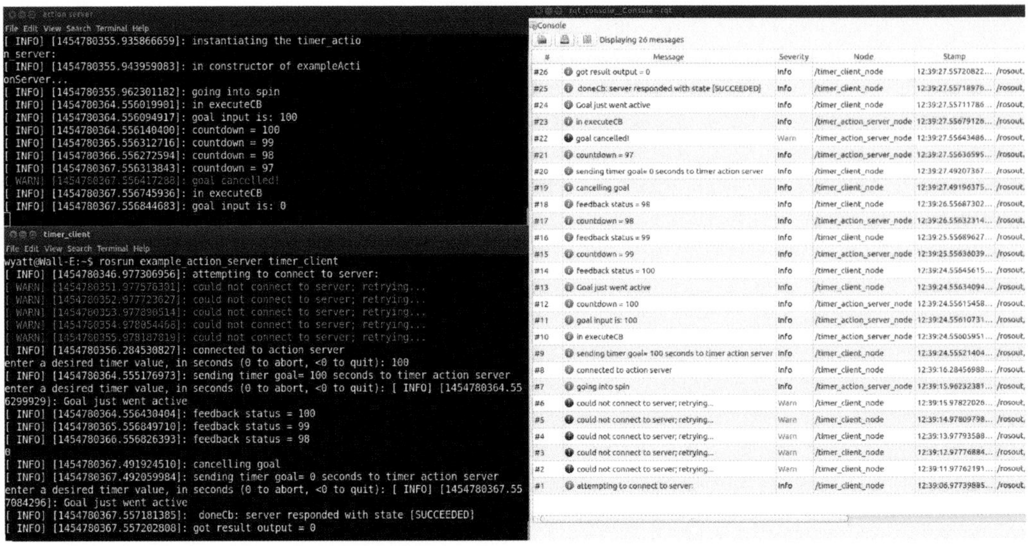

Figure 2.1: Output of `rqt_console` and action client and action server terminals with timer example

2.6 INTRODUCTION TO PARAMETER SERVER

Sending a message, a service request or an action server goal are all transactions that are typically very fast. At the other end of the spectrum, the ROS parameter server is useful for sharing values that are infrequently changed. From `http://wiki.ros.org/Parameter% 20Server`:

> A parameter server is a shared, multi-variate dictionary that is accessible via network APIs. Nodes use this server to store and retrieve parameters at runtime. As it is not designed for high-performance, it is best used for static, non-binary data such as configuration parameters.

While a publisher broadcasts its messages and service and action server communications are peer-to-peer, the parameter server is more like a shared memory. Via code within nodes, via terminal commands or via launch files, any process can can read, write or alter parameter values on the parameter server. This communication mechanism can be very convenient, but it must also be used appropriately and with care.

The parameter server is commonly used to set configuration parameters or specifications, including:

- Controller gains for joint servos

- Coordinate transform parameters, *e.g.* for tool transforms

- Sensor intrinsic and extrinsic calibration parameters

- Robot models

It is frequently convenient to store parameter settings in YAML files, which can be loaded into the parameter server via a command line, or (more commonly) as procedures within a launch file. With this approach, one can make parameter changes and retest a system without touching or recompiling the source code. If a sensor mount is changed, the modified coordinate transform data can be incorporated in the system by editing the corresponding configuration file and re-launching the system. If a robot's end-effector is changed, the corresponding new tool transform can be incorporated by changing its corresponding configuration file. Parameter changes to a robot's model can also be introduced without changing any source code.

On the other hand, the parameter server should not be used for dynamically changing values. Unlike a subscriber, a service or an action service, incoming data is not constantly detected. Rather, nodes typically consult the parameter server once, upon start-up, and they are subsequently unaware of changes to the parameter server.

The `rosparam` command is a tool for getting and setting ROS parameters on the parameter server using YAML-encoded information. (See `http://wiki.ros.org/rosparam`.) Options for `rosparam` are shown by running:

```
rosparam
```

which results in displaying:

```
rosparam is a command-line tool for getting, setting, and deleting parameters from
the ROS Parameter Server.

Commands:
rosparam set set parameter
rosparam get get parameter
rosparam load load parameters from file
rosparam dump dump parameters to file
rosparam delete delete parameter
rosparam list list parameter names
```

A common example would be to set proportional, derivative and integral-error gains for a servo control algorithm. Values of interest could be put on the parameter server manually, *e.g.* setting P, I and D values to 1,2 and 3, respectively, with the command:

```
rosparam set /gains "p: 1.0
i: 2.0
d: 3.0"
```

The result of this action can be seen by entering:

```
rosparam list
```

which produces the output:

```
/gains/d
/gains/i
/gains/p
/rosdistro
/roslaunch/uris/host_wall_e__54538
/rosversion
/run_id
```

The command:

```
rosparam get /gains/d
```

displays the expected result of 3.0. Alternatively, all of the gains can be seen by commanding:

```
rosparam get /gains
```

which displays:

{d: 3.0, i: 2.0, p: 1.0}

The entire set of parameter values can be obtained by dumping it to a file with, *e.g.*:

```
rosparam dump param_dump
```

where **param_dump** is a chosen file name in which to dump the data. The contents of **param_dump** at this point are:

```
gains: {d: 3.0, i: 2.0, p: 1.0}
rosdistro: 'indigo

  '
roslaunch:
  uris: {host_wall_e__54538: 'http://Wall-E:54538/'}
rosversion: '1.11.13

  '
run_id: f8504ae0-b747-11e5-af4f-c48508582a82
```

Alternatively, the parameter values may be loaded from a file. As an example, the file **test_param.yaml** with contents:

joint1_gains: {p: 4.0, i: 5.0, d: 6.0}

can be loaded onto the parameter server with the command:

```
rosparam load jnt1_gains.yaml
```

After loading this data, it can be confirmed that the desired values are present on the parameter server with the command:

```
rosparam get /joint1_gains
```

which results in displaying:

{d: 6.0, i: 5.0, p: 4.0}

To automate loading configuration files, ROS launch files accept a parameter tag. As an example, a launch file called **load_gains.launch** can be composed with the contents:

```
<launch>
<rosparam command="load" file="jnt1_gains.yaml" />
</launch>
```

and run with the command:

```
roslaunch load_gains.launch
```

The effect of this command is identical to running: **rosparam load jnt1_gains.yaml**.

While it is convenient to be able to load parameters onto the parameter server automatically through launch files, the value of the parameter server is realized when nodes are able to access the data. Means to access parameter data using C++ code are described in http://wiki.ros.org/roscpp/Overview/Parameter%20Server. An example package illustrating accessing parameter data is in the accompanying repository under example_parameter_server. This package contains src and launch subdirectories. The launch directory contains the launch file load_gains.launch, introduced above. The YAML file jnt1_gains.yaml also resides in this directory. The desired data can be loaded into the parameter server with the command:

```
roslaunch example_parameter_server load_gains.launch
```

Within the package example_parameter_server, the src folder contains a file named read_param_from_node.cpp with contents:

Listing 2.28: read_param_from_node.cpp: C++ code to illustrate pulling parameters from the parameter server

```
1  #include <ros/ros.h>#include <ros/ros.h>
2
3  int main(int argc, char **argv) {
4    ros::init(argc, argv, "param_reader"); // name of this node will be "↩
         minimal_publisher"
5    ros::NodeHandle nh; // two lines to create a publisher object that can talk to ROS
6    double P_gain,D_gain,I_gain;
7
8    if (nh.getParam("/joint1_gains/p", P_gain)) {
9    ROS_INFO("proportional gain set to %f",P_gain);
10   }
11   else
12   {
13   ROS_WARN("could not find parameter value /joint1_gains/p on parameter server");
14   }
15   if (nh.getParam("/joint1_gains/d", D_gain)) {
16   ROS_INFO("proportional gain set to %f",D_gain);
17   }
18   else
19   {
20   ROS_WARN("could not find parameter value /joint1_gains/d on parameter server");
21   }
22   if (nh.getParam("/joint1_gains/i", I_gain)) {
23   ROS_INFO("proportional gain set to %f",I_gain);
24   }
25   else
26   {
27   ROS_WARN("could not find parameter value /joint1_gains/i on parameter server");
28   }
29  }
```

The code is compiled with the CMakeLists.txt line cs_add_executable(read_param_from_nodesrc/read_param_from_node.cpp). To test the resulting executable, first make sure the parameters are loaded by running:

```
roslaunch example_parameter_server load_gains.launch
```

Then, run the test node with:

```
rosrun example_parameter_server read_param_from_node
```

The resulting output displayed is:

```
[ INFO] [1452404560.596815928]: proportional gain set to 7.000000
[ INFO] [1452404560.598031455]: proportional gain set to 9.000000
```

```
[ INFO] [1452404560.599246423]: proportional gain set to 8.000000
```

To see what happens when the data is not present on the parameter server, delete the gain values with the command:

```
rosparam delete /joint1_gains
```

Re-running `rosrun example_parameter_server read_param_from_node` results in the output:

```
[ WARN] [1452404731.031529362]: could not find parameter value /joint1_gains/p
on parameter server
[ WARN] [1452404731.032923909]: could not find parameter value /joint1_gains/d
on parameter server
[ WARN] [1452404731.034251408]: could not find parameter value /joint1_gains/i
on parameter server
```

The function `ROS_WARN()` displays output in a yellow color and carries the tag `WARN`. Such messages are also highlighted in `rqt_console`. The line:

```
if (nh.getParam("/joint1_gains/p", P_gain))
```

tests for the existence of the parameter `/joint1_gains/p` on the parameter server. This capability is useful to make sure start-up is consistent, with all required parameters specified before they are used in operation.

Parameters can also be set programmatically, although such use is less common.

Although the example provided here is (deliberately) trivial, the parameter server is capable of handling much larger and more complex data. A parameter used in most ROS applications is `robot_description`, which contains a full description of a robot model. In Section 3.3, a unified robot model description format will be introduced. Such models typically have corresponding launch files. An example of such a file contains the lines:

```
<!-- send robot urdf to param server -->
<param name="robot_description"
textfile="$(find minimal_robot_description)/minimal_robot_w_sensor.urdf"/>
```

This induces `roslaunch` to search for the package `minimal_robot_description`, find the file `minimal_robot_w_sensor.urdf` within this package, and load the contents onto the parameter server with the associated parameter name `robot_description`. The robot model is then available both for control algorithms and for visualization displays.

2.7 WRAP-UP

This concludes our introduction to ROS foundations. The material presented in Section I describes the philosophy of ROS, including packages, nodes, messages, and services. With this framework, large systems can be built collaboratively and more easily. Contributions can be broken up by packages containing messages, libraries and/or nodes. By defining the interface options among nodes (publish–subscribe, service–client, action server–action client or parameter server), collaborators can make their work compatible with large systems, simplifying integration. Even within a single-person project, breaking up a design task in terms of nodes promotes faster design with simpler integration and supports growth, extensions, upgrades and software re-use. These attributes make ROS attractive for robot software design.

Standardization on the ROS infrastructure is compelling enough to make it a preferred approach to robot software design. In addition, ROS has extensive modeling, simulation and visualization capabilities that further enhance its utility. Modeling, simulation and visualization in ROS will be introduced in Section II.

II

Simulation and Visualization in ROS

INTRODUCTION

Valuable characteristics of ROS include its simulation and visualization capabilities. When software development can be performed only on specific hardware (*e.g.* a particular robot), productivity plummets. A physical robot is typically a bottleneck for development, because it is a limited resource presumably shared by a team. Further, some robots, including vehicles and humanoids, require teams to run experiments. Most often, the robot and team are idling while a programmer is seeking errors in code under test.

With a suitable robot simulator, code can be developed under simulation, allowing work to proceed in parallel and without imposing on a support staff for experiments. If the simulator has high fidelity (providing a good approximation of the actual robot behavior), few or no changes may be required to run the same code on the physical system.

One of the common barriers is that code written for simulation may need to be modified significantly to run on the physical system. This is an issue addressed by ROS. Typically, code written for simulation in ROS requires no changes (not even recompilation) to run on a target physical system. Using ROS's messaging system, nodes are indifferent to where other nodes reside in the system, as long as the message types, topic names, and server names are consistent.

A ROS-enabled robot should publish its sensory data to named topics with defined message types, and the corresponding simulator should publish the same (simulated) information on the same topics with the same message types. If this is achieved, switching a perceptual processing system from simulation to a physical system requires only that the physical system (rather than the simulator) start publishing the information.

In the other direction, one controls a robot by publications to topics (or ROS service requests or ROS action server goal requests). This interface is defined in terms of topic names, message types and server names. If the simulator responds to these publications or requests in the same manner as the physical system, the controller code that runs on the simulation would run identically using the actual robot.

Similarly, human-machine interfaces may be developed with respect to a robot (and environment) simulator, and the developed code may be used verbatim with the physical system. Although the simulator becomes unnecessary when running the physical hardware on the target system, the human interface developed under simulation can be re-used without change from simulation to hardware.

The success of developing software in simulation depends on the quality of the simulator. Fortunately, ROS has powerful simulation capabilities, both for robot dynamics and for sensor emulation in virtual worlds. These capabilities will be introduced in this section.

Simulation in ROS

CONTENTS

INTRODUCTION

This chapter introduces simulation in ROS, starting with a simple two-dimensional mobile-robot simulator and extending to Gazebo–a powerful dynamic simulator. The Unified Robot Description Format (URDF) is introduced for modeling robots suitable for simulation in Gazebo. Controlling wheels or joints of a robot in Gazebo is illustrated in simplified examples.

3.1   SIMPLE TWO-DIMENSIONAL ROBOT SIMULATOR

A helpful place to start the introduction to simulation in ROS is with the Simple Two-Dimensional Robot simulator (STDR). This package is described at `http://wiki.ros.org/stdr_simulator`, with accompanying tutorials at `http://wiki.ros.org/stdr_simulator/Tutorials`. If the STDR package is not already installed, it may be installed with the command:

```
sudo apt-get install ros-$ROS_DISTRO-stdr-simulator
```

(STDR should already be installed when using the recommended installation scripts accompanying this text). STDR may be launched with the command:

```
roslaunch stdr_launchers server_with_map_and_gui_plus_robot.launch
```

Figure 3.1: Display from STDR launch

This results in a screen display as shown in Fig 3.1. A small circle in the lower left corner of the maze represents an abstracted mobile robot. The red rays illustrate simulated laser lines from a hypothetical LIDAR sensor. (See `https://en.wikipedia.org/wiki/Lidar` for a description of LIDAR sensors.) The LIDAR rays extend from the robot out to the first point of reflection in the environment subject to a maximum sensing range. The green cones represent simulated sonar signals. The distance reported by a sonar sensor corresponds to the range to the first sound-reflecting object in the environment within the sensing angle of the sonar (again, subject to a sensing range limitation). Sensor values from the robot simulator are published to topics. The robot subscribes to a command topic, via which the user can control motion of the simulated robot. In fact, running:

```
rostopic list
```

shows more than 30 different topics active from the STDR launch. Additionally,

```
rosservice list
```

shows 24 services running.

With this simple simulator, one can develop code that can interpret the sensor signals, both to avoid collisions and to help identify the robot's location in space. Based on sensor interpretation, one can command incremental robot motions to achieve a desired behavior, such as navigating to specified coordinates in the map. We will defer sensory processing for now and focus on controlling the robot's motion (open loop, without benefit of sensor information). The topic of interest for controlling the robot is `/robot0/cmd_vel`. The name `cmd_vel` is (by convention, though not requirement) the topic name used for controlling robots in velocity-command mode. This topic lives within the **namespace** of `/robot0`. The `cmd_vel` topic has been made a subset of `/robot0` to allow for launching multiple robots within the same simulation. If (as is typical) a ROS system is dedicated to control a single robot, the `/robot0 namespace` may be omitted.

We can examine the topic `/robot0/cmd_vel` with:

```
rostopic info /robot0/cmd_vel
```

which displays:

```
Type: geometry_msgs/Twist

Publishers: None

Subscribers:
 * /robot_manager (http://Wall-E:58336/)
 * /stdr_gui_node_Wall_E_15095_8750187360825501198 (http://Wall-E:60301/)
```

This reveals that the `/robot_manager` node is listening to this topic, but at present there are no publishers to this topic. Further, we see that messages on this topic are of type `geometry_msgs/Twist`. We can examine the `Twist` message with:

```
rosmsg show geometry_msgs/Twist
```

which displays:

```
geometry_msgs/Vector3 linear
   float64 x
   float64 y
   float64 z
geometry_msgs/Vector3 angular
   float64 x
   float64 y
   float64 z
```

The `Twist` message contains the equivalent of two vectors: a speed vector (with x, y and z components) and an angular-velocity (rotation rate) vector, also with x, y and z components. This message type is sufficient for specifying arbitrary velocities in space, both translating and spinning or tumbling. It is often useful to have full 6-D command capability, for example for aerial vehicles, submersible vehicles or the end-effector of an arm.

For our two-D mobile robot, only two of these six components are viable. Constraining the motion of the robot to a plan, the robot can only move forward (relative to its own heading) and/or spin about its center. These speed and spin components correspond to the x component of linear velocity and the z component of angular velocity.

We can command a velocity to the robot manually from a terminal by entering:

```
rostopic pub -r 2 /robot0/cmd_vel geometry_msgs/Twist  '{linear:  {x: 0.5, y: 0.0, z: ↵
    0.0}, angular: {x: 0.0,y: 0.0,z: 0.0}}'
```

The above command instructs the robot to move forward at a velocity of 0.5 meters per second (m/s) with constant heading (zero angular velocity). (Note: by default, all units in ROS messages are international standard, meters-kilograms-seconds, or MKS.) As a result of this command, the robot moves forward until it hits an obstacle, as shown in Fig 3.2. Note that the LIDAR rays and sonar cones have changed to be consistent with the new relative positions of reflective surfaces in the robot's virtual world.

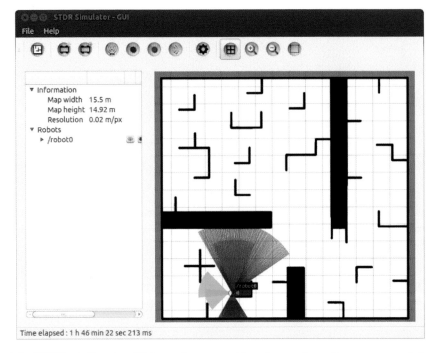

Figure 3.2: STDR stalls at collision after executing forward motion command from start position

The manually entered **Twist** command is reiterated at 2 Hz, via the option **-r 2**. However, if the **rostopic pub** command is halted (with control-C), the robot will continue to try to move with the last **Twist** command received. We can return the **Twist** command to zero with:

```
rostopic pub -r 2 /robot0/cmd_vel geometry_msgs/Twist  '{linear:  {x: 0.0, y: 0.0, z: ↩
    0.0}, angular: {x: 0.0,y: 0.0,z: 0.0}}'
```

To get the robot unstuck and allow it to move forward again, a pure rotation command may be issued, *e.g.* as:

```
rostopic pub -r 2 /robot0/cmd_vel geometry_msgs/Twist  '{linear:  {x: 0.0, y: 0.0, z: ↩
    0.0}, angular: {x: 0.0,y: 0.0,z: 0.1}}'
```

The only non-zero component of the above command is the z component of the angular velocity. The robot's z axis points up (out of the plan view) and a positive angular-velocity command corresponds to counter-clockwise rotation. The commanded angular velocity is 0.1 radians per second. Allowing this command to run for approximately 15 seconds corresponds to a rotation of 1.5 radians, or approximately 90 degrees. The resulting pose of the robot is shown in Fig 3.3. The robot has not translated after this command, but it has rotated (as is apparent from the new sensor visualization) to point approximately toward the upper boundary of the map. If the command:

```
rostopic pub -r 2 /robot0/cmd_vel geometry_msgs/Twist  '{linear:  {x: 0.5, y: 0.0, z: ↩
    0.0}, angular: {x: 0.0,y: 0.0,z: 0.0}}'
```

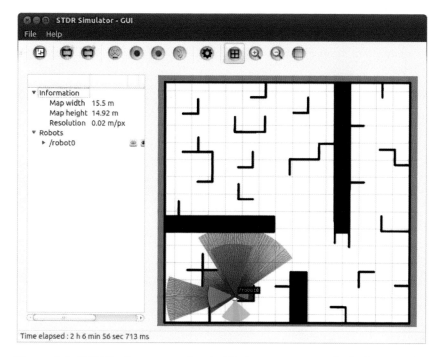

Figure 3.3: STDR after approximate 90-degree counter-clockwise rotation

is issued, the robot will again move forward until it encounters a new barrier, as shown in Fig 3.4. While the manual command-line twist publications illustrate how to control the robot, such commands should be issued programmatically (*i.e.* from a ROS node). An example node to control the STDR mobile robot appears in the accompanying repository within package **stdr_control** (under Part II of the repository). This package was created using:

```
cs_create_pkg stdr_control roscpp geometry_msgs/Twist
```

which specifies a dependency on **geometry_msgs/Twist**, which is needed to publish messages on topic **robot0/cmd_vel**.

A command publisher node, **stdr_open_loop_commander.cpp**, was written, starting from a copy of **minimal_publisher**. The contents of this program appear in Listing 3.1. Line 2 includes the required header file to use the **geometry_msgs/Twist** message.

```
#include <geometry_msgs/Twist.h>
```

A publisher is instantiated on line 7, specifying the command topic and compatible message type:

```
ros::Publisher twist_commander = n.advertise<geometry_msgs::Twist>("/robot0/cmd_vel", ↵
    1);
```

A sample period is defined on line 9:

```
double sample_dt = 0.01; //specify a sample period of 10ms
```

Listing 3.1: `stdr_open_loop_commander.cpp`: C++ code to command velocities to the STDR simulator

```cpp
#include <ros/ros.h>
#include <geometry_msgs/Twist.h>
//node to send Twist commands to the Simple 2-Dimensional Robot Simulator via cmd_vel
int main(int argc, char **argv) {
    ros::init(argc, argv, "stdr_commander");
    ros::NodeHandle n; // two lines to create a publisher object that can talk to ROS
    ros::Publisher twist_commander = n.advertise<geometry_msgs::Twist>("/robot0/↵
        cmd_vel", 1);
    //some "magic numbers"
    double sample_dt = 0.01; //specify a sample period of 10ms
    double speed = 1.0; // 1m/s speed command
    double yaw_rate = 0.5; //0.5 rad/sec yaw rate command
    double time_3_sec = 3.0; // should move 3 meters or 1.5 rad in 3 seconds


    geometry_msgs::Twist twist_cmd; //this is the message type required to send twist ↵
        commands to STDR
    // start with all zeros in the command message; should be the case by default, but↵
        just to be safe..
    twist_cmd.linear.x=0.0;
    twist_cmd.linear.y=0.0;
    twist_cmd.linear.z=0.0;
    twist_cmd.angular.x=0.0;
    twist_cmd.angular.y=0.0;
    twist_cmd.angular.z=0.0;

    ros::Rate loop_timer(1/sample_dt); //create a ros object from the ros Rate class; ↵
        set 100Hz rate
    double timer=0.0;
    //start sending some zero-velocity commands, just to warm up communications with ↵
        STDR
    for (int i=0;i<10;i++) {
        twist_commander.publish(twist_cmd);
        loop_timer.sleep();
    }
    twist_cmd.linear.x=speed; //command to move forward
    while(timer<time_3_sec) {
        twist_commander.publish(twist_cmd);
        timer+=sample_dt;
        loop_timer.sleep();
        }
    twist_cmd.linear.x=0.0; //stop moving forward
    twist_cmd.angular.z=yaw_rate; //and start spinning in place
    timer=0.0; //reset the timer
    while(timer<time_3_sec) {
        twist_commander.publish(twist_cmd);
        timer+=sample_dt;
        loop_timer.sleep();
        }

    twist_cmd.angular.z=0.0; //and stop spinning in place
    twist_cmd.linear.x=speed; //and move forward again
    timer=0.0; //reset the timer
    while(timer<time_3_sec) {
        twist_commander.publish(twist_cmd);
        timer+=sample_dt;
        loop_timer.sleep();
        }
    //halt the motion
    twist_cmd.angular.z=0.0;
    twist_cmd.linear.x=0.0;
    for (int i=0;i<10;i++) {
        twist_commander.publish(twist_cmd);
        loop_timer.sleep();
    }
    //done commanding the robot; node runs to completion
}
```

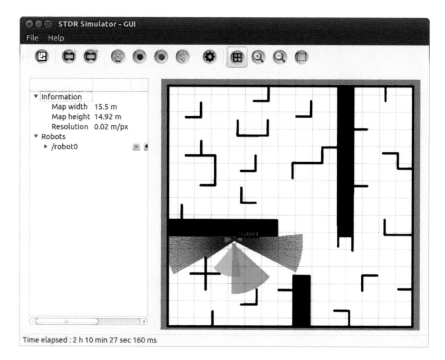

Figure 3.4: STDR stalls again at collision after executing another forward motion command

and this sample period is used to keep a timer consistent with `sleep()` calls of a ROS `Rate` object. A `Twist` object is defined on line 15:

```
geometry_msgs::Twist twist_cmd; //this is the message type required to send twist ↩
    commands to STDR
```

and lines 17 through 22 show how to access all six components of this message object.

The robot is commanded to move forward 3 m, spin counter-clockwise 90 degrees, then move forward again 3 m. This is done open-loop by commanding speeds for computed durations in lines 32 through 36, 39 through 44, and 48 through 53, thus:

```
timer=0.0; //reset the timer
while(timer<time_3_sec) {
    twist_commander.publish(twist_cmd);
    timer+=sample_dt;
    loop_timer.sleep();
    }
```

As a result, the robot moves as desired, ending at the final pose shown in Fig 3.5.

The simple two-dimensional robot simulator may be used to develop more intelligent control code by interpreting sensor values and commanding computed twists that achieve effective reactive behaviors. Approaches to such control will be deferred until sensory-processing techniques have been introduced.

For designing sensor-based behaviors, this level of simulation abstraction may be appropriate, since it easy to operate, requires little computational power and allows the designer to focus on a limited context. This level of abstraction, however, does not include a variety of realistic effects, and thus code developed using this simulator may require additional work with a more general simulator before the code is suitable for testing on a real robot.

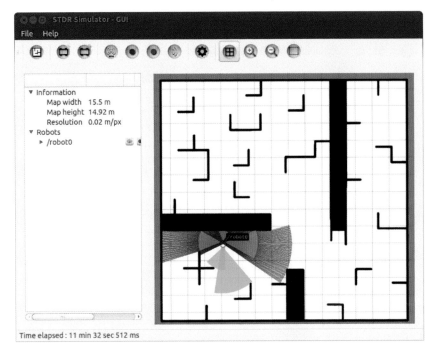

Figure 3.5: STDR final pose after executing programmed speed control

In addition to the 2-D abstraction, the STDR simulator does not account for sensor noise, force and torque disturbances (including wheel slip), or realistic vehicle dynamics.

The STDR dynamic response can be viewed using `rqt_plot`. The topic `/robot0/odom` is useful for analysis, since messages on this topic (of type `nav_msgs/Odometry`) include the simulated robot's velocity and angular velocity. (By convention in ROS, a mobile robot should publish its state to an `odom` topic with message type `nav_msgs/Odometry`, where the state is based on sensor measurements. Details of the `/odom` topic will be covered later, in the context of mobile robots.) Figure 3.6 shows the time history of the commanded and actual robot speed and yaw rate.

Both the speed and spin are commanded with step changes, persisting for 3 seconds within each of the control blocks of the `stdr_open_loop_commander.cpp` code. The corresponding actual robot speed and yaw rate follow the commands very closely, with almost imperceptible errors due to jitter between when the command is issued and when the robot responds. In reality, the influences of inertia, angular inertia, actuator saturation, wheel friction and servo-controller response would result in very different behavior. Expected non-idealities include ramp-up slew rate limitations, overshoot, and wheel slip. The robot might even fall over due to commanded instantaneous braking. A more realistic controller would ramp up and ramp down the velocity commands to respect dynamic limitations. Appropriate ramping can be implemented to control the STDR simulation, but the necessity of such ramping will not be revealed by STDR, nor can we expect that ramping implemented on STDR will be appropriately tuned for a real robot.

More sophisticated robot simulation requires a sufficiently detailed robot model, including geometry, mass properties, surface contact properties and actuator dynamics. These can be specified using the unified robot description format introduced next.

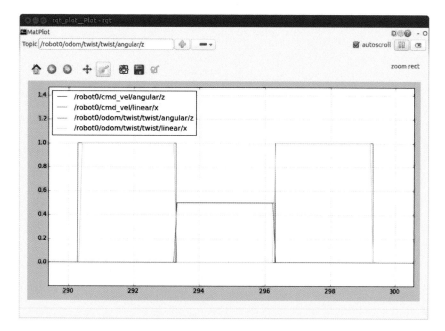

Figure 3.6: STDR commanded and actual speed and yaw rate versus time

3.2 MODELING FOR DYNAMIC SIMULATION

To perform physically realistic dynamic simulations, objects must be specified with sufficient detail. Modeling details are organized within three broad categories: a dynamic model, a collision model and a visual model. (For robots with controlled joints, a kinematic model is also specified.)

Dynamic objects must always include specification of inertial properties. This constitutes the necessary description for an abstract physics model. With an inertial model alone, one can already perform useful simulations. As an example, one could compute the dynamics of a satellite acted on by gravity and thrusters, producing a simulation that incorporates the influences of gravity, centrifugal and Coriolis effects acting on an arbitrary inertia tensor.

To compute the dynamics of interacting bodies including colliding objects or a robot grabbing or pushing an object, it is also necessary to compute the forces and moments due to contacts. Simulation of contact dynamics is challenging, and different physics engines address this problem with different methods. The default physics engine used in ROS is Open Dynamics Engine (ODE). (See `http://ode-wiki.org/wiki`.) This physics engine uses energy and conservation of momentum to deduce the outcome of collisions rather than attempt to detail the very fast dynamics of force profiles during impacts. (As a consequence, this physics engine does a good job of modeling brief collisions, but it suffers from artifacts in simulating sustained contact forces, including wheels on the ground and fingers grasping objects.) To include the dynamic effects of collisions, the simulator must be able to deduce when and where (on each body) contact occurs. It is thus necessary to describe the envelope (boundary description) of 3-D objects in a manner suitable for efficient computation of contact sites. This part of the robot description is a collision model. It can be as simple as a primitive 3-D object description (*e.g.* a rectangular prism or a cylinder), or it can be a high-fidelity model of a complex surface, typically originating from a CAD model. The surface descriptions are translated to faceted approximations (equivalent to a stereolithography, or STL model, comprised of triangular facets), which is convenient for computing intersections (collisions) between model boundaries. A pragmatic concern with the collision model is that a large number of facets results in slower simulations.

The third category of model description is the visual model, which is used for graphical display purposes. Having a graphical display of the computed dynamics can be very helpful in interpreting results and debugging software development. The visual model is often identical to the collision model, since the visual model also requires a boundary description of each object. It is common, however, for the visual model to be of higher fidelity (*e.g.* incorporating more facets) than the collision model. Displaying a high-fidelity model is less demanding than computing collisions using a model with a large number of facets.

In the simplest case, the visual model can be identical to the collision model. However, these two categories can include their own options. The collision model, for example, can include specification of surface properties including friction and resilience (which are irrelevant to visual appearance). The visual model may include specifications of color, reflectivity and transparency (which are irrelevant to the collision model).

Some simple model descriptions are contained in the accompanying repository under the package (ROS folder) `exmpl_models`. Within the subfolder `rect_prism` of the `exmpl_models` package, the file `model-1_4.sdf` is shown in Listing 3.2.

Listing 3.2: `rect_prism/model-1_4.sdf`: model description of a simple rectangular prism

```
1  <?xml version='1.0'?>
2  <sdf version='1.4'>
3    <model name="rect_prism">
4      <link name='link'>
5        <inertial>
6        <mass>2000</mass>
7        <inertia>
8            <ixx>3000</ixx>
9            <ixy>0</ixy>
10           <ixz>0</ixz>
11           <iyy>3000</iyy>
12           <iyz>0</iyz>
13           <izz>1000</izz>
14       </inertia>
15       </inertial>
16       <collision name='collision'>
17         <geometry>
18             <box>
19                 <size> 2 2 4 </size>
20             </box>
21         </geometry>
22       </collision>
23
24       <visual name='visual'>
25         <geometry>
26             <box>
27                 <size> 2 2 4 </size>
28             </box>
29         </geometry>
30       </visual>
31     </link>
32   </model>
33 </sdf>
```

This file contains a model description using the Simulation Description Format (SDF). (SDF; see `SDF;http://sdformat.org`.) The model description is in XML format and contains three fields: inertial, collision, and visual. The inertial properties include a specification of both the mass (in kilograms) and the moments of inertia (in kilograms-m^2); explanation of inertial properties will be detailed later in this chapter. The collision and visual models are identical, in this simple case. They both specify a simple box (a rectangular prism) with x, y and z dimensions of 2, 2 and 4 meters, respectively. The rectangular prism has an associated coordinate frame defined to have its origin in the middle of the box, and with x, y, and z axes parallel to the respective specified dimensions.

The simple rectangular prism model is sufficient for performing interesting dynamic simulations. Modeling robots with joint controllers requires additional detail.

3.3 UNIFIED ROBOT DESCRIPTION FORMAT

A robot model can be specified using SDF. However, the older unified robot description format (URDF; see http://wiki.ros.org/urdf) is closely related, and most existing open-source robot models are expressed in URDF. Further, SDF models end up being translated into URDF for use with ROS. This section will thus describe robot modeling in URDF.

A minimal example is introduced here. A good tutorial for more detail can be found at http://gazebosim.org/tutorials?tut=ros_urdf for the rrbot (a 2-degree-of-freedom robot arm). A URDF model can be used as input for dynamic simulation. The dynamic simulator used with ROS is Gazebo (see http://gazebosim.org/), which offers alternative choices for physics engines.

3.3.1 Kinematic model

A minimal kinematic model of a single-degree-of-freedom robot in the URDF style is given in Listing 3.3, which is links_and_joints.urdf file in the minimal_robot_description package of our accompanying repository.

Listing 3.3: links_and_joints.urdf: one-DOF robot kinematic description

```xml
 1  <?xml version="1.0"?>
 2  <robot  name="one_DOF_robot">
 3
 4    <!-- Base Link -->
 5    <link name="link1" />
 6    <!--distal link -->
 7    <link name="link2" />
 8
 9    <joint name="joint1" type="continuous">
10      <parent link="link1"/>
11      <child link="link2"/>
12      <origin xyz="0 0 0.5" rpy="0 0 0"/>
13      <axis xyz="0 1 0"/>
14    </joint>
15  </robot>
```

In Listing 3.3, a robot called **one_DOF_robot** is defined, consisting of two links and one joint. Descriptions of links and joints are the minimal requirements of a robot definition. These elements are described in XML syntax and may be described in any order in the URDF. It is necessary only that the result is kinematically consistent.

A restriction on URDF files is that one cannot describe closed chains. A URDF will have a single, unique base link, and all other links form a tree relative to this base. This may be a single, open chain (like a conventional robot arm), or the tree may be more complex (such as a humanoid robot with legs, neck, and arms sprouting from a base link on the torso). Closed chains, such as a four-bar linkage, can be "spoofed" in a URDF by making some links kinematically dependent and by introducing virtual actuators; this is adequate for visualization but inaccurate in terms of dynamics. The URDF robot description is not suitable for supporting dynamic simulation of closed-chain mechanisms.

An abstraction of a robot model requires only specifying spatial relationships among (solid) links. In the conventional Denavit-Hartenberg representation, this requires only one frame definition per link, and spatial relations between successive link frames are implied by only four values (three fixed parameters and one joint variable) from parent link to child link. The URDF format departs from this compact description, using up to ten values (nine fixed parameters and one joint variable) instead of the minimum four values. These parameters are described in on-line ROS tutorials; an alternative description of the URDF conventions is offered here.

Each link within a URDF (except for the base link) has a single parent link. However, a link may have multiple child links, *e.g.* such as multiple fingers extending from the base of a hand, or arms and legs extending from a torso.

Each link has a single reference frame, which is rigidly associated with the link and moves with the link as it moves through space. (The link frame may or may not be located within some part of the physical body of the link, but it nonetheless moves with the link conceptually, as though rigidly attached to some part of the physical body of the link.) In addition to the link frame, there may be one or more joint frames that help to describe how children of the link are kinematically related. (Note: in Denavit-Hartenberg notation, there is no such thing as a joint frame; the joint frame is a URDF construct, which may be convenient in some instances, but also introduces confusing redundancy.)

In Listing 3.3, `link1` is the base frame, and `link2` is a child of `link1`. The link frame for the base link is arbitrary—typically a consequence of whatever reference is chosen when describing this link in a CAD system. The respective link frames associated with all other (non-base) links, however, are constrained in their placements by URDF conventions.

For our base link, a joint frame is defined by the lines:

```
<joint name="joint1" type="continuous">
  <parent link="link1"/>
  <child link="link2"/>
  <origin xyz="0 0 1" rpy="0 0 0"/>
  <axis xyz="0 1 0"/>
</joint>
```

The joint frame is named `joint1`. This frame is also associated with `link1` (the parent link) and moves rigidly with `link1`. The position and location of this frame are specified by describing the displacement and rotation of the joint frame relative to the parent (`link1`) reference frame. This spatial relationship is specified by the line:

```
  <origin xyz="0 0 0.5" rpy="0 0 0"/>
```

This specification declares that the origin of the `joint1` frame is offset from the `link1` frame by the vector $[0,0,1]$, *i.e.* aligned with the x and y coordinates of the `link1` frame, but offset by 1 m in the `link1` z direction. (By default, all units in a URDF are MKS.)

We must also specify the orientation of the `joint1` frame. In this case, the `joint1` frame is orientationally aligned with the `link1` reference frame. (*i.e.* the respective x, y and z axes of these two frames are parallel). This is declared via the specification `rpy="0 0 0"`, which says that the `joint1` frame orientation relative to the `link1` reference frame is describable with zero roll, pitch and yaw angles.

The `joint1` frame is useful as an intermediate frame for describing how `link2` is related to `link1`. The line:

```
  <joint name="joint1" type="continuous">
```

declares that the pose of `link2` is related to `link1` through a revolute joint. If unspecified, the joint axis of this revolute joint lies colinear with the x axis of the defined joint frame.

(Note: this is in contrast to Denavit-Hartenberg convention, in which a joint axis is always defined as a z axis.) In our example, the line:

```
<axis xyz="0 1 0"/>
```

specifies that the joint axis is not aligned with the joint-frame x axis but is instead colinear with the joint-frame's y axis. The three components of the `<axis xyz="0 1 0"/>` specification allow for defining an arbitrary joint-axis orientation relative to the joint frame (where the three components specify a unit vector direction in space). Although the joint axis can be oriented arbitrarily, the URDF convention requires that the joint axis pass through the joint-frame's origin. (Note: orientation of the joint axis specifies more than a line in space; as a direction vector, it also implies a positive sense of rotation for the child link relative to the parent link about a joint axis colinear with this vector.)

Having defined a joint, the reference frame for the child link follows implicitly. This is an important conceptual constraint (similar to the Denavit-Hartenberg convention for assigning link frames). In the URDF convention, a child link's reference frame must have its origin coincident with the joint frame that connects the child to the parent. In the simple example provided, the origin of the `link2` frame is defined to be coincident with the origin of the specified `joint1` frame.

The definition of the orientation of the `link2` frame is also constrained by convention, based on the `joint1` frame and definition of a home angle for `joint1`. Since a revolute joint joins `link2` to `link1`, `link2` can move only by rotating about the joint axis defined within the `joint1` frame. When the robot moves with this single degree of freedom, the variable angle is known as the joint angle of `joint1`. The home angle (or zero angle) for `joint1` is a definition, typically chosen to be something convenient (*e.g.* the 0 reading of the associated rotational sensor, or a convenient alignment that is easy to visualize). When a home angle is chosen, the child link's reference frame follows. It is defined to be coincident with its parent's joint frame. That is, if one puts `link2` in the defined home position relative to `link1`, the value of the `joint1` angle will be defined to be zero at this pose, and the `link2` reference frame will be identical to the `joint1` reference frame at this pose. At any other (*i.e.* non-zero, non-periodic) values of the `joint1` angle, the `joint1` and `link2` frames will not be aligned (although their respective origins will remain coincident).

Our example URDF file thus defines a link frame for each link and how these link frames are constrained to move relative to each other. We can check whether our URDF file is consistent by running:

```
check_urdf links_and_joints.urdf
```

We can either run this from the `minimal_robot_description` directory, or provide a path as part of the filename argument to **check_urdf**. The output produced is:

```
robot name is: one_DOF_robot
---------- Successfully Parsed XML ---------------
root Link: link1 has 1 child(ren)
    child(1):  link2
```

This confirms that our file has correct XML syntax and that the robot definition is kinematically consistent.

To this point, our URDF is consistent, but it only defines the constrained spatial relationship between two frames. For simulation purposes, we need to provide more information: a visual model, a collision model and a dynamic model.

3.3.2 Visual model

Our minimal kinematic model is augmented with visual information in Listing 3.4 (which also appears in the package `minimal_robot_description` as file `one_link_description.urdf` in our associated code repository).

Listing 3.4: `one_link_description.urdf`: one-link model with visual description

```
 1   <?xml version="1.0"?>
 2   <robot  name="static_robot">
 3
 4   <!-- Used for fixing robot to the simulator's world frame -->
 5     <link name="world"/>
 6
 7     <joint name="glue_robot_to_world" type="fixed">
 8       <parent link="world"/>
 9       <child link="link1"/>
10     </joint>
11
12   <!-- Base Link -->
13     <link name="link1">
14       <visual>
15         <origin xyz="0 0 0.5" rpy="0 0 0"/>
16         <geometry>
17           <box size="0.2 0.2 1"/>
18         </geometry>
19       </visual>
20     </link>
21   </robot>
```

For simulation purposes, a world frame with a ground plane will be defined, and we can place `link1` fixed in the world by defining a static joint, lines 7 through 10:

```
<joint name="glue_robot_to_world" type="fixed">
  <parent link="world"/>
  <child link="link1"/>
</joint>
```

In defining joint `glue_robot_to_world`, the joint type is specified as `fixed`, which means our `link1` will have a static relationship with respect to the world frame. Further, since we did not specify the x,y,z or r,p,y coordinates of the joint frame, the default values (all 0's) are applied, and our `link1`-frame is thus defined to be identical to the world frame.

Within the `link1` definition, in the sub-section inside the `<visual>` tags, lines 16 through 18

```
<geometry>
  <box size="0.2 0.2 1"/>
</geometry>
```

define a 3-D box entity with dimensions 0.2 by 0.2 by 1.0 meters. For this simple 3-D primitive (essentially identical to our earlier `rect_prism` model), a model frame is defined with origin at the center of the box, and x ,y ,z axes along the specified dimensions 0.2, 0.2, 1.0. This entity can be used to define the visual appearance of `link1`. (More commonly an entire CAD file is referenced to define each robot element, but simple 3-D primitives such as box are useful for quick models and for simple illustration of concepts.)

Although we have defined a `link1` frame (which is coincident with the world frame), we also need to define a visual frame for `link1`, such that our visual model (the box) appears in the correct location. Our `link1` frame has its origin at the level of the ground plane, whereas our box entity has its origin in the center of the box. We wish to place the appearance of the box such that it appears to sit on one (square) face (*i.e.* with this face coplanar with the ground plane) and with the long dimension of the box (the visual frame *z* axis) pointing up (*i.e.* normal to the ground plane and parallel to the *z* axis of the world frame). This is accomplished with line 15:

```
<origin xyz="0 0 0.5" rpy="0 0 0"/>
```

which specifies that the box frame axes are parallel to the respective link1 frame axes, but that the box frame origin is elevated (offset in the link1 frame *z* direction) by 0.5 m.

We cannot test our visualization in our simulator yet until we augment the model with additional, dynamic information.

3.3.3 Dynamic model

For our robot model to be consistent with a physics engine, we must define mass properties for every link in the system. (Although it should not be necessary conceptually, a dynamic model is required even for link1, which we intend to fix rigidly to the ground plane.) Our one-link model, augmented with mass-property information, is shown in Listing 3.5.

Listing 3.5: **one_link_w_mass.urdf**: one-link model with visual and inertial descriptions

```
 1  <?xml version="1.0"?>
 2  <robot   name="static_robot">
 3
 4  <!-- Used for fixing robot to the simulator's world frame -->
 5    <link name="world"/>
 6
 7    <joint name="glue_robot_to_world" type="fixed">
 8      <parent link="world"/>
 9      <child link="link1"/>
10    </joint>
11
12  <!-- Base Link -->
13    <link name="link1">
14      <visual>
15        <origin xyz="0 0 0.5" rpy="0 0 0"/>
16        <geometry>
17          <box size="0.2 0.2 1"/>
18        </geometry>
19      </visual>
20      <inertial>
21        <origin xyz="0 0 0.5" rpy="0 0 0"/>
22        <mass value="1"/>
23        <inertia
24          ixx="1.0" ixy="0.0" ixz="0.0"
25          iyy="1.0" iyz="0.0"
26          izz="1.0"/>
27      </inertial>
28    </link>
29    </link>
30  </robot>
```

Within the \link tags of the link1 description, the additional lines 20 through 27 describe the mass properties:

```
    <inertial>
      <origin xyz="0 0 0.5" rpy="0 0 0"/>
      <mass value="1"/>
      <inertia
        ixx="1.0" ixy="0.0" ixz="0.0"
        iyy="1.0" iyz="0.0"
        izz="1.0"/>
    </inertial>
```

In the `<inertial>` field, one specifies the mass of the link (here, set to 1 kg) and the coordinates of the center of mass (here, `xyz="0 0 0.5"`). The coordinates of the center of mass are specified in the `link1` frame. For our box example, if the box has uniform density, the center of mass will be in the center of the box, which is 0.5 m above the `link1` frame.

Specifying rotational inertia properties is more complex. Specification of rotational inertia requires defining a coordinate frame with respect to which the inertial properties are computed. With respect to this inertial frame, the components of a 3×3 matrix inertia tensor may be computed as:

$$\mathbf{I} = \int_V \rho(x, y, z) \begin{bmatrix} x^2 + y^2 & -xy & -xz \\ -xy & z^2 + x^2 & -yx \\ -xz & -yz & x^2 + y^2 \end{bmatrix} dx dy dz \qquad (3.1)$$

where $\rho(x, y, z)$ is the density of material at location (x, y, z), and the integral over volume V is defined as the volume of all material comprising the rigid link of interest. The matrix $\mathbf{I}$ is always symmetric, and thus there are only six values to specify. For simple shapes, this moment of inertia tensor is easy to compute. Many common shapes have published tabulated values. The inertia tensor can be difficult to compute for complex link shapes. If the link is detailed in a CAD program, the CAD program typically can compute the inertia tensor numerically.

Note that the matrix $\mathbf{I}$ will have different numerical components for different positions and/or orientations of the inertial reference frame. In a URDF, the origin of the inertial frame is always coincident with the center of mass (which is typically convenient in dynamics). It is sometimes convenient to define the orientation of this inertial frame to be aligned with some axes of symmetry, which simplifies specification of inertial components. In the present case (and often in URDFs) the inertial frame is simply chosen to be parallel to the link frame (as specified by the rpy values in `<origin xyz="0 0 0.5" rpy="0 0 0"/>`).

With respect to the defined reference frame, the rotational inertial components can be specified in the URDF as in lines 23 through 27. The terms of $\mathbf{I}$ are labeled in the URDF as ixx $= \mathbf{I}_{1,1}$, ixy $= \mathbf{I}_{1,2} = \mathbf{I}_{2,1}$, ixz $= \mathbf{I}_{1,3} = \mathbf{I}_{3,1}$, iyy $= \mathbf{I}_{2,2}$, iyz $= \mathbf{I}_{2,3} = \mathbf{I}_{3,2}$ and izz $= \mathbf{I}_{3,3}$.

Ideally, these values will be a close approximation to the true inertial components of the actual links. One should attempt to at least estimate these values roughly. It is important, though, that neither the mass nor the diagonal components of inertia be assigned a value of 0. The physics engine will have divide-by-zero numerical problems trying to simulate the dynamics of massless or inertialess objects.

For our single-link URDF, the mass properties have been assigned to unity (1 kg mass, and 1 $m^2 kg$ rotational inertia about each of the x, y and z principal axes). The inertial values are not realistic for a uniform rectangular prism. However, since this link will be fixed to the ground plane, the accurate mass values will not be a concern.

At this point, our model is boring (a static, rectangular prism). Nonetheless, it is a viable URDF that may be loaded into a simulator. This will be deferred, though, until our model becomes more interesting by adding a movable link. Listing 3.6 combines our initial kinematic model with visual and inertial properties.

Listing 3.6: `minimal_robot_description_wo_collision.urdf`: one-DOF robot URDF description

```
1  <?xml version="1.0"?>
2  <robot  name="one_DOF_robot">
3
4  <!-- Used for fixing robot to the simulator's world frame -->
5    <link name="world"/>
```

```
6
7      <joint name="glue_robot_to_world" type="fixed">
8        <parent link="world"/>
9        <child link="link1"/>
10     </joint>
11
12  <!-- Base Link -->
13     <link name="link1">
14       <visual>
15         <origin xyz="0 0 0.5" rpy="0 0 0"/>
16         <geometry>
17           <box size="0.2 0.2 1"/>
18         </geometry>
19       </visual>
20
21       <inertial>
22         <origin xyz="0 0 0.5" rpy="0 0 0"/>
23         <mass value="1"/>
24         <inertia
25           ixx="1.0" ixy="0.0" ixz="0.0"
26           iyy="1.0" iyz="0.0"
27           izz="1.0"/>
28       </inertial>
29     </link>
30
31  <!-- Moveable Link -->
32     <link name="link2">
33       <visual>
34         <origin xyz="0 0 0.5" rpy="0 0 0"/>
35         <geometry>
36           <cylinder length="1" radius="0.1"/>
37         </geometry>
38       </visual>
39
40       <inertial>
41         <origin xyz="0 0 0.5" rpy="0 0 0"/>
42         <mass value="1"/>
43         <inertia
44           ixx="0.1" ixy="0.0" ixz="0.0"
45           iyy="0.1" iyz="0.0"
46           izz="0.005"/>
47       </inertial>
48     </link>
49
50     <joint name="joint1" type="continuous">
51       <parent link="link1"/>
52       <child link="link2"/>
53       <origin xyz="0 0 1" rpy="0 0 0"/>
54       <axis xyz="0 1 0"/>
55     </joint>
56  </robot>
```

In Listing 3.6, visual and dynamic models are defined for two links. The joint between these links, `joint1`, could include additional properties, *e.g.* to model viscous or Coulomb friction. Further, joint limits and actuator torque limits may be specified. However, the minimal model that we have is adequate for simulating robot dynamics subject to specified joint (actuator) torques and the influence of gravity.

The visual model for `link2` is defined on line 36:

```
<cylinder length="1" radius="0.1"/>
```

as a cylinder of length 1 m and radius 0.1 m. The inertial properties for `link2` are specified as mass = 1 kg, ix = iyy = 0.1 and izz = 0.005. These values are reasonable approximations. For a thin rod of uniform density, mass m and length l, the rotational inertia about its x or y axis is $I_{xx} = I_{yy} = (1/12)ml^2$. The assigned values have been rounded up to 0.1 m^2kg. The inertia of this rod spinning about its cylindrical axis is $I_{zz} = (1/2)mr^2$, which evaluates to the assignment of izz = 0.005 m^2kg.

As with the box entity, the cylinder entity's reference frame has its origin in the middle of the cylinder. The z axis of this visualization frame points along the cylinder's major axis.

As introduced in Subsection 3.3.1, the `link2` frame is defined to be coincident with the `joint1` frame when the angle of `joint1` is in its home ($q = 0$) position. The reference frame of our visual model is specified relative to the `link2` frame in line 34: `<origin xyz="0 0 0.5" rpy="0 0 0"/>`. That is, the visual frame is aligned with the `link2` frame (`rpy="0 0 0"`), but the origin of the visual frame of `link2` is offset from the `link2` reference from by 0.5 m along the `link2` z axis (`xyz="0 0 0.5"`). As with our box visual description, this 0.5 m offset locates the geometric (visual) model such that one endcap of the model coincides with a joint origin (in this case, `joint1`).

In the home position, the `link2` frame is aligned with the `joint1` frame. With our choices, the home position of our one-DOF, two-link robot corresponds to `link2` pointing straight up.

3.3.4 Collision model

Our minimal robot description so far has included a kinematic model, an inertial model and a visual model. Motion of our robot will depend on forces and moments exerted on the robot. The dynamics engine of our simulator will enforce the kinematic constraints (*e.g.* how `link2` can move with respect to `link1`) and will compute angular accelerations of `link2` about `joint1`. These accelerations will be due to effects including gravity, joint torque exerted by an actuator (once we have defined an actuator), and possible collisions with other bodies. To include the influence of contact forces (*e.g.* due to collisions), we must include a collision model in our URDF.

A `<collision>` tag defines the region in which the collision model is defined in the URDF. Our one-DOF URDF with collision properties appears in Listing 3.7 (which also appears in the package `minimal_robot_description` as file `minimal_robot_description.urdf` in our associated code repository).

Listing 3.7: `minimal_robot_description.urdf`: one-DOF robot URDF description

```
1   <?xml version="1.0"?>
2   <robot   name="one_DOF_robot">
3
4   <!-- Used for fixing robot to the simulator's world frame -->
5     <link name="world"/>
6
7     <joint name="glue_robot_to_world" type="fixed">
8       <parent link="world"/>
9       <child link="link1"/>
10    </joint>
11
12  <!-- Base Link -->
13    <link name="link1">
14      <collision>
15        <origin xyz="0 0 0.5" rpy="0 0 0"/>
16        <geometry>
17          <box size="0.2 0.2 0.7"/>
18        </geometry>
19      </collision>
20
21      <visual>
22        <origin xyz="0 0 0.5" rpy="0 0 0"/>
23        <geometry>
24          <box size="0.2 0.2 1"/>
25        </geometry>
26      </visual>
27
28      <inertial>
```

```
29          <origin xyz="0 0 0.5" rpy="0 0 0"/>
30          <mass value="1"/>
31          <inertia
32            ixx="1.0" ixy="0.0" ixz="0.0"
33            iyy="1.0" iyz="0.0"
34            izz="1.0"/>
35         </inertial>
36       </link>
37
38   <!-- Moveable Link -->
39     <link name="link2">
40       <collision>
41          <origin xyz="0 0 0.5" rpy="0 0 0"/>
42          <geometry>
43            <cylinder length="1" radius="0.1"/>
44            <!--box size="0.15 0.15 0.8"-->
45          </geometry>
46       </collision>
47
48       <visual>
49          <origin xyz="0 0 0.5" rpy="0 0 0"/>
50          <geometry>
51            <cylinder length="1" radius="0.1"/>
52          </geometry>
53       </visual>
54
55       <inertial>
56          <origin xyz="0 0 0.5" rpy="0 0 0"/>
57          <mass value="1"/>
58          <inertia
59            ixx="0.1" ixy="0.0" ixz="0.0"
60            iyy="0.1" iyz="0.0"
61            izz="0.005"/>
62       </inertial>
63     </link>
64
65     <joint name="joint1" type="continuous">
66       <parent link="link1"/>
67       <child link="link2"/>
68       <origin xyz="0 0 1" rpy="0 0 0"/>
69       <axis xyz="0 1 0"/>
70     </joint>
71   </robot>
```

In Listing 3.7, the collision models define the geometric detail corresponding to a "skin" on the links. The collision model is used to compute intersections between solids within a world model; such collisions lead to interaction forces and torques at the points of contact.

Often, the collision model is identical to the visual model (and this is the case in Listing 3.7). However, collision checking can be a computationally intensive process, and therefore the collision model should be as sparse as possible. This can be done by reducing the number of triangles in a tessellated surface model, or by creating a primitive collision model based on geometric solids, *e.g.* rectangular prisms, cylinders or spheres.

Another concern is that a collision model that does not offer adequate clearance between the links can result in simulation instability, as the simulator repeatedly determines that the links are colliding with each other. For our crude model, we will set the collision model of link2 to be identical to the visual model of link2—a simple cylinder. However, we will set the collision model of link1 to be a shorter box, thus providing clearance for link2.

Since link1 is stationary, fidelity of its collision model is not a concern for computing dynamics of this minimal robot. However, if this model were part of a finger, one would care about how this link contacts objects to be grasped. Alternatively, if there were additional robots in the virtual world, one would care about how these robots might collide with link1 of this minimal robot, as such collisions would affect the dynamics of the other robots.

Our one-DOF minimal robot URDF now contains kinematic, inertial, visual and collision information. We next introduce the Gazebo simulator, which can perform dynamic simulations of robots based on URDF specifications.

3.4 INTRODUCTION TO GAZEBO

Gazebo is the simulator used with ROS (see `http://gazebosim.org/`). Gazebo offers options for alternative physics engines, with a default of ODE (Open Dynamics Engine). The Gazebo simulator consists of two parts: a server (which runs as process `gzserver`) and a client (which runs as process `gzclient`). The client process presents a graphical display and human interface. However, Gazebo can be run "headless" if a visual display is not needed.

An impressive and valuable capability of Gazebo is that it can simulate sensors as well as dynamics, including force sensors, accelerometers, sonar, LIDARs, color cameras, and 3-D point-cloud sensors. Description of camera simulation will be deferred to Part III.

Gazebo simulations require a world model in addition to one or more robot models. The world model may contain details of terrain, buildings, barriers, tables, graspable objects, and additional active entities, including swarms of robots. To start simple, however, we consider our minimal robot in a minimal world.

A common default world model consists only of a flat ground plane oriented perpendicular to the direction of gravity. To start Gazebo with this empty world model, run:

```
roslaunch gazebo_ros empty_world.launch
```

This starts a launch file called `empty_world.launch` from the ROS package `gazebo_ros`. The result of this command is to start both the `gzserver` and the `gzclient`, presenting a Gazebo display of an empty world (except for a ground plane). The Gazebo window will appear as in Fig 3.7. Along the bottom bar of the Gazebo simulator are some controls and

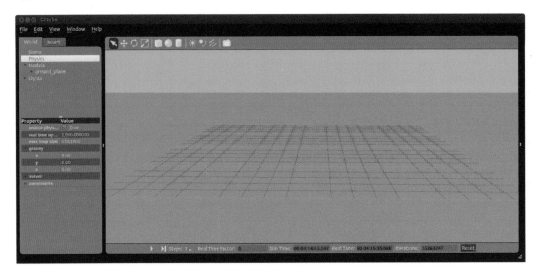

Figure 3.7: Gazebo display of empty world with gravity set to 0

displays. Run and pause buttons allow the user to suspend a simulation and resume it at will. The real time factor display indicates the efficiency of the simulation. Gazebo will (by default) try to simulate the robot(s) in the virtual world in real time. If the host computer is not able to keep up with the required computations to achieve real time, the simulator will slow down, resulting in an apparent slow-motion output. If the real time factor is unity, the simulation is equivalent to real-time dynamics; if this factor is *e.g.* 0.5, the simulation will take twice as long as reality.

On the left side of the Gazebo display, as shown in Fig 3.7, are tabs labeled "World" and "Insert." The World tab contains the option Physics. This element has been expanded and displays various properties, including the gravity item. The gravity item has also been expanded, revealing the numerical values for its x, y and z components. In this example, the z component was changed from its initial value of -9.8 to 0 (which is done by clicking on the the displayed value and editing it). With gravity set to zero, models can float in space indefinitely, whereas when gravity has a negative z component, models will fall to the ground plane.

Also within the Gazebo window's World tab is a Models menu, which allows one to inspect the spatial and dynamic properties of all models within the simulation. At this point, the only model in the simulation is a ground plane. We can add models to the simulation in multiple ways. Using the Insert tab, one can select any of the pre-defined models displayed in the model list to be inserted in the simulation. The list of available models will include on-line models in Gazebo's database, as well as models defined locally that reside within the (hidden) directory `~/.gazebo`.

Alternatively, one can load a model into Gazebo manually by invoking a Gazebo node with a path to a model. For example, first navigate to the directory `~/ros_ws/src/learning_ros/exmpl_models/rect_prism`, then enter the following command:

```
rosrun gazebo_ros spawn_model -file model-1_4.sdf -sdf -model rect_prism
```

Our simple, rectangular prism model will be loaded into Gazebo, and the display will appear as in Fig 3.8. We can remove this model by clicking on it, then pressing the delete key.

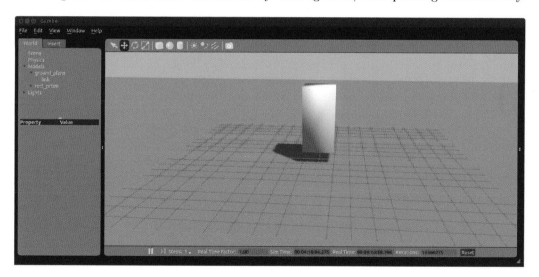

Figure 3.8: Gazebo display after loading rectangular prism model

Alternatively, the prism can be loaded with specified world coordinates. For example, the command:

```
rosrun gazebo_ros spawn_model -file model-1_4.sdf -sdf -model rect_prism -x 0 -y 0 -z ↩
    4
```

loads the prism with its coordinate frame located 4 m above the ground plane.

Instead of navigating to the model directory to load models, one can include the full path to the model location. This is made somewhat more general with use of the environment variable `ROS_WORKSPACE`. The following command loads the model using an explicit path:

```
rosrun gazebo_ros spawn_model -file $ROS_WORKSPACE/src/learning_ros/exmpl_models/↩
    rect_prism/model-1_4.sdf -sdf -model rect_prism
```

It can be more convenient to ask ROS to find the package in which the model resides, which can be done with `$(rospack find package_name)`, *e.g.* as with the following command:

```
rosrun gazebo_ros spawn_model -file $(rospack find exmpl_models)/rect_prism/model-1_4.↩
    sdf -sdf -model rect_prism
```

Since direct commands can be tedious to type in, it is typically more convenient to enter the commands in a launch file, then invoke the launch file to load models. An example is in `exmpl_models/launch/add_rect_prism.launch`. The contents of this file are:

```
<launch>
<node name="spawn_sdf" pkg="gazebo_ros" type="spawn_model" args="-file $(find ↩
    exmpl_models)/rect_prism/model-1_4.sdf -sdf -model rect_prism -x 0 -y 0 -z 5" />
</launch>
```

The syntax is similar to the command line, though with variations. Within a launch file, one finds ROS packages using `$(find package_name)` instead of `$(rospack find package_name)`. This launch file can be invoked from any directory with:

```
roslaunch exmpl_models add_rect_prism.launch
```

This command finds the package `exmpl_models`, implicitly looks in the subdirectory `launch`, and invokes the launch file `add_rect_prism.launch`. This launch file loads the rectangular prism model into Gazebo with initial coordinates of $(x, y, z) = (0, 0, 5)$. With gravity off, this model remains stationary, floating above the ground plane.

A second simple model (a cylinder) can be added with a similar launch file:

```
roslaunch exmpl_models add_cylinder.launch
```

which results in the Gazebo display in Fig 3.9. Note that the `Models` menu now includes `rect_prism` and `cylinder`.

At this point, we have confirmed that our models are syntactically correct, present the expected visual displays, and contain the minimum requirements to be compatible with Gazebo simulation. More interestingly, dynamic simulation can be observed by giving these models an initial velocity that results in collision, from which we can observe behavior of the physics engine. To initialize our models with a non-zero velocity and angular velocity, we can use a service of Gazebo. Running:

```
rosservice list
```

shows 30 services running, one of which is `/gazebo/set_model_state`. Examining this service with

```
rosservice info gazebo/set_model_state
```

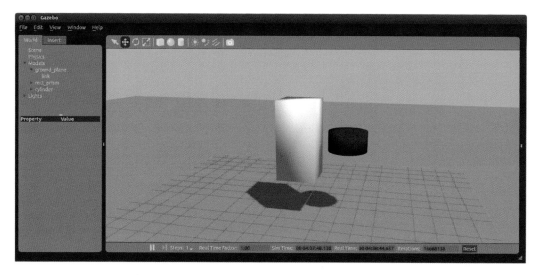

Figure 3.9: Gazebo display after loading rectangular prism and cylinder

shows that the service message expects an argument, `model_state`, of type `gazebo_msgs/SetModelState`. Looking in the `gazebo_msgs/srv` directory, one finds the service message description `SetModelState.srv`, the contents of which is:

```
gazebo_msgs/ModelState model_state
---
bool success              # return true if setting state successful
string status_message     # comments if available
```

which confirms that the `request` field of this service message contains a field called `model_state` of type `gazebo_msgs/ModelState`.

The message type `gazebo_msgs/ModelState` can be examined with:

```
rosmsg show gazebo_msgs/ModelState
```

which displays details of this message type to be:

```
string model_name
geometry_msgs/Pose pose
  geometry_msgs/Point position
    float64 x
    float64 y
    float64 z
  geometry_msgs/Quaternion orientation
    float64 x
    float64 y
    float64 z
    float64 w
geometry_msgs/Twist twist
  geometry_msgs/Vector3 linear
    float64 x
    float64 y
    float64 z
  geometry_msgs/Vector3 angular
    float64 x
    float64 y
    float64 z
string reference_frame
```

A model state can be set using a manual command. An example is:

```
rosservice call /gazebo/set_model_state '{model_state: {model_name: rect_prism, twist:↩
    {angular:{z: 1.0}}}}'
```

This command specifies the model name to be our rectangular prism, and it commands the z component of the angular velocity to be 1.0 rad/sec. Implicitly, all other components of position, orientation, linear velocity and angular velocity are set to 0. The syntax for specifying components of a message type is YAML. (See http://wiki.ros.org/ROS/YAMLCommandLine for details on using YAML within a ROS command line.)

Since the YAML syntax can become tedious (and error prone for direct typing), it can be more convenient to set the model state programmatically. An example of how to do this is contained in the package example_gazebo_set_state in the source code (node) example_gazebo_set_prism_state.

Key lines of this node include:

```
ros::ServiceClient set_model_state_client =
        nh.serviceClient<gazebo_msgs::SetModelState>("/gazebo/set_model_state");
```

A compatible service message is instantiated with the line:

```
gazebo_msgs::SetModelState model_state_srv_msg;
```

Components of this message are filled in, *e.g.* as:

```
model_state_srv_msg.request.model_state.model_name = "rect_prism";
```

which specifies the model for which the state is to be set, and:

```
model_state_srv_msg.request.model_state.twist.angular.z= 1.0;
```

which specifies a z component of angular velocity to be 1.0 rad/sec. Similarly, all components of position, orientation, translational velocity and angular velocity can be specified. After the service message is populated, it is sent to the Gazebo service to set the specified model state with:

```
set_model_state_client.call(model_state_srv_msg);
```

This program can be run by entering:

```
rosrun example_gazebo_set_state example_gazebo_set_prism_state
```

which causes the prism to rotate about its z axis and translate slowly in the x direction. After a single service call, this program concludes. The objects subsequently evolve in time according to their twist vectors.

By using the initial conditions within the launch files for loading a prism and a cylinder, then invoking example_gazebo_set_prism_state, the prism will begin spinning and translating, eventually colliding with the cylinder. This collision results in a change in momentum (including angular momentum) of both objects. The states of the models can be observed with:

```
rostopic echo gazebo/model_states
```

Before the models collide, this will show that the cylinder has zero twist, but the prism has an angular velocity about its z axis of 1 rad/sec, as well as a translational velocity in the x direction of 0.02 m/sec. After the models collide, they both have altered translational and rotational velocities. However, it can be shown that the total system linear momentum and system angular momentum are preserved. (Momentum before collision equals momentum after collision.) This illustrates that the physics engine behaves as expected. Although the objects may be translating and tumbling in a complex way after collision, the system momentum is conserved.

Robot models can be inserted and dynamically modeled in a similar fashion, although the model and its stimulus are more complex. To import our minimal robot model, in a separate terminal, navigate to the `minimal_robot_description` package and enter:

```
rosrun gazebo_ros spawn_model -urdf -file minimal_robot_description.urdf -model ↩
    one_DOF
```

This invokes the `spawn_model` node from the `gazebo_ros` package with three arguments: declaration that the input file is in URDF format, specification of the URDF file name to be loaded, and a model name to assign to the loaded file in Gazebo. This node will run to completion, resulting in inserting the named URDF file into the simulator. The result appears as in Fig 3.10. The figure verifies that we have an upright, rectangular prism for

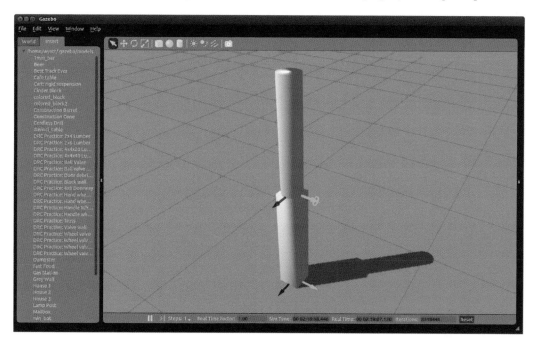

Figure 3.10: Gazebo display of two-link, one-DOF robot URDF model

`link1` and a cylindrical `link2`. The coordinate frames for `link1` and `link2` are illustrated as well, where the red axes are x axes and the green axes are y axes. The green circle about the `link2` y axis indicates that this vector is also a joint axis (the axis of `joint1`). This display was enabled via the top menu bar of the Gazebo display by enabling `view->Joints`. The Gazebo menus offer a variety of additional options that can be convenient for visualizing simulations, including centers of mass and contact forces.

The command `rosrun gazebo_ros spawn_model ...` can be run alternatively from a launch file. For future interaction with ROS nodes and sensor displays, it will be convenient to first load the robot model onto the parameter server, then spawn the model into Gazebo from the parameter server. This will assure that ROS nodes and Gazebo refer to the same details of the robot model. This is accomplished with the example launch file in package `minimal_robot_description`, `minimal_robot_description.launch`, which can be run with:

```
roslaunch minimal_robot_description minimal_robot_description.launch
```

(If the robot has already been spawned in Gazebo, this will generate an error, due to spawning two identical models.) The contents of this launch file are:

```
<launch>
  <param name="robot_description"
    textfile="$(find minimal_robot_description)/minimal_robot_description.urdf "/>

  <!-- Spawn a robot into Gazebo -->
  <node name="spawn_urdf" pkg="gazebo_ros" type="spawn_model"
    args="-param robot_description -urdf -model one_DOF_robot" />
</launch>
```

The first subfield in this launch file puts a copy of the robot model on the parameter server, and the second field spawns the model into Gazebo using the model drawn from the parameter server. After running this launch file, it can be verified that the robot model is on the parameter server. Running `rosparam list` shows an item called `robot_description` on the parameter server. Running `rosparam get /robot_description` displays the contents of this item, which contains the entire URFD specifications from `minimal_robot_description.urdf`. In the future, it will also be convenient to include start-up of joint controllers in the same launch file, which may include specifications of joint control parameters placed on the parameter server as well.

Models can also be inserted into Gazebo interactively via the Insert tab in the Gazebo GUI, which can be convenient for constructing variations on the virtual world for experiments with the robot. A variety of models are available on-line to import. Additionally, Gazebo will look in the (hidden) user directory called `.gazebo` for available models; the `.gazebo` directory typically resides in the user's home directory. Robots can be spawned into Gazebo in this fashion as well, although it is more useful to do so via the parameter server for better ROS integration.

With Gazebo running, entering:

```
rostopic list
```

reveals that the active topics include:

```
/clock
/gazebo/link_states
/gazebo/model_states
/gazebo/parameter_descriptions
/gazebo/parameter_updates
/gazebo/set_link_state
/gazebo/set_model_state
```

Entering:

```
rosservice list
```

shows that the following Gazebo services are active:

```
/gazebo/apply_body_wrench
/gazebo/apply_joint_effort
/gazebo/clear_body_wrenches
/gazebo/clear_joint_forces
/gazebo/delete_model
/gazebo/get_joint_properties
/gazebo/get_link_properties
/gazebo/get_link_state
/gazebo/get_loggers
/gazebo/get_model_properties
/gazebo/get_model_state
/gazebo/get_physics_properties
/gazebo/get_world_properties
/gazebo/pause_physics
/gazebo/reset_simulation
/gazebo/reset_world
/gazebo/set_joint_properties
/gazebo/set_link_properties
/gazebo/set_link_state
/gazebo/set_logger_level
/gazebo/set_model_configuration
/gazebo/set_model_state
/gazebo/set_parameters
/gazebo/set_physics_properties
/gazebo/spawn_gazebo_model
/gazebo/spawn_sdf_model
/gazebo/spawn_urdf_model
/gazebo/unpause_physics
```

Running:

```
rostopic echo gazebo/link_states
```

displays updates of the 6-D pose and 6-D velocity of each link in the system. With the minimal robot spawned into Gazebo, the initial part of the echo display looks like:

```
name: ['ground_plane::link', 'one_DOF::link1', 'one_DOF::link2']
pose:
  -
    position:
      x: 0.0
      y: 0.0
      z: 0.0
    orientation:
      x: 0.0
      y: 0.0
      z: 0.0
      w: 1.0
```

The output declares that there are three links in the system (including the ground plane). The position and orientation of every link and the linear and angular velocity vectors are given. For the simulation at present, these values are boring. All velocities are zero, and all links are aligned with the world frame (except for link2 being elevated by 1.0 m).

The initial pose in this scene is the home pose, corresponding to `link2` straight up. In Fig 3.10, `link2` is precariously balanced at an unstable equilibrium point. There is no joint controller keeping it upright, and it may unexpectedly tip over, tilting about the `joint1` axis.

To make our robot more interesting, a joint controller is needed. The joint controller can integrate with Gazebo, obtaining joint angles and inducing joint torques, simulating a servoed actuator.

3.5   MINIMAL JOINT CONTROLLER

An important connection between Gazebo and ROS is how controls are exerted using joint actuator commands and joint displacement sensors. For purposes of illustrating this interaction, a minimal joint controller ROS node is presented. The source code `minimal_joint_controller.cpp` (in the `minimal_joint_controller` package) is described here, which interacts with Gazebo and creates a ROS interface (similar to constructing the bridges necessary to interact with a real robot).

Please note that the example controller described here normally would not be used. For a physical robot, the proportional-plus-derivative (PD) controller would be contained within dedicated control hardware. Similarly, for Gazebo, there are pre-defined Gazebo plug-ins that perform the equivalent of the present joint controller example. (See `http://gazebosim.org/tutorials?tut=ros_plugins` for a tutorial on how to write Gazebo plug-ins.) Thus, you would not need to use the `minimal_joint_controller` package; this is presented only to illustrate the equivalent of what a Gazebo controller plug-in performs, using concepts already introduced. A disadvantage of using the present controller code relative to a Gazebo plug-in is that the example code incurs additional computational and bandwidth loads of serialization, de-serialization and corresponding latency in message passing, which can be critical in high-performance control implementations.

The contents of `minimal_joint_controller.cpp` appear in Listings 3.8 (preamble and helper functions) and 3.9 (main program).

Listing 3.8: `minimal_joint_controller.cpp`: minimal joint controller via Gazebo services (preamble and helper functions)

```
1   #include <ros/ros.h> //ALWAYS need to include this
2   #include <gazebo_msgs/GetModelState.h>
3   #include <gazebo_msgs/ApplyJointEffort.h>
4   #include <gazebo_msgs/GetJointProperties.h>
5   #include <sensor_msgs/JointState.h>
6   #include <string.h>
7   #include <stdio.h>
8   #include <std_msgs/Float64.h>
9   #include <math.h>
10
11  //some "magic number" global params:
12  const double Kp = 10.0; //controller gains
13  const double Kv = 3;
14  const double dt = 0.01;
15
16  //a simple saturation function; provide saturation threshold, sat_val, and arg to be ↩
          saturated, val
17
18  double sat(double val, double sat_val) {
19      if (val > sat_val)
20          return (sat_val);
21      if (val< -sat_val)
22          return (-sat_val);
23      return val;
24
```

```
25   }
26
27   double min_periodicity(double theta_val) {
28       double periodic_val = theta_val;
29       while (periodic_val > M_PI) {
30           periodic_val -= 2 * M_PI;
31       }
32       while (periodic_val< -M_PI) {
33           periodic_val += 2 * M_PI;
34       }
35       return periodic_val;
36   }
37
38   double g_pos_cmd = 0.0; //position command input-- global var
39
40   void posCmdCB(const std_msgs::Float64& pos_cmd_msg) {
41       ROS_INFO("received value of pos_cmd is: %f", pos_cmd_msg.data);
42       g_pos_cmd = pos_cmd_msg.data;
43   }
44
45   bool test_services() {
46       bool service_ready = false;
47       if (!ros::service::exists("/gazebo/apply_joint_effort", true)) {
48           ROS_WARN("waiting for apply_joint_effort service");
49           return false;
50       }
51       if (!ros::service::exists("/gazebo/get_joint_properties", true)) {
52           ROS_WARN("waiting for /gazebo/get_joint_properties service");
53           return false;
54       }
55       ROS_INFO("services are ready");
56       return true;
```

Listing 3.9: `minimal_joint_controller.cpp`: minimal joint controller via Gazebo services (main program)

```
59   int main(int argc, char **argv) {
60       //initializations:
61       ros::init(argc, argv, "minimal_joint_controller");
62       ros::NodeHandle nh;
63       ros::Duration half_sec(0.5);
64
65       // make sure services are available before attempting to proceed, else node will ←
                crash
66       while (!test_services()) {
67           ros::spinOnce();
68           half_sec.sleep();
69       }
70
71       ros::ServiceClient set_trq_client =
72               nh.serviceClient<gazebo_msgs::ApplyJointEffort>("/gazebo/←
                    apply_joint_effort");
73       ros::ServiceClient get_jnt_state_client =
74               nh.serviceClient<gazebo_msgs::GetJointProperties>("/gazebo/←
                    get_joint_properties");
75
76       gazebo_msgs::ApplyJointEffort effort_cmd_srv_msg;
77       gazebo_msgs::GetJointProperties get_joint_state_srv_msg;
78
79       ros::Publisher trq_publisher = nh.advertise<std_msgs::Float64>("jnt_trq", 1);
80       ros::Publisher vel_publisher = nh.advertise<std_msgs::Float64>("jnt_vel", 1);
81       ros::Publisher pos_publisher = nh.advertise<std_msgs::Float64>("jnt_pos", 1);
82       ros::Publisher joint_state_publisher = nh.advertise<sensor_msgs::JointState>("←
                joint_states", 1);
83       ros::Subscriber pos_cmd_subscriber = nh.subscribe("pos_cmd", 1, posCmdCB);
84
85       std_msgs::Float64 trq_msg, q1_msg, q1dot_msg;
86       double q1, q1dot, q1_err, trq_cmd;
87       sensor_msgs::JointState joint_state_msg;
88       ros::Duration duration(dt);
89       ros::Rate rate_timer(1 / dt);
```

```
90
91      effort_cmd_srv_msg.request.joint_name = "joint1";
92      effort_cmd_srv_msg.request.effort = 0.0;
93      effort_cmd_srv_msg.request.duration = duration;
94      get_joint_state_srv_msg.request.joint_name = "joint1";
95
96      // set up the joint_state_msg fields to define a single joint,
97      // called joint1, and initial position and vel values of 0
98      joint_state_msg.header.stamp = ros::Time::now();
99      joint_state_msg.name.push_back("joint1");
100     joint_state_msg.position.push_back(0.0);
101     joint_state_msg.velocity.push_back(0.0);
102
103     //here is the main controller loop:
104     while (ros::ok()) {
105         get_jnt_state_client.call(get_joint_state_srv_msg);
106         q1 = get_joint_state_srv_msg.response.position[0];
107         q1_msg.data = q1;
108         pos_publisher.publish(q1_msg); //republish his val on topic jnt_pos
109
110         q1dot = get_joint_state_srv_msg.response.rate[0];
111         q1dot_msg.data = q1dot;
112         vel_publisher.publish(q1dot_msg);
113
114         joint_state_msg.header.stamp = ros::Time::now();
115         joint_state_msg.position[0] = q1;
116         joint_state_msg.velocity[0] = q1dot;
117         joint_state_publisher.publish(joint_state_msg);
118
119         q1_err = min_periodicity(g_pos_cmd - q1); //jnt angle err; watch for ←
                periodicity
120
121         trq_cmd = Kp * (q1_err) - Kv*q1dot;
122         trq_msg.data = trq_cmd;
123         trq_publisher.publish(trq_msg);
124
125         effort_cmd_srv_msg.request.effort = trq_cmd; // send torque command to Gazebo
126         set_trq_client.call(effort_cmd_srv_msg);
127         //make sure service call was successful
128         bool result = effort_cmd_srv_msg.response.success;
129         if (!result)
130             ROS_WARN("service call to apply_joint_effort failed!");
131         ros::spinOnce();
132         rate_timer.sleep();
133     }
134 }
```

The example controller in `minimal_joint_controller.cpp` interacts with Gazebo via service clients of the services `/gazebo/get_joint_properties` and `/gazebo/apply_joint_effort`. In lines 36 through 41, the helper function `test_services()` (lines 45 through 57) is used to make sure the Gazebo services are available, after which corresponding service clients are instantiated (lines 71 through 74). Compatible service messages are instantiated in lines 76 and 77. We can examine the operation of these respective Gazebo services manually.

To examine Gazebo's services, make sure the minimal robot is running, by first starting Gazebo:

```
roslaunch gazebo_ros empty_world.launch
```

and loading the robot model with:

```
roslaunch minimal_robot_description minimal_robot_description.launch
```

The Gazebo service `/gazebo/get_joint_properties` can be examined manually by entering:

```
rosservice call /gazebo/get_joint_properties "joint1"
```

which results in the following example output:

```
type: 0
damping: []
position: [0.0]
rate: [0.0]
success: True
status_message: GetJointProperties: got properties
```

From this service, we can obtain the state of `joint1`, which includes the joint position and the joint (angular) velocity. The service client of this service, `get_jnt_state_client`, makes such calls repeatedly in the control loop to obtain the joint position and velocity from the dynamic simulator.

The service client of `/gazebo/apply_joint_effort`, `set_trq_client`, sets fields in the corresponding service message for `joint_name` to `joint1` and `effort` to a desired joint torque, enabling the controller node to impose joint torques on `joint1` of the simulated robot.

A subscriber to the topic `pos_cmd` is also set up (line 83), ready to accept user input for desired `joint1` position values via callback function `posCmdCB` (lines 40 through 43).

The main loop of our controller node does the following:

- Obtains the current joint position and velocity from Gazebo and republishes these on topic `joint_states` (lines 104 through 117)

- Compares the (virtual) joint sensor value to the commanded joint angle (from the `pos_cmd` callback), accounting for periodicity (line 119)

- Computes a PD torque response (line 121)

- Sends this effort to Gazebo via the `apply_joint_effort` service (lines 125 and 126)

The minimal controller node can be started with the command:

```
rosrun minimal_joint_controller minimal_joint_controller
```

Initially, there is no noticeable effect on Gazebo, since the start-up desired angle is 0, and the robot is already at 0 angle. However, we can command a new desired angle manually from the command line (and later, under program control) with the command:

```
rostopic pub pos_cmd std_msgs/Float64 1.0
```

which commands a new joint angle of 1.0 radians. The robot then moves to the position shown in Fig 3.11. We can record the dynamic response of an input command by plotting the published values of joint torque, velocity and position, using `rqt_plot`. Enter the command:

```
rqt_plot
```

then add topics of `/jnt_pos/data`, `/jnt_trq/data` and `/jnt_vel/data`. The plot in Fig 3.12 shows the transient response, starting from a position command of 1.0, then responding to a new command of 2.0. As shown, the initial joint angle is larger than the commanded 1.0. This is due to a low feedback proportional gain and the influence of gravity, causing droop relative to the desired angle. At approximately $t = 52.8$, the input position command is changed to 2.0 rad, resulting in a transient in joint torque as the link accelerates toward

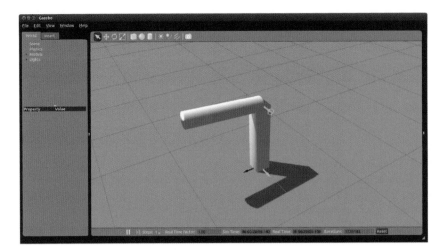

Figure 3.11: Gazebo display of minimal robot with minimal controller

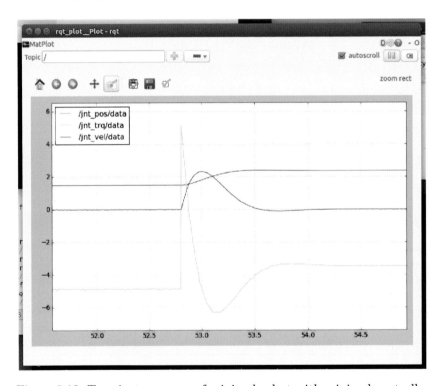

Figure 3.12: Transient response of minimal robot with minimal controller

the new goal. The goal of 2.0 is overshot, again due to gravity load. As the link settles, a sustained torque of approximately $-3.5Nm$ is required to hold the link against gravity.

To illustrate the influence of contact dynamics, an additional model is added to Gazebo, as shown in Fig 3.13. Here, a cafe table (from a list of pre-defined models) is added using the Insert menu of Gazebo. Gazebo offers the user the capability of moving the table to a desired location, which was chosen here to be within reach of the robot.

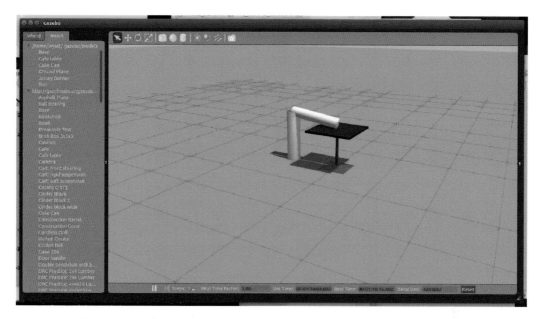

Figure 3.13: Gazebo display of minimal robot contacting rigid object

Next, the robot was commanded to position 0 (straight up), then commanded to position 2.5, which is unreachable with the table in the way. The transient dynamics from this command are shown in the Fig 3.14. The collision model of the link and the collision model of the table are used by Gazebo to detect that contact has occurred. This results in a reaction force from the table to the robot (and from the robot to the table). As a result, the robot does not reach its goal angle, and equilibrium occurs with the robot's joint actuator exerting an effort downward on the table, while struggling to reach the desired angle of 2.5 rad.

The ability to move models in Gazebo under program control can have useful applications in performing simulations in dynamic virtual worlds. As noted, though, simulating joint controllers in Gazebo is better performed using plug-ins, as described next.

3.6 USING GAZEBO PLUG-IN FOR JOINT SERVO CONTROL

A Gazebo plug-in can run at the full rate of the Gazebo simulator (1 kHz, by default), without the overhead and associated latency of message passing. ROS controllers in Gazebo are intended to be constructed with interfaces identical to actual hardware as well as dynamic behavior in simulation that emulates dynamic behavior of actual hardware. Further, the ROS packages and example controllers were constructed with anticipation of growth and generalization. Unfortunately, this resulted in a fair amount of complexity in creating and simulating models of joint controllers. Further, multiple additional packages are required (including, for example, with the Indigo ROS release: `ros-indigo-controller-interface`, `ros-indigo-gazebo-ros-control`, `ros-indigo-joint-state-controller`, and `ros-indigo-effort-controllers`). These packages are not installed automatically, even with the `desktop-full` ROS installation. If you installed ROS using the setup scripts accompanying this text (at `https://github.com/wsnewman/learning_ros_setup_scripts`), these control packages will al-

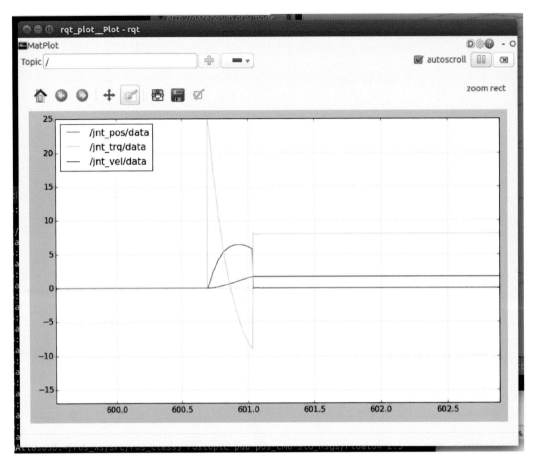

Figure 3.14: Contact transient when colliding with table

ready be installed. If you installed ROS manually with the `desktop-full` option, you will need to install the ros-controller packages (as admin) with:

```
sudo apt-get install ros-indigo-controller-interface ros-indigo-gazebo-ros-control ros
    -indigo-joint-state-controller ros-indigo-effort-controllers
```

A good example to follow regarding ROS control is the `rrbot` tutorial, which can be found at `http://gazebosim.org/tutorials/?tut=ros_control`.

The process for incorporating ROS joint controllers as a Gazebo plug-in requires the following steps:

- Edit the `joint` fields to specify torque and joint limits in the URDF file

- Add a `transmission` field to the URDF, one for each joint to be controlled

- Add a `gazebo` block in the URDF to bring in the `libgazebo_ros_control.so` controller plug-in library

- Create a controller parameter YAML file that declares the controller gains

- Modify the launch file to put the control parameters on the parameter server and start the controllers running

These steps are illustrated here, revisiting our minimal robot specification. Within the package `minimal_robot_description`, a modified URDF file, `minimal_robot_description_w_jnt_ctl.urdf`, contains the additions necessary for ROS joint control. This model file is largely identical to our previous minimal robot URDF, Listing 3.7, and thus it is not repeated in full here. The important additions within `minimal_robot_description_w_jnt_ctl.urdf` exist within three blocks.

First, the previous `joint1` block is modified as follows:

```
<joint name="joint1" type="revolute">
  <parent link="link1"/>
  <child link="link2"/>
  <origin xyz="0 0 1" rpy="0 0 0"/>
  <axis xyz="0 1 0"/>
  <limit effort="10.0" lower="0.0" upper="2.0" velocity="0.5"/>
  <dynamics damping="1.0"/>
</joint>
```

This block shows a new type of joint and additional parameters. The `revolute` joint type is more appropriate than `continuous` for a robot arm, since robot joints typically have limited range of motion (in contrast to joints for wheels). Another common joint type is `prismatic`, for joints that extend and retract. (For more detail on joint types and specifications, see http://wiki.ros.org/urdf/XML/joint.) Joint limit values are expressed in the `limit` tag, including upper and lower joint range of motion limits (constrained to the range 0 to 2.0, for this example). Additionally, one can express actuator dynamic limits, including a velocity limit (here set to 0.5 rad/s) and a torque limit (set to 10.0 Nm). The expression "effort" is used instead of "torque", since an effort can be either a force or a torque, depending on whether the joint is prismatic or revolute. The joint might also have inherent damping (linear friction). This is included in our example with the expression `<dynamics damping="1.0"/>`, which imposes joint friction of 1.0 $(Nm)/(rad/s)$.

A second addition to the URDF file is a `transmission` block (see http://wiki.ros.org/urdf/XML/Transmission). In the present example, this declares a transmission associated with `joint1`:

```
<transmission name="tran1">
  <type>transmission_interface/SimpleTransmission</type>
  <joint name="joint1">
    <hardwareInterface>EffortJointInterface</hardwareInterface>
  </joint>
  <actuator name="motor1">
    <hardwareInterface>EffortJointInterface</hardwareInterface>
    <mechanicalReduction>1</mechanicalReduction>
  </actuator>
</transmission>
```

The above block is required, and the joint name must be associated with a corresponding joint name detailed in the URDF—"joint" in this case. (For multiple joints to be controlled, a corresponding transmission block must be inserted for each controlled joint.) While future options are anticipated, the lines

```
<type>transmission_interface/SimpleTransmission</type>
```

and

```
<hardwareInterface>EffortJointInterface</hardwareInterface>
```

are, at the time of this writing, the only available options for these required elements.

In the example **transmission** block, the transmission ratio is set to unity. For realistic robot dynamics, transmission ratios on the order of 100 to 1000 are common, and the reflected inertia of the motor can contribute a significant influence of the link inertia. At present, incorporating this in the Gazebo model would require that the modeler incorporate an estimated reflected motor inertia augmentation of the corresponding link inertia. Correspondingly, the actuator parameters defining velocity saturation, torque saturation and control gains must be represented consistently. If the default unity transmission ratio is used, the gains must be expressed in joint-space (transmission output) values, as though the motor were a low-speed, high-torque, direct-drive actuator.

A second block must be inserted in the URDF file for use of Gazebo plug-in controllers, as follows:

```
<gazebo>
  <plugin name="gazebo_ros_control" filename="libgazebo_ros_control.so">
    <robotNamespace>/one_DOF_robot</robotNamespace>
  </plugin>
</gazebo>
```

This block is an instruction to the Gazebo simulator. It brings in the plug-in library **libgazebo_ros_control.so**. The line

```
<robotNamespace>/one_DOF_robot</robotNamespace>
```

sets a **namespace** for the controller. Since multiple robots may be present in the simulator, it is useful to separate their interfaces into separate namespaces. The name chosen, **one_DOF_robot**, must be consistent verbatim with naming in two other files: the control-parameter YAML file and the launch file (described below).

A file that must be created for use with Gazebo plug-in controllers is a controller parameter file in YAML syntax. Such files typically reside within a subdirectory for configuration files. In the the package **minimal_robot_description**, a subdirectory **control_config** was created, which contains the file **one_dof_ctl_params.yaml**. The contents of this file are:

Listing 3.10: **one_dof_ctl_params.yaml**: control gains file

```
%\begin{lstlisting}[numbers=none]
one_DOF_robot:
  # Publish all joint states -------------------------------
  joint_state_controller:
    type: joint_state_controller/JointStateController
    publish_rate: 50

  # Position Controllers ----------------------------------
  joint1_position_controller:
    type: effort_controllers/JointPositionController
    joint: joint1
    pid: {p: 10.0, i: 10.0, d: 10.0, i_clamp_min: -10.0, i_clamp_max: 10.0}
```

The control parameter file starts with a robot name, in this case **one_DOF_robot:**. This name must agree with the namespace name in the **gazebo** tag within the URDF file.

The control parameter file associates a controller name with each controlled joint. In the present case, a controller named **joint1_position_controller** is associated with **joint1**. For additional controlled joints, this block should be replicated, assigning a unique controller name associated with each controlled joint name in the corresponding URDF file.

The type of controller specified for this example is a PID joint position controller, which is a type of servo controller with proportional, derivative and integral-error gains. The gain values are specified by the line:

```
pid: {p: 10.0, i: 10.0, d: 10.0, i_clamp_min: -10.0, i_clamp_max: 10.0}
```

which assigns values for the proportional gain (10.0 $(Nm)/rad$), the derivative gain (10.0 $(Nm)/(rad/s)$) and the integral-error gain (10.0 $(Nm)/rads$). The suggested values are not well tuned, but they are valid.

Finding good gains for joint controllers can be challenging. This can be done interactively with graphical assistance using the command:

```
rosrun rqt_reconfigure rqt_reconfigure
```

Use of this graphical tool for joint control parameter tuning is described at http://wiki.ros.org/rqt_reconfigure.

Tuning the integral-error gain can be particularly challenging. This control gain can easily lead to instability. Using a gain of 0 is a good place to start. If this gain is non-zero, anti-windup constraints should be imposed on the integral-error computation. These are specified by the values of i_clamp_min and i_clamp_max. If integral-error feedback is not used (*i.e.* by setting the i: term to zero), it is not necessary to specify these values.

For the control parameter file to be associated with the corresponding real-time control code, the YAML file is first loaded onto the parameter server. This is conveniently done within a launch file.

A launch file that performs the necessary start-up functions for our example is minimal_robot_w_jnt_ctl.launch, contained within the minimal_robot_description package. The contents of this launch file appear in Listing 3.11.

Listing 3.11: minimal_robot_w_jnt_ctl.launch: launch file for minimal robot using ROS control plug-ins

```
1  <launch>
2    <!-- Load joint controller configurations from YAML file to parameter server -->
3    <rosparam file="$(find minimal_robot_description)/control_config/one_dof_ctl_params.↩
         yaml" command="load"/>
4    <param name="robot_description"
5        textfile="$(find minimal_robot_description)/minimal_robot_description_w_jnt_ctl.↩
           urdf"/>
6
7    <!-- Spawn a robot into Gazebo -->
8    <node name="spawn_urdf" pkg="gazebo_ros" type="spawn_model"
9      args="-param robot_description -urdf -model one_DOF_robot" />
10
11   <!--start up the controller plug-ins via the controller manager -->
12   <node name="controller_spawner" pkg="controller_manager" type="spawner" respawn="↩
         false"
13     output="screen" ns="/one_DOF_robot" args="joint_state_controller ↩
           joint1_position_controller"/>
14
15  </launch>
```

Line 3 of Listing 3.11 loads the control parameter file on the parameter server, where it will be accessed when the joint controllers are started. Lines 4 and 5 load the (modified) robot model URDF file onto the parameter server. Lines 8 and 9 load the robot model into the Gazebo simulator. This task was formerly done with a separate terminal command, but this operation is now automated by including it in the launch file. Also, the robot model in this instance is spawned into Gazebo by accessing it from the parameter server instead of from a file. Lines 12 and 13 use the spawner node from the controller_manager package to bring in the PID controller(s) and start them up. In this command, the argument joint1_position_controller refers to the controller name specified in the control parameter YAML file. If there are more joint controllers to be used, each controller's name should be included in the list of arguments in this command.

Note that the controller launch command specifies ns="/one_DOF_robot". This namespace assignment must be identical to the name assigned in the control parameter YAML file as well as the namespace specified in the gazebo tag of the URDF file.

A roslaunch option used in the controller launch command is: output="screen". With this option specified, printed output from the node being launched will appear within the terminal from which roslaunch is invoked. This is the case even when there are multiple nodes launched from the same launch file. (Formerly, we used rqt_console to view such messages when launching minimal_nodes, since launching multiple nodes from a single terminal suppressed their ROS_INFO displays within that terminal.)

With the preceding changes, our controlled robot can be launched as follows. First, bring up Gazebo with an empty world from a terminal with the command:

```
roslaunch gazebo_ros empty_world.launch
```

In a second terminal, launch our robot model, complete with controllers:

```
roslaunch minimal_robot_description minimal_robot_w_jnt_ctl.launch
```

Terminal output from this launch ends with:

```
Loading controller: joint_state_controller
Loading controller: joint1_position_controller
Controller Spawner: Loaded controllers: joint_state_controller,
    joint1_position_controller
Started controllers: joint_state_controller, joint1_position_controller
```

Also, the terminal from which Gazebo was launched displays:

```
Loading gazebo_ros_control plugin
Starting gazebo_ros_control plugin in namespace: /one_DOF_robot
gazebo_ros_control plugin is waiting for model URDF in
    parameter [/robot_description] on the ROS param server.
Loaded gazebo_ros_control.
```

Our one-DOF robot appears in the Gazebo graphical display, identical to the initial case presented in Section 3.4. However, we now have additional topics. Running

```
rostopic list
```

shows the following additional topics:

```
/one_DOF_robot/joint1_position_controller/command
/one_DOF_robot/joint1_position_controller/pid/parameter_descriptions
/one_DOF_robot/joint1_position_controller/pid/parameter_updates
/one_DOF_robot/joint1_position_controller/state
/one_DOF_robot/joint_states
```

These topics all appear under the namespace one_DOF_robot. The topic /one_DOF_robot/ joint1_position_controller/command is subscribed to by Gazebo, and the message type is std_msgs/Float64. This topic is used by the joint position controller as a desired setpoint, equivalent to pos_command in our previous example minimal controller. We can use this topic to command our robot manually by entering, e.g.:

```
rostopic pub -r 10 /one_DOF_robot/joint1_position_controller/command std_msgs/Float64 ↵
    1.0
```

which commands the joint to a desired angle of 1.0 rad. The resulting response is shown in Fig 3.15. The response is slow and has overshoot, calling for tuning of the control param-

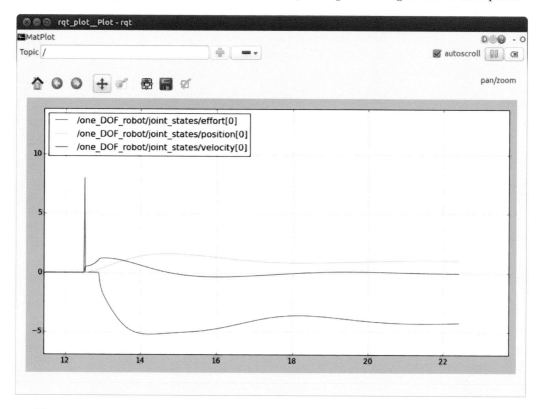

Figure 3.15: Transient response to step position command with ROS PD controller

eters. With the integral-error term active, however, the link ultimately converges virtually perfectly on the commanded setpoint of 1.0 rad.

With the preceding introductory material, we next consider a slightly more complex model in the context of a mobile robot.

3.7   BUILDING MOBILE-ROBOT MODEL

To extend the URDF modeling introduced in Section 3.3, modeling of a simple mobile robot is presented here, introducing some additional modeling and Gazebo capabilities. An on-line tutorial for building mobile robots can be found at http://wiki.ros.org/urdf/Tutorials/Building%20a%20Visual%20Robot%20Model%20with%20URDF%20from%20Scratch. (See http://gazebosim.org/tutorials?tut=build_robot for a mobile-robot modeling tutorial using SDF, which is a similar, but richer modeling format that is becoming more popular).

Extensions of URDF modeling to be introduced here include the use of xacro to help simplify URDF models, and a Gazebo plug-in for inclusion of a differential-drive controller.

The illustrative mobile-robot model, called mobot.xacro, is contained in the accompanying repository within the package mobot_urdf. The contents of mobot.xacro are given in Listing 3.12. A `xacro` file, (see `http://wiki.ros.org/xacro`), can be converted to a URDF file, but using `xacro` can help simplify model specification. Notably, `mobot.xacro` takes advantage of the `xacro` instruction `<xacro:include />` to import five other `xacro` files. This allows distributing the modeling details over multiple smaller files with restricted contexts.

Listing 3.12: `mobot.xacro`: simple mobile robot xacro model file

```
1  <?xml version="1.0"?>
2  <robot
3      xmlns:xacro="http://www.ros.org/wiki/xacro" name="mobot">
4      <xacro:include filename="$(find mobot_urdf)/urdf/mobot_xacro_defs.xacro" />
5      <xacro:include filename="$(find mobot_urdf)/urdf/mobot_static_links.xacro" />
6      <xacro:include filename="$(find mobot_urdf)/urdf/mobot_wheels.xacro" />
7      <xacro:include filename="$(find mobot_urdf)/urdf/casters.xacro" />
8      <xacro:include filename="$(find mobot_urdf)/urdf/gazebo_tags.xacro" />
9  </robot>
```

An interpretation of the Listing 3.12 follows. Line 1, `<?xml version="1.0"?>`, must be the first line of all `xacro` files. Line 3,

```
xmlns:xacro="http://www.ros.org/wiki/xacro" name="mobot">
```

establishes that this file will use the `xacro` package for defining and using macros. Subsequent lines include five other `xacro` files (and these files may include other files, hierarchically).

A useful `xacro` capability is defining properties and macros. Examples are included in the file `mobot_xacro_defs.xacro`, which appears in Listing 3.13.

Listing 3.13: `mobot_xacro_defs.xacro`: xacro file that contains parameter values

```
1  <?xml version="1.0"?>
2  <robot
3      xmlns:xacro="http://www.ros.org/wiki/xacro" >
4      <!-- define the base-link origin to lie at floor level, between the drive wheels↩
           -->
5      <!--main body is a simple box; origin is a center of box-->
6      <xacro:property name="bodylen" value="0.5461" />
7      <xacro:property name="bodywidth" value="0.4572" />
8      <xacro:property name="bodyheight" value="0.2" />
9      <xacro:property name="bodyclearance" value="0.4" />   <!--clearance from bottom of ↩
           box to ground-->
10     <!-- derived values -->
11     <xacro:property name="half_bodylen" value="${bodylen/2.0}" />
12     <xacro:property name="half_bodyheight" value="${bodyheight/2.0}" />
13     <!-- placement of main body relative to base link frame -->
14     <xacro:property name="bodyOX" value="${-half_bodylen}" />
15     <xacro:property name="bodyOY" value="0" />
16     <xacro:property name="bodyOZ" value="0.45" />
17
18     <!-- define the drive-wheel dimensions-->
19     <xacro:property name="tirediam" value="0.3302" />
20     <xacro:property name="tirerad" value="${tirediam/2.0}" />
21     <xacro:property name="tirewidth" value="0.06985" />
22     <!-- "track" is the distance between the drive wheels -->
23     <xacro:property name="track" value=".56515" />
24
25     <!-- battery box dimensions -->
26     <xacro:property name="batterylen" value="0.381" />
27     <xacro:property name="batterywidth" value="0.3556" />
28     <xacro:property name="batteryheight" value="0.254" />
29     <!-- placement of battery box relative to base frame -->
30     <xacro:property name="batOX" value="-0.05" />
```

```
31        <xacro:property name="batOY" value="0" />
32        <xacro:property name="batOZ" value="0.22" />
33
34
35        <xacro:property name="M_PI" value="3.1415926535897931" />
36        <xacro:property name="boschwidth" value="0.0381" />
37        <xacro:property name="casterdrop" value="0.125" />
38        <xacro:property name="bracketwidth" value="0.1175" />
39        <xacro:property name="bracketheight" value="0.16" />
40        <xacro:property name="bracketthick" value="0.0508" />
41        <xacro:property name="bracketangle" value="0.7854" />
42        <xacro:property name="casterwidth" value="0.0826" />
43        <xacro:property name="casterdiam" value="0.2286" />
44
45        <!--here is a default inertia matrix with small, but legal values; use this when ↩
            don't need accuracy for I -->
46        <!--model will assign inertia matrix dominated by main body box -->
47        <xacro:macro name="default_inertial" params="mass">
48            <inertial>
49                <mass value="${mass}" />
50                <inertia ixx="0.01" ixy="0.0" ixz="0.0"
51            iyy="0.01" iyz="0.0"
52            izz="0.01" />
53            </inertial>
54        </xacro:macro>
55 </robot>
```

Lines 6 through 43 use **xacro** properties to define mnemonic names to represent numerical values. This technique allows for defining URDF elements with symbolic parameters. Use of xacro properties helps to keep numerical values consistent throughout the file (which is particularly important when these values are used in multiple instances). It also makes the file easier to modify to change dimensions to tune a model to a physical system.

In addition to defining parameters, **xacro** allows for defining macros. For example, lines 47 through 54 define the macro **default_inertial**:

```
<xacro:macro name="default_inertial" params="mass">
    <inertial>
        <mass value="${mass}" />
        <inertia ixx="0.01" ixy="0.0" ixz="0.0"
    iyy="0.01" iyz="0.0"
    izz="0.01" />
    </inertial>
</xacro:macro>
```

This macro is used six times in the model listing, *e.g.* as in line 19 of **casters.xacro**:

```
<xacro:default_inertial mass="0.2"/>
```

Macros can be used within the definition of other macros. In fact, use of the macro **default_inertial** on line 19 of **casters.xacro** is such an instance, since it is embedded within the macro **caster**.

The next three files included use box and cylinder geometric objects to define the 14 links in this system, both for their visual representation and for their collision boundaries. Inertial components are defined for each link in the system. All fields within the link and joint definitions are of the same style as introduced in the **minimal_robot** URDF, except for the use of named parameters in place of numerical values. Details are distributed over multiple files, the first of which is **mobot_static_links.xacro**. This file appears in Listing 3.14.

Listing 3.14: `mobot_static_links.xacro`: specification of static links in mobot model

```
1   <?xml version="1.0"?>
2   <robot
3       xmlns:xacro="http://www.ros.org/wiki/xacro" name="static_links">
4       <link name="base_link">
5           <visual>
6               <geometry>
7                   <box size="${bodylen} ${bodywidth} ${bodyheight}"/>
8               </geometry>
9               <origin xyz="${bodyOX} ${bodyOY} ${bodyOZ}" rpy="0 0 0"/>
10          </visual>
11          <collision>
12              <geometry>
13                  <box size="${bodylen} ${bodywidth} ${bodyheight}"/>
14              </geometry>
15              <origin xyz="${bodyOX} ${bodyOY} ${bodyOZ}" rpy="0 0 0"/>
16          </collision>
17          <inertial>
18              <!--assign almost all the mass to the main body box; set m= 100kg; treat I↵
                    as approx m*r^2 -->
19              <mass value="100" />
20              <inertia ixx="10" ixy="0" ixz="0"
21                  iyy="10" iyz="0"
22                  izz="10" />
23          </inertial>
24      </link>
25
26      <link name="batterybox">
27          <visual>
28              <geometry>
29                  <box size="${batterylen} ${batterywidth} ${batteryheight}"/>
30              </geometry>
31              <origin xyz="0 0 0" rpy="0 0 0"/>
32          </visual>
33          <collision>
34              <geometry>
35                  <box size="${batterylen} ${batterywidth} ${batteryheight}"/>
36              </geometry>
37              <origin xyz="0 0 0" rpy="0 0 0"/>
38          </collision>
39          <xacro:default_inertial mass="1"/>
40      </link>
41      <joint name="batterytobase" type="fixed">
42          <parent link="base_link"/>
43          <child link="batterybox"/>
44          <origin xyz="${batOX} ${batOY} ${batOZ}" rpy="0 0 0"/>
45      </joint>
46   </robot>
```

The `mobot_static_links.xacro` file contains visual, collision and inertial properties for two links: a `base_link` and a `batterybox`. Every robot model has a root link, conventionally called `base_link`, that is the root of the transformation tree of all other frames of the robot. Arbitrarily many additional links (including sensor links) may be joined to the base link. The file `mobot_static_links.xacro` also describes a `batterybox` link, which is joined to the `base_link` through a defined fixed joint.

The third included file is `mobot_wheels.xacro`. This file, displayed in Listing 3.15, describes the drive wheels of the mobile robot. This description includes visual, collision and inertial properties of the wheels, along with effort and velocity limits of the named drive-wheel joints.

Listing 3.15: `mobot_wheels.xacro`: description of drive wheels

```
1   <?xml version="1.0"?>
2   <robot
3       xmlns:xacro="http://www.ros.org/wiki/xacro" name="wheels">
4
```

```
5    <xacro:macro name="wheel" params="prefix reflect">
6        <link name="${prefix}_wheel">
7            <visual>
8                <geometry>
9                    <cylinder radius="${tirerad}" length="${tirewidth}"/>
10               </geometry>
11           </visual>
12           <collision>
13               <geometry>
14                   <cylinder radius="${tirerad}" length="${tirewidth}"/>
15               </geometry>
16           </collision>
17           <inertial>
18           <!--assign inertial properties to drive wheels -->
19           <mass value="1" />
20           <inertia ixx="0.1" ixy="0" ixz="0"
21             iyy="0.1" iyz="0"
22             izz="0.1" />
23       </inertial>
24       </link>
25       <joint name="${prefix}_wheel_joint" type="continuous">
26           <axis xyz="0 0 1"/>
27           <parent link="base_link"/>
28           <child link="${prefix}_wheel"/>
29           <origin xyz="0 ${reflect*track/2} ${tirerad}" rpy="0 ${M_PI/2} ${M_PI/2}"↵
                /> 
30           <limit effort="100" velocity="15" />
31           <joint_properties damping="0.0" friction="0.0" />
32       </joint>
33   </xacro:macro>
34   <xacro:wheel prefix="left" reflect="1"/>
35   <xacro:wheel prefix="right" reflect="-1"/>
36 </robot>
```

The `wheel` macro is used twice in the code to define symmetric left and right drive wheels through use of the `prefix` parameter. This allows defining links with names `right_wheel` and `left_wheel` with associated joints `right_wheel_joint` and `left_wheel_joint`. These are declared in the model file on lines 34 through 35:

```
<xacro:wheel prefix="left" reflect="1"/>
<xacro:wheel prefix="right" reflect="-1"/>
```

Through use of this macro, it is assured that the left and right wheels will be identical, except for their placement within the model. Changes to parameter values used within this macro will still result in left and right wheels that are identical.

The file `casters.xacro`, which is included in `mobot.xacro`, contains detailed modeling of a pair of passive casters attached to the robot. Each caster has two degrees of freedom: caster swivel and caster wheel rotation. The modeling file specifies the visual, collision and inertial properties of the caster components, as well as the (passive) joint properties. This file is displayed in Listing 3.16.

Listing 3.16: `casters.xacro`: description of passive caster wheels

```
1  <?xml version="1.0"?>
2  <robot
3      xmlns:xacro="http://www.ros.org/wiki/xacro" name="casters">
4
5      <xacro:macro name="caster" params="prefix reflect">
6          <link name="castdrop_${prefix}">
7              <visual>
8                  <geometry>
9                      <box size="${boschwidth} ${boschwidth} ${casterdrop}"/>
10                 </geometry>
11                 <origin xyz="0 0 0" rpy="0 0 0"/>
12             </visual>
```

```
13          <collision>
14              <geometry>
15                  <box size="${boschwidth} ${boschwidth} ${casterdrop}"/>
16              </geometry>
17              <origin xyz="0 0 0" rpy="0 0 0"/>
18          </collision>
19          <xacro:default_inertial mass="0.2"/>
20      </link>
21      <joint name="cast2base_${prefix}" type="fixed">
22          <parent link="base_link"/>
23          <child link="castdrop_${prefix}"/>
24          <origin xyz="${-bodylen/2+bodyOX+boschwidth/2} ${reflect*bodywidth/2-↩
                  reflect*boschwidth/2} ${-casterdrop/2-bodyheight/2+bodyOZ}" />
25      </joint>
26      <link name="brackettop_${prefix}">
27          <visual>
28              <geometry>
29                  <box size="${bracketwidth} ${bracketthick} .005"/>
30              </geometry>
31              <origin xyz="0 0 0" rpy="0 0 0"/>
32          </visual>
33          <collision>
34              <geometry>
35                  <box size="${bracketwidth} ${bracketthick} .005"/>
36              </geometry>
37              <origin xyz="0 0 0" rpy="0 0 0"/>
38          </collision>
39          <xacro:default_inertial mass="0.2"/>
40      </link>
41      <joint name="cast2bracket_${prefix}" type="continuous">
42          <axis xyz="0 0 1"/>
43          <parent link="castdrop_${prefix}"/>
44          <child link="brackettop_${prefix}"/>
45          <origin xyz="0 0 ${-casterdrop/2}" rpy="0 0 ${M_PI/2}"/>
46          <joint_properties damping="0.0" friction="0.0" />
47      </joint>
48      <link name="bracketside1_${prefix}">
49          <visual>
50              <geometry>
51                  <box size="${bracketthick} ${bracketheight} .005"/>
52              </geometry>
53              <origin xyz="0 0 0" rpy="${M_PI/2} ${-bracketangle} ${M_PI/2}"/>
54          </visual>
55          <collision>
56              <geometry>
57                  <box size="${bracketthick} ${bracketheight} .005"/>
58              </geometry>
59              <origin xyz="0 0 0" rpy="${M_PI/2} ${-bracketangle} ${M_PI/2}"/>
60          </collision>
61          <xacro:default_inertial mass="0.2"/>
62      </link>
63      <joint name="brack2top1_${prefix}" type="fixed">
64          <parent link="brackettop_${prefix}"/>
65          <child link="bracketside1_${prefix}"/>
66          <origin xyz="${bracketwidth/2} .04 -${bracketheight/2-.02}" rpy="0 0 0" />
67      </joint>
68      <link name="bracketside2_${prefix}">
69          <visual>
70              <geometry>
71                  <box size="${bracketthick} ${bracketheight} .005"/>
72              </geometry>
73              <origin xyz="0 0 0" rpy="${M_PI/2} ${-bracketangle} ${M_PI/2}"/>
74          </visual>
75          <collision>
76              <geometry>
77                  <box size="${bracketthick} ${bracketheight} .005"/>
78              </geometry>
79              <origin xyz="0 0 0" rpy="${M_PI/2} ${-bracketangle} ${M_PI/2}"/>
80          </collision>
81          <xacro:default_inertial mass="0.2"/>
82      </link>
83      <joint name="brack2top2_${prefix}" type="fixed">
84          <parent link="brackettop_${prefix}"/>
85          <child link="bracketside2_${prefix}"/>
86          <origin xyz="${-bracketwidth/2} .04 -${bracketheight/2-.02}" rpy="0 0 0" ↩
                  />
```

```
87          </joint>
88          <link name="${prefix}_casterwheel">
89              <visual>
90                  <geometry>
91                      <cylinder radius="${casterdiam/2}" length="${casterwidth}"/>
92                  </geometry>
93              </visual>
94              <collision>
95                  <geometry>
96                      <cylinder radius="${casterdiam/2}" length="${casterwidth}"/>
97                  </geometry>
98              </collision>
99              <!-- accept default inertial properties for caster wheels-->
100             <xacro:default_inertial mass="0.5"/>
101         </link>
102         <joint name="${prefix}_caster_joint" type="continuous">
103             <axis xyz="0 0 1"/>
104             <parent link="bracketside1_${prefix}"/>
105             <child link="${prefix}_casterwheel"/>
106             <origin xyz="${-casterwidth/2-.02} .053 -.053" rpy="0 ${M_PI/2} 0"/>
107             <limit effort="100" velocity="15" />
108             <joint_properties damping="0.0" friction="0.0" />
109         </joint>
110     </xacro:macro>
111     <xacro:caster prefix="left" reflect="1"/>
112     <xacro:caster prefix="right" reflect="-1"/>
113 </robot>
```

As with the drive wheels, a macro that supports creating identical, albeit reflected caster models is defined.

Finally, `mobot.xacro` includes the file `gazebo_tags.xacro`, which is shown in Listing 3.17. As with the minimal robot arm example, it is useful to include a Gazebo plug-in for joint control. For the mobot model, this is done as shown in file `gazebo_tags.xacro`.

Listing 3.17: `gazebo_tags.xacro`: gazebo block to include differential-drive plug-in

```
1  <?xml version="1.0"?>
2  <robot
3       xmlns:xacro="http://www.ros.org/wiki/xacro" name="gazebo_tags">
4
5   <gazebo>
6    <plugin name="differential_drive_controller" filename="libgazebo_ros_diff_drive.so">
7     <alwaysOn>true</alwaysOn>
8     <updateRate>100</updateRate>
9     <leftJoint>right_wheel_joint</leftJoint>
10    <rightJoint>left_wheel_joint</rightJoint>
11    <wheelSeparation>${track}</wheelSeparation>
12    <wheelDiameter>${tirediam}</wheelDiameter>
13    <torque>200</torque>
14    <commandTopic>cmd_vel</commandTopic>
15    <odometryTopic>odom</odometryTopic>
16    <odometryFrame>odom</odometryFrame>
17    <robotBaseFrame>base_link</robotBaseFrame>
18    <publishWheelTF>true</publishWheelTF>
19    <publishWheelJointState>true</publishWheelJointState>
20   </plugin>
21  </gazebo>
22 </robot>
```

The XML code between the `<gazebo>` tags brings in the library `libgazebo_ros_diff_drive.so`, which is useful for differential-drive control similar to that of the previous simple two-dimensional robot simulator. To use the differential-drive plug-in, several parameters must be defined, including

- Names of the drive-wheel joints, `right_wheel_joint` and `left_wheel_joint`

- Wheel separation, `track`

- Name of the root of the URDF tree, which is `base_link`

- Topic name to be used to command speed and spin, set (per convention) to `cmd_vel`

Further details can be found at `http://gazebosim.org/tutorials?tut=ros_gzplugins` and `http://www.theconstructsim.com/?p=3332`.

To use a xacro file, one converts it to a URDF file using the `xacro` executable within the `xacro` package. For example, to convert the file `mobot.xacro` to a URDF with filename `mobot.urdf`, run the following command:

```
rosrun xacro xacro mobot.xacro > mobot.urdf
```

This action creates a URDF file with all of the substitutions defined by the `xacro` macros. The URDF file produced can be checked for consistency with the command:

```
check_urdf mobot.urdf
```

which produces the output:

```
robot name is: mobot
---------- Successfully Parsed XML ---------------
root Link: base_link has 5 child(ren)
    child(1):  batterybox
    child(2):  castdrop_left
        child(1):  brackettop_left
            child(1):  bracketside1_left
                child(1):  left_casterwheel
            child(2):  bracketside2_left
    child(3):  castdrop_right
        child(1):  brackettop_right
            child(1):  bracketside1_right
                child(1):  right_casterwheel
            child(2):  bracketside2_right
    child(4):  left_wheel
    child(5):  right_wheel
```

This output shows that the URDF file can be parsed logically, and it displays 14 links in the model. Relationships among the links are shown in an outline style. The tree of links also can be visualized graphically using the command:

```
urdf_to_graphiz mobot.urdf
```

which produces the file `mobot_graphiz.pdf`, which is displayed in Fig 3.16. This figure illustrates the connectivities and spatial relationships among the links.

The mobot model can be loaded into Gazebo to visualize and simulate. As before, this is done by first loading the robot description file into the parameter server, then executing the `spawn_model` node within the `gazebo_ros` package, referencing the robot model on the parameter server. This is accomplished via the launch file `mobot.launch` in the `urdf` subdirectory of the package `mobot_urdf`. First, Gazebo is started with the command:

```
roslaunch gazebo_ros empty_world.launch
```

Then the mobot model is inserted into the Gazebo simulation by running the corresponding launch file:

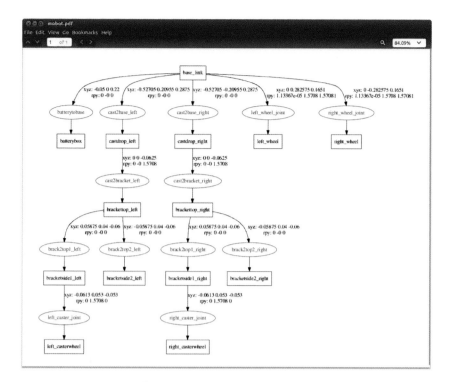

Figure 3.16: Graphical display of URDF tree for mobot

```
roslaunch mobot_urdf mobot.launch
```

The resulting Gazebo display is shown in Fig 3.17. The Gazebo option `view->joints` has been enabled. Note that we can see 6 movable joints: two joints corresponding to the two large drive wheels, and four joints associated with the passive casters.

3.8 SIMULATING MOBILE-ROBOT MODEL

As noted earlier, the robot model can be introduced in the Gazebo simulator by

```
roslaunch gazebo_ros empty_world.launch
```

after which the mobot model is inserted into the Gazebo simulation by running the corresponding launch file:

```
roslaunch mobot_urdf mobot.launch
```

At this point, the mobile robot can be controlled via the `cmd_vel` topic, identical to what was done with the simple two-dimensional robot simulator. For example, the command:

```
rostopic pub cmd_vel geometry_msgs/Twist  '{linear:  {x: 0.5, y: 0.0, z: 0.0}, angular←
    : {x: 0.0,y: 0.0,z: 0.3}}'
```

causes the robot to move in a counter-clockwise circle.

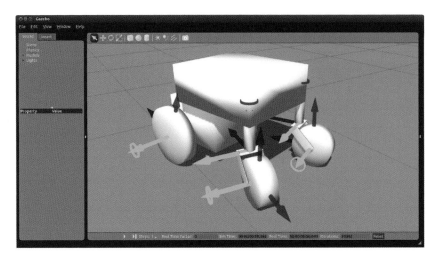

Figure 3.17: Gazebo view of mobot in empty world

We can also command the robot under program control, as formerly illustrated with the STDR simulator, or with a teleoperation program. Use of such command nodes with different robots is made simpler through a ROS feature of topic remapping. This can be seen by re-using the `stdr_open_loop_commander` node from the package `stdr_control` introduced in Section 3.1. Note that the `stdr_open_loop_commander.cpp` code publishes to the topic `/robot0/cmd_vel`, whereas our differential-drive controller of the mobot model expects commands on topic `cmd_vel`. We can re-use the `stdr_open_loop_commander` node nonetheless if we remap the output topic name. From the command line, this can be invoked by entering:

```
rosrun stdr_control stdr_open_loop_commander /robot0/cmd_vel:=cmd_vel
```

By using the option `/robot0/cmd_vel:=cmd_vel`, when this command is executed, the output from `stdr_open_loop_commander` is redirected from `/robot0/cmd_vel` to `cmd_vel`, and our mobot responds.

Topic remapping can also be performed within launch files, as illustrated in `open_loop_squarewave_commander.launch` within the `launch` subdirectory of package `mobot_urdf`.

Listing 3.18: `open_loop_squarewave_commander.launch`

```
<launch>
<!-- original node publishes to /robot0/cmd_vel; direct this instead to topic /cmd_vel↩
    -->
<node pkg="stdr_control" type="stdr_open_loop_commander" name="commander">
 <remap from="/robot0/cmd_vel" to="cmd_vel" />
</node>
</launch>
```

Running this launch file has the same effect as the previous command line execution with topic remapping.

An alternative, more convenient means to send Twist commands to the robot interactively is by running a keyboard teleoperation node, *e.g.*:

```
rosrun teleop_twist_keyboard teleop_twist_keyboard.py
```

(This package should be installed by the provided installation scripts, or it can be installed manually. See http://wiki.ros.org/teleop_twist_keyboard.) This node takes keyboard input to command linear and angular velocities, and these commands are published as a `Twist` message to the `cmd_vel` topic, resulting in robot motion. This interface works as well with the STDR simulator or with any mobile robot that subscribes to a velocity-command interface with a `Twist` message type (which is the *de facto* standard in ROS). By default, this node publishes `Twist` commands to the topic `cmd_vel`, although this can be remapped (*e.g.* to control the STDR simulator).

An important difference of the URDF model in Gazebo versus the STDR simulator is that the Gazebo simulation includes physics. Inertias, friction, controller dynamics and actuator saturation are taken into account. With the Gazebo simulation of the mobot running, invoking the launch file:

```
roslaunch mobot_urdf open_loop_squarewave_commander.launch
```

commands the robot to move, and the resulting response can be plotted with `rqt_plot`. The mobot simulation in Gazebo publishes its state to the `odom` topic, from which we can plot the robot forward velocity in response to step velocity commands. As seen in Fig 3.18, the robot does not change velocities instantaneously.

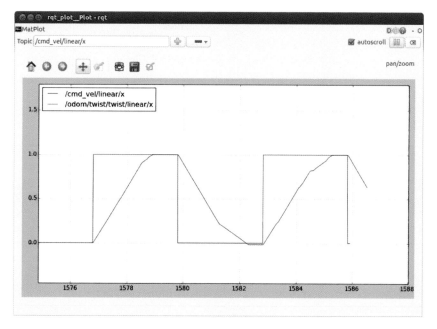

Figure 3.18: Response of mobot to step velocity commands in Gazebo simulation

One of the idiosyncrasies of the default physics simulator, open dynamics engine, is that it does not do a good job of modeling sustained stiff contact between separate models. This is noticeable with wheels or feet on the ground, as well as gripper fingers grasping objects. This can be observed with the simple mobile robot model. After starting Gazebo and inserting the mobot model, the robot will slowly slip to its left. Actually, the robot wheels are chattering against the ground. This can be seen in the Gazebo viewer by enabling

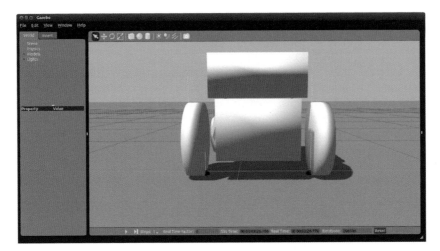

Figure 3.19: Gazebo simulation of mobot with contacts display enabled

`view->contacts` from the top menu bar. An example screenshot with contacts displayed appears in Fig 3.19. The blue markers are points of contact and the green lines show the direction and (by length) the magnitude of the contact forces. At this instant, the contact forces on wheels are stronger on its left side than on its right. However, if the display is viewed dynamically, the contact forces can be seen to toggle on and off and to shift around due to numerical instability of modeling stiff contact. This problem is exacerbated by the fact that the robot has four wheels, and simultaneous contact of four points with a flat plane is numerically challenging. As a result, the robot seems to constantly vibrate on the ground.

A few adjustments can be made to try to improve the numerical stability. Within the URDF (or `xacro` file), insert the lines:

```
<gazebo reference="left_wheel">
  <mu1>100000.0</mu1>
  <mu2>100000.0</mu2>
  <material>Gazebo/Black</material>
</gazebo>
```

and repeat the block for `right_wheel` as well. This Gazebo tag passes parameters to the physics engine, declaring a custom friction property. In fact, the `mu` coefficients are not realistically physical. They do influence the simulation, but these values cannot be treated like genuine dimensionless Coulomb friction components. Some experimentation may be required to achieve acceptable simulation results for sustained contact simulations, including wheels, feet and fingers.

The line `<material>Gazebo/Black</material>` was introduced above to show how to set colors for links in Gazebo. This property must also be included within `<gazebo>` delimiters. The result of the example insertion (repeated for left and right wheels) is that the model appears with black wheels.

Another variation that can be introduced is to tell the simulator to run more iterations of internal computations each time step (where the time step is, by default, 1 ms). This can result in better fidelity of simulation, including less chatter with sustained, stiff contacts. But this is done at the price of more demanding computation, potentially resulting in a lower real-time factor. The `iterations` parameter can be changed with a Gazebo tag in the URDF or interactively from the Gazebo window. For the latter, on the left pane, choose the World tab, then under `physics->solver->iterations`, edit the numerical value from its default value (50) to something larger (*e.g.* 200). This change can slow the side slip of the mobot model to be imperceptible.

The mobile-robot simulation can be made more interesting by introducing elements in the world model. This can be done by manually inserting existing models or by inserting custom-designed models. An interesting world model that can be inserted from the Gazebo Insert tab is `Starting Pen`. After selecting this model from the Insert tab, the user has the opportunity to move the corresponding model around in the simulated world to a desired location (via the mouse), then click to complete the model placement. Figure 3.20 shows the mobot model within a virtual world that includes the starting pen model.

Figure 3.20: Gazebo display of a mobot in starting pen

3.9 COMBINING ROBOT MODELS

Since robot URDF files can get quite long, it is useful to be able to include subsystems within a single model file. Similarly, it is convenient to include launch files within launch files.

To illustrate, we will mount our minimal robot arm model onto our simple mobile base. Revisiting the minimal arm model in the `minimal_robot_description` package, we make a simple change to the URDF to create the file `minimal_robot_description_unglued.urdf`. Importantly, we delete the joint `glue_robot_to_world`, so our arm can be attached to the mobile base instead. To make the arm more attractive, the following material tags are added to colorize the links:

```
<gazebo reference="link1">
  <material>Gazebo/Blue</material>
</gazebo>
<gazebo reference="link2">
  <material>Gazebo/Red</material>
</gazebo>
```

We can combine the mobile-platform and minimal arm models as illustrated by the file `mobot_w_arm.xacro`, the contents of which are shown in Listing 3.19.

Listing 3.19: `mobot_w_arm.xacro`: model file combining mobot and minimal arm

```
1   <?xml version="1.0"?>
2   <robot
3       xmlns:xacro="http://www.ros.org/wiki/xacro" name="mobot">
4     <xacro:include filename="$(find mobot_urdf)/urdf/mobot2.xacro" />
5     <xacro:include filename="$(find minimal_robot_description)/↵
          minimal_robot_description_unglued.urdf" />
6
7     <!-- attach the simple arm to the mobile robot -->
8     <joint name="arm_base_joint" type="fixed">
9       <parent link="base_link" />
10      <child link="link1" />
11      <origin rpy="0 0 0 " xyz="${-bodylen/2} 0 ${bodyOZ+bodyheight/2}"/>
12    </joint>
13  </robot>
```

The xacro file of Listing 3.19 resides in the `mobot_urdf` package. Lines 4 and 5:

```
<xacro:include filename="$(find mobot_urdf)/urdf/mobot2.xacro" />
<xacro:include filename="$(find minimal_robot_description)/↵
    minimal_robot_description_unglued.urdf" />
```

use the `include` feature of `xacro` to bring in two files verbatim. The named file paths start with `$(find mobot_urdf)`. This syntax informs the launcher to search for the named files by stated package name (`mobot_urdf` in this example). Specifying directories in this manner simplifies specifying the search path, and it also makes the launcher more robust. Packages installed in a different system under different directories can still be found using this syntax.

After bringing in both the mobile base and the simple arm models, the arm is attached to the base by declaring a new (fixed) joint, `arm_base_joint`:

```
<joint name="arm_base_joint" type="fixed">
    <parent link="base_link" />
    <child link="link1" />
    <origin rpy="0 0 0 " xyz="${-bodylen/2} 0 ${bodyOZ+bodyheight/2}"/>
</joint>
```

This new joint specifies how `link1` (the first link of the arm model) is to be attached to its parent, `base_link` (the base link of the mobile-platform model). The arm model is offset in y and z, using `xacro` parameters, to center the base of `link1` on the top of the main link of the mobile robot.

In addition to combining URDF (or `xacro`) models, it is also convenient to combine launch files. Our launch file for the robot arm, `minimal_robot_w_jnt_ctl.launch`, performed three functions: it loaded the arm URDF onto the parameter server, it spawned the robot model from the parameter server into Gazebo, and it started the ROS joint position controller(s). Since our integrated mobile base with arm is a new model, we wish to separate the arm-specific launch commands from the model loading and spawning. For this purpose, a smaller launch file in the `minimal_robot_description` package, called `minimal_robot_ctl.launch`, is created. The contents of this file appear in Listing 3.20.

Listing 3.20: `minimal_robot_ctl.launch`: launch file for starting arm controller

```
1  <launch>
2    <!-- Load joint controller configurations from YAML file to parameter server -->
3    <rosparam file="$(find minimal_robot_description)/control_config/one_dof_ctl_params.↵
         yaml" command="load"/>
4
5    <!--start up the controller plug-ins via the controller manager -->
6    <node name="controller_spawner" pkg="controller_manager" type="spawner" respawn="↵
         false"
7      output="screen" ns="/one_DOF_robot" args="joint_state_controller ↵
           joint1_position_controller"/>
8  </launch>
```

A launch file for the integrated robot model resides in the **mobot_urdf** package within the **launch** subdirectory, named **mobot_w_arm.launch**. The contents of this launch file appear in Listing 3.21.

Listing 3.21: launch file combining mobot and minimal arm

```
1  <launch>
2  <!-- Convert xacro model file and put on parameter server -->
3  <param name="robot_description" command="$(find xacro)/xacro.py '$(find mobot_urdf)/↵
       urdf/mobot_w_arm.xacro'" />
4
5  <!-- Spawn the robot from parameter server into Gazebo -->
6  <node name="spawn_urdf" pkg="gazebo_ros" type="spawn_model" args="-param ↵
       robot_description -urdf -model mobot" />
7
8  <!-- load the controller parameter yaml file and start the ROS controllers for the arm↵
       -->
9  <include file="$(find minimal_robot_description)/minimal_robot_ctl.launch">
10 </include>
11
12 </launch>
```

This launch file loads the combined **xacro** file onto the parameter server and spawns this model into Gazebo. The line:

```
<include file="$(find minimal_robot_description)/minimal_robot_ctl.launch">
```

searches for the file **minimal_robot_ctl.launch** in the **minimal_robot_description** package and includes this launch file verbatim.

After launching Gazebo, the integrated launch file can be invoked with:

```
roslaunch mobot_urdf mobot_w_arm.launch
```

The resulting Gazebo display appears in Fig 3.21.

Running:

```
rostopic list
```

shows that the following topics are active:

```
/joint_states
/odom
/cmd_vel
/one_DOF_robot/joint1_position_controller/command
/one_DOF_robot/joint1_position_controller/pid/parameter_descriptions
/one_DOF_robot/joint1_position_controller/pid/parameter_updates
```

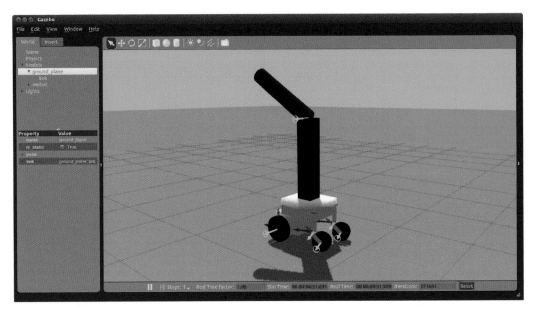

Figure 3.21: Gazebo display of combined mobile base and minimal arm models

```
/one_DOF_robot/joint1_position_controller/state
/one_DOF_robot/joint_states
```

We thus see that command interface topics are available for both the mobile platform and the arm. Feedback values of joint states for both the arm and mobile base are available as well.

3.10   WRAP-UP

Our robot modeling so far has been quite crude in utilizing only primitive boxes and cylinders. Nonetheless, we have seen the basic elements of modeling, including kinematic, dynamic, visual and collision properties, as well as use of plug-ins for physically realistic controllers. The simple examples illustrate modeling both of articulated mechanisms and mobile vehicles.

Making more sophisticated models involves extending the same techniques to more joints and using CAD-file descriptions that offer more detailed and realistic modeling of visual and collision properties. A few examples examples of more realistic robot models are shown in Figs 3.22 through 3.24.

The Baxter [9] robot model in Fig 3.22 includes control over 15 joints (7 per arm plus a neck pan motion) plus grippers (which can be substituted). Sensor outputs include the dynamic state of every joint, streaming images from three color cameras, distance sensors from the wrists, and sonar sensors around the head. This model is realistic in terms of its kinematics (including joint limits), actuator behavior (including torque saturations), visual appearance and collision characteristics. The robot is highly dexterous and well instrumented. Further, the model is publicly available [37]. Consequently, this model will be used in this text for further examples in use of ROS for sensing and manipulation.

The model in Fig 3.23 is of a dual-arm DaVinci surgical robot [41]. The CAD descriptions have been posted by Johns Hopkins University [18], [16]. Inertial parameters and ROS controller plug-ins were added to make this model Gazebo compliant. The resulting sys-

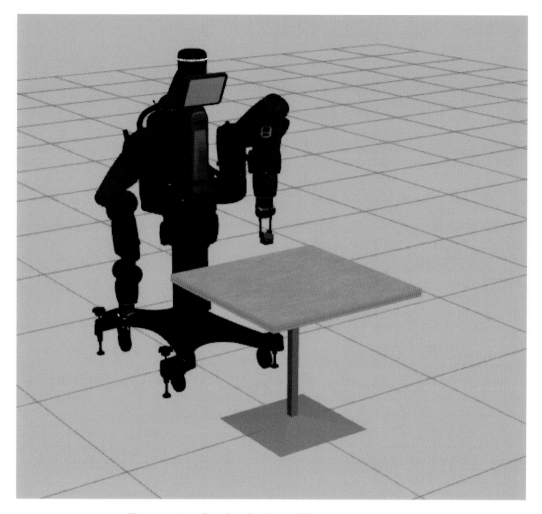

Figure 3.22: Gazebo display of Baxter robot model

tem is useful for testing extensions to computer-assisted robotic surgery and potentially development of a surgical training system.

Figure 3.24 shows the Boston-Dynamics, Inc. Atlas robot [6]; the model was developed by the Open Source Robotics Foundation (OSRF) [1] for the DARPA[1] Robotics Challenge [30]. This model was used by teams to develop code for the DARPA competition tasks, including both bipedal motion control and object manipulations (*e.g.* valve turning). Given the risk, difficulty and crew size required to perform physical experiments with Atlas, this model helped teams develop code more quickly, enabling individuals to make progress in parallel and to perform extensive software debugging and testing in simulation before moving to physical trials.

It should be appreciated that robot models in Gazebo include realistic physical interactions, including collisions, effects of gravity and inertial effects. Simulators that lack these properties can be used to evaluate viability of kinematic trajectories in terms of reachability,

---

[1]DARPA: Defense Advanced Research Projects Agency of the U.S. Department of Defense.

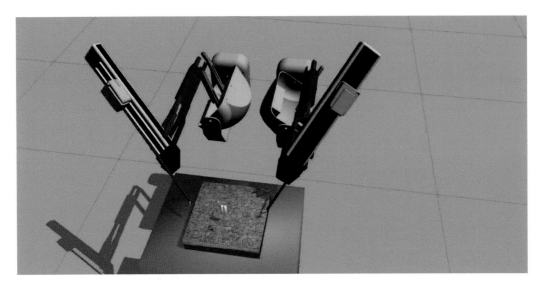

Figure 3.23: Gazebo display of DaVinci robot model

but simulators that lack physics engines cannot evaluate dynamic effects, including walking or object manipulation.

An even stronger motivation for using a physics-based simulator is incorporation of sensor simulation. Gazebo is capable of simulating physical sensors acting in a virtual environment. Plug-ins for sensors include color cameras, LIDARs, depth cameras (including the Microsoft Kinect [38]), stereo cameras, accelerometers, force sensors, and sonar sensors. With emulation of sensors, one can develop sensor-based behaviors that are testable in Gazebo. Since such development can be time consuming, having a suitable simulator is a valuable productivity tool.

Integrating sensory data requires reconciling all model and sensor information to a common reference frame, which requires extensive use of coordinate transforms in ROS, which is introduced next.

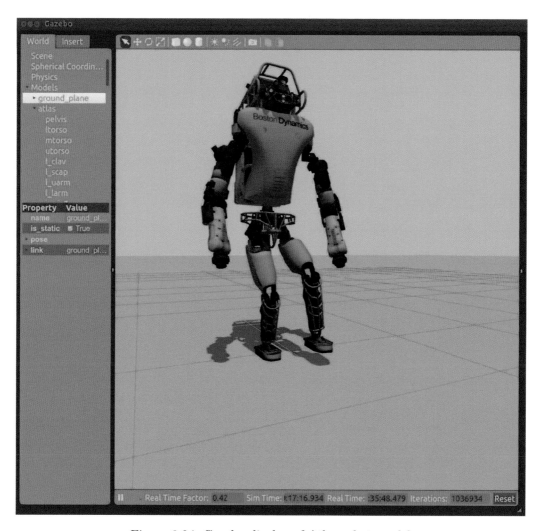

Figure 3.24: Gazebo display of Atlas robot model

Coordinate Transforms in ROS

CONTENTS

I NTRODUCTION
Reconciliation of coordinates is necessary to take advantage of sensor-driven behaviors, whether in simulation or in physical control. Coordinating data from multiple models and sensors is accomplished through coordinate transformations. Fortunately, ROS provides extensive support for coordinate transformations. Since transforms are ubiquitous in ROS and in robotics in general, it is important to have an understanding of how coordinate transforms are handled in ROS.

4.1   INTRODUCTION TO COORDINATE TRANSFORMS IN ROS

Coordinate frame transformations are fundamental to robotics. For articulated robot arms, full six-DOF transforms are required to compute gripper poses as a function of joint angles. Multiplication of sequential transforms, link by link, performs such computations. Sensor data, *e.g.* from cameras or LIDARs, is acquired in terms of the sensor's own frame, and this data must be interpreted in terms of alternative frames (*e.g.* world frame or robot frame).

Any introductory robotics text will cover coordinate-frame assignments and transformations (*e.g.* [36], [3]). A brief introduction to coordinate-frame transformations is given here, leading to ROS's treatment of coordinate transforms.

A frame is defined by a point in 3-D space, **p**, that is the frame's origin, and three vectors: **n**, **t** and **b** (which define the local **x**, **y** and **z** axes, respectively). The axis vectors are normalized (have unit length), and they form a right-hand triad, such that **n** crossed into **t** equals **b**: $\mathbf{b} = \mathbf{n} \times \mathbf{t}$.

These three directional axes can be stacked side-by-side as column vectors, comprising a 3×3 matrix, **R**.

$$\mathbf{R} = \begin{bmatrix} \mathbf{n} & \mathbf{t} & \mathbf{b} \end{bmatrix} = \begin{bmatrix} n_x & t_x & b_x \\ n_y & t_y & b_y \\ n_z & t_z & b_z \end{bmatrix} \tag{4.1}$$

We can include the origin vector as well to define a 3×4 matrix as:

$$\begin{bmatrix} \mathbf{n} & \mathbf{t} & \mathbf{b} & \mathbf{p} \end{bmatrix} = \begin{bmatrix} n_x & t_x & b_x & p_x \\ n_y & t_y & b_y & p_y \\ n_z & t_z & b_z & p_z \end{bmatrix} \tag{4.2}$$

A useful trick to simplify mathematical operations is to define a homogeneous transformation matrix by converting the above 3×4 matrix into a square 4×4 matrix by adding a fourth row consisting of $[\,0\ 0\ 0\ 1\,]$. We will refer to this augmented matrix as a $\mathbf{T}$ matrix:

$$\mathbf{T} = \begin{bmatrix} n_x & t_x & b_x & p_x \\ n_y & t_y & b_y & p_y \\ n_z & t_z & b_z & p_z \\ 0 & 0 & 0 & 1 \end{bmatrix} \tag{4.3}$$

Conveniently, matrices constructed as above (consistent with valid frame specifications) are always invertible. Further, computation of the inverse of a $\mathbf{T}$ matrix is efficient.

Abstractly, one can refer to an origin and a set of orientation axes (vectors) without having to specify numerical values. However, to perform computations, numerical values are required. When numerical values are given, one must define the coordinate system in which the values are measured. For example, to specify the origin of frame B (*i.e.* point $\mathbf{p}$) with respect to frame A, we can measure from the origin of frame A to the origin of frame B along three directions: the $\mathbf{x}$ axis of frame A, the $\mathbf{y}$ axis of frame A, and the $\mathbf{z}$ axis of frame A. These measurements can be referred to explicitly as $p_{x/A}$, $p_{y/A}$, and $p_{z/A}$, respectively. If we provided coordinates for the origin of frame B from any other viewpoint, the numerical values of the components of $\mathbf{p}$ would be different.

Similarly, we can describe the components of frame B's $\mathbf{n}$ axis (correspondingly, $\mathbf{t}$ and $\mathbf{b}$ axes) with respect to frame A by measuring the x, y and z components along the respective axes of frame A.

Equivalently, this representation can be interpreted as the following operation. First, translate frame B such that its origin coincides with frame A, but do not change the orientation of frame B. The three axis vectors of frame B have tips that are unit length from the (common) origin. These define three distinct points in space, and each of these axis tips of frame B can be expressed as 3-D points, *e.g.* as measured in frame A. Equivalently, the tip of the $\mathbf{n}$ axis of frame B is simply $\begin{bmatrix} 1 & 0 & 0 \end{bmatrix}^T$ in the B frame and $\begin{bmatrix} n_{x/A} & n_{y/A} & n_{z/A} \end{bmatrix}^T$ with respect to the A frame.

We can then state the position and orientation of frame B with respect to frame A as:

$$^{A}\mathbf{T_B} = \mathbf{T_{B/A}} = \begin{bmatrix} n_{x/A} & t_{x/A} & b_{x/A} & p_{x/A} \\ n_{y/A} & t_{y/A} & b_{y/A} & p_{y/A} \\ n_{z/A} & t_{z/A} & b_{z/A} & p_{z/A} \\ 0 & 0 & 0 & 1 \end{bmatrix} \tag{4.4}$$

We can refer to the above matrix as "frame B with respect to frame A."

Having labeled frames A and B, providing values for the elements of $^{A}\mathbf{T_B}$ fully specifies the position and orientation of frame B with respect to frame A.

In addition to providing a means to explicitly declare the position and orientation of a frame (with respect to some named frame, *e.g.* frame B with respect to frame A), $\mathbf{T}$ matrices can also be interpreted as operators. For example, if we know the position and orientation of frame C with respect to frame B, *i.e.* $^{B}\mathbf{T_C}$, and if we also know the position and orientation of frame B with respect to frame A, $^{A}\mathbf{T_B}$, we can compute the position and orientation of frame C with respect to frame A as follows:

$$^{A}\mathbf{T_C} = {}^{A}\mathbf{T_B}\,{}^{B}\mathbf{T_C} \tag{4.5}$$

That is, a simple matrix multiplication yields the desired transform. This process can be extended, *e.g.*:

$$^A\mathbf{T_F} = {}^A\mathbf{T_B}{}^B\mathbf{T_C}{}^C\mathbf{T_D}{}^D\mathbf{T_E}{}^E\mathbf{T_F} \tag{4.6}$$

In the above, the 4×4 on the left-hand side can be interpreted column by column. For example, the fourth column (rows 1 through 3), contains the coordinates of the origin of frame F as measured with respect to frame A.

The notation of prefix superscripts and post subscripts provides a visual mnemonic to aid logical compatibility. The super and sub-scripts act like Lego™blocks, such that a subscript of a leading $\mathbf{T}$ matrix must match the pre-superscript of the trailing $\mathbf{T}$ matrix. Following this convention helps to keep transform operations consistent.

Since coordinate transforms are so common in robotics, ROS provides a powerful `tf` package for handling transforms. (See `http://wiki.ros.org/tf`.) Use of this package can be confusing, because `tf` performs considerable work.

To examine the `tf` topic, we can start Gazebo with an empty world:

```
roslaunch gazebo_ros empty_world.launch
```

and then bring up our mobile robot with one-DOF arm:

```
roslaunch mobot_urdf mobot_w_arm.launch
```

This results in 18 topics published, including the topic `tf`. Issuing:

```
rostopic info tf
```

shows that the topic `tf` carries messages of type `tf2_msgs/TFMessage`.
Running:

```
rosmsg show tf2_msgs/TFMessage
```

reveals that the `tf2_msgs/TFMessage` is organized as follows:

```
geometry_msgs/TransformStamped[] transforms
  std_msgs/Header header
    uint32 seq
    time stamp
    string frame_id
  string child_frame_id
  geometry_msgs/Transform transform
    geometry_msgs/Vector3 translation
      float64 x
      float64 y
      float64 z
    geometry_msgs/Quaternion rotation
      float64 x
      float64 y
      float64 z
      float64 w
```

That is, this message contains a vector (variable-length array) of messages of type `geometry_msgs/TransformStamped`.

The ROS transform datatype is not identical to a 4×4 homogeneous transformation matrix, but it carries equivalent information (and more). The ROS transform datatype contains a 3-D vector (equivalent to the fourth column of a 4×4 transform) and a quaternion (an alternative representation of orientation). In addition, transform messages have time stamps, and they explicitly name the child frame and the parent frame (as text strings).

There can be (and typically are) many publishers to the `tf` topic. Each publisher expresses a transform relationship, describing a named child frame with respect to a named parent frame. In the present example, Gazebo publishes to `tf` because, within the differential-drive plug-in, this was requested with the `publishWheelTF` option:

```
<gazebo>
  <plugin name="differential_drive_controller" filename="libgazebo_ros_diff_drive.so">
    <publishWheelTF>true</publishWheelTF>
    <publishWheelJointState>true</publishWheelJointState>
```

(Note that `publishWheelJointState` was also enabled for this plug-in).

Running:

```
rostopic hz tf
```

shows that the `tf` topic is updated at 300 Hz. Examining the output of `tf` with:

```
rostopic echo tf
```

shows that the **transforms** component (a variable-length array) of this message includes multiple, individual transform relationships. An (abbreviated) excerpt from the `tf` echo output is:

```
frame_id: base_link
child_frame_id: left_wheel
transform:
  translation:
    x: -9.1739781936e-08
    y: 0.282574370084
    z: 0.165117569901
  rotation:
    x: -0.497142106739
    y: 0.502848355739
    z: 0.502843417149
    w: 0.497133538064

frame_id: base_link
child_frame_id: right_wheel
transform:
  ...
```

The `left_wheel` link has a defined reference frame, as does the `base_link`. From the `tf` output, we see that a transform is expressed between the `base_link` frame and the `left_wheel` frame (and similarly between the `base_link` frame and the `right_wheel` frame). The transform from the `base_link` frame to the `left_wheel` frame has a translational part and a rotational part.

The translation from the origin of the base frame to the origin of the left-wheel frame is approximately [0, 0.283, 0.165]. These values are interpretable in terms of our robot model. Our base frame has its x axis pointing forward, its y axis pointing to the left, and its z axis pointing up. The origin of the base frame is at ground level, immediately below the point between the two wheels. Consequently, the x value of the left-wheel origin is 0 (neither in front of nor behind the base-frame origin). The y value of the left-wheel origin is positive 0.283 (half the track width of the vehicle). The z value of the left-wheel origin is equal to the wheel radius (*i.e.*, the axle is above ground level by one wheel radius).

The `left_wheel` frame is also rotated relative to the `base_link` frame. Its z axis points to the left of the base frame, parallel to the base-frame's y axis. However, the wheel's x and y axes will change their orientations as the wheel rotates (*i.e.* they are functions of the `left_wheel_joint` rotation value).

In ROS, orientations are commonly expressed as unit quaternions. The quaternion representation of orientation is an alternative to rotation matrices. Quaternions can be converted to 3×3 rotation matrices (and used within 4×4 coordinate transform matrices), or quaternions can be used directly with mathematically defined quaternion operations for coordinate transforms. Quaternions are more compact than 3×3 rotation matrices and they have attractive mathematical properties. However, they are not as intuitive as 3×3 rotation matrices for visualizing orientations in terms of **n**, **t** and **b** axes.

Details of representations and mathematical operations with quaternions will not be covered here but can be found in many robotics textbooks and publications (*e.g.* [25], [11]). At this point, it is sufficient to know that there is a correspondence between quaternions and rotation matrices (and ROS functions to perform such conversions will be introduced later), and that there are corresponding mathematical operations for coordinate transformations with quaternions.

Another topic to which Gazebo is publishing is `joint_states`, which carries messages of type `sensor_msgs/JointState`. Running:

```
rostopic echo joint_states
```

shows output such as the (abbreviated) following:

```
name: ['left_wheel_joint', 'right_wheel_joint']
position: [0.005513943571622271, 0.007790399443280194]
```

These messages list the names of joints (*e.g.* `left_wheel_joint`), then list states of these joints (including positions) in the same order as the list of joint names (*e.g.* the left wheel joint angle is 0.0055 rad, per the example message). This information is needed as part of the computation of transforms. For example, the transform of the `left_wheel` frame with respect to the base frame cannot be computed without knowing the value of the wheel rotation.

The wheel rotation values are published on `joint_states` because the URDF model contains the Gazebo option: `<publishWheelJointState>true</publishWheelJointState>`. However, the `joint_states` topic does not contain other angles, including the caster wheel angles nor the arm joint angle (`joint1`).

The caster wheel angles are known to Gazebo, and these joint angles can be published to a ROS topic by including another Gazebo plug-in in the model. The model file `mobot_w_jnt_pub.xacro` (in the `mobot_urdf` package) is the same as `mobot2.xacro`, except for the addition of the following lines:

```
<gazebo>
  <plugin name="joint_state_publisher" filename="↩
      libgazebo_ros_joint_state_publisher.so">
    <jointName>cast2bracket_right, cast2bracket_left, right_caster_joint, ↩
        left_caster_joint </jointName>
  </plugin>
</gazebo>
```

This Gazebo plug-in accesses the internal Gazebo dynamic model to obtain the joint values for the named joints, and it publishes these values to the `joint_states` topic.

Similarly, the model file `minimal_robot_w_jnt_pub.urdf` in the `minimal_robot_description` package is identical to `minimal_robot_description_unglued.urdf`, except for the addition of the following lines:

```
<gazebo>
    <plugin name="joint_state_publisher" filename="←
        libgazebo_ros_joint_state_publisher.so">
      <jointName>joint1</jointName>
    </plugin>
</gazebo>
```

which invokes the `libgazebo_ros_joint_state_publisher.so` Gazebo plug-in to publish the values of `joint1` angles to the `joint_states` topic.

These two modified files are included in a common model file, `mobot_w_arm_and_jnt_pub.xacro` (in the `mobot_urdf` package), as follows:

Listing 4.1: `mobot_w_arm_and_jnt_pub.xacro`

```
<?xml version="1.0"?>
<robot
    xmlns:xacro="http://www.ros.org/wiki/xacro" name="mobot">
  <xacro:include filename="$(find mobot_urdf)/urdf/mobot_w_jnt_pub.xacro" />
  <xacro:include filename="$(find minimal_robot_description)/minimal_robot_w_jnt_pub.←
      urdf" />

  <!-- attach the simple arm to the mobile robot -->
  <joint name="arm_base_joint" type="fixed">
    <parent link="base_link" />
    <child link="link1" />
    <origin rpy="0 0 0 " xyz="${-bodylen/2} 0 ${bodyOZ+bodyheight/2}"/>
  </joint>
</robot>
```

The `mobot_w_arm_and_jnt_pub.xacro` model file is identical to `mobot_w_arm.xacro`, except that it includes the modified mobot and arm model files. The `mobot_w_arm_and_jnt_pub.xacro` model file is referenced in a launch file, `mobot_w_arm_and_jnt_pub.launch`:

Listing 4.2: `mobot_w_arm_and_jnt_pub.launch`

```
<launch>
<!-- Convert xacro model file and put on parameter server -->
<param name="robot_description" command="$(find xacro)/xacro.py '$(find mobot_urdf)/←
    urdf/mobot_w_arm_and_jnt_pub.xacro'" />

<!-- Spawn the robot from parameter server into Gazebo -->
<node name="spawn_urdf" pkg="gazebo_ros" type="spawn_model" args="-param ←
    robot_description -urdf -model mobot" />

<!-- load the controller parameter yaml file and start the ROS controllers for the arm←
    -->
<include file="$(find minimal_robot_description)/minimal_robot_ctl.launch">
</include>

</launch>
```

To observe the effects of the joint-state publisher Gazebo plug-in, kill and restart Gazebo:

```
roslaunch gazebo_ros empty_world.launch
```

and then invoke the new launch file:

```
roslaunch mobot_urdf mobot_w_arm_and_jnt_pub.launch
```

This again brings up the mobile robot with attached one-DOF arm. Now, however, the `joint_states` topic is richer. Example output of:

```
rostopic echo joint_states
```

(excerpted) is:

```
name: ['left_wheel_joint', 'right_wheel_joint']
position: [-0.0010584164794300577, -0.0007484677653710747]

name: ['cast2bracket_right', 'cast2bracket_left', 'right_caster_joint',
       'left_caster_joint']
position: [0.12480710950733798, 0.090932225345286, 0.21368044937114838,
          0.2107250800969398]

name: ['joint1']
position: [0.17308318317281302]
```

We see that now there are seven joint values being published. These values are needed to compute transforms for link poses. Given the model description, which prescribes how connected links relate to each other through joint displacements, it is possible to compute all individual transforms for pair-wise connected links. This could be done manually from the information available. However, this is such a common need that ROS has a package designed to do this: the `robot_state_publisher` (see `http://wiki.ros.org/robot_state_publisher`).

With the simulation running, invoking:

```
rosrun robot_state_publisher robot_state_publisher
```

results in the `tf` topic being much richer as well. Excerpts from the output of:

```
rostopic echo tf
```

follow:

```
    frame_id: base_link
  child_frame_id: castdrop_right
    ...(static)

    frame_id: castdrop_right
  child_frame_id: brackettop_right
      transform:
    translation:
      x: 0.0
      y: 0.0
      z: -0.0625
    rotation:
      x: 0.0
      y: 0.0
      z: 0.947030895892
      w: 0.321142464065
```

```
    frame_id: brackettop_right
child_frame_id: bracketside1_right
  ...(static)

   frame_id: bracketside1_right
child_frame_id: right_casterwheel
transform:
  translation:
    x: -0.0613
    y: 0.053
    z: -0.053
  rotation:
    x: 0.673542175535
    y: -0.215269453877
    z: 0.673542175539
    w: -0.215269453878

  frame_id: base_link
child_frame_id: link1
transform:
  translation:
    x: -0.27305
    y: 0.0
    z: 0.55
  rotation:
    x: 0.0
    y: 0.0
    z: 0.0
    w: 1.0

    frame_id: link1
child_frame_id: link2
transform:
  translation:
    x: 0.0
    y: 0.0
    z: 1.0
  rotation:
    x: 0.0
    y: 0.0864419753897
    z: 0.0
    w: 0.996256886998
```

From the `tf` output, one can follow chains of transforms. One kinematic branch starts at `base_link` and has successive parent-child connections through `castdrop_right` (a static transform) to `brackettop_right` (which swivels about its z axis, as evidenced by the quaternion values), to `bracketside1_right` (a static transform) to `right_casterwheel` (which spins about its axle). The individual transforms can be multiplied together (an operation defined in the transform class, equivalent to multiplying $\mathbf{T}$ matrices) to find the pose of any link frame with respect to any desired reference frame (as long as there is a complete chain between the two frames).

Note also the transform between `base_frame` and `link1` (a static transform) and the transform between `link1` and `link2` (which depends on the `joint1` angle). From these transforms, one can obtain the frame of link2 of the one-DOF robot with respect to the base link. ROS also provides facilities to compute transforms between any two (connected) frames within a user's program with a `transform_listener`, details of which are introduced in Section 4.2.

4.2   TRANSFORM LISTENER

Performing transforms from one frame to another is enabled by the `tf` library in ROS (see: `http://wiki.ros.org/tf/Tutorials`). A very useful capability in the `tf` library is the `tf_listener`. We have seen that transforms are published to the topic `tf`, where each such message contains a detailed description of how a child frame is spatially related to its parent frame. (In ROS, a parent can have multiple children, but a child must have a unique parent, thus guaranteeing a tree of geometric relationships.) The `tf_listener` is typically started as an independent thread (thus not dependent on `spin()` or `spinOnce()` invocations from a main program). This thread subscribes to the `tf` topic and assembles a kinematic tree from individual parent-child transform messages. Since the `tf_listener` incorporates all transform publications, it is able to address specific queries, such as "where is my right-hand palm frame relative to my left-camera optical frame?" The response to such a query is a transform message that can be used to reconcile different frames (*e.g.* for hand-eye coordination). As long as a complete tree is published connecting frames of interest, the transform listener can be used to transform all sensor data into a common reference frame, thus allowing for display of sensory data from multiple sources in a common view.

Use of the tf listener is illustrated with example code in the package `example_tf_listener` of the accompanying code repository. This illustrative package specifically assumes reference to our mobot model, notably with respect to frames `base_link`, `link1` and `link2`. The example code is comprised of three files: `example_tf_listener.h`, which defines a class `DemoTfListener`; `example_tf_listener_fncs.cpp`, which contains the implementation of the class methods of `DemoTfListener`; and `example_tf_listener.cpp`, which contains a `main()` program illustrating operations using a transform listener. The contents of the main program are shown in Listing 4.3.

Listing 4.3: `example_tf_listener.cpp`: example code illustrating transform listener

```
1   //example_tf_listener.cpp:
2   //wsn, March 2016
3   //illustrative node to show use of tf listener, with reference to the simple mobile-↩
        robot model
4   // specifically, frames: odom, base_frame, link1 and link2
5
6   // this header incorporates all the necessary #include files and defines the class "↩
        DemoTfListener"
7   #include "example_tf_listener.h"
8   using namespace std;
9
10  //main pgm to illustrate transform operations
11
12  int main(int argc, char** argv) {
13      // ROS set-ups:
14      ros::init(argc, argv, "demoTfListener"); //node name
15      ros::NodeHandle nh; // create a node handle; need to pass this to the class ↩
            constructor
16      ROS_INFO("main: instantiating an object of type DemoTfListener");
17      DemoTfListener demoTfListener(&nh); //instantiate an ExampleRosClass object and ↩
            pass in pointer to nodehandle for constructor to use
18
19      tf::StampedTransform stfBaseToLink2, stfBaseToLink1, stfLink1ToLink2;
20      tf::StampedTransform testStfBaseToLink2;
21
22      tf::Transform tfBaseToLink1, tfLink1ToLink2, tfBaseToLink2, altTfBaseToLink2;
23
```

```
24      demoTfListener.tfListener_->lookupTransform("base_link", "link1", ros::Time(0), ←
            stfBaseToLink1);
25      cout << endl << "base to link1: " << endl;
26      demoTfListener.printStampedTf(stfBaseToLink1);
27      tfBaseToLink1 = demoTfListener.get_tf_from_stamped_tf(stfBaseToLink1);
28
29      demoTfListener.tfListener_->lookupTransform("link1", "link2", ros::Time(0), ←
            stfLink1ToLink2);
30      cout << endl << "link1 to link2: " << endl;
31      demoTfListener.printStampedTf(stfLink1ToLink2);
32      tfLink1ToLink2 = demoTfListener.get_tf_from_stamped_tf(stfLink1ToLink2);
33
34      demoTfListener.tfListener_->lookupTransform("base_link", "link2", ros::Time(0), ←
            stfBaseToLink2);
35      cout << endl << "base to link2: " << endl;
36      demoTfListener.printStampedTf(stfBaseToLink2);
37      tfBaseToLink2 = demoTfListener.get_tf_from_stamped_tf(stfBaseToLink2);
38      cout << endl << "extracted tf: " << endl;
39      demoTfListener.printTf(tfBaseToLink2);
40
41      altTfBaseToLink2 = tfBaseToLink1*tfLink1ToLink2;
42      cout << endl << "result of multiply tfBaseToLink1*tfLink1ToLink2: " << endl;
43      demoTfListener.printTf(altTfBaseToLink2);
44
45      if (demoTfListener.multiply_stamped_tfs(stfBaseToLink1, stfLink1ToLink2, ←
            testStfBaseToLink2)) {
46          cout << endl << "testStfBaseToLink2:" << endl;
47          demoTfListener.printStampedTf(testStfBaseToLink2);
48      }
49      cout << endl << "attempt multiply of stamped transforms in wrong order:" << endl;
50      demoTfListener.multiply_stamped_tfs(stfLink1ToLink2, stfBaseToLink1, ←
            testStfBaseToLink2);
51
52      geometry_msgs::PoseStamped stPose, stPose_wrt_base;
53      stPose = demoTfListener.get_pose_from_transform(stfLink1ToLink2);
54      cout << endl << "pose link2 w/rt link1, from stfLink1ToLink2" << endl;
55      demoTfListener.printStampedPose(stPose);
56
57      demoTfListener.tfListener_->transformPose("base_link", stPose, stPose_wrt_base);
58      cout << endl << "pose of link2 transformed to base frame:" << endl;
59      demoTfListener.printStampedPose(stPose_wrt_base);
60
61      return 0;
62  }
```

In Listing 4.3, an object of class `DemoTfListener` is instantiated (line 17). This object has
a transform listener, a pointer to which is `tf::TransformListener* tfListener_;`. The
transform listener is used in four places in the main program: lines 24, 29, 34 and 57.

The first instance, line 24, is:

```
demoTfListener.tfListener_->lookupTransform("base_link", "link1", ros::Time(0), ←
    stfBaseToLink1);
```

The transform listener subscribes to the `tf` topic and it constantly attempts to assemble
the most current chain of transforms possible from all published parent-child spatial relationships. The transform listener, once it has been instantiated and given a brief time to
start collecting published transform information, offers a variety of useful methods. In the
above case, the method `lookupTransform` finds a transformation between the named frames
(`base_link` and `link1`). Equivalently, this transform tells us where is the `link1` frame with
respect to the `base_link` frame. The `lookupTransform` method fills in `stfBaseToLink1`,
which is an object of type `tf::StampedTransform`. A stamped transform contains both an
origin (a 3-D vector) and an orientation(a 3×3 matrix). The example code prints components of this transform, using accessor functions of transform objects.

The `lookupTransform()` function is called with arguments to define the frame of interest
(`link1`) and the desired reference frame (`base_link`). The argument `ros::Time(0)` specifies

that the current transform is desired. (Optionally, one can request a transform a historical transform corresponding to some specified time of interest in the past.)

The object stfBaseToLink1 has a timestamp, labels for the reference frame (frame_id) and for the child frame (child_frame_id), and a tf::Transform object. Objects of type tf::Transform have a variety of member methods and defined operators. Extracting the tf::Transform from a tf::StampedTransform object is not as simple as would be expected. The function get_tf_from_stamped_tf() is defined within the class DemoTfListener to assist. On line 27,

```
tfBaseToLink1 = demoTfListener.get_tf_from_stamped_tf(stfBaseToLink1);
```

the tfBaseToLink1 transform is extracted from stfBaseToLink1. This transform describes the position and orientation of frame link1 with respect to frame base_frame.

On line 29, the stamped-transform stfLink1ToLink2 is obtained using the transform listener, but with specified frame of interest link2 with respect to frame link1. The transform is extracted from the stamped transform into the object tfLink1ToLink2.

Lines 34 and 37 perform this operation again, this time to populate tfBaseToLink2, which is the transform of link2 with respect to base_frame.

The operator * is defined for tf::Transform objects. Thus, the objects tfBaseToLink1 and tfLink1ToLink2 can be multiplied together, as in line 41:

```
altTfBaseToLink2 = tfBaseToLink1*tfLink1ToLink2;
```

The meaning of this operation is to cascade these transforms, equivalent to multiplying 4×4 transforms:

$$^A\mathbf{T}_C = {}^A\mathbf{T}_B {}^B\mathbf{T}_C \qquad (4.7)$$

The result in altTfBaseToLink2 is the same as the result of

```
demoTfListener.tfListener_->lookupTransform("base_link", "link2", ros::Time(0), ←
    stfBaseToLink2);
```

and extracting the transform from stfBaseToLink2.

The example code displays the various transforms using the member functions printTf() and printStampedTf() of class DemoTfListener.

Line 57 shows use of another member method of the transform listener:

```
demoTfListener.tfListener_->transformPose("base_link", stPose, stPose_wrt_base);
```

With this function, one can transform an object of type geometry_msgs::PoseStamped. The position and orientation of this pose are expressed with respect to the named frame_id. With the transformPose() function, the input PoseStamped object, stPose, is re-expressed as an output pose, stPose_wrt_base, expressed with respect to the named, desired frame (base_link, in this example).

To see the results of the example code, start Gazebo with:

```
roslaunch gazebo_ros empty_world.launch
```

and invoke the mobot launch file:

```
roslaunch mobot_urdf mobot_w_arm_and_jnt_pub.launch
```

Running:

```
rostopic echo tf
```

one can see published transforms that include relationships among `odom`, `base_link`, `left_wheel` and `right_wheel`, which are published courtesy of the differential-drive Gazebo plug-in. To get more transforms, including the minimal robot arm, run:

```
rosrun robot_state_publisher robot_state_publisher
```

The `tf` topic then carries many additional transform messages, including `link2` to `link1` (of the minimal robot arm) and `link1` to `base_link`.

With these nodes running, start the example transform listener:

```
rosrun example_tf_listener example_tf_listener
```

The example transform listener output begins with:

```
[ INFO] [1457913167.079553126]: main: instantiating an object of type DemoTfListener
[ INFO] [1457913167.079639652]: in class constructor of DemoTfListener
[ INFO] [1457913167.097079718]: waiting for tf between link2 and base_link...
[ WARN] [1457913167.097393435]: "base_link" passed to lookupTransform argument
    target_frame does not exist. ; retrying...
[ WARN] [1457913167.843127914, 414.927000000]: Lookup would require extrapolation into the pa
    Requested time 414.914000000 but the earliest data is at time 414.934000000,
        when looking up transform from frame [link2] to frame [base_link]; retrying...
[ INFO] [1457913168.344223832, 415.427000000]: tf is good
```

Initially, the transform listener does not have knowledge of all of the incremental transforms of the kinematic tree. The transform listener call thus fails. This failure is trapped, and the call is re-attempted. By the time of the next try, the connecting transforms are all known and the call is successful. Ordinarily, successive attempts to find transforms between any two named frames will be successful. However, it is still good practice to use "try and catch" in case of future missing transforms.

The first result displayed is:

```
base to link1:
frame_id: base_link
child_frame_id: link1
vector from reference frame to to child frame: -0.27305,0,0.55
orientation of child frame w/rt reference frame:
1,0,0
0,1,0
0,0,1
quaternion: 0, 0, 0, 1
```

This shows that the transform between frame `base_link` and frame `link1` is relatively simple. The orientation is merely the identity, indicating that the `link1` frame is aligned with the `base_link` frame. The vector from the `base_link` frame origin to the `link1` frame origin has a negative x component, a 0 y component and a positive z component. This makes intuitive sense, as the `link1` frame origin is above the `base_link` origin (and thus positive z component) and behind the `base_link` frame origin (and thus the negative x component). The quaternion corresponding to the identity matrix orientation is (0,0,0,1).

The next part of the display output is:

```
link1 to link2:
frame_id: link1
child_frame_id: link2
vector from reference frame to to child frame: 0,0,1
orientation of child frame w/rt reference frame:
 0.978314,  0, 0.207128
 0,         1, 0
-0.207128,  0, 0.978314
quaternion: 0, 0.10413, 0, 0.994564
```

The vector from the `link1` frame origin to the `link2` frame origin, $(0, 0, 1)$, is simply a 1 m displacement in the z direction. The `link2` frame is nearly aligned with the `link1` frame. The y axis (second column of the rotation matrix) is $(0,1,0)$, which implies the `link2` y axis is identically parallel to the `link1` y axis. However, the x and z axes of the `link2` frame are not identical to the corresponding `link1` axes. Since `link2` is leaning slightly forward, the `link2` z axis, $(0.207, 0, 0.078)$, has a slight positive x component (as expressed with respect to the `link1` frame), and the `link2` x axis, $(0.978, 0, -0.207)$, points slightly down, and thus has a negative z component (with respect to the `link1` frame). The next display output corresponds to the transform between `link2` and the base link.

```
base to link2:
frame_id: base_link
child_frame_id: link2
vector from reference frame to to child frame: -0.27305,0,1.55
orientation of child frame w/rt reference frame:
0.978314,0,0.207128
0,1,0
-0.207128,0,0.978314
quaternion: 0, 0.10413, 0, 0.994564
```

For comparison, the output of the product `tfBaseToLink1` and `tfLink1ToLink2` is:

```
result of multiply tfBaseToLink1*tfLink1ToLink2:
vector from reference frame to to child frame: -0.27305,0,1.55
orientation of child frame w/rt reference frame:
0.978314,0,0.207128
0,1,0
-0.207128,0,0.978314
quaternion: 0, 0.10413, 0, 0.994564
```

The components of this result are identical to those of the transform lookup directly from the base link to `link2`, demonstrating that products of transforms behave equivalent to 4×4 matrix transform multiplications (although the result of `tf::Transform` multiplications are objects of type `tf::Transform`, not merely matrices).

Multiplication of `tf::StampedTransform` objects is not defined. However, a member function of class `DemoTfListener` performs the equivalent operation. Line 45:

```
if (demoTfListener.multiply_stamped_tfs(stfBaseToLink1, stfLink1ToLink2, ↵
    testStfBaseToLink2))
```

performs the expected operation. The transform components of the stamped transforms are extracted, multiplied together, and used to populate the transform component of a resulting stamped transform, `testStfBaseToLink2`. The example function gets assigned the `frame_id` of the first stamped transform and the `child_id` of the second stamped transform. However, for the multiplication to make sense, the `child_id` of the first stamped transform must be identical to the `frame_id` of the second stamped transform. If this condition is not satisfied, the multiplication function returns `false` to indicate a logic error. (See line 50.)

Finally, lines 52 through 59 illustrate how to transform a pose to a new frame. The output display is:

```
pose link2 w/rt link1, from stfLink1ToLink2
frame id = link1
origin: 0, 0, 1
quaternion: 0, 0.10413, 0, 0.994564

pose of link2 transformed to base frame:
frame id = base_link
origin: -0.27305, 0, 1.55
quaternion: 0, 0.10413, 0, 0.994564
```

Note that the transformed `link2` pose has a translation and a rotation identical to the corresponding components of the stamped transform `stfBaseToLink2` obtained using `tfListener_->lookupTransform`.

Another check on this transform can be obtained with use of a command-line tool within the `tf` package. Running:

```
rosrun tf tf_echo base_link link2
```

results in output to the screen displaying the transform between the named frames. The order of naming matters. The above command displays the frame of `link2` with respect to frame `base_link`. Example output from this command is:

```
At time 23.585
- Translation: [-0.273, 0.000, 1.550]
- Rotation: in Quaternion [0.000, 0.103, 0.000, 0.995]
            in RPY (radian) [0.000, 0.207, 0.000]
            in RPY (degree) [0.000, 11.850, 0.000]
```

which agrees with the transform-listener result.

Converting between `geometry_msgs` types and `tf` types can be tedious. The code in Listing 4.4, extracted from `example_tf_listener_fncs.cpp`, illustrates how to extract a `geometry_msgs::PoseStamped` from a `tf::StampedTransform`.

Listing 4.4: example conversion from `tf` to `geometry_msgs` type

```
1   geometry_msgs::PoseStamped DemoTfListener::get_pose_from_transform(tf::↵
        StampedTransform tf) {
2     //clumsy conversions--points, vectors and quaternions are different data types in tf↵
            vs geometry_msgs
3     geometry_msgs::PoseStamped stPose;
4     geometry_msgs::Quaternion quat;  //geometry_msgs object for quaternion
5     tf::Quaternion tfQuat; // tf library object for quaternion
6     tfQuat = tf.getRotation(); // member fnc to extract the quaternion from a transform
7     quat.x = tfQuat.x(); // copy the data from tf-style quaternion to geometry_msgs-↵
            style quaternion
8     quat.y = tfQuat.y();
9     quat.z = tfQuat.z();
10    quat.w = tfQuat.w();
11    stPose.pose.orientation = quat; //set the orientation of our PoseStamped object from↵
            result
12
13    // now do the same for the origin--equivalently, vector from parent to child frame
14    tf::Vector3 tfVec;  //tf-library type
15    geometry_msgs::Point pt; //equivalent geometry_msgs type
16    tfVec = tf.getOrigin(); // extract the vector from parent to child from transform
17    pt.x = tfVec.getX(); //copy the components into geometry_msgs type
18    pt.y = tfVec.getY();
```

```
19    pt.z = tfVec.getZ();
20    stPose.pose.position= pt; //and use this compatible type to set the position of the ←
          PoseStamped
21    stPose.header.frame_id = tf.frame_id_; //the pose is expressed w/rt this reference ←
          frame
22    stPose.header.stamp = tf.stamp_; // preserve the time stamp of the original ←
          transform
23    return stPose;
24  }
```

Although vectors and quaternions are defined within `geometry_msgs` and `tf`, these types are not compatible. The code in Listing 4.4 shows how these can be converted.

When instantiating a transform listener, it was noted that the lookup function should be tested for return errors. This is done in the constructor of `DemoTfListener`, as shown in Listing 4.5, extracted from `example_tf_listener_fncs.cpp`.

Listing 4.5: constructor of `DemoTfListener` illustrating `tfListener`

```
1   DemoTfListener::DemoTfListener(ros::NodeHandle* nodehandle):nh_(*nodehandle)
2   {
3       ROS_INFO("in class constructor of DemoTfListener");
4       tfListener_ = new tf::TransformListener;  //create a transform listener and assign←
            its pointer
5       //here, the tfListener_ is a pointer to this object, so must use -> instead of "."←
            operator
6       //somewhat more complex than creating a tf_listener in "main()", but illustrates ←
            how
7       // to instantiate a tf_listener within a class
8
9       // wait to start receiving valid tf transforms between base_link and link2:
10      // this example is specific to our mobot, which has a base_link and a link2
11      // lookupTransform will through errors until a valid chain has been found from ←
            target to source frames
12      bool tferr=true;
13      ROS_INFO("waiting for tf between link2 and base_link...");
14      tf::StampedTransform tfLink2WrtBaseLink;
15      while (tferr) {
16          tferr=false;
17          try {
18                  //try to lookup transform, link2-frame w/rt base_link frame; this will←
                      test if
19              // a valid transform chain has been published from base_frame to link2
20                  tfListener_->lookupTransform("base_link", "link2", ros::Time(0), ←
                      tfLink2WrtBaseLink);
21              } catch(tf::TransformException &exception) {
22                  ROS_WARN("%s; retrying...", exception.what());
23                  tferr=true;
24                  ros::Duration(0.5).sleep(); // sleep for half a second
25                  ros::spinOnce();
26              }
27      }
28      ROS_INFO("tf is good");
29      // from now on, tfListener will keep track of transforms; do NOT need ros::spin(),←
            since
30      // tf_listener gets spawned as a separate thread
31  }
```

The try and catch construct is used to trap errors from the `lookupTransform()` function of the transform listener. When using a transform-listener lookup function, it is advisable to always have a try and catch test. Otherwise, if the lookup function fails, the main program will crash.

Another concern with the transform listener is clock synchronization of multiple computers running nodes within a common ROS system. ROS supports distributed processing. However, with a network of computers, each computer has its own clock. This can cause time stamps within published transforms to be out of synchronization. The transform listener may complain that some transforms appear to be posted for times in the future. Resolution of this problem may require using `chrony` (see `http://chrony.tuxfamily.org/`) or some alternative network time protocol clock synchronization.

Additional functions in `example_tf_listener_fncs.cpp` include `multiply_stamped_tfs()`, `get_tf_from_stamped_tf()`, `get_pose_from_transform()`, `printTf()`, `printStampedTf()` and `printStampedPose()`. These will not be covered in detail here, but viewing the source code can be useful in understanding how to access or populate components of transform types. The conversions introduced here are incorporated within a useful library called `XformUtils`, which will be described in the next section. In addition to tf operations, an alternative library known as `Eigen` can be more convenient for performing linear-algebra operations, as introduced next.

4.3 USING EIGEN LIBRARY

ROS messages are designed for efficient serialization for network communication. These messages can be inconvenient to work with when one desires to perform operations on data. One common need is to perform linear algebra operations. A useful C++ library for linear algebra is the Eigen library (see `http://eigen.tuxfamily.org`). The Eigen open-source project is independent of ROS. However, one can still use Eigen with ROS.

To use Eigen with ROS, one must include the associated header files in the `*.cpp` source code and add lines to `CmakeLists.txt`. Our custom `cs_create_pkg` script already includes the necessary lines in `CmakeLists.txt`; they only need to be uncommented. Specifically, uncomment the following lines in the `CMakeLists.txt` file:

```
#uncomment the following 4 lines to use the Eigen library
find_package(cmake_modules REQUIRED)
find_package(Eigen3 REQUIRED)
include_directories(${EIGEN3_INCLUDE_DIR})
add_definitions(${EIGEN_DEFINITIONS})
```

(see `https://github.com/ros/cmake_modules/blob/0.3-devel/README.md#usage` for an explanation of `cmake_modules` in ROS). In the source code, include the header files for the functionality desired. The following steps include much functionality:

```
#include <Eigen/Eigen> //for the Eigen library
#include <Eigen/Dense>
#include <Eigen/Geometry>
#include <Eigen/Eigenvalues>
```

For access to additional capabilities in Eigen, more header files may be included, as described in `http://eigen.tuxfamily.org/dox/group__QuickRefPage.html#QuickRef_Headers`.

A program that illustrates some Eigen capabilities is `example_eigen_plane_fit.cpp` in the package `example_eigen`. Lines from this program are explained here.

An example vector may be defined as follows:

```
Eigen::Vector3d normal_vec(1,2,3); // here is an arbitrary 3x1 vector, initialized to ↩
    (1,2,3) upon instantiation
```

This instantiates an Eigen object that is a column vector comprised of three double-precision values. The object is named `normal_vec`, initialized to the values [1;2;3].

One of the member functions of `Eigen::Vector3d` is `norm()`, which computes the Euclidean length of the vector (square root of the sum of squares of the components). The vector can be coerced to unit length (if it is a non-zero vector!) as follows:

```
normal_vec/=normal_vec.norm(); // make this vector unit length
```

Note that, although `normal_vec` is an object, the operators ∗ and / are defined to scale the components of the vector as expected. Thus, vector-times-scalar operations behave as expected (where `normal_vec.norm()` returns a scalar value).

Here is an example of instantiating a 3×3 matrix object comprised of double-precision values:

```
Eigen::Matrix3d Rot_z;
```

With the following notation, one can fill the matrix with data, one row at a time:

```
Rot_z.row(0)<<0,1,0;   // populate the first row--shorthand method
Rot_z.row(1)<<1,0,0;   //second row
Rot_z.row(2)<<0,0,1;   // third row
```

There are a variety of other methods for initializing or populating matrices and vectors. For example, one can fill a vector or matrix with zeros using:

```
Eigen::Vector3d centroid;
// here's a convenient way to initialize data to all zeros; more variants exist
centroid = Eigen::MatrixXd::Zero(3,1); // http://eigen.tuxfamily.org/dox/↵
    AsciiQuickReference.txt
```

The arguments (3,1) specify 3 rows, 1 column. (A vector is simply a special case of a matrix, for which there is either a single row or a single column.)

Alternatively, one may specify initial values as arguments to the constructor upon instantiation. To initialize a vector to values of 1:

```
Eigen::VectorXd  ones_vec= Eigen::MatrixXd::Ones(npts,1);
```

In the illustrative example code, a set of points is generated that lie on (or near) a predetermined plane. Eigen methods are invoked to discover what was the original plane, using only the data points. This operation is valuable in point-cloud processing, where we may wish to find flat surfaces of interest, *e.g.* tables, walls, floors, doors, etc.

In the example code, the plane of interest is defined to have a surface normal called `normal_vec`, and the plane is offset from the origin by a distance `dist`. A plane has a unique definition of distance from the origin. If one cares about positive and negative surfaces of a plane, the distance from the origin can be a signed number, where the offset is measured from origin to plane in the direction of the plane's normal (and thus can result in a negative offset).

To generate sample data, we construct a pair of vectors perpendicular to the plane's normal vector. We can do so starting with vector `v1` that is not colinear with the plane normal. In the example code, this vector is generated by rotating `normal_vec` 90 degrees about the z axis, which is accomplished with the following matrix*vector multiply:

```
v1 = Rot_z*normal_vec; //here is how to multiply a matrix times a vector
```

(Note: if `normal_vec` is parallel to the z axis, v1 will be equal to `normal_vec`, and the subsequent operations will not work.)

For display purposes, Eigen types are nicely formatted for `cout`. Matrices are formatted with new lines for each row, *e.g.* using:

```
cout<<Rot_z<<endl;
```

Rather than use `cout`, it is preferable to use `ROS_INFO_STREAM()`, which is more versatile than `ROS_INFO()`. This function outputs the data via network communication, and it is thus visible via `rqt_console` and loggable. To display `Rot_z` as a formatted matrix with `ROS_INFO_STREAM()`, one can use:

```
ROS_INFO_STREAM(endl<<Rot_z); // start w/ endl, so get a clean first line of data ↩
    display
```

For short column vectors, it is more convenient to display the values on a single line. To do so, one can output the transpose of the vector, *e.g.* as follows:

```
ROS_INFO_STREAM("v1: "<<v1.transpose()<<endl);
```

Two common vector operations are the dot product and the cross product. From the example code, here are some excerpts:

```
double dotprod = v1.dot(normal_vec); //using the "dot()" member function
double dotprod2 = v1.transpose()*normal_vec;// alt: turn v1 into a row vector, ↩
    then multiply times normal_vec
```

and for the cross product, `v1` crossed into `normal_vec`:

```
v2 = v1.cross(normal_vec);
```

Note that the result of `v1 x normal_vec` must be mutually orthogonal to both `v1` and `normal_vec`. Since the result, `v2`, is perpendicular to `normal_vec`, it is parallel to the plane under construction.

A second vector in the plane can be computed as:

```
v1 = v2.cross(normal_vec);   // re-use v1; make it the cross product of v2 into ↩
    normal_vec
```

Using the vectors `v1`, `v2` and `normal_vec`, one can define any point in the desired plane as:

```
p = a*v1 + b*v2 + c*normal_vec
```

with the constraint $c = $ `dist`, but a and b can be any scalar values.

Random points in the plane are generated and stored as column vectors in a $3 \times N$ matrix. The matrix is instantiated with the line:

```
Eigen::MatrixXd points_mat(3,npts); //create a matrix, double-precision values, 3 rows↩
    and npts cols
```

The example generates random points within the desired plane as follows:

```
Eigen::Vector2d rand_vec; //a 2x1 vector
//generate random points that all lie on plane defined by distance and normal_vec
for (int ipt = 0;ipt<npts;ipt++) {
    // MatrixXd::Random returns uniform random numbers in the range (-1, 1).
    rand_vec.setRandom(2,1);  // populate 2x1 vector with random values
    //cout<<"rand_vec: "<<rand_vec.transpose()<<endl; //optionally, look at these ↩
        random values
```

```
                //construct a random point ON the plane normal to normal_vec at distance "dist↵
                    " from origin:
                // a point on the plane is a*x_vec + b*y_vec + c*z_vec, where we may choose
                // x_vec = v1, y_vec = v2 (both of which are parallel to our plane) and z_vec ↵
                    is the plane normal
                // choose coefficients a and b to be random numbers, but "c" must be the plane↵
                    's distance from the origin, "dist"
                point =  rand_vec(0)*v1 + rand_vec(1)*v2 + dist*normal_vec;
                //save this point as the i'th column in the matrix "points_mat"
                points_mat.col(ipt) = point;
                }
```

Random noise can be added to the (ideal) data on the plane using:

```
// add random noise to these points in range [-0.1,0.1]
Eigen::MatrixXd Noise = Eigen::MatrixXd::Random(3,npts);
// add two matrices, term by term.  Also, scale all points in a matrix by a scalar: ↵
    Noise*g_noise_gain
points_mat = points_mat + Noise*g_noise_gain;
```

The matrix `points_mat` now contains points, column by column, that approximately lie on a plane distance `dist` from the origin and with surface-normal vector `normal_vec`. This dataset can be used to illustrate the operation of plane fitting. The first step is to compute the centroid of all of the points, which can be computed as follows:

```
    // first compute the centroid of the data:
    // here's a handy way to initialize data to all zeros; more variants exist
    // see http://eigen.tuxfamily.org/dox/AsciiQuickReference.txt
    Eigen::Vector3d centroid = Eigen::MatrixXd::Zero(3,1);

    //add all the points together:
    npts = points_mat.cols(); // number of points = number of columns in matrix; check↵
        the size
    cout<<"matrix has ncols = "<<npts<<endl;
    for (int ipt =0;ipt<npts;ipt++) {
    centroid+= points_mat.col(ipt); //add all the column vectors together
    }
    centroid/=npts; //divide by the number of points to get the centroid
```

The (approximately) planar points are then offset by subtracting the centroid from each point.

```
// subtract this centroid from all points in points_mat:
Eigen::MatrixXd points_offset_mat = points_mat;
for (int ipt =0;ipt<npts;ipt++) {
    points_offset_mat.col(ipt)  = points_offset_mat.col(ipt)-centroid;
}
```

The resulting matrix (three rows and npts columns) can be used to compute a covariance matrix by multiplying this matrix times its transpose, resulting in a 3 × 3 matrix:

```
Eigen::Matrix3d CoVar;
CoVar = points_offset_mat*(points_offset_mat.transpose());  //3xN matrix times Nx3 ↵
    matrix is 3x3
```

One of the more advanced Eigen options is computation of eigenvectors and associated eigenvalues. This can be invoked with:

```
    // here is a more complex object: a solver for eigenvalues/eigenvectors;
    // we will initialize it with our covariance matrix, which will induce computing ↵
        eval/evec pairs
    Eigen::EigenSolver<Eigen::Matrix3d> es3d(CoVar);
    Eigen::VectorXd evals; //we'll extract the eigenvalues to here
```

```
// in general, the eigenvalues/eigenvectors can be complex numbers
//however, since our matrix is self-adjoint (symmetric, positive semi-definite), ←
    we expect
// real-valued evals/evecs; we'll need to strip off the real parts of the ←
    solution
evals= es3d.eigenvalues().real(); // grab just the real parts
```

The three eigenvalues are all non-negative. The smallest of the eigenvalues corresponds to the eigenvector that is the best-fit approximation to the plane normal. (The direction of smallest variance is perpendicular to the best-fit plane.) If the smallest eigenvalue corresponds to index `ivec`, the corresponding eigenvector is:

```
est_plane_normal = es3d.eigenvectors().col(ivec).real();
```

The distance of the plane from the origin can computed as the projection (dot product) of the vector from the origin to the centroid onto the plane normal:

```
double est_dist = est_plane_normal.dot(centroid);
```

The program `example_eigen_plane_fit.cpp` illustrates a variety of Eigen capabilities, and the specific algorithm illustrated is an efficient and robust technique for fitting planes to data.

Another Eigen object that will be useful is `Eigen::Affine3d`, which has the equivalent capabilities of a coordinate-transform operator. A ROS `tf` transform can be converted to an Eigen affine object as follows:

```
Eigen::Affine3d transformTFToEigen(const tf::Transform &t) {
    Eigen::Affine3d e;
    // treat the Eigen::Affine as a 4x4 matrix:
    for (int i = 0; i < 3; i++) {
        e.matrix()(i, 3) = t.getOrigin()[i]; //copy the origin from tf to Eigen
        for (int j = 0; j < 3; j++) {
            e.matrix()(i, j) = t.getBasis()[i][j]; //and copy 3x3 rotation matrix
        }
    }
    // Fill in (0,0,0,1) in the last row
    for (int col = 0; col < 3; col++)
        e.matrix()(3, col) = 0;
    e.matrix()(3, 3) = 1;
    return e;
}
```

Subsequently, Eigen affine objects can be multiplied together, or premultiplied by Eigen `Vector3d` objects to transform points to new coordinates frames, as in the following example code snippet:

```
//let's say we have a point, "p", as detected in the sensor frame;
// arbitrarily, initialize this to [1;2;3]
Eigen::Vector3d p_wrt_sensor(1,2,3);
//create an affine object that defines the transform between the sensor frame and ←
    the world frame:
Eigen::Affine3d affine_sensor_wrt_world;
//assume "tfTransform" has been filled in by a transform listener for sensor frame←
    w/rt world frame
affine_sensor_wrt_world =transformTFToEigen(tfTransform); //convert tf to Eigen::←
    Affine
// here's how to convert a sensor point to the world frame:
Eigen::Vector3d p_wrt_world;
p_wrt_world = affine_sensor_wrt_world*p_wrt_sensor; //point is now expressed in ←
    world-frame coordinates

//we can transform in the opposite direction with the transform inverse:
```

```
Eigen::Vector3d p_back_in_sensor_frame;
p_back_in_sensor_frame = affine_sensor_wrt_world.inverse()*p_wrt_world;
```

Eigen-style affine transforms are very useful both in computing poses of kinematic chains and transforming sensor data into useful reference frames, such as a robot's base frame.

4.4   TRANSFORMING ROS DATATYPES

As we have seen, ROS nodes communicate with each other through defined message types. A ROS publisher deconstructs the data components of a message object, serializes the data and transmits it via a topic. A complementary subscriber receives the serial data and reconstructs the components of the corresponding message object. This process is convenient for communications, but message types can be cumbersome for performing mathematical operations. For example, a `geometry_msgs::Pose` object can be inspected in terms of its individual elements, but it cannot be used directly in linear algebraic operations.

The package `xform_utils` contains a library of convenient conversion functions. For example, `transformPoseToEigenAffine3d()`, which takes a `geometry_msgs::Pose` and returns an equivalent `Eigen::Affine3d`, is given below.

```
Eigen::Affine3d XformUtils::transformPoseToEigenAffine3d(geometry_msgs::Pose pose) {
    Eigen::Affine3d affine;
    Eigen::Vector3d Oe;
    Oe(0) = pose.position.x;
    Oe(1) = pose.position.y;
    Oe(2) = pose.position.z;
    affine.translation() = Oe;

    Eigen::Quaterniond q;
    q.x() = pose.orientation.x;
    q.y() = pose.orientation.y;
    q.z() = pose.orientation.z;
    q.w() = pose.orientation.w;
    Eigen::Matrix3d Re(q);
    affine.linear() = Re;

    return affine;
}
```

To use these utility functions in one's own package, the package should include `xform_utils` dependency in `package.xml`, and the package source code should include the header file `#include <xform_utils/xform_utils.h>`. Within the source code, an object of type `XformUtils` can be instantiated, after which the member functions can be used.

Example usage is illustrated in the source code `example_xform_utils.cpp` in the `xform_utils` package. This example anticipates a forthcoming need to transform from an object pose to a gripper pose. In this instance, the object pose is hard coded, with position and orientation components of a `geometry_msgs::Pose` object called `object_pose`. A gripper approach pose is to be derived in which the origin of the gripper frame is identical to the origin of the object frame, the x axis of the gripper frame is coincident with the object frame, and the z axis of the gripper frame is anti-parallel to the z axis of the object frame. This is constructed as follows. Given the populated object frame, it is transformed into an equivalent `Eigen::Affine3d` object using `XformUtils`, and the components of the affine are displayed:

```
Eigen::Affine3d object_affine, gripper_affine;
//use XformUtils to convert from pose to affine:
object_affine = xformUtils.transformPoseToEigenAffine3d(object_pose);
cout << "object_affine origin: " << object_affine.translation().transpose() << ↩
    endl;
```

```
cout << "object_affine R matrix: " << endl;
cout << object_affine.linear() << endl;
```

The rotation matrix of the **Affine3d** object is interpreted in terms of x, y and z direction axes. These are used to construct the corresponding axes of the gripper frame. The x axis of the gripper frame is identical to the x axis of the object frame, and the z axis of the gripper frame is the negative of the z axis of the gripper frame. The y axis is constructed as the cross product of z into x to create a right-hand triad.

```
Eigen::Vector3d x_axis, y_axis, z_axis;
Eigen::Matrix3d R_object, R_gripper;
R_object = object_affine.linear(); //get 3x3 matrix from affine
x_axis = R_object.col(0); //extract the x axis
z_axis = -R_object.col(2); //define gripper z axis anti-parallel to object z-axis
y_axis = z_axis.cross(x_axis); // construct a right-hand coordinate frame
```

From the object origin and the constructed orientation, the gripper affine is populated:

```
R_gripper.col(0) = x_axis; //populate orientation matrix from axis directions
R_gripper.col(1) = y_axis;
R_gripper.col(2) = z_axis;
gripper_affine.linear() = R_gripper; //populate affine w/ orientation
gripper_affine.translation() = object_affine.translation(); //and origin
cout << "gripper_affine origin: " << gripper_affine.translation().transpose() << ↩
    endl;
cout << "gripper_affine R matrix: " << endl;
cout << gripper_affine.linear() << endl;
```

The resulting **Affine3d** object is converted into a pose (using a **XformUtils** function), and this pose is used in populating a **PoseStamped** object, and the result is displayed (using another **XformUtils** function).

```
//use XformUtils fnc to convert from Affine to a pose
gripper_pose = xformUtils.transformEigenAffine3dToPose(gripper_affine);
gripper_pose_stamped.pose = gripper_pose;
gripper_pose_stamped.header.stamp = ros::Time::now();
gripper_pose_stamped.header.frame_id = "torso";
ROS_INFO("desired gripper pose: ");
xformUtils.printStampedPose(gripper_pose_stamped); //display the output
```

This illustrates how object types in ROS can be transformed to be compatible with additional libraries (notably, the Eigen library) to exploit additional library functions. The specific example used here will be useful in computing poses for manipulation, as discussed in Section 15.4.

4.5   WRAP-UP

Coordinate transforms are ubiquitous in robotics, and correspondingly, ROS has extensive tools for handling coordinate transforms. These can be somewhat confusing, though, since there are multiple representations. The **tf** library offers a **tf** listener that can be used conveniently within nodes to listen for all transform publications and assemble these into coherent chains. The **tf** listener can be queried for spatial relationships between any two connected frames. The listener responds with the most current transform information available. In contrast, a planner would consider hypothetical relationships. As such, **tf** is less useful for planning purposes, and thus one may need to compute kinematic transforms independently for planning. Additionally, although **tf** is updated dynamically, a potential limitation is that the assembled chains of transforms may be slightly time delayed. Consequently, **tf** results should not be used in time-critical feedback loops (*e.g.* for force control, for which it may be necessary to compute one's own kinematic transforms).

The Eigen library, which is independent of ROS, can be integrated within ROS nodes to perform fast and sophisticated linear algebra computations. Eigen transforms can use 4×4 matrices or the Eigen datatype `Eigen::Affine3`. As with other independent libraries, conversions are required between ROS message types and (in this case) Eigen objects. Such conversions require some extra overhead, but the benefits are worth the effort.

With an understanding of transforms in ROS, we are ready to introduce the valuable ROS visualization tool known as `rviz`.

Sensing and Visualization in ROS

CONTENTS

I NTRODUCTION
    The primary tool for visualization in ROS is `rviz` (see http://wiki.ros.org/rviz).
With this tool, one can display sensor values, real or emulated, as well as robot pose (as
inferred from a robot model and from published robot joint values). One can also superim-
pose graphics (*e.g.* to display results from perceptual processing nodes) and provide operator
inputs (*e.g.* via a mouse).

Figure 5.1 shows an example `rviz` display for the Atlas robot. The colored points correspond to simulated data from a rotating Hokuyo LIDAR (see [22]) on a Carnegie Robotics
sensor head (see [35]). The LIDAR data has been colorized in the `rviz` display to indicate
z height of each point. From the `rviz` display, we can interpret points on the floor (red),
points on a table (yellow) and multiple cylinders on the table. There are also colored points
on the robot model itself, illustrating that the simulated sensor is aware of the robot model
as well, appropriately displaying how the sensor would see the robot's arm.

Figure 5.2 is an `rviz` display of data from an actual (physical) Boston Dynamics Atlas
robot in a lab at University of Hong Kong. In this display, the LIDAR data points have
been processed to classify specific entities, and the display colors indicate class membership,
including the floor, a wall in front of the robot, a wall to the robot's left, and a door. Points
not colorized are not classified in one of the defined categories, including clutter near walls, a
gantry, and the ceiling. The LIDAR display from the physical robot is a good approximation
to simulated LIDAR from a Gazebo model of the robot. This supports developing perceptual
processing software in simulation, which is then applicable to the physical system.

Another example is shown in Fig 5.3. The left scene is an actual Baxter™robot in a lab
at Case Western Reserve University, with a Microsoft Kinect™and a Yale OpenHand [31].
The right scene has a similar composition (a cylinder on a table in front of the robot). It

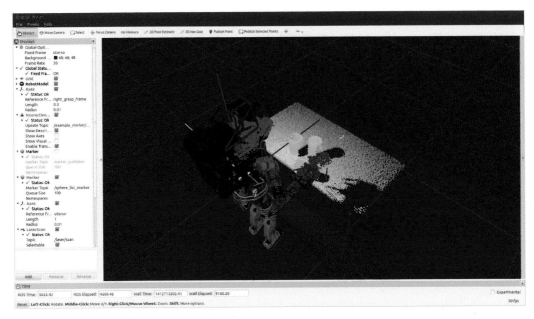

Figure 5.1: An `rviz` view of Atlas robot model with LIDAR sensor display

is an `rviz` view of a model of the Baxter robot and a display of colorized points from a simulated Kinect sensor mounted on the robot. The points are colorized by z height.

In the `rviz` display, a robot model's pose is rendered consistent with the robot's current joint angles. The position and orientation of each link of the robot are knowable from published coordinates. If the coordinates are published by a real robot, the `rviz` display will be synchronized to render the robot model to mimic the real robot. Alternatively, the source of coordinates could be a Gazebo model, or simply a rosbag playback of previously recorded poses.

Similarly, display of sensor values (*e.g.* 3-D points from a depth sensor) can be from physical sensors streaming in real time, from simulated sensors within a Gazebo model, or from playback of pre-recorded data.

As seen in Figs 5.1 and 5.3, data from a real or simulated sensor may include sensing of the robot's own body. Figures 5.1 and 5.3 show that sensory data is appropriately colocated with respective surfaces on the robot model. Such reconciliation of coordinates is necessary to take advantage of sensor-driven behaviors in simulation. Coordinating data from multiple models and sensors is accomplished through coordinate transformations.

Via the parameter server (which holds the kinematic and visual model of our robot under the name `robot_description`), and publications of transforms on `tf`, our system is aware of all visual models of all links, as well as all transforms necessary to compute the pose of any link with respect to any other link. With this much information, it is possible for an independent node to access the parameter server, subscribe to the `tf` topic and render all links in a consistent frame of reference. Transformations and rendering are performed in ROS using `rviz`. By running:

```
rosrun rviz rviz
```

(and after some interactive configuration), a view of the robot model can be seen, as in Fig 5.4.

Figure 5.2: An `rviz` view of Atlas robot in lab with interpretation of actual LIDAR data

The view from `rviz` is, in fact, less appealing than the view from Gazebo. One issue with visualizing the link poses in `rviz` is that the material color information specified for Gazebo display is not compatible with colors specified for `rviz` (and thus all links appear red by default). This can be corrected by augmenting the model file to include color specifications to be used by `rviz`. The `rviz` view also does not show a ground plane, shading, nor additional modeling that might be brought into the virtual world. The unique value of `rviz` will not be apparent until we introduce additional features, notably, display of sensor values.

Our primitive `rviz` view nonetheless already illustrates multiple important points. First, the display of the robot's configuration does not generally require Gazebo. As long as a robot model has been loaded into the parameter server and all the link transforms are published, `rviz` can display a rendering of the robot in a pose consistent with its joint angles. Publication of link transforms could be accomplished, *e.g.* from playback of a rosbag recording, without the need to run Gazebo.

More importantly, the same robot model can be rendered using data from a real robot. Such rendering requires satisfying a few requirements. First, there must be an adequate visual model of the physical robot (as in the examples in the introduction to this chapter), and this model must be loaded onto the parameter server. Second, the real robot must publish its joint displacements on the `joint_states` topic. And third, one must run the `robot_state_publisher` node, which combines URDF information from the parameter server with joint state information from the robot to compute and publish transform information on the `tf` topic. With these conditions satisfied, `rviz` can display a rendering of the robot that mimics the actual robot pose, updating continuously as the robot moves. This display can be used, for example, to help debug hardware problems. With the physical robot's motors disabled, one could move the joints manually and observe whether the `rviz` model follows along. If so, this test would confirm that the corresponding joint sensors are alive and reporting correct joint displacement values.

 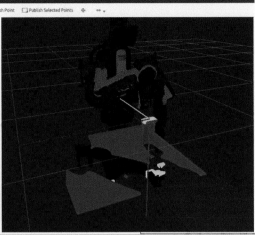

(a) Baxter robot approaching an object (Photo by: Russell Lee)

(b) `rviz` model of Baxter robot with Kinect sensor

Figure 5.3: Physical and simulated Baxter robots. `rviz` view (b) displays points from simulated Kinect sensor with points colorized by z height

A second valuable debugging capability of `rviz` is to visualize playback of records (rosbags) recorded from tests performed on a real robot. Such replay can be useful for identifying clues for debugging, *e.g.* did a robot fall due to instability of balance or did it trip? Did the robot controller exhibit oscillations at different poses? Did joint limits or inertial effects result in undesirable motion?

An `rviz` display can also be useful in understanding a robot model. In Fig 5.4, `display items` includes the robot model, as well as four axis frames: `base_link`, `left_wheel`, `link2` and `left_casterwheel`. Coordinate frame displays are color coded: red for x, green for y and blue for z. These displays should reconcile with our expectations. The base frame has its x axis forward, its y axis to the left, and its z axis up. Its origin is at ground level, immediately below the point between the drive wheels. The left-wheel frame has its z axis pointing to the left, coincident with the wheel axle. The y and x axes are orthogonal to the z axis, and they rotate about the wheel z axis.

The left caster wheel frame also has its z axis coincident with the wheel's axle, though the caster is rotated in this scene so the wheel axle is not parallel to the base-frame y axis. The `link2` frame shows that the `joint1` axis is coincident with the y axis of the `link2` frame.

This display shows that `rviz` has full information of all poses of all links of the robot.

The true value of `rviz` will become apparent with the introduction of sensor displays, graphical overlays and operator interaction.

Before diving deeper into `rviz`, a subtle point is worth mentioning. For `rviz` to display all links, it must have access to all transforms. All joint angles are known to Gazebo, since Gazebo runs the physics simulation. The example models included a joint-state publisher, with which Gazebo makes all joint angles available to ROS on the `joint_states` topic. On a real robot that corresponds to our mobot model, motion of the caster wheels would be similar to that of Gazebo simulation. However, it would be unusual to have joint sensors on passive joints. Thus, the real robot would be unable to publish joint values for the caster brackets and caster wheels. Consequently, the robot state publisher would be unable to

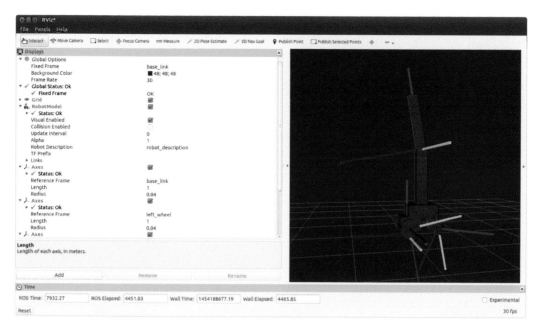

Figure 5.4: `rviz` view of simple mobile robot with one-DOF arm

compute the corresponding transforms, and `rviz` would be unable to render the casters and caster wheels. A work-around for this disparity is to run a node that publishes 0 for all four of the caster joints. In `rviz`, the casters and caster wheels would appear to be static, but at least `rviz` would be able to display their models without complaint.

A second subtle issue with `rviz` concerns pose of the robot in the world. Within Gazebo, a world frame is defined, and all models (and links within models) have poses that are known to Gazebo with respect to the world (a necessary requirement to perform computation of dynamics in an inertial frame). However, the pose of the robot relative to the world frame is not published to ROS. Consequently, `rviz`cannot know where the robot is in the world. In order to produce a consistent rendering, a (non-world) reference frame must be declared. In `rviz` (within the `displays` panel, `globaloptions` item), the `FixedFrame` can be set. When this is set to `base_link`, all rendering will be with respect to the robot's base frame. As a consequence, when sensor values are displayed in `rviz`, it will appear that the robot is stationary and the world moves towards the robot–as though the observer were riding on the robot.

Rendering the robot in a world frame requires additional steps, including defining a map of the world and running localization routines that help establish the robot's position and orientation with respect to the map. Localization for mobile robots will be introduced in Section 9.2.

5.1   MARKERS AND INTERACTIVE MARKERS IN RVIZ

It is frequently useful to display computed graphical overlays in a scene. These can be used, for example, to help highlight an object of interest, to show a robot's focus of attention, display a perception module's belief in the pose of an object model with respect to sensory data, or indicate kinematically reachable regions. A feature to provide graphical overlays in `rviz` is the publishing of `markers`. A more sophisticated type is the `Interactive Marker`, which allows the operator to move a marker in `rviz` with a mouse (or other input device) and publish the resulting coordinates in ROS.

5.1.1 Markers in `rviz`

One can display markers in the `rviz` view by publishing to appropriate topics. An example is in the accompanying repository in package `example_rviz_marker`. The source code, `example_rviz_marker.cpp`, is shown in Listings 5.1 (preamble and helper functions) and 5.2 (the main program).

Listing 5.1: `example_rviz_marker.cpp`: C++ code to publish grid of markers for `rviz` display, preamble and helper functions

```
1   #include <ros/ros.h>
2   #include <visualization_msgs/Marker.h> // need this for publishing markers
3   #include <geometry_msgs/Point.h> //data type used for markers
4   #include <string.h>
5   #include <stdio.h>
6   #include <example_rviz_marker/SimpleFloatSrvMsg.h> //a custom message type defined in ↩
        this package
7   using namespace std;
8
9   //set these two values by service callback, make available to "main"
10  double g_z_height = 0.0;
11  bool g_trigger = true;
12
13  //a service to prompt a new display computation.
14  // E.g., to construct a plane at height z=1.0, trigger with:
15  // rosservice call rviz_marker_svc 1.0
16
17  bool displaySvcCB(example_rviz_marker::SimpleFloatSrvMsgRequest& request,
18          example_rviz_marker::SimpleFloatSrvMsgResponse& response) {
19      g_z_height = request.request_float32;
20      ROS_INFO("example_rviz_marker: received request for height %f", g_z_height);
21      g_trigger = true; // inform "main" a new computation is desired
22      response.resp = true;
23      return true;
24  }
25
26  void init_marker_vals(visualization_msgs::Marker &marker) {
27      marker.header.frame_id = "/world"; // reference frame for marker coords
28      marker.header.stamp = ros::Time();
29      marker.ns = "my_namespace";
30      marker.id = 0;
31      // use SPHERE if you only want a single marker
32      // use SPHERE_LIST for a group of markers
33      marker.type = visualization_msgs::Marker::SPHERE_LIST; //SPHERE;
34      marker.action = visualization_msgs::Marker::ADD;
35      // if just using a single marker, specify the coordinates here, like this:
36
37      //marker.pose.position.x = 0.4;
38      //marker.pose.position.y = -0.4;
39      //marker.pose.position.z = 0;
40      //ROS_INFO("x,y,z = %f %f, %f",marker.pose.position.x,marker.pose.position.y,
41      //          marker.pose.position.z);
42      // otherwise, for a list of markers, put their coordinates in the "points" array, ↩
            as below
43
44      //whether a single marker or list of markers, need to specify marker properties
45      // these will all be the same for SPHERE_LIST
46      marker.pose.orientation.x = 0.0;
47      marker.pose.orientation.y = 0.0;
48      marker.pose.orientation.z = 0.0;
49      marker.pose.orientation.w = 1.0;
50      marker.scale.x = 0.02;
51      marker.scale.y = 0.02;
52      marker.scale.z = 0.02;
53      marker.color.a = 1.0;
```

```
54        marker.color.r = 1.0;
55        marker.color.g = 0.0;
56        marker.color.b = 0.0;
57  }
```

Listing 5.2: `example_rviz_marker.cpp`: C++ code to publish grid of markers for `rviz` display, main program

```
59  int main(int argc, char **argv) {
60      ros::init(argc, argv, "example_rviz_marker");
61      ros::NodeHandle nh;
62      ros::Publisher vis_pub = nh.advertise<visualization_msgs::Marker>("↩
            example_marker_topic", 0);
63      visualization_msgs::Marker marker; // instantiate a marker object
64      geometry_msgs::Point point; // points will be used to specify where the markers go
65
66      //set up a service to compute marker locations on request
67      ros::ServiceServer service = nh.advertiseService("rviz_marker_svc", displaySvcCB);
68
69      init_marker_vals(marker);
70
71      double z_des;
72
73      // build a wall of markers; set range and resolution
74      double x_min = -1.0;
75      double x_max = 1.0;
76      double y_min = -1.0;
77      double y_max = 1.0;
78      double dx_des = 0.1;
79      double dy_des = 0.1;
80
81      while (ros::ok()) {
82          if (g_trigger) { // did service get request for a new computation?
83              g_trigger = false; //reset the trigger from service
84              z_des = g_z_height; //use z-value from service callback
85              ROS_INFO("constructing plane of markers at height %f", z_des);
86              marker.header.stamp = ros::Time();
87              marker.points.clear(); // clear out this vector
88
89              for (double x_des = x_min; x_des < x_max; x_des += dx_des) {
90                  for (double y_des = y_min; y_des < y_max; y_des += dy_des) {
91                      point.x = x_des;
92                      point.y = y_des;
93                      point.z = z_des;
94                      marker.points.push_back(point);
95                  }
96              }
97          }
98          ros::Duration(0.1).sleep();
99          //ROS_INFO("publishing...");
100         vis_pub.publish(marker);
101         ros::spinOnce();
102     }
103     return 0;
104 }
```

The source code of this node defines a service called `rviz_marker_svc`. This service uses the service message, `SimpleFloatSrvMsg` defined in the `example_rviz_marker` package, and which expects the client to provide a single floating-point value. This service can be invoked manually, *e.g.* with the terminal command:

```
rosservice call rviz_marker_svc 1.0
```

When this service is invoked, the `example_rviz_marker` node will compute a grid of points within a horizontal plane at height 1.0 (as specified in the above example command).

This package lists dependency on `visualization_msgs` and `geometry_msgs` in the `package.xml` file. Correspondingly, the source code includes the headers:

```
#include <visualization_msgs/Marker.h>
```

and

```
#include <geometry_msgs/Point.h>
```

Within `main()` of `example_rviz_marker.cpp`, line 62, a publisher object is instantiated using the message type:

```
ros::Publisher vis_pub = nh.advertise<visualization_msgs::Marker>("←
    example_marker_topic", 0);
```

which publishes to the chosen topic `example_marker_topic`.

The message type `visualization_msgs::Marker` is fairly complex, as can be seen by entering:

```
rosmsg show visualization_msgs/Marker
```

This message type contains 15 fields, many of which contain subfields. Additional details on definition and use of this message type can be found at `http://wiki.ros.org/visualization_msgs` and `http://wiki.ros.org/rviz/DisplayTypes/Marker`. A helper function within the example code, `init_marker_vals()`, populates the fields: header (including `frame_id` and time stamp), type, action, pose, scale, color and the vector of points (with x,y,z coordinates).

In the main loop of the main program of this example node, the x, y, z coordinates of a list of points are populated and published (lines 89 through 96) with the chosen shape, color and size held constant.

To run the example node, first start a `roscore` running, then enter:

```
rosrun example_rviz_mark example_rviz_marker
```

Running:

```
rostopic list
```

shows `example_marker_topic` and

```
rostopic info example_marker_topic
```

shows that the node `example_rviz_marker` publishes messages of type `visualization_msgs/Marker` to this topic.

Upon start-up, this node will populate points (markers) to be displayed at elevation zero. `rviz` will be able to display these markers graphically after adding the display type and topic name. To do so, start `rviz`:

```
rosrun rviz rviz
```

Note that it is not necessary to have Gazebo running.

In the `rviz` display, choose a `fixed frame` of `world`.

To see the computed markers, one must add a display item `Marker` to the `rviz displays` list, as shown in Fig 5.5, by clicking `Add` and choosing `Marker` from the pop-up options. After clicking "OK," the display list will include the item `Marker`. Clicking on

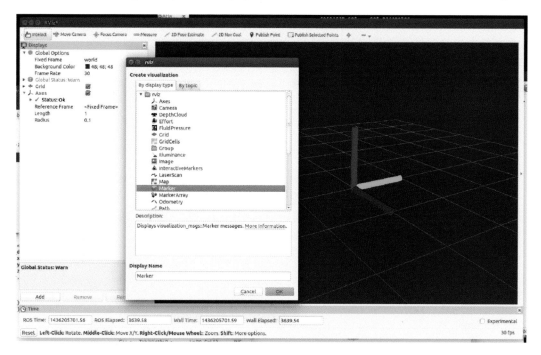

Figure 5.5: Adding marker display in rviz

this item to expand it, gives one the option of editing the topic to which this display should subscribe. As shown in Fig 5.6, the topic should be edited by typing in the text `example_marker_topic` (or selecting it from drop-down list options), which is the topic name chosen in our source code for `example_rviz_marker`. At this point, `rviz` starts receiving messages from our `example_rviz_marker` node, resulting in a display of red spheres within a planar patch at height zero about the origin. The markers will appear at height 1.0 if the `example_rviz_marker` node responds to a service request of:

```
rosservice call rviz_marker_svc 1.0
```

as shown in Fig 5.7. Ordinarily, the service client would originate from another node, but manually stimulating the service from the command line is useful for incremental testing.

5.1.2 Triad display example

A second illustrative example in the package `example_rviz_marker` is `triad_display.cpp`. This node listens for publications of type `geometry_msgs::PoseStamped` on topic `triad_display_pose`. The source code is given in Listings 5.3 through 5.6. Lines 29 through 31 instantiate markers for x, y and z axes. The function `init_markers()`, lines 84 through 128, sets parameters that are unchanging for these markers, including marker type (`ARROW`), scale factors, colors, and marker identification numbers. The function `update_arrows()` (lines 35 through 80) updates coordinate values for these markers, based on poses received

Listing 5.3: `triad_display.cpp`: C++ code to publish triad of axes in rviz display, preamble

```cpp
1   // triad_display.cpp
2   // Wyatt Newman, 8/16
3   // node to assist display of triads (axes) in rviz
4   // this node subscribes to topic "triad_display_pose", from which it receives ←
        geometry_msgs/PoseStamped poses
5   // it uses this info to populate and publish axes, using whatever frame_id is in the ←
        pose header
6   // To see the result, add a "Marker" display in rviz and subscribe to the marker topic←
        "/triad_display"
7   // Can test this display node with the test node: "triad_display_test_node", which ←
        generates moving poses
8   // corresponding to a marker origin spiralling up in z
9
10  #include <ros/ros.h>
11  #include <visualization_msgs/Marker.h>
12  //#include <visualization_msgs/InteractiveMarkerFeedback.h>
13  #include <geometry_msgs/Point.h>
14  #include <geometry_msgs/PointStamped.h>
15  #include <geometry_msgs/PoseStamped.h>
16  #include <math.h>
17  #include <Eigen/Eigen>
18  #include <Eigen/Core>
19  #include <Eigen/Geometry>
20  #include <Eigen/Dense>
21  #include <tf_conversions/tf_eigen.h>
22
23  //some globals...
24  geometry_msgs::Point vertex1;
25  geometry_msgs::PoseStamped g_stamped_pose;
26  Eigen::Affine3d g_affine_marker_pose;
27
28  // create arrow markers; do this 3 times to create a triad (frame)
29  visualization_msgs::Marker arrow_marker_x; //this one for the x axis
30  visualization_msgs::Marker arrow_marker_y; //this one for the y axis
31  visualization_msgs::Marker arrow_marker_z; //this one for the y axis
```

Listing 5.4: `triad_display.cpp`: C++ code to publish triad of axes in rviz display, helper function to update arrows

```
33
34  //udpdate_arrows() set the frame and
35
36  void update_arrows() {
37      geometry_msgs::Point origin, arrow_x_tip, arrow_y_tip, arrow_z_tip;
38      Eigen::Matrix3d R;
39      Eigen::Quaterniond quat;
40      quat.x() = g_stamped_pose.pose.orientation.x;
41      quat.y() = g_stamped_pose.pose.orientation.y;
42      quat.z() = g_stamped_pose.pose.orientation.z;
43      quat.w() = g_stamped_pose.pose.orientation.w;
44      R = quat.toRotationMatrix();
45      Eigen::Vector3d x_vec, y_vec, z_vec;
46      double veclen = 0.2; //make the arrows this long
47      x_vec = R.col(0) * veclen;
48      y_vec = R.col(1) * veclen;
49      z_vec = R.col(2) * veclen;
50
51      //update the arrow markers w/ new pose:
52      origin = g_stamped_pose.pose.position;
53      arrow_x_tip = origin;
54      arrow_x_tip.x += x_vec(0);
55      arrow_x_tip.y += x_vec(1);
56      arrow_x_tip.z += x_vec(2);
57      arrow_marker_x.points.clear();
58      arrow_marker_x.points.push_back(origin);
59      arrow_marker_x.points.push_back(arrow_x_tip);
60      arrow_marker_x.header = g_stamped_pose.header;
61
62      arrow_y_tip = origin;
63      arrow_y_tip.x += y_vec(0);
64      arrow_y_tip.y += y_vec(1);
65      arrow_y_tip.z += y_vec(2);
66
67      arrow_marker_y.points.clear();
68      arrow_marker_y.points.push_back(origin);
69      arrow_marker_y.points.push_back(arrow_y_tip);
70      arrow_marker_y.header = g_stamped_pose.header;
71
72      arrow_z_tip = origin;
73      arrow_z_tip.x += z_vec(0);
74      arrow_z_tip.y += z_vec(1);
75      arrow_z_tip.z += z_vec(2);
76
77      arrow_marker_z.points.clear();
78      arrow_marker_z.points.push_back(origin);
79      arrow_marker_z.points.push_back(arrow_z_tip);
80      arrow_marker_z.header = g_stamped_pose.header;
81  }
```

Listing 5.5: `triad_display.cpp`: C++ code to publish triad of axes in rviz display, helper function to initialize markers

```
82    //init persistent params of markers, then variable coords
83
84    void init_markers() {
85        //initialize stamped pose for at a legal (if boring) pose
86        g_stamped_pose.header.stamp = ros::Time::now();
87        g_stamped_pose.header.frame_id = "world";
88        g_stamped_pose.pose.position.x = 0;
89        g_stamped_pose.pose.position.y = 0;
90        g_stamped_pose.pose.position.z = 0;
91        g_stamped_pose.pose.orientation.x = 0;
92        g_stamped_pose.pose.orientation.y = 0;
93        g_stamped_pose.pose.orientation.z = 0;
94        g_stamped_pose.pose.orientation.w = 1;
95
96        //the following parameters only need to get set once
97        arrow_marker_x.type = visualization_msgs::Marker::ARROW;
98        arrow_marker_x.action = visualization_msgs::Marker::ADD; //create or modify marker
99        arrow_marker_x.ns = "triad_namespace";
100       arrow_marker_x.lifetime = ros::Duration(); //never delete
101       // make the arrow thin
102       arrow_marker_x.scale.x = 0.01;
103       arrow_marker_x.scale.y = 0.01;
104       arrow_marker_x.scale.z = 0.01;
105       arrow_marker_x.color.r = 1.0; // red, for the x axis
106       arrow_marker_x.color.g = 0.0;
107       arrow_marker_x.color.b = 0.0;
108       arrow_marker_x.color.a = 1.0;
109       arrow_marker_x.id = 0;
110       arrow_marker_x.header = g_stamped_pose.header;
111
112       //y and z arrow params are the same, except for colors
113       arrow_marker_y = arrow_marker_x;
114       arrow_marker_y.color.r = 0.0;
115       arrow_marker_y.color.g = 1.0; //green for y axis
116       arrow_marker_y.color.b = 0.0;
117       arrow_marker_y.color.a = 1.0;
118       arrow_marker_y.id = 1;
119
120       arrow_marker_z = arrow_marker_x;
121       arrow_marker_z.id = 2;
122       arrow_marker_z.color.r = 0.0;
123       arrow_marker_z.color.g = 0.0;
124       arrow_marker_z.color.b = 1.0; //blue for z axis
125       arrow_marker_z.color.a = 1.0;
126       //set the poses of the arrows based on g_stamped_pose
127       update_arrows();
128   }
```

Listing 5.6: `triad_display.cpp`: C++ code to publish triad of axes in rviz display, callback function and main program

```
130   void poseCB(const geometry_msgs::PoseStamped &pose_msg) {
131       ROS_DEBUG("got pose message");
132
133       g_stamped_pose.header = pose_msg.header;
134       g_stamped_pose.pose = pose_msg.pose;
135
136   }
137
138   int main(int argc, char** argv) {
139       ros::init(argc, argv, "triad_display"); // this will be the node name;
140       ros::NodeHandle nh;
141
142       // subscribe to stamped-pose publications
143       ros::Subscriber pose_sub = nh.subscribe("triad_display_pose", 1, poseCB);
144       ros::Publisher vis_pub = nh.advertise<visualization_msgs::Marker>("triad_display",↩
                1);
145       init_markers();
146
147       ros::Rate timer(20); //timer to run at 20 Hz
148
149
150
151       while (ros::ok()) {
152           update_arrows();
153           vis_pub.publish(arrow_marker_x); //publish the marker
154           ros::Duration(0.01).sleep();
155           vis_pub.publish(arrow_marker_y); //publish the marker
156           ros::Duration(0.01).sleep();
157           vis_pub.publish(arrow_marker_z); //publish the marker
158           ros::spinOnce(); //let callbacks perform an update
159           timer.sleep();
160       }
161   }
```

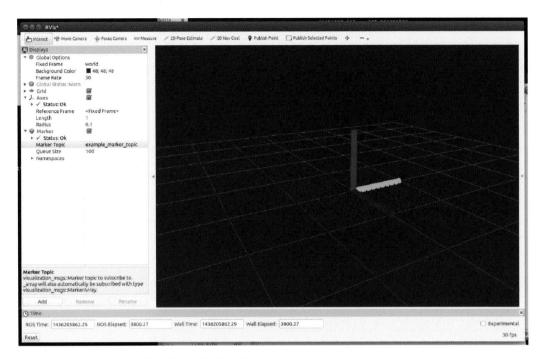

Figure 5.6: Markers displayed in `rviz` from `example_marker_topic`

by a subscriber. The origin of each arrow (tail of arrow) is set to match the origin of the received pose. Directions of the arrows are based on the received orientation of the pose. However, this requires some conversions. The Eigen library (introduced in Section 4.3) is used to convert from a quaternion to a rotation matrix, the columns of which are the three desired axis directions (lines 37 through 48). The arrow markers require specifying points for the tail (*e.g.* line 57, assigning the tail to the pose origin) and the head (*e.g.* line 58, which is an offset of `veclen` in the direction of `x_vec`).

The unchanging parameters are set upon start-up (line 145). Within the main loop (lines 151 through 160), the arrow coordinates are assigned based on the most current value of desired pose (line 152), and the marker messages are published (lines 153, 155, 157), making these marker specifications available to `rviz` for display.

A callback function (lines 130 through 136) receives updates of desired pose on topic `triad_display_pose`.

The node `triad_display_test_node` illustrates use of `triad_display`. This node generates and publishes poses corresponding to an origin that spirals upward, orientation with z axis pointing up, and x axis pointing tangent to the spiral. To run this test, first start up a `roscore` and start `rviz`. Within `rviz`, set the fixed frame to `world` and add a `Marker` display item with topic set to `triad_display`. In a terminal, run:

```
rosrun example_rviz_marker triad_display
```

and in another terminal run:

```
rosrun example_rviz_marker triad_display_test_node
```

In `rviz`, a triad will appear, rotating and spiralling upward. A snapshot of this is shown in

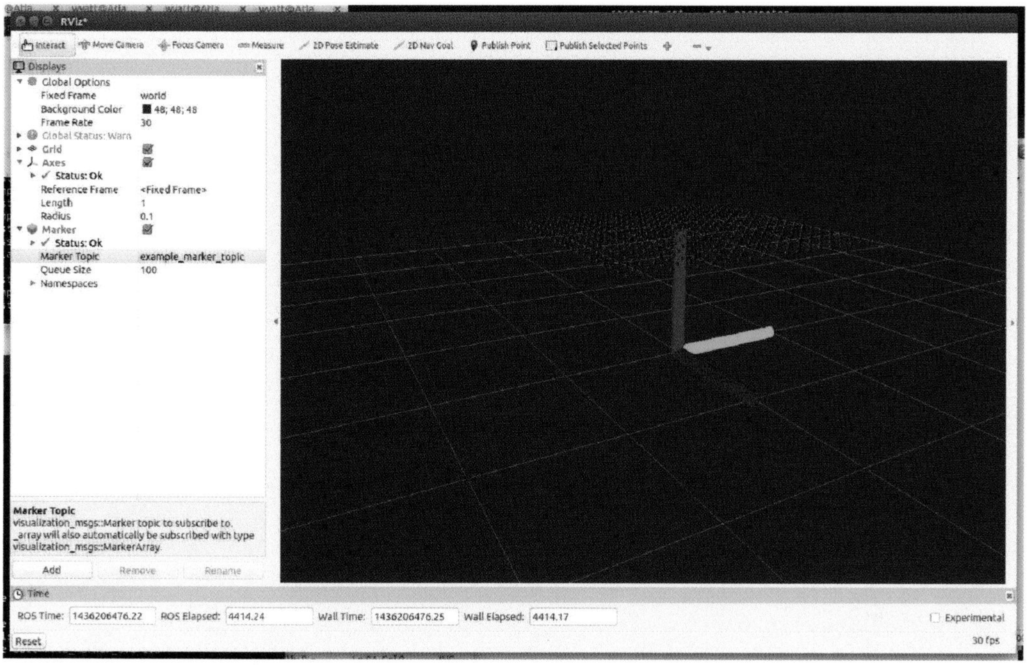

Figure 5.7: Markers at height 1.0 after `rosservice call`

Fig 5.8. This capability can be convenient in visualizing computed frames, *e.g.* to test the results of image processing or point-cloud processing.

5.1.3 Interactive markers in `rviz`

A more complex marker type is the `Interactive Marker`. (See `http://wiki.ros.org/rviz/Tutorials/InteractiveMarkers:GettingStarted`.) Example code for using interactive markers is in the accompanying repository package `example_interactive_marker`. The node `example_interactive_marker` is the executable name compiled from the source code in `IM_6DOF.cpp`. The (lengthy) source code is shown in Listings 5.7 through 5.11. Most of the code simply assigns parameter values to visualization of arrows and instantiates controls for three translations and three rotations.

The interactive-marker code relies on the `interactive_markers` package, and thus it includes (line 5):

```
#include <interactive_markers/interactive_marker_server.h>
```

A custom service message, `ImNodeSvcMsg` is defined in the `example_interactive_marker` package. This message includes a `geometry_msgs/PoseStamped` both for the request and the response. The service message is used by a service, `IM6DofSvc`, implemented in the callback function `IM6DofSvcCB` (lines 21 through 57). This callback function has different behaviors depending on the command mode in the service message. If the command mode is `IM_GET_CURRENT_MARKER_POSE`, the current marker pose is returned in the service response (lines 26 through 33). If the command mode is `IM_SET_NEW_MARKER_POSE`, an object of type `geometry_msgs::PoseStamped`, `poseStamped_IM_desired` (line 37) is populated with data

Listing 5.7: `IM_6DOF.cpp`: C++ for a six-DOF interactive marker, preamble and callback functions

```cpp
// IM_6DOF.cpp
// Wyatt Newman, based on ROS tutorial 4.2 on Interactive Markers
#include <ros/ros.h>
#include <iostream>
#include <interactive_markers/interactive_marker_server.h>
#include <geometry_msgs/Point.h>
#include <example_interactive_marker/ImNodeSvcMsg.h>

const int IM_GET_CURRENT_MARKER_POSE=0;
const int IM_SET_NEW_MARKER_POSE= 1;

geometry_msgs::Point g_current_point;
geometry_msgs::Quaternion g_current_quaternion;
ros::Time g_marker_time;

interactive_markers::InteractiveMarkerServer *g_IM_server; //("rt_hand_marker");
visualization_msgs::InteractiveMarkerFeedback *g_IM_feedback;

//service:  return pose of marker from above globals;
// depending on mode, move IM programmatically,
bool IM6DofSvcCB(example_interactive_marker::ImNodeSvcMsgRequest& request, ↩
        example_interactive_marker::ImNodeSvcMsgResponse& response) {
    //if busy, refuse new requests;

    // for a simple status query, handle it now;
    if (request.cmd_mode == IM_GET_CURRENT_MARKER_POSE) {
        ROS_INFO("IM6DofSvcCB: rcvd request for query--GET_CURRENT_MARKER_POSE");
        response.poseStamped_IM_current.header.stamp = g_marker_time;
        response.poseStamped_IM_current.header.frame_id = "world";
        response.poseStamped_IM_current.pose.position = g_current_point;
        response.poseStamped_IM_current.pose.orientation = g_current_quaternion;
        return true;
    }

    //command to move the marker to specified pose:
    if (request.cmd_mode == IM_SET_NEW_MARKER_POSE) {
        geometry_msgs::PoseStamped poseStamped_IM_desired;
        ROS_INFO("IM6DofSvcCB: rcvd request for action--SET_NEW_MARKER_POSE");
        g_current_point = request.poseStamped_IM_desired.pose.position;
        g_current_quaternion = request.poseStamped_IM_desired.pose.orientation;
        g_marker_time = ros::Time::now();
        poseStamped_IM_desired = request.poseStamped_IM_desired;
        poseStamped_IM_desired.header.stamp = g_marker_time;
        response.poseStamped_IM_current = poseStamped_IM_desired;
        //g_IM_feedback->pose = poseStamped_IM_desired.pose;

        response.poseStamped_IM_current.header.stamp = g_marker_time;
        response.poseStamped_IM_current.header.frame_id = "torso";
        response.poseStamped_IM_current.pose.position = g_current_point;
        response.poseStamped_IM_current.pose.orientation = g_current_quaternion;
        g_IM_server->setPose("des_hand_pose",poseStamped_IM_desired.pose); //↩
                g_IM_feedback->marker_name,poseStamped_IM_desired.pose);
        g_IM_server->applyChanges();
        return true;
    }
    ROS_WARN("IM6DofSvcCB: case not recognized");
    return false;
}

void processFeedback(
        const visualization_msgs::InteractiveMarkerFeedbackConstPtr &feedback) {
    ROS_INFO_STREAM(feedback->marker_name << " is now at "
            << feedback->pose.position.x << ", " << feedback->pose.position.y
            << ", " << feedback->pose.position.z);
    g_current_quaternion = feedback->pose.orientation;
    g_current_point = feedback->pose.position;
    g_marker_time = ros::Time::now();
}
```

Listing 5.8: IM_6DOF.cpp: C++ for a six-DOF interactive marker, marker visualization
parameter initialization functions

```cpp
void init_arrow_marker_x(visualization_msgs::Marker &arrow_marker_x) {
    geometry_msgs::Point temp_point;

    arrow_marker_x.type = visualization_msgs::Marker::ARROW; //ROS example was a CUBE;←
        changed to ARROW
    // specify/push-in the origin point for the arrow
    temp_point.x = temp_point.y = temp_point.z = 0;
    arrow_marker_x.points.push_back(temp_point);
    // Specify and push in the end point for the arrow
    temp_point = g_current_point;
    temp_point.x = 0.2; // arrow is this long in x direction
    temp_point.y = 0.0;
    temp_point.z = 0.0;
    arrow_marker_x.points.push_back(temp_point);

    // make the arrow very thin
    arrow_marker_x.scale.x = 0.01;
    arrow_marker_x.scale.y = 0.01;
    arrow_marker_x.scale.z = 0.01;

    arrow_marker_x.color.r = 1.0; // red, for the x axis
    arrow_marker_x.color.g = 0.0;
    arrow_marker_x.color.b = 0.0;
    arrow_marker_x.color.a = 1.0;
}

void init_arrow_marker_y(visualization_msgs::Marker &arrow_marker_y) {
    geometry_msgs::Point temp_point;
    arrow_marker_y.type = visualization_msgs::Marker::ARROW;
    // Push in the origin point for the arrow
    temp_point.x = temp_point.y = temp_point.z = 0;
    arrow_marker_y.points.push_back(temp_point);
    // Push in the end point for the arrow
    temp_point.x = 0.0;
    temp_point.y = 0.2; // points in the y direction
    temp_point.z = 0.0;
    arrow_marker_y.points.push_back(temp_point);

    arrow_marker_y.scale.x = 0.01;
    arrow_marker_y.scale.y = 0.01;
    arrow_marker_y.scale.z = 0.01;

    arrow_marker_y.color.r = 0.0;
    arrow_marker_y.color.g = 1.0; // color it green, for y axis
    arrow_marker_y.color.b = 0.0;
    arrow_marker_y.color.a = 1.0;
}

void init_arrow_marker_z(visualization_msgs::Marker &arrow_marker_z) {
    geometry_msgs::Point temp_point;

    arrow_marker_z.type = visualization_msgs::Marker::ARROW; //CUBE;
    // Push in the origin point for the arrow
    temp_point.x = temp_point.y = temp_point.z = 0;
    arrow_marker_z.points.push_back(temp_point);
    // Push in the end point for the arrow
    temp_point.x = 0.0;
    temp_point.y = 0.0;
    temp_point.z = 0.2;
    arrow_marker_z.points.push_back(temp_point);

    arrow_marker_z.scale.x = 0.01;
    arrow_marker_z.scale.y = 0.01;
    arrow_marker_z.scale.z = 0.01;

    arrow_marker_z.color.r = 0.0;
    arrow_marker_z.color.g = 0.0;
    arrow_marker_z.color.b = 1.0;
    arrow_marker_z.color.a = 1.0;
}
```

Listing 5.9: `IM_6DOF.cpp`: C++ for a six-DOF interactive marker, marker translation and rotation control initialization functions

```cpp
139  void init_translate_control_x(visualization_msgs::InteractiveMarkerControl &↵
         translate_control_x) {
140      translate_control_x.name = "move_x";
141      translate_control_x.interaction_mode =
142          visualization_msgs::InteractiveMarkerControl::MOVE_AXIS;
143  }
144
145  void init_translate_control_y(visualization_msgs::InteractiveMarkerControl &↵
         translate_control_y) {
146      translate_control_y.name = "move_y";
147      translate_control_y.interaction_mode =
148          visualization_msgs::InteractiveMarkerControl::MOVE_AXIS;
149      translate_control_y.orientation.x = 0; //point this in the y direction
150      translate_control_y.orientation.y = 0;
151      translate_control_y.orientation.z = 1;
152      translate_control_y.orientation.w = 1;
153  }
154
155  void init_translate_control_z(visualization_msgs::InteractiveMarkerControl &↵
         translate_control_z) {
156      translate_control_z.name = "move_z";
157      translate_control_z.interaction_mode =
158          visualization_msgs::InteractiveMarkerControl::MOVE_AXIS;
159      translate_control_z.orientation.x = 0; //point this in the y direction
160      translate_control_z.orientation.y = 1;
161      translate_control_z.orientation.z = 0;
162      translate_control_z.orientation.w = 1;
163  }
164
165  void init_rotx_control(visualization_msgs::InteractiveMarkerControl &rotx_control) {
166      rotx_control.always_visible = true;
167      rotx_control.interaction_mode = visualization_msgs::InteractiveMarkerControl::↵
             ROTATE_AXIS;
168      rotx_control.orientation.x = 1;
169      rotx_control.orientation.y = 0;
170      rotx_control.orientation.z = 0;
171      rotx_control.orientation.w = 1;
172      rotx_control.name = "rot_x";
173  }
174
175  void init_roty_control(visualization_msgs::InteractiveMarkerControl &roty_control) {
176      roty_control.always_visible = true;
177      roty_control.interaction_mode = visualization_msgs::InteractiveMarkerControl::↵
             ROTATE_AXIS;
178      roty_control.orientation.x = 0;
179      roty_control.orientation.y = 0;
180      roty_control.orientation.z = 1;
181      roty_control.orientation.w = 1;
182      roty_control.name = "rot_y";
183  }
184
185  void init_rotz_control(visualization_msgs::InteractiveMarkerControl &rotz_control) {
186      rotz_control.always_visible = true;
187      rotz_control.interaction_mode = visualization_msgs::InteractiveMarkerControl::↵
             ROTATE_AXIS;
188      rotz_control.orientation.x = 0;
189      rotz_control.orientation.y = 1;
190      rotz_control.orientation.z = 0;
191      rotz_control.orientation.w = 1;
192      rotz_control.name = "rot_z";
193  }
```

Listing 5.10: `IM_6DOF.cpp`: C++ for a six-DOF interactive marker, marker control initialization functions

```
195  void init_IM_control(visualization_msgs::InteractiveMarkerControl &IM_control,
196          visualization_msgs::Marker &arrow_marker_x,
197          visualization_msgs::Marker &arrow_marker_y, visualization_msgs::Marker &↩
             arrow_marker_z) {
198      init_arrow_marker_x(arrow_marker_x); //set up arrow params for x
199      init_arrow_marker_y(arrow_marker_y); //set up arrow params for y
200      init_arrow_marker_z(arrow_marker_z); //set up arrow params for z
201      IM_control.always_visible = true;
202
203      IM_control.markers.push_back(arrow_marker_x);
204      IM_control.markers.push_back(arrow_marker_y);
205      IM_control.markers.push_back(arrow_marker_z);
206  }
207
208  void init_int_marker(visualization_msgs::InteractiveMarker &int_marker) {
209      int_marker.header.frame_id = "world"; //base_link"; ///world"; // the reference ↩
             frame for pose coordinates
210      int_marker.name = "des_hand_pose"; //name the marker
211      int_marker.description = "Interactive Marker";
212
213      /** Scale Down: this makes all of the arrows/disks for the user controls smaller ↩
             than the default size */
214      int_marker.scale = 0.2;
215
216      /** specify/push-in the origin for this marker */
217      //let's pre-position the marker, else it will show up at the frame origin by ↩
             default
218      int_marker.pose.position.x = g_current_point.x;
219      int_marker.pose.position.y = g_current_point.y;
220      int_marker.pose.position.z = g_current_point.z;
221  }
```

Listing 5.11: `IM_6DOF.cpp`: C++ for a six-DOF interactive marker, main program

```
223  int main(int argc, char** argv) {
224      ros::init(argc, argv, "simple_marker"); // this will be the node name;
225      ros::NodeHandle nh; //standard ros node handle
226      // create an interactive marker server on the topic namespace simple_marker
227      interactive_markers::InteractiveMarkerServer server("rt_hand_marker");
228      g_IM_server = &server;
229      ros::ServiceServer IM_6dof_interface_service = nh.advertiseService("IM6DofSvc", &↩
             IM6DofSvcCB);
230      // look for resulting pose messages on the topic: /rt_hand_marker/feedback,
231      // which publishes a message of type visualization_msgs/InteractiveMarkerFeedback,↩
             which
232      // includes a full "pose" of the marker.
233      // Coordinates of the pose are with respect to the named frame
234      g_current_point.x = 0.5; //init these global values
235      g_current_point.y = -0.5; //will be used in subsequent init fncs
236      g_current_point.z = 0.2;
237
238      // create an interactive marker for our server
239      visualization_msgs::InteractiveMarker int_marker;
240      init_int_marker(int_marker);
241
242      // arrow markers; 3 to create a triad (frame)
243      visualization_msgs::Marker arrow_marker_x, arrow_marker_y, arrow_marker_z;
244      // create a control that contains the markers
245      visualization_msgs::InteractiveMarkerControl IM_control;
246      //initialize values for this control
247      init_IM_control(IM_control, arrow_marker_x, arrow_marker_y, arrow_marker_z);
248      // add the control to the interactive marker
249      int_marker.controls.push_back(IM_control);
250
251      // create a control that will move the marker
252      // this control does not contain any markers,
253      // which will cause RViz to insert two arrows
254      visualization_msgs::InteractiveMarkerControl translate_control_x,
255              translate_control_y, translate_control_z;
256      init_translate_control_x(translate_control_x);
257      init_translate_control_y(translate_control_y);
258      init_translate_control_z(translate_control_z);
259
260      // add x,y,and z-rotation controls
261      visualization_msgs::InteractiveMarkerControl rotx_control, roty_control,
262              rotz_control;
263      init_rotx_control(rotx_control);
264      init_roty_control(roty_control);
265      init_rotz_control(rotz_control);
266
267      // add the controls to the interactive marker
268      int_marker.controls.push_back(translate_control_x);
269      int_marker.controls.push_back(translate_control_y);
270      int_marker.controls.push_back(translate_control_z);
271      int_marker.controls.push_back(rotx_control);
272      int_marker.controls.push_back(rotz_control);
273      int_marker.controls.push_back(roty_control);
274
275      // add the interactive marker to our collection &
276      // tell the server to call processFeedback() when feedback arrives for it
277      //server.insert(int_marker, &processFeedback);
278      g_IM_server->insert(int_marker, &processFeedback);
279      // 'commit' changes and send to all clients
280      //server.applyChanges();
281      g_IM_server->applyChanges();
282
283      // start the ROS main loop
284      ROS_INFO("going into spin...");
285      ros::spin();
286  }
```

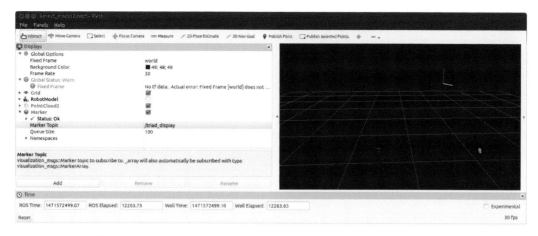

Figure 5.8: Screenshot of triad display node with `triad_display_test_node`

from the service request `poseStamped_IM_desired` field. The interactive marker is then moved programmatically by lines 51 and 52:

```
g_IM_server->setPose("des_hand_pose",poseStamped_IM_desired.pose);
g_IM_server->applyChanges();
```

Invoking motion of the interactive marker uses an interactive marker server, defined in the main program on lines 227 and 228:

```
interactive_markers::InteractiveMarkerServer server("rt_hand_marker");
g_IM_server = &server;
```

A (global) pointer to the interactive-marker server is defined so that callback functions can access the server (as in lines 51 and 52). The server is associated with the callback function `processFeedback` in `main()` on line 278:

```
g_IM_server->insert(int_marker, &processFeedback);
```

A limitation of interactive markers is that one cannot query them for their current pose. Instead, one must monitor publications when the marker is moved, then remember these values for future use. Marker pose memory is implemented via the callback function `processFeedback` (lines 59 through 67). When the marker is moved via its graphical handles, the new pose is received by this function, and these values are saved to global variables. Subsequently, a request to the service `IM6DofSvc` can be sent at any time to get the most recently received pose of the interactive marker.

The bulk of the `IM_6DOF.cpp` code (lines 69 through 137) describes the type, size and color of the six components of the interactive marker and associates type and direction of the six interactive motion controls (lines 139 through 193). After these initializations, the main program goes into a spin (line 285), and all further interactions are handled by the service and the callback function.

To run the example interactive marker, with `roscore` and `rviz` running (optionally, additional nodes, such as `example_rviz_marker`, displaying a robot model, etc.), start this node with:

```
rosrun  example_interactive_marker example_interactive_marker_node
```

In `rviz`, one must add a display and enter the appropriate topic name to visualize the interactive marker. To do so, in `rviz` click `Add` and choose the item `InteractiveMarkers` as shown in Fig 5.9.

The new `InteractiveMarker` item is expanded, and the topic `rt_hand_marker/update` is selected from the drop-down menu. An interactive marker then appears, as in Fig 5.10.

This display has nine interactive handles for moving the marker $+/- x$, y, z, and $+/-$ rotation about x, y and z. In `rviz`, if one hovers the mouse over one of these controls, the color of the handle will bolden, and by clicking and dragging, the marker will change its displacement or orientation. As the marker is moved interactively, its pose (6-D position and orientation) in space changes. The new values are published, invoking a callback response from the `processFeedback` function.

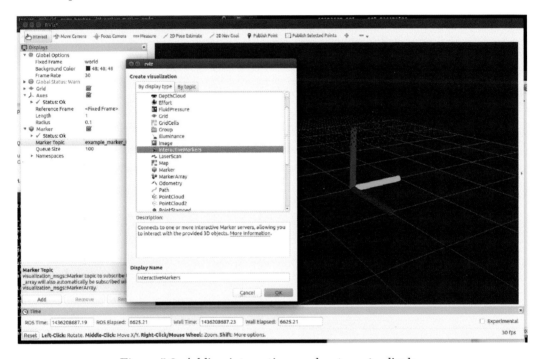

Figure 5.9: Adding interactive marker to `rviz` display

Interactive markers can be used to input full 6-D poses of interest. Such inputs can be used, *e.g.* to specify a desired hand pose or to direct attention to an object of interest. Moving markers under program control can be used to illustrate computed or proposed poses to achieve or to indicate interpretations of poses of objects of interest.

Markers in `rviz` are particularly useful when they can be placed with respect to sensory data. We next consider how sensor data can be displayed in `rviz`.

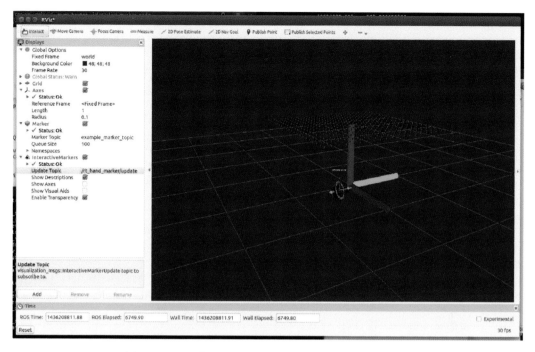

Figure 5.10: Display of interactive marker in `rviz`

5.2 DISPLAYING SENSOR VALUES IN RVIZ

One of the most valuable aspects of Gazebo (together with `rviz`) is the ability to perform realistic emulation of common sensors. With a virtual world, simulated sensors provide corresponding data that can be used for off-line program development of sensor-driven behaviors. This section will show how a few common sensors that can be incorporated in Gazebo and interpreted in `rviz`. It should be noted that the `rviz` display of emulated sensors is performed the same way as `rviz` display of real sensors. The `rviz` display is unaware of the origin of the sensor signals, whether physical, rosbag logs played back, or simulated sensors.

5.2.1 Simulating and displaying LIDAR

One of the most common sensors used in robots (including autonomous vehicles) is LIDAR (LIght Detection And Ranging). LIDAR sensors send out very brief pulses of laser light and measure the time of flight of reflected light to infer distance. Most commonly, LIDAR uses a spinning mirror, resulting in samples of the environment in a single plane. Popular manufacturers include Sick [23] and Hokuyo [22]. Data from these devices is obtained at regular angular intervals (typically between 1 and 0.25 degrees). The output is streamed in a format consisting of a list of radii, one list per revolution of the LIDAR's mirror. For example, a Sick LMS200 LIDAR can provide 181 radial distances sampled over a 180-degree semicircle with a 1-degree sample resolution repeated at 75 Hz. With known start angle, end angle and angular resolution, it is only necessary to transmit a list of radii, and these distances can be inferred to associate with a corresponding angle, thus providing samples in polar coordinates. LIDAR sensors also are used to obtain 3-D panoramic data. In the DARPA Urban Challenge [5] and subsequently with Google cars [14], 3-D Velodyne sensors

[42] were used, essentially equivalent to 64 LIDARS in parallel. Lower-cost LIDARs have been used to acquire 3-D data by adding a mechanism to change the LIDAR's viewpoint, either with a wobbler (oscillating the mirror's spin axis) or by spinning the LIDAR about an axis orthogonal to the mirror's spin axis, as used on the sensor heads [35] of the Boston-Dynamics Atlas robots in the DARPA Robotics Challenge [30].

The simple two-dimensional robot simulator (STDR) introduced in Section 3.1 illustrated the concept of LIDAR graphically with red rays emanating from an abstracted mobile robot. As shown in Fig 3.1, each LIDAR line originates from the sensor and pings a point in the environment. By knowing the length of this line (as deduced from time of flight) and its angle, a single point in the environment is sampled. With careful attention to transforms, the 3-D vector of the line of sight of this pulse is known in some reference frame, from which one can compute corresponding 3-D coordinates of a point in the environment with respect to this reference frame.

Messages from LIDAR can use the ROS message type `sensor_msgs/LaserScan`. The example code in package `lidar_alarm` in the accompanying repository includes the file `lidar_alarm.cpp`. This code shows how to interpret messages of type `sensor_msgs/LaserScan`. Although these LIDAR messages are generated by STDR, the format is identical for LIDAR sources from physical sensors or from Gazebo emulation of LIDAR.

To create simulated LIDAR data, we need to augment our robot model by adding a Gazebo plug-in for LIDAR emulation. This is illustrated by the model `mobot_w_lidar.xacro` in the package `mobot_urdf`. This model is identical to the model `mobot_w_jnt_pub.xacro` described in Section 4.1 (and not repeated here), except for the following two inserted blocks. First, a visual, collision, and dynamic model of a LIDAR (a simple box) is defined as a new link, and this link is attached to the robot with a static joint, as per Listing 5.12.

Listing 5.12: Link and joint modeling for adding LIDAR to mobot model

```
1  <!-- add a simulated lidar, including visual, collision and inertial properties, and ↩
       physics simulation-->
2    <link name="lidar_link">
3        <collision>
4            <origin xyz="0 0 0" rpy="0 0 0"/>
5            <geometry>
6                <!-- coarse LIDAR model; a simple box -->
7                <box size="0.2 0.2 0.2"/>
8            </geometry>
9        </collision>
10
11       <visual>
12           <origin xyz="0 0 0" rpy="0 0 0" />
13           <geometry>
14               <box size="0.2 0.2 0.2" />
15           </geometry>
16           <material name="sick_box">
17               <color rgba="0.7 0.5 0.3 1.0"/>
18           </material>
19       </visual>
20
21       <inertial>
22           <mass value="4.0" />
23           <origin xyz="0 0 0" rpy="0 0 0"/>
24           <inertia ixx="0.01" ixy="0" ixz="0" iyy="0.01" iyz="0" izz="0.01" />
25       </inertial>
26   </link>
27   <!--the above displays a box meant to imply Lidar-->
28
29   <joint name="lidar_joint" type="fixed">
30       <axis xyz="0 1 0" />
```

```
31        <origin xyz="0.1 0 0.56" rpy="0 0 0"/>
32        <parent link="base_link"/>
33        <child link="lidar_link"/>
34      </joint>
```

In listing 5.12, lines 11 through 19 describe the visual appearance of a simple box meant to represent a LIDAR sensor. Within this block, lines 16 through 18 show how one can set color for `rviz` display. Recall that `rviz` and Gazebo use different color representations. An additional `<gazebo>` field could be added to describe the color to Gazebo as well, but lacking this, the Gazebo appearance will default to light gray.

In fact, defining visual, collision and inertial properties for a sensor seem to be overkill, since we are primarily concerned with emulating the sensor physics. To compute transforms consistently, however, we must associate the sensor with a link in the model, and a link must be attached to the model via a joint. The physics engine of Gazebo also insists that every link includes inertial properties. The collision and visual blocks may be ignored, but including them allows the model to be more realistic.

To include computations of an equivalent LIDAR, a Gazebo plug-in is used. Following the tutorial at `http://gazebosim.org/tutorials?tut=ros_gzplugins#GPULaser`, and modifying values lightly to apply to a Sick LMS200 LIDAR, emulation is enabled by including the block of code in Listing 5.13 (extracted from `mobot_w_lidar.xacro`).

Listing 5.13: Gazebo block to include LIDAR emulation plug-in for mobot model

```
1    <!-- here is the gazebo plug-in to simulate a lidar sensor -->
2    <gazebo reference="lidar_link">
3      <sensor type="gpu_ray" name="sick_lidar_sensor">
4        <pose>0 0 0 0 0 0</pose>
5        <visualize>false</visualize>
6        <update_rate>40</update_rate>
7        <ray>
8          <scan>
9            <horizontal>
10             <samples>181</samples>
11             <resolution>1</resolution>
12             <min_angle>-1.570796</min_angle>
13             <max_angle>1.570796</max_angle>
14           </horizontal>
15         </scan>
16         <range>
17           <min>0.10</min>
18           <max>80.0</max>
19           <resolution>0.01</resolution>
20         </range>
21         <noise>
22           <type>gaussian</type>
23           <mean>0.0</mean>
24           <stddev>0.01</stddev>
25         </noise>
26       </ray>
27       <plugin name="gazebo_ros_lidar_controller" filename="libgazebo_ros_gpu_laser.so"
                >
28         <topicName>/scan</topicName>
29         <frameName>lidar_link</frameName>
30       </plugin>
31     </sensor>
32   </gazebo>
```

In Listing 5.13, lines 2 through 4 declare that the sensor is to be located coincident with the `lidar_link` frame. That is, relative to the `lidar_link` frame, the sensor frame transform is $(x, y, z) = (0, 0, 0)$, and $(R, P, Y) = (0, 0, 0)$, *i.e.* identical to the `lidar_link` frame.

Lines 6 through 26 set various parameters of the LIDAR to be emulated, including scan repetition rate, start angle, end angle, angular resolution (angle increments between samples), min and max range, and an option for adding noise to the computed result (to more realistically simulate an actual LIDAR).

Line 27,

```
<plugin name="gazebo_ros_lidar_controller" filename="libgazebo_ros_gpu_laser.so">
```

references the Gazebo library that contains the code for simulating a LIDAR sensor. Importantly, this library assumes use of a Graphical Processing Unit (GPU) on the host computer. Use of a GPU makes this computation much faster. However, it also imposes constraints on the hardware. If a GPU is not present, the LIDAR simulator will attempt to run, but it will output meaningless range values; all range values will be set to the minimum LIDAR range. (If you run into this problem, it may help to install **bumblebee** and launch Gazebo with **optirun** to direct the GPU-based code to run appropriately on available graphics chips; search on these keywords for possible solutions, if needed.)

With the LIDAR additions to our mobile-robot model, we can see its behavior by starting Gazebo, loading the robot model onto the parameter server, spawning the model into Gazebo, bringing an interesting virtual world into Gazebo (something for the LIDAR to sense), starting a **robot_state_publisher**, starting **rviz**, and configuring **rviz** to display the LIDAR sensor topic. This process can become tedious. Fortunately it can be automated with launch files. But for now, to illustrate the steps, we will start them separately with the following commands. First, launch Gazebo (optionally, with **optirun**):

```
(optirun) roslaunch gazebo_ros empty_world.launch
```

A launch file to bring in the modified robot is in the **mobot_urdf** package (in the **launch** subdirectory), called **mobot_w_lidar.launch**. The contents are shown in Listing 5.14.

Listing 5.14: Launch file for mobot with LIDAR

```
1  <launch>
2  <!-- Convert xacro model file and put on parameter server -->
3  <param name="robot_description" command="$(find xacro)/xacro.py '$(find mobot_urdf)/↵
       urdf/mobot_w_lidar.xacro'" />
4
5  <!-- Spawn the robot from parameter server into Gazebo -->
6  <node name="spawn_urdf" pkg="gazebo_ros" type="spawn_model" args="-param ↵
       robot_description -urdf -model mobot" />
7
8  <!-- start a robot_state_publisher -->
9  <node name="robot_state_publisher" pkg="robot_state_publisher" type="↵
       robot_state_publisher" />
10 </launch>
```

This launch file locates the modified robot-with-lidar model, puts it on the parameter server, and spawns it in Gazebo. Additionally, this launch file starts the robot state publisher node.

A more interesting Gazebo environment can be utilized. In the Gazebo display, under the **Insert** tab, one can select **Starting Pen**. The entire starting-pen model will move around in the Gazebo scene until the mouse is clicked, which establishes the model location in the world. When doing this, be careful not to place the model such that the robot is embedded in a wall, or the physics simulation will blow up.

A view of the Gazebo display with the modified robot in the starting pen is shown in Fig 5.11. At this point, Gazebo is computing simulated LIDAR. A **rostopic echo scan** will display output like the following (truncated):

```
frame_id: lidar_link
```

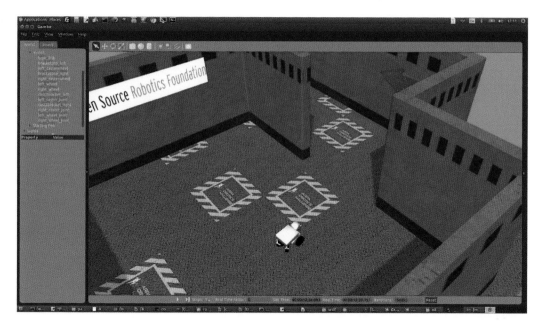

Figure 5.11: Gazebo view of simple mobile robot with LIDAR sensor in a virtual world

```
angle_min: -1.57079994678
angle_max: 1.57079994678
angle_increment: 0.0174533333629
time_increment: 0.0
scan_time: 0.0
range_min: 0.10000000149
range_max: 80.0
ranges: [1.4379777908325195, 1.458155632019043, 1.430367350578308, ...
```

The values for frame id, max angle, min angle, angle increment, range min, and range max correspond to the values in our URDF model within the LIDAR Gazebo block. The vector of range values contains 181 radii (in meters) corresponding to individual LIDAR rays. Note also that `scan` is the `topicName` value set in the Gazebo plug-in.

Next, bring up `rviz`, with:

```
rosrun rviz rviz
```

In the `rviz` display, `Add` a display item called `LaserScan`. Expand this item in the `displays` window. Beside the `Topic` field, click to show the drop-down menu of options and choose `/scan`, which is the topic to which our emulated LIDAR instrument publishes its data.

With these settings, the `rviz` display appears as in Fig 5.12. The `rviz` view is still not as interesting as the Gazebo view. However, what we can see in the `rviz` view is the information available to the robot to perceive its environment. Lacking other visual sensors, the robot cannot know the detail of its world (in contrast to Gazebo).

A software developer can interpret the `rviz` display to determine what signal processing would be appropriate for the robot to function usefully in the world. Code can be written, *e.g.* to make maps of the environment (at least at the height of the LIDAR's slice plane) and reconcile sensor data with such maps to estimate the robot's pose in the world. Alternatively,

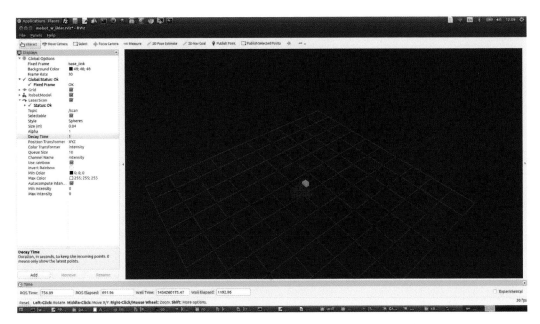

Figure 5.12: Rviz view of simple mobile robot with LIDAR sensor data being displayed

the LIDAR data could be used to attempt path planning without a map, *e.g.* by following walls or seeking corridors of suitable clearance.

The `rviz` view can be customized with more than a dozen display options to make the sensor display easier to interpret. Once the settings are as desired (including display items, topics, colorization, etc.), one can save the `rviz` settings for future use. In `rviz`, on the top menu, under `file->save config as`, the user has the option to save the current `rviz` settings with a name and directory of choice. In the present case, `rviz` settings were saved to a file called `mobot_w_lidar.rviz` in a subdirectory `rviz_config` within the package `mobot_urdf`.

An `rviz` display can be launched automatically from a launch file and directed to use a desired configuration file. The launch file `robot_w_lidar_and_rviz.launch` within the `launch` subdirectory of package `mobot_urdf` is identical to `mobot_w_lidar.launch` except for one additional line:

```
<node pkg="rviz" type="rviz" name="rviz" args="-d $(find mobot_urdf)/rviz_config/↵
    mobot_w_lidar.rviz"/>
```

This addition to the launch file starts `rviz` running, specifically directed to start up with a specified configuration file.

Although the `rviz` view in Fig 5.12 is relatively impoverished, it is appropriate and realistic. If a physical robot with a LIDAR sensor publishes its LIDAR data, the result can be visualized in `rviz`, and it would look essentially the same as Fig 5.12, except that individual points would correspond to samples in a real environment.

It can be illuminating to see the LIDAR data change dynamically in `rviz` as the robot moves around. The robot can be commanded to move (open-loop) in circles from a command line by entering:

```
rostopic pub -r 2 cmd_vel geometry_msgs/Twist  '{linear:  {x: 0.5, y: 0.0, z: 0.0}, ↩
    angular: {x: 0.0,y: 0.0,z: 0.2}}'
```

The resulting **rviz** view will show the LIDAR points refreshing as the robot changes its perspective. The **rviz** view, with its fixed frame set to **base_link**, shows a stationary robot with sensor data translating and rotating with respect to the robot, *i.e.* from the perspective of the robot. In contrast, the Gazebo view shows the robot moving within a stationary (virtual) world. Although **rviz** sensor data is displayed with respect to the robot, the viewpoint can be translated and rotated, which can help the observer get a better sense of three dimensions.

The value of rotating a **rviz** viewpoint to get a 3-D sense is more striking when displaying richer sets of 3-D data, such as from cameras.

5.2.2 Simulating and displaying color-camera data

An impressive capability of Gazebo is simulation of color cameras. As with the LIDAR, we can use a Gazebo plug-in to emulate a color camera. The format to do so is similar.

One option is to edit the **mobot_w_lidar.xacro** file to add details to include emulation of a camera. However, it is more convenient to model the camera separately, then include the camera model in an integrated robot model. A camera model, **example_camera.xacro**, is included in the **mobot_urdf** package. The contents of this model file are shown in Listing 5.15.

Listing 5.15: **example_camera.xacro**: example model file for camera

```
1  <?xml version="1.0"?>
2  <robot
3      xmlns:xacro="http://www.ros.org/wiki/xacro" name="mobot_camera">
4
5    <!-- add a simulated camera, including visual, collision and inertial properties, ↩
          and physics simulation-->
6    <link name="camera_link">
7        <!-- here is the physical body (case) of the camera-->
8        <collision>
9            <origin xyz="0 0 0" rpy="0 0 0"/>
10           <geometry>
11               <box size="0.1 0.02 0.02"/>
12           </geometry>
13       </collision>
14
15       <visual>
16           <origin xyz="0 0 0" rpy="0 0 0" />
17           <geometry>
18               <box size="0.1 0.02 0.02"/>
19           </geometry>
20           <material name="camera_case">
21               <color rgba="0.7 0.0 0.0 1.0"/>
22           </material>
23       </visual>
24
25       <inertial>
26           <mass value="0.1" />
27           <origin xyz="0 0 0" rpy="0 0 0"/>
28           <inertia ixx="0.0001" ixy="0" ixz="0" iyy="0.0001" iyz="0" izz="0.0001" />
29       </inertial>
30   </link>
31
32   <!-- here is the gazebo plug-in to simulate a color camera -->
33   <!--must refer to the above-defined link to place the camera in space-->
34   <gazebo reference="camera_link">
35       <!--optionally, displace/rotate the optical frame relative to the enclosure-->
36       <pose>0.1 00 0.0 0 0 0</pose>
37       <sensor type="camera" name="example_camera">
```

```
38      <update_rate>30.0</update_rate>
39      <camera name="example_camera">
40        <!--describe some optical properties of the camera-->
41        <!--field of view is expressed as an angle, in radians-->
42        <horizontal_fov>1.0</horizontal_fov>
43        <!--set resolution of pixels of image sensor, e.g. 640x480-->
44        <image>
45          <width>640</width>
46          <height>480</height>
47          <format>R8G8B8</format>
48        </image>
49        <clip>
50          <!--min and max range of camera-->
51          <near>0.01</near>
52          <far>100.0</far>
53        </clip>
54        <!--optionally, add noise, to make images more realistic-->
55        <noise>
56          <type>gaussian</type>
57          <mean>0.0</mean>
58          <stddev>0.007</stddev>
59        </noise>
60      </camera>
61      <!--here is the plug-in that does the work of camera emulation-->
62      <plugin name="camera_controller" filename="libgazebo_ros_camera.so">
63        <alwaysOn>true</alwaysOn>
64        <updateRate>10.0</updateRate> <!--can set the publication rate-->
65        <cameraName>example_camera</cameraName> <!--topics will be example_camera/... ↩
              -->
66        <!--listen to the following topic name to get streaming images-->
67        <imageTopicName>image_raw</imageTopicName>
68        <!--the following topic carries info about the camera, e.g. 640x480, etc-->
69        <cameraInfoTopicName>camera_info</cameraInfoTopicName>
70        <!--frameName must match gazebo reference name...seems redundant-->
71        <!-- this name will be the frame_id name in header of published frames-->
72        <frameName>camera_link</frameName>
73        <!-- optionally, add some lens distortion -->
74        <distortionK1>0.0</distortionK1>
75        <distortionK2>0.0</distortionK2>
76        <distortionK3>0.0</distortionK3>
77        <distortionT1>0.0</distortionT1>
78        <distortionT2>0.0</distortionT2>
79      </plugin>
80    </sensor>
81  </gazebo>
82
83 </robot>
```

In Listing 5.15, a **robot** model is defined, although this pseudo-robot consists only of a single link. As usual, the link is defined to have visual, collision and inertial properties. This link is defined to be a simple box, meant to suggest the enclosure of a camera.

The listing is more interesting starting with line 32. A Gazebo tag introduces a sensor, specifically a camera, and the frame rate for publications is set to 30 Hz (line 38). Camera parameters are defined on lines 39 through 60, including the array dimensions (640×480), the optics (equivalently, a pin-hole camera with a field-of-view angle of 1.0 rad projecting onto the image plane 640 pixels wide). When images are broadcast, they will be encoded as 8-bit values each of red, green, blue (in that order), as specified on line 47.

Minimum and maximum ranges for the camera are set (lines 49 through 53). The maximum range is a matter of computational pragmatism rather than physics. Since synthetic images are computed from ray tracing in the simulated environment, one must put an upper bound on how far to extend the rays to make this computation practical.

Lines 55 through 59 add noise to the image. (see `http://gazebosim.org/tutorials?` `tut=sensor_noise` for details). Introducing noise helps make the synthetic images more realistic. Image processing developed using such images would be more robust by avoiding the pitfall of depending on unrealistic assumptions regarding image quality.

Lines 62 through 79 introduce the Gazebo plug-in for camera emulation. The camera software library computes ray tracing in the simulated world to evaluate intensities of colors for each pixel in the camera's image plane, updated at the specified frequency (if this update rate can be achieved on the target simulation computer). The conventional name chosen in ROS for transmitting camera images is `image_raw` (line 67). This topic carries messages of type `sensor_msgs/Image`. The topic `camera_info` is the conventional name of the topic for messages describing camera parameters, via message type `sensor_msgs/CameraInfo`. As specified by line 65, the camera topic names are pre-pended with the namespace `example_camera`, yielding topics `/example_camera/image_raw` and `/example_camera/camera_info`.

The Gazebo plug-in computes synthetic images and publishes them to the `image_raw` topic. These messages will have a header with `frame_id` set to `camera_link`, as specified on line 72. Note that the frame name on line 72 must agree with the Gazebo reference name on line 34, and the named frame must be associated with a corresponding link in the model file (`camera_link`, in this case).

In addition to noise, one can also introduce lens distortion effects (lines 74 through 78). (see `http://gazebosim.org/tutorials?tut=camera_distortion` for details). Typically, camera calibration is performed to find these parameters. Given calibration coefficients, an additional node is run that subscribes to the raw images, undistorts the images, and republishes them as rectified images. This process, however, is not part of the Gazebo simulation. Rather, the Gazebo simulation attempts to create and publish streaming images with realism, including noise and distortion, attempting to emulate physical cameras.

Our camera model in `example_camera.xacro` can be added to our mobile robot model, as was done when adding an arm to the mobile base. To do so, one must specify a joint that connects the camera link to the base link. A xacro file that combines the base and camera is `mobot_w_lidar_and_camera.xacro`, given in Listing 5.16.

Listing 5.16: `mobot_w_lidar_and_camera.xacro`: model file combining base and camera

```
1  <?xml version="1.0"?>
2  <robot
3      xmlns:xacro="http://www.ros.org/wiki/xacro" name="mobot">
4    <xacro:include filename="$(find mobot_urdf)/urdf/mobot_w_lidar.xacro" />
5    <xacro:include filename="$(find mobot_urdf)/urdf/example_camera.xacro" />
6
7    <!-- attach the camera to the mobile robot -->
8    <joint name="camera_joint" type="fixed">
9      <parent link="base_link" />
10     <child link="camera_link" />
11     <origin rpy="0 0 0 " xyz="0.1 0 0.7"/>
12   </joint>
13 </robot>
```

Listing 5.16 includes the prior mobile-robot model (including its LIDAR and joint publications) and also includes the example camera model. The two models are joined together by declaring `camera_joint` to establish the `camera_link` (the base of the camera model) as a child of `base_link` (the base of the mobile platform). It is not necessary that the parent link be a base link. Any defined link on the mobile robot will work (including a frame on an attached arm if desired).

The combined model can be launched with `mobot_w_lidar_and_camera.launch`, given in Listing 5.17.

Listing 5.17: `mobot_w_lidar_and_camera.launch`: launch file for combined base and camera

```
1   <launch>
2   <!-- Convert xacro model file and put on parameter server -->
3   <param name="robot_description" command="$(find xacro)/xacro.py '$(find mobot_urdf)/↩
        urdf/mobot_w_lidar_and_camera.xacro'" />
4
5   <!-- Spawn the robot from parameter server into Gazebo -->
6   <node name="spawn_urdf" pkg="gazebo_ros" type="spawn_model" args="-param ↩
        robot_description -urdf -model mobot" />
7
8   <!-- start a robot_state_publisher -->
9   <node name="robot_state_publisher" pkg="robot_state_publisher" type="↩
        robot_state_publisher" />
10
11  <!-- launch rviz using a specific config file -->
12   <node pkg="rviz" type="rviz" name="rviz" args="-d $(find mobot_urdf)/rviz_config/↩
        mobot_w_lidar.rviz"/>
13
14  </launch>
```

To launch the new, combined model, first start Gazebo:

```
roslaunch gazebo_ros empty_world.launch
```

then load the robot model onto the parameter server, spawn the robot model into Gazebo, start up a robot state publisher and start `rviz`:

```
roslaunch mobot_urdf mobot_w_lidar_and_camera.launch
```

One can display camera views in `rviz` by adding a `camera` item. However, a separate node can be used for this (which is somewhat more convenient and stable). By running:

```
rosrun image_view image_view image:=example_camera/image_raw
```

we start the node `image_view` from the package `image_view` (see http://wiki.ros.org/image_view). As specified by the command-line argument, this will subscribe to the `example_camera/image_raw` topic and display images published on this topic. Initially, the output will be bland, because the robot is in an empty world. From Gazebo, one can insert existing world models, such as the **Starting Pen**. The resulting displays of Gazebo and `image_view` appear as in Fig 5.13. Figure 5.13 shows the model robot in the starting pen world, as well as a display of the simulated camera. It can be seen that the camera view makes sense in terms of the pose of the robot in the world. Directions, colors and perspective are appropriate for this pose. As the robot drives around in the world, images transmitted will continue to update to reflect the robot's viewpoint. Image-processing code can be written that subscribes to this sensor topic and interprets the data. Such code can be developed and tested in simulation, then applied to a physical system with few changes. In practice, one would need to calibrate the physical camera for its intrinsic parameters (focal length, central pixel and distortion coefficients) and its extrinsic parameters (the true values for the `camera_joint` transform, specifying precisely how the camera is mounted to the robot). The Gazebo model should be reconciled with the corresponding physical system to ensure that all camera parameters are in agreement. Subsequently, code developed in simulation should behave well on the real system, depending on fidelity of the virtual world model representing a real setting, although the image-processing code would likely need additional tuning on the real system.

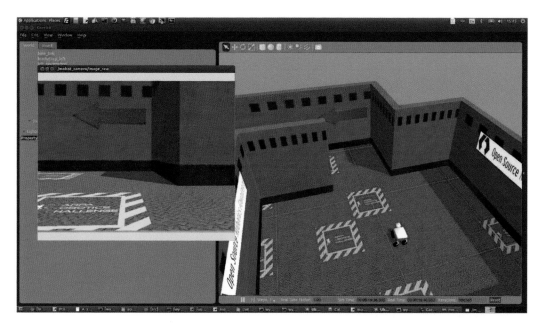

Figure 5.13: Gazebo view of simple mobile robot in virtual world and display of emulated camera

5.2.3 Simulating and displaying depth-camera data

Another valuable sensor type is the depth camera. Various sensors, including stereo vision systems, the Kinect™camera, and some LIDARs, are able to sense 3-D coordinates of points in the environment. Some sensors, including the Kinect, also associate color with each 3-D point, constituting an RGBD (red-green-blue-depth) camera. Similar to LIDAR, line-of-sight trigonometry can be applied to vectors associated with image-plane pixels (and passing through a focal point), such that pixel coordinates augmented with depth information imply 3-D (and each such pixel may have associated RGB color values). Performing such computations, one can express the result as a point cloud (see http://pointclouds.org/).

To simulate a Kinect camera, we can construct a similar model file, provided in the mobot_urdf package within example_kinect.xacro. The contents of this file appear in Listing 5.18.

Listing 5.18: example_kinect.xacro: example model file for Kinect sensor

```
1  <?xml version="1.0"?>
2  <robot
3      xmlns:xacro="http://www.ros.org/wiki/xacro" name="example_kinect">
4
5    <!-- add a simulated Kinecct camera, including visual, collision and inertial ↵
         properties, and physics simulation-->
6    <link name="kinect_link">
7      <!-- here is the physical body (case) of the camera-->
8      <collision>
9        <origin xyz="0 0 0" rpy="0 0 0"/>
10       <geometry>
11         <box size="0.02 0.1 0.02"/>
12       </geometry>
13     </collision>
```

```
14
15          <visual>
16              <origin xyz="0 0 0" rpy="0 0 0" />
17              <geometry>
18                  <box size="0.02 0.1 0.02"/>
19              </geometry>
20              <material name="camera_case">
21                  <color rgba="0.0 0.0 0.7 1.0"/>
22              </material>
23          </visual>
24
25          <inertial>
26              <mass value="0.1" />
27              <origin xyz="0 0 0" rpy="0 0 0"/>
28              <inertia ixx="0.0001" ixy="0" ixz="0" iyy="0.0001" iyz="0" izz="0.0001" />
29          </inertial>
30      </link>
31
32  <!-- here is the gazebo plug-in to simulate a color camera -->
33  <!--must refer to the above-defined link to place the camera in space-->
34      <gazebo reference="kinect_link">
35          <sensor type="depth" name="openni_camera_camera">
36              <always_on>1</always_on>
37              <visualize>true</visualize>
38              <camera>
39                  <horizontal_fov>1.047</horizontal_fov>
40                  <image>
41                      <width>640</width>
42                      <height>480</height>
43                      <format>R8G8B8</format>
44                  </image>
45                  <depth_camera>
46
47                  </depth_camera>
48                  <clip>
49                      <near>0.1</near>
50                      <far>100</far>
51                  </clip>
52              </camera>
53          <!--here is the plug-in that does the work of kinect emulation-->
54              <plugin name="camera_controller" filename="libgazebo_ros_openni_kinect.so">
55                  <alwaysOn>true</alwaysOn>
56                  <updateRate>10.0</updateRate>
57                  <cameraName>kinect</cameraName>
58                  <frameName>kinect_depth_frame</frameName>
59              <imageTopicName>rgb/image_raw</imageTopicName>
60              <depthImageTopicName>depth/image_raw</depthImageTopicName>
61              <pointCloudTopicName>depth/points</pointCloudTopicName>
62              <cameraInfoTopicName>rgb/camera_info</cameraInfoTopicName>
63              <depthImageCameraInfoTopicName>depth/camera_info</↵
                    depthImageCameraInfoTopicName>
64              <pointCloudCutoff>0.4</pointCloudCutoff>
65                  <hackBaseline>0.07</hackBaseline>
66                  <distortionK1>0.0</distortionK1>
67                  <distortionK2>0.0</distortionK2>
68                  <distortionK3>0.0</distortionK3>
69                  <distortionT1>0.0</distortionT1>
70                  <distortionT2>0.0</distortionT2>
71              <CxPrime>0.0</CxPrime>
72              <Cx>0.0</Cx>
73              <Cy>0.0</Cy>
74              <focalLength>0.0</focalLength>
75              </plugin>
76          </sensor>
77      </gazebo>
78
79  </robot>
```

As with the example camera, this model file is called a **robot**, even though it has no degrees of freedom. A link is defined to represent the housing of the Kinect sensor (lines 6 through 30). The Gazebo tag references this link (line 34). The Kinect includes both depth information (from an infra-red camera) and color information (from an RGB camera). Many of the specifications for the Kinect camera are similar to the previous example camera model, including field of view, dimensions of image array (in pixels), range clipping, optional noise and distortion coefficients, and update rate.

The plug-in library used to emulate the Kinect is referenced on line 54 (`libgazebo_ros_openni_kinect.so`).

The value of tag `cameraName` is set to `kinect`, and thus all topics published by this Gazebo plug-in will be in the namespace `kinect`. The topic for the RGB camera, `imageTopicName` is set on line 59 to `rgb/image_raw`. To display images from the RGB camera, the appropriate topic is thus `kinect/rgb/image_raw`.

The reference frame for Kinect topics, tag `frameName`, is set on line 58 to `kinect_depth_frame`. Note that this is different from the previous camera example, in which this frame was identical to the Gazebo reference. The Kinect model, inconveniently, requires an additional transform to align the sensor correctly relative to its mounting link. This issue is addressed in the launch file, described later.

The Kinect model is incorporated with a robot using the same technique as before by including it hierarchically in another xacro file. The file `mobot_w_lidar_and_kinect.xacro` appears in Listing 5.19.

Listing 5.19: `mobot_w_lidar_and_kinect.xacro`: model file combining robot and kinect sensor

```
1   <?xml version="1.0"?>
2   <robot
3        xmlns:xacro="http://www.ros.org/wiki/xacro" name="mobot">
4     <xacro:include filename="$(find mobot_urdf)/urdf/mobot_w_lidar.xacro" />
5     <xacro:include filename="$(find mobot_urdf)/urdf/example_camera.xacro" />
6     <xacro:include filename="$(find mobot_urdf)/urdf/example_kinect.xacro" />
7     <!-- attach the camera to the mobile robot -->
8     <joint name="camera_joint" type="fixed">
9       <parent link="base_link" />
10      <child link="camera_link" />
11      <origin rpy="0 0 0 " xyz="0.1 0 0.7"/>
12    </joint>
13    <!-- attach the kinect to the mobile robot -->
14    <joint name="kinect_joint" type="fixed">
15      <parent link="base_link" />
16      <child link="kinect_link" />
17      <origin rpy="0 0 0 " xyz="0.1 0 0.72"/>
18    </joint>
19    <!-- kinect depth frame has a different viewpoint; publish it separately-->
20  </robot>
```

Most of Listing 5.19 is the same as our prior camera model (Listing 5.16). Line 6 brings in the example Kinect xacro file. Lines 14 through 18 define a static joint attaching the `kinect_depth_frame` link in the Kinect model to the base link of the mobile-robot model.

Launching the combined model is achieved by the launch file `mobot_w_lidar_and_kinect.launch` in Listing 5.20.

Listing 5.20: `mobot_w_lidar_and_kinect.launch`: launch file for combined robot and kinect sensor

```
1   <launch>
2   <!-- Convert xacro model file and put on parameter server -->
3   <param name="robot_description" command="$(find xacro)/xacro.py '$(find mobot_urdf)/↵
        urdf/mobot_w_lidar_and_kinect.xacro'" />
4
5   <!-- Spawn the robot from parameter server into Gazebo -->
6   <node name="spawn_urdf" pkg="gazebo_ros" type="spawn_model" args="-param ↵
        robot_description -urdf -model mobot" />
7
```

```
8  <node pkg="tf" type="static_transform_publisher" name="kinect_broadcaster" args="0 0 0↩
       -0.500 0.500 -0.500 0.500 kinect_link kinect_depth_frame 100" />
9
10 <!-- start a robot_state_publisher -->
11 <node name="robot_state_publisher" pkg="robot_state_publisher" type="↩
       robot_state_publisher" />
12
13 <!-- launch rviz using a specific config file -->
14  <node pkg="rviz" type="rviz" name="rviz" args="-d $(find mobot_urdf)/rviz_config/↩
       mobot_w_lidar_and_kinect.rviz"/>
15
16 <!-- launch image_view as well -->
17  <node pkg="image_view" type="image_view" name="image_view">
18    <remap from="image" to="/kinect/rgb/image_raw" />
19  </node>
20
21 </launch>
```

This launch file illustrates two new features. First, line 8:

```
<node pkg="tf" type="static_transform_publisher" name="kinect_broadcaster" args="0 0 0↩
    -0.500 0.500 -0.500 0.500 kinect_link kinect_depth_frame 100" />
```

starts a node from the `tf` package called `static_transform_publisher`. This node publishes a transform relationship between specified frames on the `tf` topic. Our Kinect model file defined a frame for the image data, called `kinect_depth_frame`, but our URDF contains no information regarding how this frame relates to any other frame in the model. The `kinect_depth_frame` is not associated with any physical link and connects only with a reference frame defined with respect to the sensor itself. The static transform publisher node is provided with the names of a child frame, `kinect_depth_frame`, and a parent frame, `kinect_link`, and transform parameters between them. The arguments $(0, 0, 0)$ state that the depth frame origin is coincident with the Kinect-link frame origin. The arguments $(-0.5, 0.5, -0.5, 0.5)$ describe a quaternion orientation transformation involving 90 degree rotations about each of the x, y and z axes. By starting this static transform publisher, `rviz` is able to fully connect frames from the Kinect depth frame to the mobile robot's base frame. (See `http://wiki.ros.org/tf#static_transform_publisher` for more details on the static transform publisher.)

A second new feature of the launch file in Listing 5.20 appears in lines 17 through 19:

```
<node pkg="image_view" type="image_view" name="image_view">
  <remap from="image" to="/kinect/rgb/image_raw" />
</node>
```

This instruction automates start-up of the `image_view` node and instructs it to subscribe to the `/kinect/rgb/image_raw` topic. Note that from the command line, this was done with:

```
rosrun image_view image_view image:=example_camera/image_raw
```

but when started from a launch file, the syntax for topic assignment uses the `<remap>` tag.

To try the newly augmented robot model, start Gazebo with:

```
roslaunch gazebo_ros empty_world.launch
```

then run the launch script, which loads the robot model onto the parameter server, spawns the robot model into Gazebo, starts up a robot state publisher, starts a static transform publisher, and starts `rviz` (with a new `config` file):

```
roslaunch mobot_urdf mobot_w_lidar_and_kinect.launch
```

The Gazebo model is made more interesting by bringing in the world model gas station. Figure 5.14 shows three displays (overlapping on the screen). The Gazebo view shows the model robot within a gas-station virtual world. Recall that Gazebo is a stand-in for reality. It is useful for developing code in simulation, but it is ultimately replaced by a real robot, real sensors and a real environment.

The image_view view shows the synthetic image computed by Gazebo equivalent to the viewpoint of the Kinect's color camera viewing a gas pump. The image display appears virtually identical to the Gazebo display, although it should be remembered that the Gazebo display has 3-D information that can be inspected by moving the observer around relative to the virtual world. In contrast, the camera display only contains 2-D information, equivalent to values from an actual camera's image sensor.

The most interesting addition to Fig 5.14 is in the rviz view. A new display item has been added: PointCloud2. Within this display item, the topic is set to kinect/depth/ points. This topic carries the Kinect 3-D points. These are rendered in rviz with consistent transformations, resulting in points displayed at an appropriate distance from the robot. Further, these points align virtually perfectly with the red markers indicating pings from the robot's LIDAR. Unlike the image_view scene, the rviz scene is fully 3-D. The observer can rotate this view to observe the data from alternative viewpoints. Even though Gazebo gets replaced with reality, rviz views, such as in Fig 5.14, can still be displayed. This capability can give the operator perception of the robot's surroundings, thus providing potential for either teleoperation or supervisory control.

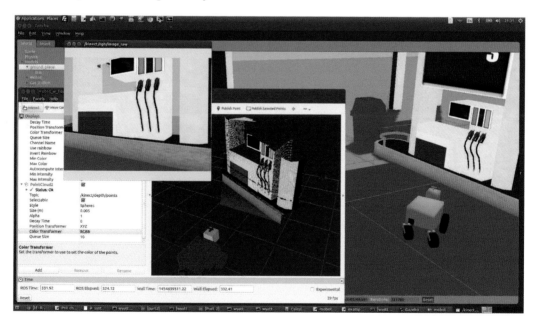

Figure 5.14: Gazebo, rviz and image-view displays of simple mobile robot with LIDAR, camera and Kinect sensor in virtual world

A nuisance inconsistency is that colors in the rviz display are different from the colors in the Gazebo display. This is because rviz and Gazebo are separate developments that use different color representations. In practice, this is not a problem. An rviz display from real cameras (including the Kinect) displays color correctly. Also, many 3-D sensors do not have associated color, making this a non-issue even for Gazebo simulations.

A very useful capability of point-cloud displays in `rviz` is that points (or collections of points) can be selected interactively and published for use by nodes. This capability is described next.

5.2.4 Selection of points in `rviz`

We have seen that `rviz` is a useful tool for visualization of sensor data, allowing the operator to view a display of values with respect to a robot model. As introduced in Section 5.1, one can also overlay graphics (markers) to draw attention to specific regions of the display.

In addition to using `rviz` for sensor visualization, `rviz` can also be used as a remote operator interface for either teleoperation or supervisory control of robots. We have already seen that one can interact with `rviz` via interactive markers (Section 5.1.3). A valuable additional input option is the ability to select points of interest in the `rviz` display, and publish these point coordinates for consumption by perceptual processing nodes. The tool `PublishSelectedPoints` is a plug-in of `rviz` that provides this capability.

The `rviz` plug-in `selected_points_publisher` (from Technical University of Berlin, Robotics and Biology Laboratory [33]; see also [32]) is contained in the accompanying repository **learning_ros_external_packages** within the package **rviz_plugin_selected_points_topic**. With initial installation of `rviz`, this tool is not yet present. There are installation instructions within the README file in the **selected_points_publisher** package. It may be necessary to run:

```
catkin_make install
```

for ROS to be able to find this plug-in. From `rviz`, on the title bar is a blue + sign. Upon clicking this symbol, a menu will pop up, as shown in Fig 5.15. In the example of Fig 5.15,

Figure 5.15: View of `rviz` during addition of plug-in tool

`PublishSelectedPoints` is grayed out, since this tool was already installed in `rviz`.

The Publish Selected Points tool can be enabled by clicking its icon on the `rviz` title bar. One may then click and drag within the `rviz` scene to select and publish points of interest.

Figure 5.16 shows the LIDAR display with the mobile robot closer to the exit from the starting pen. The individual LIDAR points are displayed using relatively large red spheres as markers (selectable in `rviz` within the `LaserScan` display item). Within the `LaserScan` display item, the option `selectable` is checked (enabled). Consequently, when one clicks and drags to enclose one or more of these points, the corresponding coordinates are published. Figure 5.16 shows a single LIDAR point selected (indicated by being enclosed by a light-blue wireframe box).

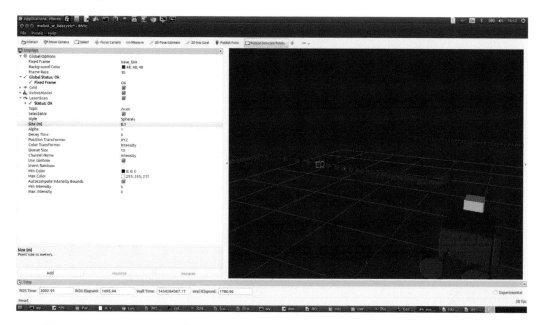

Figure 5.16: `rviz` view showing selection of single LIDAR point to be published

The `PublishSelectedPoints` tool publishes information about user-selected points to the topic `selected_points`. This topic carries messages of type `sensor_msgs/PointCloud2`. With `rostopic echo selected_points`, we can see the result of the selected-points publisher. After selecting the LIDAR point shown in Fig 5.16, the selected-points topic echo displays:

```
fields:
  -
    name: x
    offset: 0
    datatype: 7
    count: 1
  -
    name: y
    offset: 4
    datatype: 7
    count: 1
  -
    name: z
```

```
     offset: 8
     datatype: 7
     count: 1
is_bigendian: False
point_step: 12
row_step: 12
data: [230, 233, 118, 63, 26, 249, 52, 64, 41, 92, 15, 63]
```

This format is not as obvious as previous sensor messages. Briefly, it declares in a header that the points will be represented by x, y and z coordinates, represented with 4 bytes per coordinate (*i.e.* single-precision floating point). The `data` component of the message carries a potentially large number of bytes, but in this example there are only 12 bytes (3 coordinates at 4 bytes each). The format for `PointCloud` messages is less convenient than most ROS messages since it is had to be designed to carry large amounts of data to be processed efficiently. More interpretation of `PointCloud` messages will be covered in Chapter 8 in the context of point cloud processing.

Figure 5.17 shows both Gazebo and `rviz` displays of our simple robot with the simulated gas station world. The Kinect camera can see gas pumps. The `rviz` display is zoomed on the pumps to emphasize the pump handles. Note on the title bar of `rviz` that `PublishSelectedPoints` is highlighted. With this tool selected, one can click and drag on the `rviz` scene to select a set of point-cloud points. Although the `rviz` view is a 2-D display, the display is generated by 3-D data. By choosing points in 2-D, the underlying 3-D source of this data can be referenced, and thus the corresponding 3-D coordinates of the selected points can be published. In Fig 5.17, a small patch of points on the pump handle, highlighted in blue, has been selected.

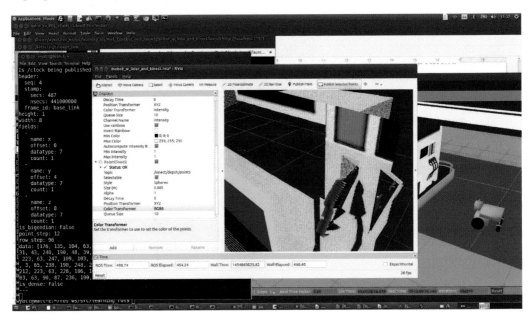

Figure 5.17: Gazebo view of mobot in simulated gas station, and `rviz` display of simulated Kinect data. A small, light-blue patch of points on the pump handle displays user-selected points.

Once points are selected in rviz, a message (of type `PointCloud2`) is published on the topic `selected_points`. Each point is described in terms of x, y and z coordinates, each of which is encoded as four bytes (corresponding to datatype 7, which is defined in `sensor_msgs/PointCloud2` as a `FLOAT32`). Via this topic, one can obtain 3-D data corresponding to interacting with displayed sensor values in `rviz`. The resulting publications can be received and interpreted by ROS nodes. Further details of point-cloud messages and interpretation will be covered in Chapter 8.

5.3 WRAP-UP

This concludes our introduction to sensing and visualization in ROS. It should be appreciated that Gazebo simulations can and should be designed such that the interfaces are identical to corresponding physical robot systems. With attention to this commonality, it is possible to perform extensive software design and debugging in simulation, then apply the results to corresponding physical robots productively. Inevitably, some tuning is required to account for modeling imperfections, but the vast majority of programming can be done in simulation.

The `rviz` interface is highly useful for interpreting data from a robot, whether physical or simulated. If the simulated robot is designed to be consistent with the physical robot, the `rviz` display should be realistic in simulation. With the addition of user-designed markers displayed in `rviz`, one can help visualize the results of perceptual processing and/or path planning, which is highly useful for development and debugging. In addition, the same display can be used as an operator interface to interpret sensory data from a remote robot, including fusion of LIDAR, point cloud and camera data, together with displays of any available *a priori* models.

With the addition of interactive markers and publication of selected points, the `rviz` display can also perform as an intuitive operator interface for supervisory control of robots.

Given this introduction to the foundations of ROS, we are ready to illustrate use of ROS for robot programming, including both re-use of existing packages and design of new capabilities.

III

Perceptual Processing in ROS

INTRODUCTION

Intelligent robot behavior depends on performing actions appropriately in the context of its environment. For example, a mobile robot should avoid collisions and avoid navigating over impassable or dangerous terrain. Robot manipulators should perceive and interpret objects of interest, including object identification and localization, and should plan collision-free trajectories for part acquisition and manipulation. By perceiving and interpreting the environment, a robot can locate objects of interest or deduce appropriate actions (*e.g.* putting dishes in a dishwasher or fetching a specified article from a warehouse), as well as generate viable grasp and manipulation plans. Realizing sensor-based behaviors requires perceptual processing of sensory data.

In general, understanding one's environment based on sensory data is an enormous challenge encompassing multiple fields. Nonetheless, some useful sensory-driven behaviors are currently practical, and ROS tools exist to to assist in such design. Perceptual processing (*e.g.* computer vision) has a much longer history than ROS, and it is important that ROS be compatible with existing open-source libraries. Notably, OpenCV and the Point Cloud Library offer powerful tools to interpret sensory data from cameras, stereo cameras, 3-D LIDAR and depth cameras.

The next three chapters will introduce using cameras in ROS, depth imaging and point clouds, and point-cloud processing. It should be appreciated that this introduction is not a substitute for learning image processing in general, nor OpenCV or PCL in particular. A recommended guide for using OpenCV is [4]. Use of the Point-Cloud Library is not, at the time of this writing, presented in textbook style. However, there are on-line tutorials at `http://pointclouds.org/`.

Using Cameras in ROS

CONTENTS

INTRODUCTION

Cameras are commonly used with robots. To interpret camera data for purposes of navigation or grasp planning, one needs to interpret patterns of pixel values (intensities and/or colors) and associate labels and coordinates with these interpretations. This chapter will introduce conventions for camera coordinate frames in ROS, camera calibration, and low-level image-processing operations with OpenCV. Extensions to 3-D imaging are deferred to Chapter 7.

6.1   PROJECTIVE TRANSFORMATION INTO CAMERA COORDINATES

A necessary step toward associating images with spatial coordinates is camera calibration. This includes *intrinsic* properties and *extrinsic* properties. The intrinsic properties of a camera remain constant independent of the way the camera is mounted or how it moves in space. Intrinsic properties include image sensor dimensions (numbers of rows and columns of pixels); the central pixel of the image plane (which depends on how the image plane is mounted relative to the idealized focal point); distance from the image plane to the focal point; and coefficients of a model of lens distortion. Extrinsic properties describe how the camera is mounted, which is specified as a transform between a frame associated with the camera and a frame of interest in the world (*e.g.* the robot's base frame). The camera's intrinsic properties should be identified before attempting to calibrate the extrinsic properties.

Calibrating a camera requires defining coordinate frames associated with the camera. An illustration of the standard coordinate system of a camera is shown in Fig 6.1.

The point "C" in Fig 6.1 corresponds to the focal point of a pinhole camera model (*i.e.*, "C" is the location of the pinhole). This point is referred to as the projection center, and it will constitute the origin of our camera frame.

The camera will have a sensor plane (typically, a CCD array) behind the focal point (between the sensor plane and objects in the environment viewed by the sensor). This sensor plane has a surface-normal direction vector. We can define the optical axis as the unique vector perpendicular to the sensor plane passing through point "C." The optical axis defines the z axis of our camera coordinate frame. The sensor plane will also have rows and columns of pixels. We define the camera frame x axis as parallel to sensor rows, and the y axis as parallel to sensor columns, consistent with a right-hand frame (X,Y,Z). Choice of positive direction of X depends on how the sensor defines increasing pixel column numbers.

The projection of light rays from the environment to the sensor plane results in inverted images. It is mathematically convenient to avoid consideration of image inversion by assuming a virtual image plane that lies in *front* of the focal point (between the focal point and objects viewed). This image plane, or principal plane, is defined parallel to the physical sensor device, but offset by distance "f" (the focal length) in front of the projection center. This principal plane has two local coordinate system: (x, y) and (u, v). The directions of the x and u axes are parallel, as are the directions of the y and v axes. The origin of the (u, v) frame is one corner of the image plane, whereas the origin of the (x, y) frame is near the center of the principal plane. To be precise, the origin of the (x, y) system is at point "c" in Fig 6.1, which corresponds to the intersection of the optical axis with the principal plane. Additionally, the frames (x, y) and (u, v) have different units. The (x, y) dimensions are metric, whereas the (u, v) coordinates are measured in pixels. The u coordinate ranges from 0 to NCOLS-1, where NCOLS is the number of columns in the sensor; v ranges from 0 to NROWS-1, where NROWS is the number of rows. (A common sensor dimension is $640 \times 480 =$ NCOLS $\times$ NROWS). Data from the camera will be available in terms of light intensities indexed by (u, v) (in each of three planes, for color cameras).

The location of point "c" ideally is near $u =$ NCOLS$/2$ and $v =$ NROWS$/2$. In practice, the actual location of point "c" on the principal plane will depend on assembly tolerances. The true coordinates of point "c" are (u_c, v_c), which defines a central pixel known as the optical center. The coordinates of the central pixel are two of the intrinsic parameters to be determined.

Conversion from (x, y) coordinates to (u, v) coordinates requires knowledge of the physical dimensions of the pixel sensor elements in the detector. These elements, in general, are not square. Conversion from (x, y) to (u, v) can be expressed as: $u = u_c + x/w_{pix}$ and $v = v_c + y/h_{pix}$, where w_{pix} is the width of a pixel and h_{pix} is the height of a pixel.

In Fig 6.1, a point "M" in the environment is shown, where light emanating from this point (presumably background light reflecting off the object) is received by our detector. Following a ray from point "M", passing through the focal point (the projection center) results in intersection with our principal plane at image point "m". Point M in the environment has coordinates $M(X, Y, Z)$, as measured with respect to our defined camera frame. The projected image point "m" has coordinates $m(x, y)$ in the principal-frame coordinate system. If the coordinates of M are known with respect to the camera frame, the coordinates of m can be computed as a projection. Given the focal length "f", projection may be computed as $x = fX/Z$ and $y = fY/Z$ (where x, f, X, Y and Z are all in consistent units, *e.g.* meters). Additionally, with knowledge of the optical center in pixel coordinates (u_c, v_c) and knowledge of the pixel dimensions w_{pix} and h_{pix} we can compute the coordinates (u, v) of the projection of this optical ray onto our sensor. This projection computation thus requires knowledge of five parameters: f, u_c, v_c, w_{pix} and h_{pix}. This set of parameters can be further reduced by defining $f_x = f/h_{pix}$ and $f_y = f/w_{pix}$, where f_x and f_y are treated as separate focal lengths measured in units of pixels. Using only four parameters, f_x, f_y, u_c, and v_c, it is possible to compute projection from the environment onto (u, v) coordinates in the principal plane. These four parameters are intrinsic, since they do not change as the camera moves in space. Identifying these parameters is the first step in camera calibration.

Beyond a linear camera model, one can also account for lens distortion. Commonly, this is done by finding parameters for an analytic approximation of radial and tangential distortion as follows. Given $M(X, Y, Z)$, one can compute the projection $m(x, y)$. The values of (x, y) can be normalized by the focal length: $(x', y') = (x/f, y/f)$. In these di-

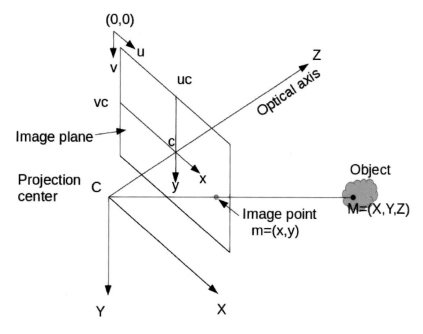

Figure 6.1: Standard camera frame definition

mensionless units, distortion can be modeled to map (x', y') linear projection onto (x'', y'') projection subject to distortion, and the resulting prediction of light impingement can then be converted to (u, v) pixel coordinates. A distortion model to map (x', y') onto (x'', y'') is (see http://docs.opencv.org/2.4/modules/calib3d/doc/camera_calibration_and_3d_reconstruction.html):

$$x'' = x' \frac{1 + k_1 r^2 + k_2 r^4 + k_3 r^6}{1 + k_4 r^2 + k_5 r^4 + k_6 r^6} + 2p_1 x' y' + p_2 (r^2 + 2x'^2) \tag{6.1}$$

$$y'' = y' \frac{1 + k_1 r^2 + k_2 r^4 + k_3 r^6}{1 + k_4 r^2 + k_5 r^4 + k_6 r^6} + 2p_2 x' y' + p_1 (r^2 + 2y'^2) \tag{6.2}$$

where $r^2 = x'^2 + y'^2$. Note that when all coefficients are zero, the distortion formula degenerates to simply $x'' = x'$ and $y'' = y'$. Also, for pixels close to the optical center "c", the (dimensionless) value of r will be small, and thus the higher order terms in Equations 6.1 and 6.2 will be negligible. The distortion model often includes $k_1, k_2, p_1, p2$ and disregards higher order terms. Discovery of distortion model coefficients is also part of the intrinsic parameter calibration process, which will be described next.

6.2 INTRINSIC CAMERA CALIBRATION

ROS provides support for intrinsic camera calibration through the `camera_calibration` package (which consists of a ROS wrapper on camera calibration code originating from openCV). Theory and details of the process can be found in the OpenCV documentation (see http://docs.opencv.org/2.4/doc/tutorials/calib3d/camera_calibration/camera_calibration.html).

The calibration routine assumes use of a checkerboard-like vision target with known numbers of rows and columns and known dimensions. Calibration can be as simple as waving the calibration target in front of the camera while running the calibration process. The process will acquire snapshots of the target, inform the user when there are a sufficient number of good images for identification, then compute the intrinsic parameters from the acquired images.

This process can be illustrated in simulation, given a Gazebo emulation of a camera, a model for the vision target, and means to move the vision target in front of the camera.

Emulation of cameras in Gazebo was introduced in Section 5.2.2. In the `simple_camera_model` package (in Part 3 of the associated repository), emulation of a camera in Gazebo is specified with only a static robot acting as a camera holder. The xacro file appears in listing 6.3.

Listing 6.1: `simple_camera_model.xacro`: simple camera model

```
1   <?xml version="1.0" ?>
2   <robot name="simple_camera" xmlns:xacro="http://www.ros.org/wiki/xacro">
3
4     <link name="world">
5         <origin xyz="0.0 0.0 0.0"/>
6     </link>
7
8     <joint name="camera_joint" type="fixed">
9         <parent link="world"/>
10        <child link="camera_link"/>
11        <origin rpy="0.0 1.5708 1.5708" xyz="0 0.0 0.5"/>
12    </joint>
13
14    <link name="camera_link">
15      <visual>
16        <origin xyz="0 0 0.0" rpy="0 0 0"/>
17        <geometry>
18          <box size="0.03 0.01 0.01"/>
19        </geometry>
20      </visual>
21
22      <inertial>
23        <mass value="1e-5" />
24        <origin xyz="0 0 0" rpy="0 0 0"/>
25        <inertia ixx="1e-6" ixy="0" ixz="0" iyy="1e-6" iyz="0" izz="1e-6" />
26      </inertial>
27    </link>
28
29    <!-- camera simulator plug-in -->
30    <gazebo reference="camera_link">
31      <sensor type="camera" name="camera">
32        <update_rate>30.0</update_rate>
33        <camera name="camera">
34          <horizontal_fov>0.6</horizontal_fov>
35          <image>
36            <width>640</width>
37            <height>480</height>
38            <format>R8G8B8</format>
39          </image>
40          <clip>
41            <near>0.005</near>
42            <far>0.9</far>
43          </clip>
44          <noise>
45            <type>gaussian</type>
46            <mean>0.0</mean>
47            <stddev>0.000</stddev>
48          </noise>
49        </camera>
50        <plugin name="camera_controller" filename="libgazebo_ros_camera.so">
51          <alwaysOn>true</alwaysOn>
52          <updateRate>0.0</updateRate>
53          <cameraName>simple_camera</cameraName>
54          <imageTopicName>image_raw</imageTopicName>
55          <cameraInfoTopicName>camera_info</cameraInfoTopicName>
```

```
56        <frameName>camera_link</frameName>
57        <distortionK1>0.0</distortionK1>
58        <distortionK2>0.0</distortionK2>
59        <distortionK3>0.0</distortionK3>
60        <distortionT1>0.0</distortionT1>
61        <distortionT2>0.0</distortionT2>
62      </plugin>
63    </sensor>
64  </gazebo>
65
66 </robot>
```

This model file specifies a camera located 0.5 m above the ground plane, oriented facing downward (looking at the ground plane). The model camera has 640 × 480 pixels and publishes images encoded with eight bits each for the red, green and blue channels. Noise can be modeled, but it has been set to zero in this simple model. Radial and tangential distortion coefficients can be specified in the model as well; in the present case, these have been set to zero to emulate an ideal (linear) camera. The focal length is specified indirectly by specifying a field-of-view angle of 0.6 rad. Emulated images will be published to the topic /simple_camera/image_raw.

Our robot can be launched into Gazebo with the command:

```
roslaunch simple_camera_model simple_camera_simu_w_checkerboard.launch
```

This command invokes the simple_camera_simu_w_checkerboard.launch file within the launch subdirectory of the package simple_camera_model. This launch file performs multiple steps, including starting Gazebo, setting gravity to 0, loading the simple camera model onto the parameter server and spawning it into Gazebo, and finding and loading a model called small_checkerboard.

With Gazebo running and the camera model spawned, rostopic list shows 13 topics under /simple_camera. Running:

```
rostopic hz /simple_camera/image_raw
```

shows that images are published at 30 Hz, which is consistent with the update rate specified in the Gazebo model. Running:

```
rostopic info /simple_camera/image_raw
```

shows that the message type on this topic is sensor_msgs/Image. Running:

```
rosmsg show sensor_msgs/Image
```

shows that this message type includes a header, specification of imager dimensions (height and width in pixels is 640 × 480, in this case), a string that describes how the image is encoded (rgb8, in this case), a step parameter (number of bytes per row, which is 1920 = 3 × 640, in this case) and a single vector of unsigned 8-bit integers comprising the encoded image.

Another topic of interest is /simple_camera/camera_info. This topic carries messages of type sensor_msgs/CameraInfo, which contains intrinisic camera information, including number of rows and columns of pixels, coefficients of the distortion model, and a projection matrix that contains values of f_x, f_y, u_c, and v_c. Running:

```
rostopic echo /simple_camera/camera_info
```

displays (in part):

```
header:
  seq: 119
  stamp:
    secs: 1779
    nsecs: 333000000
  frame_id: camera_link
height: 480
width: 640
distortion_model: plumb_bob
D: [0.0, 0.0, 0.0, 0.0, 0.0]
K: [1034.4730060050647, 0.0, 320.5, 0.0, 1034.4730060050647, 240.5, 0.0, 0.0, 1.0]
R: [1.0, 0.0, 0.0, 0.0, 1.0, 0.0, 0.0, 0.0, 1.0]
P: [1034.4730060050647, 0.0, 320.5, -0.0, 0.0, 1034.4730060050647, 240.5, 0.0, 0.0,
    0.0, 1.0, 0.0]
```

This camera data includes the width and height of the sensor (640 × 480), distortion co-efficients that are all zero, and a projection matrix that sets $f_x = f_y = 1034.473$ (focal length, in pixels, for a sensor with square pixels). The focal length and horizontal width are consistent with the specified horizontal field of view (set to 0.6 rad in the model). This can be confirmed by:

$$\tan(\theta_{hfov}/2) = \tan(0.3) = 0.309336 = (NCOLS/2)/f_x = 320/1034.473 \qquad (6.3)$$

The projection matrix, P, also specifies $(u_c, v_c) = (320.5, 240.5)$, which is close to half the width and half the height of the image array. (It is surprising that these values are not $(319.5, 239.5)$, and this is possibly a Gazebo plug-in error.) The `camera_info` values reported on topic `/simple_camera/camera_info` are consistent with the model specifications for our emulated camera. Although these values are known (by construction), one can nonetheless perform virtual calibration of the camera emulating the same procedure used with physical cameras. The vision target used for this calibration is a model of a checkerboard.

The `small_checkerboard` model is located in package `exmpl_models` under subdirectory `small_checkerboard`. The checkerboard model consists of seven rows and eight columns of 1 cm squares, alternating black and white. These squares create six rows and seven columns of internal four-corner intersections, which are used as precise reference points in the calibration process. After launching `simple_camera_simu_w_checkerboard.launch`, the emulated camera images can be viewed by running:

```
rosrun image_view image_view image:=/simple_camera/image_raw
```

The node `image_view` from the `image_view` package subscribes to image messages on the topic `image` and displays received images in its own display window. To get this node to subscribe to our images on topic `/simple_camera/image_raw`, the image topic is re-mapped on the command line with the option `image:=/simple_camera/image_raw`.

With these processes running, the screen appears as in Fig 6.2. This figure shows the crude rectangular prism representing the camera body, with optical axis pointing down towards the ground plane. The checkerboard is hovering 0.2 m above the ground plane, approximately centered in the camera's field of view. A light source in the Gazebo simulation casts shadows of both the camera body and the checkerboard onto the ground plane. The display from `image_view` shows that the camera sees the checkerboard, approximately centered in the camera's field of view. Part of the shadow cast by the checkerboard is also visible in the synthetic image computed for the virtual camera.

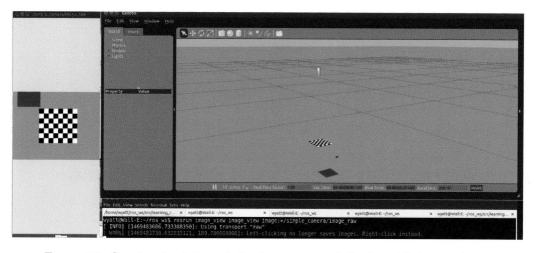

Figure 6.2: Gazebo simulation of simple camera and display with image viewer

To simulate waiving the vision target (checkerboard) in front of (below) the camera, a node is used to command Gazebo to move this model. Moving the checkerboard uses the same Gazebo interface introduced in Section 3.4. The code for moving the checkerboard is in the src directory of the example_camera_calibration package, source file name move_calibration_checkerboard.cpp. This routine generates random poses for the checkerboard, approximately constrained to remain within the camera's field of view. The checkerboard is moved vertically, horizontally, and tilted at arbitrary skew angles. The checkerboard poses are held for 0.5 sec each. Random motion generation of the checkerboard is initiated with:

```
rosrun example_camera_calibration move_calibration_checkerboard
```

The ROS camera-calibration tool can then be launched with:

```
rosrun camera_calibration cameracalibrator.py --size 7x6 --square 0.01 image:=/↩
    simple_camera/image_raw camera:=/simple_camera
```

This brings up an interactive display, as shown in Fig 6.3. The options to the camera calibrator node specify that there should be 7 × 6 internal corners in the image target, that each square of the target is 1 cm × 1 cm, that the image topic to which our camera publishes is /simple_camera/image_raw, and that the name of our camera is simple_camera. As this routine runs, it acquires snapshots of the vision target at different poses. It applies image processing to each such snapshot to identify the 42 internal corners. It displays acquired images with a graphical overlay illustrating identification of the internal corners (as seen in the colored lines and circles). As the calibration node runs, it continues to acquire sample data and informs the operator of the status of its calibration set. The horizontal bars under X, Y, size and skew in the calibration viewer indicate the range of contributions of vision target samples. When these bars are all green (as in Fig 6.3), the calibrator is suggesting that the data acquired so far is sufficient to perform calibration. When this is the case, the circle labeled "calibrate" changes from gray to green, making it an active control. Once this is enabled, the operator can click on this button to initiate parameter identification. A search algorithm is then invoked to attempt to explain all of the acquired data (with associated internal corner points) in terms of intrinsic camera coordinates. (As a

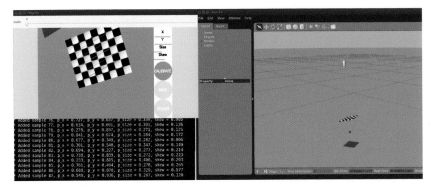

Figure 6.3: Camera calibration tool interacting with Gazebo model

by-product, the algorithm also computes an extrinsic pose estimate for each of the template snapshot poses.) Once the identification algorithm has converged, calibration results will be displayed and the Save and Commit buttons will be enabled.

Clicking Save and then clicking Commit results in camera parameters written to the directory .ros (a hidden directory under your home directory). Within the .ros directory, a subdirectory called camera_info will be created, which will contain a new file called simple_camera.yaml. For an example calibration process as above, the output yaml file contains the following:

```
image_width: 640
image_height: 480
camera_name: simple_camera
camera_matrix:
  rows: 3
  cols: 3
  data: [1035.427409570579, 0, 318.7566943335525, 0, 1035.332814088328,
         239.6921230165294, 0, 0, 1]
distortion_model: plumb_bob
distortion_coefficients:
  rows: 1
  cols: 5
  data: [-0.001604963198916835, -0.001878984799695939, 0.0001358780069028084,
         -0.0002804393593034872, 0]
rectification_matrix:
  rows: 3
  cols: 3
  data: [1, 0, 0, 0, 1, 0, 0, 0, 1]
projection_matrix:
  rows: 3
  cols: 4
  data: [1033.642944335938, 0, 318.1732831388726, 0, 0, 1033.126708984375,
         239.2189583120708, 0, 0, 0, 1, 0]
```

At this point, running rostopic echo /simple_camera/camera_info shows that the new camera parameters are being published. The camera still has identically 640 × 480 pixels (as it must), but there are now non-zero (though small) distortion parameters. These values are actually incorrect (since the simulated camera had zero distortion). However, these are the values found by the calibration algorithm. This shows that the calibration algorithm is only approximate. Better values might be obtained with more images acquired before initiating the parameter search.

Similarly, the values of f_x, f_y, u_c and v_c are slightly off. The focal-length values indicate nearly square pixels (1033.64 versus 1033.13), and these are close to the true model value (1034.47). The central coordinates of $(u_c, v_c) = (318.17, 239.22)$ are close to the model values of $(320.5, 240.5)$. These values, too, might be improved by acquiring and processing more calibration data.

This concludes the description of how to use the ROS camera calibration tool, and it provides some insight into the meaning of the camera parameters and expectations of the precision to which one can expect calibration to be achieved.

As noted earlier, calibration of a single camera is insufficient in itself to derive 3-D data from images. One must augment this information with additional assumptions or with additional sensors to infer 3-D coordinates from image data. Stereo vision is a common means to accomplish this. Calibration of a stereo vision system is described next.

6.3 INTRINSIC CALIBRATION OF STEREO CAMERAS

Stereo cameras offer the opportunity to infer 3-D from multiple images. A prerequisite is that the stereo cameras are calibrated. Commonly, stereo cameras will be mounted such that the transform between their respective optical frames is constant (*i.e.* a static transform). In this common situation, the coordinate transform from the left optical frame to the right optical frame is considered intrinsic, since this relationship should not change as the dual-camera system is moved in space. For efficiency of stereo processing, this transform is described in terms of two 3×3 rectification matrices and a single baseline dimension. These components must be identified in addition to the monocular-camera intrinsic parameters for each of the cameras.

The simple camera model is extended to dual (stereo) cameras in the model file `multi_camera_model.xacro` in package `simple_camera_model`. This model file includes two cameras, both at elevation 0.5 m. To assist with visualization, this model file specifies that the left camera body will be colored red in Gazebo. (Specification of colors in `rviz`, unfortunately, is not consistent with color specifications in Gazebo.) In this model, the left camera has its optical axis pointing anti-parallel to (and colinear with) the world z axis, and the image-plane x axis is oriented parallel to the world y axis. The right camera is offset by 0.02 m relative to the left camera. This offset is along the positive world y axis, which is also 0.02 m along the left-camera image-plane positive x axis. These frames are illustrated in Fig 6.4, which shows the left-camera optical frame, the world frame, and the locations of the left and right camera bodies. The model file includes the Gazebo elements shown below:

```
<!-- start of stereo camera plug-in -->
  <gazebo reference="left_camera_ref_frame">
  <!--gazebo reference="left_camera_optical_frame"-->
    <sensor type="multicamera" name="stereo_camera">
      <update_rate>30.0</update_rate>
      <camera name="left">
        <horizontal_fov>0.6</horizontal_fov>
        <image>
          <width>640</width>
          <height>480</height>
          <format>R8G8B8</format>
        </image>
         <clip>
          <near>0.005</near>
```

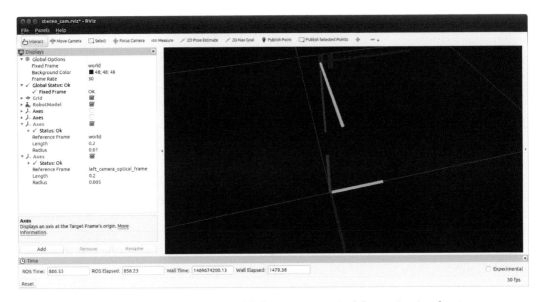

Figure 6.4: Rviz view of world frame and left-camera optical frame in simple stereo camera model

```
      <far>0.9</far>
  </clip>
  <noise>
    <type>gaussian</type>
    <mean>0.0</mean>
    <stddev>0.0</stddev>
  </noise>
</camera>
<camera name="right">
  <pose>0 -0.02  0 0 0 0</pose>
  <horizontal_fov>0.6</horizontal_fov>
  <image>
    <width>640</width>
    <height>480</height>
    <format>R8G8B8</format>
  </image>
  <clip>
    <near>0.005</near>
    <far>0.9</far>
  </clip>
  <noise>
    <type>gaussian</type>
    <mean>0.0</mean>
    <stddev>0.0</stddev>
  </noise>
</camera>
<plugin name="stereo_camera_controller" filename="libgazebo_ros_multicamera.so">
  <alwaysOn>true</alwaysOn>
  <updateRate>0.0</updateRate>
  <cameraName>stereo_camera</cameraName>
  <imageTopicName>image_raw</imageTopicName>
```

```
          <cameraInfoTopicName>camera_info</cameraInfoTopicName>
          <frameName>left_camera_optical_frame</frameName>
          <!--<rightFrameName>right_camera_optical_frame</rightFrameName>-->
          <hackBaseline>0.02</hackBaseline>
          <distortionK1>0.0</distortionK1>
          <distortionK2>0.0</distortionK2>
          <distortionK3>0.0</distortionK3>
          <distortionT1>0.0</distortionT1>
          <distortionT2>0.0</distortionT2>
      </plugin>
    </sensor>
  </gazebo>
  <!-- end of stereo plug-in -->
```

The camera specifications, right and left, are identical to the specifications used in the prior simple camera model, with zero distortion, zero noise, 640 × 480 pixels, and 0.6 rad horizontal field of view. Importantly, instead of specifying:

```
<plugin name="camera_controller" filename="libgazebo_ros_camera.so">
```

the stereo model specifies:

```
<plugin name="stereo_camera_controller" filename="libgazebo_ros_multicamera.so">
```

The reference frame for this system is the `left_camera_optical_frame`. A 0.02 m baseline is specified, meaning that the right camera is offset from the left camera by a distance of 0.02 m along the positive x axis of the left camera optical frame. [1]

The stereo camera model can be launched with:

```
roslaunch simple_camera_model multicam_simu.launch
```

Note that this launch file also starts the `robot_state_publisher` node, which publishes the model transforms, and thus enables `rviz` to display frames of interest.

With this model running in Gazebo, there are twice as many camera topics (for left and right cameras). Running:

```
rostopic echo /stereo_camera/right/camera_info
```

displays (in part):

```
  frame_id: left_camera_optical_frame
height: 480
width: 640
distortion_model: plumb_bob
D: [0.0, 0.0, 0.0, 0.0, 0.0]
K: [1034.4730060050647, 0.0, 320.5, 0.0, 1034.4730060050647, 240.5, 0.0, 0.0, 1.0]
R: [1.0, 0.0, 0.0, 0.0, 1.0, 0.0, 0.0, 0.0, 1.0]
P: [1034.4730060050647, 0.0, 320.5, -20.689460120101295, 0.0, 1034.4730060050647,
    240.5, 0.0, 0.0, 0.0, 1.0, 0.0]
```

---

[1]Specification of frames for use with the "multicamera" plug-in is non-intuitive–presumably a bug in this plug-in. Some contortions with intermediate frames were required to achieve the desired camera poses and image-plane orientations.

As with the monocular camera, the focal length (in pixels) is 1034.47, and $(u_c, v_c) = (320.5, 240.5)$. The K matrix and the R matrix are identical for the left and right cameras, and these are identical to the previous monocular camera example. The only difference is that the projection matrix (P matrix) of the right camera has a translation term (first row, third column) that is non-zero (specifically, -20.69). This translation component is $f_x b = 1034.473 \times 0.02$, where b is the baseline displacement of the right camera with respect to the left camera.

Performing stereo camera calibration incorporates the elements of monocular camera calibration to identify the intrinsic camera properties. In addition, stereo calibration finds rectification transforms for the two cameras. The intent of rectification is to transform the left and right images into corresponding virtual cameras that are perfectly aligned, *i.e.* with image planes that are coplanar, optical axes that are parallel, and x axes that are parallel and colinear. Under these conditions, the following highly useful property is obtained: for any point M in the environment that projects onto the left and right (virtual) image planes, the corresponding left and right y values (and v values) of the projected points will be identical in the left and right images. This property dramatically simplifies addressing the correspondence problem, in which one needs to identify which pixels in the left image correspond to which pixels in the right image, which is a requirement for inferring 3-D coordinates via triangulation.

To perform stereo camera calibration, the ROS node `cameracalibrator.py` can again be used, but with more parameters specified. This process is detailed on-line at `http://wiki.ros.org/camera_calibration` and illustrated at `http://wiki.ros.org/camera_calibration/Tutorials/StereoCalibration`. For our simulated stereo cameras, we can again use the model checkerboard and random pose generation, as follows. First, launch the multi-camera simulation:

```
roslaunch simple_camera_model multicam_simu.launch
```

Next, insert the checkerboard template with the command:

```
roslaunch simple_camera_model add_checkerboard.launch
```

Move the checkerboard in Gazebo to (constrained) random poses with:

```
rosrun example_camera_calibration move_calibration_checkerboard
```

The calibration tool can be started with:

```
rosrun camera_calibration cameracalibrator.py --size 7x6 --square 0.010 right:=/↵
    stereo_camera/right/image_raw left:=/stereo_camera/left/image_raw right_camera:=/↵
    stereo_camera/right left_camera:=/stereo_camera/left --fix-principal-point --fix-↵
    aspect-ratio --zero-tangent-dist
```

In this case, there are quite a few additional options and specifications. The checkerboard is again described as having 7×6 internal intersections among 0.010 m squares. The image topics `right` and `left` are remapped to match the topic names corresponding to the right and left cameras of our simulated system. The option `fix-aspect-ratio` enforces that the left and right cameras must have $f_x = f_y$, which enforces that the pixels are square. This constraint reduces the number of unknowns in the parameter search, which can improve the results (if it is known that the pixels are, in fact, square). The option `zero-tangent-dist` enforces that the camera model must assume zero tangential distortion, *i.e.* that the p_1 and p_2 coefficients are coerced to be identically zero; this removes these degrees of freedom from the calibration parameter search. The option `fix-principal-point` coerces the values of (u_c, v_c) to be in the middle of the image plane (which is true for our simple camera model, but not generally in practice). With these simplifications, the calibration process (parameter optimization) is faster and more reliable (provided the assumptions imposed are valid).

Figure 6.5 shows a screenshot during the calibration process. The camera bodies are rendered in Gazebo side by side and facing down, viewing the movable checkerboard. The graphical display for the calibration tool shows the left and right images, as well as the identification of internal corners on both images. As with monocular calibration, the graphical display shows when the data acquisition is determined adequate to perform valid calibration. Once

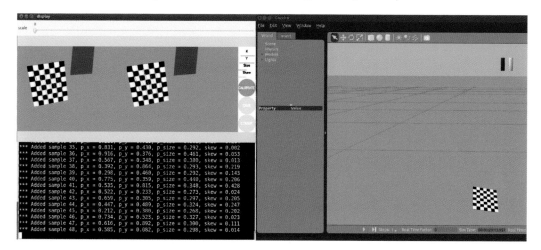

Figure 6.5: Screenshot during stereo camera calibration process, using simulated cameras

data acquisition is sufficient for calibration, the `calibrate` button changes from to green. Clicking on this button initiates the algorithmic search for optimal calibration parameters. This can take several minutes (although it is much faster with the parameter-constraints imposed in this example). After completion, the `save` and `commit` buttons are enabled, and clicking these will store the resulting calibration in `~/.ros/camera_info/stereo_camera` in files `left.yaml` and `right.yaml`. Results for an example calibration are:

```
width
640

height
480

[narrow_stereo/left]

camera matrix
1033.946820 0.000000 319.500000
0.000000 1033.946820 239.500000
0.000000 0.000000 1.000000

distortion
-0.000383 -0.006071 0.000000 0.000000 0.000000

rectification
0.999983 -0.000133 -0.005796
```

```
0.000133 1.000000 -0.000004
0.005796 0.000004 0.999983


projection
1036.332810 0.000000 326.076416 0.000000
0.000000 1036.332810 239.500460 0.000000
0.000000 0.000000 1.000000 0.000000


# oST version 5.0 parameters


[image]

width
640

height
480

[narrow_stereo/right]

camera matrix
1034.376433 0.000000 319.500000
0.000000 1034.376433 239.500000
0.000000 0.000000 1.000000

distortion
-0.001635 0.006670 0.000000 0.000000 0.000000

rectification
0.999983 -0.000136 -0.005817
0.000136 1.000000 0.000004
0.005817 -0.000005 0.999983

projection
1036.332810 0.000000 326.076416 -20.733354
0.000000 1036.332810 239.500460 0.000000
0.000000 0.000000 1.000000 0.000000
```

Rectification for the left camera is essentially a 3×3 identity matrix, meaning this camera is already virtually perfectly aligned in the epipolar frame. The right-camera rectification rotation matrix is also essentially the identity, since the Gazebo model specified the right camera to be parallel to the left camera (and thus no rotation correction is necessary). The camera matrix contains intrinsic parameters of each camera. Note that, for both cameras, $(u_c, v_c) = (319.5, 239.5)$, which is the middle of the sensor plane (as mandated by the option `fix-principal-point`). Also, $f_x = f_y$ for both left and right cameras, which was enforced as an option. However, $f_x = 1033.95$ for the left camera, and $f_x = 1036.33$ for the right camera. These should have been identical at $f_x = 1034.47$. This shows that the calibration algorithm (which performs a non-linear search in parameter space) is not perfect (although the identified values may well be adequately precise).

The left-camera projection matrix has a fourth column of all zeros. The right-camera projection matrix has $P(1, 4) = -20.733354$. This value is close to the ideal model value of $f_x * b = 20.69$.

In the present (simulated) case, the camera calibration parameters are, in fact, known identically, since these are specified in the Gazebo model. When the Gazebo simulation is restarted, the simulated cameras will use the model-specified parameters. More generally, calibration of physical cameras is a necessary step. After performing calibration (*e.g.* using the described checkerboard method), the identified calibration parameters will be stored in the `.ros` directory, and when the camera drivers are subsequently launched, these stored calibration values will be read from disk and published on the corresponding `camera_info` topic.

An alternative to the `multicamera` plug-in is the model `stereo_cam.xacro` in package `simple_camera_model`. In this model file, two individual cameras are defined, where each camera is specified as in our monocular-camera example. This alternative model can be launched with:

```
roslaunch simple_camera_model stereo_cam_simu.launch
```

This will result in separate camera publications (left and right). This model file has more intuitive frame specifications than the `multicamera` plug-in. A complication of modeling separate cameras, however, is that the image publications will have different time stamps. In performing stereo-vision analysis, right and left images must have adequate temporal correspondence to interpret dynamic data. That is, if the cameras are moving or if objects in the environment are moving, triangulation is only valid when operating on left and right images that are acquired synchronously. (For static scenes with static cameras, synchronization is not a concern.)

Defining separate camera drivers in the model file results in asynchronous image publications. As a result, the camera-calibration node and subsequent stereo image processing node will not accept the asynchronous images. If it is known that the image analysis will be performed with stationary cameras viewing stationary objects (*e.g.* interpreting objects on a table after a robot has approached the area), it is justifiable to modify the image topics to spoof synchronism. This can be done, for example, by running:

```
roslaunch simple_camera_model stereo_cam_simu.launch
```

This node simply subscribes to the raw left and right image topics, `/unsynced/left/image_raw` and `/unsynced/right/image_raw`, re-assigns the time stamp in the respective message headers as:

```
ros::Time tnow= ros::Time::now();
img_left_.header.stamp = tnow; // reset the time stamps to be identical
img_right_.header.stamp = tnow;
```

then re-publishes the images on new topics, `/stereo_sync/left/image_raw` and `/stereo_sync/right/image_raw`. The input and output topics can be re-mapped at run time to match the topic names of the actual cameras and the desired topic names for subsequent processing. With the `stereo_sync` node running, pairs of recently received left and right images will produce output image topics with identical time stamps. This will enable use of stereo calibration and stereo processing nodes.

6.4 USING OPENCV WITH ROS

OpenCV (see `http://opencv.org/`)is an open-source library of computer vision functions. It was initiated by Intel in 1999, and it has become a world-wide popular resource for computer vision programming. To leverage the capabilities of OpenCV, ROS includes bridge capabilities to translate between ROS-style messages and OpenCV-style objects (see `http://wiki.ros.org/vision_opencv`).

In addition to accommodating OpenCV classes, ROS also has customized publish and subscribe functions for working with images. The `image_transport` framework in ROS comprises both classes and nodes for managing transmission of images (see `http://wiki.ros.org/image_transport`). Since images can be very taxing on communications bandwidth, it is important to limit network burden as much as possible. There are publish and subscribe classes within `image_transport` that are used essentially identical to ROS-library publish and subscribe equivalents. The behaviors of the `image_transport` versions are different in a couple of ways. First, if there are no active subscribers to an `image_transport` publisher, images on this topic will not be published. This automatically saves bandwidth by declining to publish large messages that are not used. Second, `image_transport` publishers and subscribers can automatically perform compression and decompression (in various formats) to limit bandwidth consumption. Invoking such compression merely requires subscribing to the corresponding published topic for the chosen type of compression, thus making this process very simple for the user.

In this section, simple use of OpenCV with ROS is illustrated. This information is not intended to be a substitute for learning OpenCV; it is merely a demonstration of how OpenCV can be used with ROS.

6.4.1   Example OpenCV: finding colored pixels

The file `find_red_pixels.cpp` in the package `example_opencv` of the accompanying repository attempts to find red pixels in an image. The package `example_opencv` was created using:

```
catkin_simple example_opencv roscpp image_transport cv_bridge sensor_msgs
```

In this package, `image_transport` is used to take advantage of the efficient publish and subscribe functions customized for images. The `cv_bridge` library enables conversions between OpenCV datatypes and ROS messages. The `sensor_msgs` package describes the format of images published on ROS topics.

The `CMakeLists.txt` file auto-generated using `cs_create_pkg` contains a comment describing how to enable linking with OpenCV:

```
#uncomment next line to use OpenCV library
find_package(OpenCV REQUIRED)
```

By uncommenting `find_package(OpenCV REQUIRED)` (as above), your source code compilation within this package will be linked with the OpenCV2 library (which is the supported version for ROS Indigo and Jade).

The example source code `find_red_pixels.cpp` (in the `/src` sub-directory) computes two output images. One of these is a black and white image for which all pixels are black except for the locations of pixels in the input image that are considered to be sufficiently red (and these pixels are set to white in the output image). The second output image is a copy of the color input image, but with superimposed graphics corresponding to a small blue block displayed at the centroid of the red pixels. The code is shown in Listings 6.2 through 6.4.

Listing 6.2: `find_red_pixels.cpp`: C++ code to find centroid of red pixels, class definition

```cpp
1   //get images from topic "simple_camera/image_raw"; remap, as desired;
2   //search for red pixels;
3   // convert (sufficiently) red pixels to white, all other pixels black
4   // compute centroid of red pixels and display as a blue square
5   // publish result of processed image on topic "/image_converter/output_video"
6   #include <ros/ros.h>
7   #include <image_transport/image_transport.h>
8   #include <cv_bridge/cv_bridge.h>
9   #include <sensor_msgs/image_encodings.h>
10  #include <opencv2/imgproc/imgproc.hpp>
11  #include <opencv2/highgui/highgui.hpp>
12
13  static const std::string OPENCV_WINDOW = "Open-CV display window";
14  using namespace std;
15
16  int g_redratio; //threshold to decide if a pixel qualifies as dominantly "red"
17
18  class ImageConverter {
19      ros::NodeHandle nh_;
20      image_transport::ImageTransport it_;
21      image_transport::Subscriber image_sub_;
22      image_transport::Publisher image_pub_;
23
24  public:
25
26      ImageConverter(ros::NodeHandle &nodehandle)
27      : it_(nh_) {
28          // Subscribe to input video feed and publish output video feed
29          image_sub_ = it_.subscribe("simple_camera/image_raw", 1,
30                  &ImageConverter::imageCb, this);
31          image_pub_ = it_.advertise("/image_converter/output_video", 1);
32
33          cv::namedWindow(OPENCV_WINDOW);
34      }
35
36      ~ImageConverter() {
37          cv::destroyWindow(OPENCV_WINDOW);
38      }
39
40      //image comes in as a ROS message, but gets converted to an OpenCV type
41      void imageCb(const sensor_msgs::ImageConstPtr& msg);
42
43  }; //end of class definition
```

Listing 6.3: `find_red_pixels.cpp`: C++ code to find centroid of red pixels, method implementationlanguage

```cpp
45  void ImageConverter::imageCb(const sensor_msgs::ImageConstPtr& msg){
46      cv_bridge::CvImagePtr cv_ptr; //OpenCV data type
47      try {
48          cv_ptr = cv_bridge::toCvCopy(msg, sensor_msgs::image_encodings::BGR8);
49      } catch (cv_bridge::Exception& e) {
50          ROS_ERROR("cv_bridge exception: %s", e.what());
51          return;
52      }
53      // look for red pixels; turn all other pixels black, and turn red pixels white
54      int npix = 0; //count the red pixels
55      int isum = 0; //accumulate the column values of red pixels
56      int jsum = 0; //accumulate the row values of red pixels
57      int redval, blueval, greenval, testval;
58      cv::Vec3b rgbpix; // OpenCV representation of an RGB pixel
59      //comb through all pixels (j,i)= (row,col)
60      for (int i = 0; i < cv_ptr->image.cols; i++) {
61          for (int j = 0; j < cv_ptr->image.rows; j++) {
62              rgbpix = cv_ptr->image.at<cv::Vec3b>(j, i); //extract an RGB pixel
63              //examine intensity of R, G and B components (0 to 255)
64              redval = rgbpix[2] + 1; //add 1, to avoid divide by zero
65              blueval = rgbpix[0] + 1;
66              greenval = rgbpix[1] + 1;
67              //look for red values that are large compared to blue+green
68              testval = redval / (blueval + greenval);
69              //if red (enough), paint this white:
70              if (testval > g_redratio) {
71                  cv_ptr->image.at<cv::Vec3b>(j, i)[0] = 255;
72                  cv_ptr->image.at<cv::Vec3b>(j, i)[1] = 255;
73                  cv_ptr->image.at<cv::Vec3b>(j, i)[2] = 255;
74                  npix++; //note that found another red pixel
75                  isum += i; //accumulate row and col index vals
76                  jsum += j;
77              } else { //else paint it black
78                  cv_ptr->image.at<cv::Vec3b>(j, i)[0] = 0;
79                  cv_ptr->image.at<cv::Vec3b>(j, i)[1] = 0;
80                  cv_ptr->image.at<cv::Vec3b>(j, i)[2] = 0;
81              }
82          }
83      }
84      //cout << "npix: " << npix << endl;
85      //paint in a blue square at the centroid:
86      int half_box = 5; // choose size of box to paint
87      int i_centroid, j_centroid;
88      double x_centroid, y_centroid;
89      if (npix > 0) {
90          i_centroid = isum / npix; // average value of u component of red pixels
91          j_centroid = jsum / npix; // avg v component
92          x_centroid = ((double) isum)/((double) npix); //floating-pt version
93          y_centroid = ((double) jsum)/((double) npix);
94          ROS_INFO("u_avg: %f; v_avg: %f",x_centroid,y_centroid);
95          //cout << "i_avg: " << i_centroid << endl; //i,j centroid of red pixels
96          //cout << "j_avg: " << j_centroid << endl;
97          for (int i_box = i_centroid - half_box; i_box <= i_centroid + half_box; ↵
                 i_box++) {
98              for (int j_box = j_centroid - half_box; j_box <= j_centroid + half_box↵
                     ; j_box++) {
99                  //make sure indices fit within the image
100                 if ((i_box >= 0)&&(j_box >= 0)&&(i_box < cv_ptr->image.cols)&&(↵
                        j_box < cv_ptr->image.rows)) {
101                     cv_ptr->image.at<cv::Vec3b>(j_box, i_box)[0] = 255; //↵
                            (255,0,0) is pure blue
102                     cv_ptr->image.at<cv::Vec3b>(j_box, i_box)[1] = 0;
103                     cv_ptr->image.at<cv::Vec3b>(j_box, i_box)[2] = 0;
104                 }
105             }
106         }
107
108     }
109     // Update GUI Window; this will display processed images on the open-cv viewer↵

110     cv::imshow(OPENCV_WINDOW, cv_ptr->image);
```

```
111        cv::waitKey(3); //need waitKey call to update OpenCV image window
112
113        // Also, publish the processed image as a ROS message on a ROS topic
114        // can view this stream in ROS with:
115        //rosrun image_view image_view image:=/image_converter/output_video
116        image_pub_.publish(cv_ptr->toImageMsg());
117      }
```

Listing 6.4: `find_red_pixels.cpp`: C++ code to find centroid of red pixels, main program

```
119  int main(int argc, char** argv) {
120      ros::init(argc, argv, "red_pixel_finder");
121      ros::NodeHandle n; //
122      ImageConverter ic(n); // instantiate object of class ImageConverter
123      //cout << "enter red ratio threshold: (e.g. 10) ";
124      //cin >> g_redratio;
125      g_redratio= 10; //choose a threshold to define what is "red" enough
126      ros::Duration timer(0.1);
127      double x, y, z;
128      while (ros::ok()) {
129          ros::spinOnce();
130          timer.sleep();
131      }
132      return 0;
133  }
```

In `find_red_pixels.cpp`, lines 6 through 11 bring in the necessary headers for ROS, OpenCV, sensor messages, and ROS-OpenCV conversions. The main program (lines 119 through 133) merely instantiates an object of class `ImageConverter`, sets a global threshold parameter, then goes into a timed loop with spins. All of the computation in this node is done by callbacks defined within the `ImageConverter` class.

The `ImageConverter` class is defined in lines 18 through 43. This class has member variables (lines 20 through 22) that include a publisher and subscriber as defined in the `image_transport` library, as well as an instantiation of an `ImageTransport` object. In the constructor (lines 26 through 34), the image publisher is set to publish to the topic `image_converter/output_video`, and the subscriber is set to invoke the callback function `imageCb` upon receipt of messages on topic `simple_camera/image_raw`.

The actual work is performed in the body of the callback function, lines 44 through 117. Line 46:

```
cv_bridge::CvImagePtr cv_ptr; //OpenCV data type
```

defines a pointer to an image that is compatible with OpenCV. Data at this pointer is populated by translating the input ROS message `msg` into a compatible OpenCV image matrix in the format of an RGB image, with each color encoded as eight bits, using the code:

```
cv_ptr = cv_bridge::toCvCopy(msg, sensor_msgs::image_encodings::BGR8);
```

Lines 60 through 83 correspond to a nested loop through all rows and columns of the image, examining RGB contents of each pixel with a line of the form:

```
rgbpix = cv_ptr->image.at<cv::Vec3b>(j, i); //extract an RGB pixel
```

The redness of a pixel is chosen to be computed relative to the green and blue values, as evaluated by lines 64 through 66:

```
testval = redval / (blueval + greenval);
            //if red (enough), paint this white:
            if (testval > g_redratio) {
```

If the red component is at least 10 times stronger than the combined green and blue components, the pixel is judged to be sufficiently red. (This is an arbitrary metric and threshold; alternative metrics may be used.)

Lines 71 through 73 show how to set the RGB components of a pixel to saturation, corresponding to a maximally bright white pixel. Lines 78 through 80, conversely, set all RGB components to 0 to specify a black pixel. While combing through all pixels of the copied input image, lines 74 through 76 keep track of the total number of pixels judged to be sufficiently red, while summing the respective row and column indices. After evaluating all pixels in the image, the centroid of the red pixels is computed (lines 89 through 93).

Lines 97 through 106 alter the B/W image to display a square of blue pixels at the computed red-pixel centroid. This processed image is displayed in a native OpenCV image-display window using lines 110 and 111. The same image is published on a ROS topic using line 116:

```
image_pub_.publish(cv_ptr->toImageMsg());
```

The ROS-published image can be viewed by running:

```
rosrun image_view image_view image:=/image_converter/output_video
```

The example code can be used with any image stream. For illustration, we can re-use our stereo camera model by running:

```
roslaunch simple_camera_model multicam_simu.launch
```

Next, add a small red block for the cameras to observe with:

```
roslaunch exmpl_models add_small_red_block.launch
```

This places a small, red block at $(0, 0, 0.1)$ with respect to the left stereo camera's frame. The find_red_pixels code example can now be run using the following launch file:

```
roslaunch example_opencv find_red_pixels_left_cam.launch
```

This launch file starts the image_view node, re-mapping its input topic to the left-camera topic /stereo_camera/left/image_raw. This brings up a display window that shows the raw image from the left stereo camera. Additionally, this launch file starts up our example node find_red_pixels from package example_opencv and re-maps the input topic from simple_camera/image_raw to /stereo_camera/left/image_raw. The result shown in Fig 6.6 shows the Gazebo view with stereo cameras and a small red block centered below the left camera. The image_view node shows a display of the raw video from the left camera, which shows a red rectangle centered in the field of view. The OpenCV display window shows the result of the image processing. There is a white rectangle centered in a black background, and the centroid of this white rectangle is marked with a blue square, as expected. The terminal prints the coordinates of this centroid (319.5, 239.5), which is in the middle of the field of view of the 640×480 virtual image sensor.

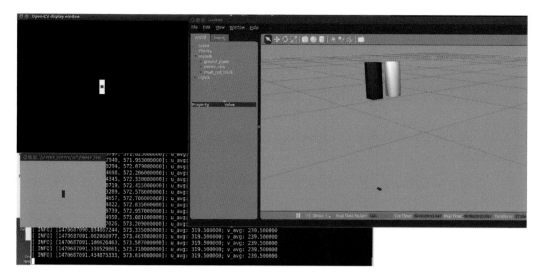

Figure 6.6: Result of running `find_red_pixels` on left-camera view of red block

The image displayed in the OpenCV display window is also published on the ROS topic `/image_converter/output_video`. This can be verified by running another instance of `image_view` with input topic remapped to `/image_converter/output_video`. Publishing intermediate steps of image processing can be useful for visualizing the behavior of an image pipeline to aid tuning and debugging.

In the example code, `ros::spinOnce()` is executed by the main program at 10 Hz. The input video rate, however, is 30 Hz. As a result, the example code only operates on every third frame. This can be a handy technique for reducing computational demand. A simpler and more effective means of reducing both bandwidth and CPU burden is throttling, which can be invoked using `rqt_reconfigure`. Within this GUI, under the topic `stereo_camera/left`, the GUI displays a slider for `imager_rate`. If the value is reduced to 1.0, frames will be published only at 1.0 Hz. Using this GUI for throttling image rates can be convenient for interactively evaluating a good compromise for image rates.

6.4.2 Example OpenCV: finding edges

The previous example performed operations on individual pixels. More commonly with OpenCV, one uses existing higher-level functions. As an example, consider the source code `find_features.cpp`, which finds edges in images using a Canny filter. This example code follows the `find_red_pixels` code, with the addition of the following lines inserted into function `imageCb()` of `find_red_pixels`.

```
cv::Mat gray_image,contours;   //two new image holders
//convert the color image to grayscale:
cv::cvtColor(cv_ptr->image, gray_image, CV_BGR2GRAY);
//use Canny filter to find edges in grayscale image;
//put result in "contours"; low and high thresh are tunable params
cv::Canny(gray_image,// gray-level image
        contours, // output contours
        125,// low threshold
        350);// high threshold
        cv::imshow(OPENCV_WINDOW, contours); //display the contours
        cv::waitKey(3); //need waitKey call to update OpenCV image window
```

This code instantiates two OpenCV image matrices, `gray_image` and `contours`. The code line:

```
cv::cvtColor(cv_ptr->image, gray_image, CV_BGR2GRAY);
```

converts the received RGB image into a gray-scale image. This is necessary because edge detection on color images is not well defined. The gray-scale image is then used to compute an output image called `contours` using the function `cv::Canny()`. This function takes two tuning parameters, a low threshold and a high threshold, which can be adjusted to find strong lines while rejecting noise. For the example image, which is black or white, the Canny filter will not be sensitive to these values, though these would be important in more realistic scenes, such as finding lines on a highway. The contour image is displayed in the OpenCV display window. (Original line 110, which displayed the processed image with blue-square centroid, has been commented out.) This example can be run much as before, with:

```
roslaunch simple_camera_model multicam_simu.launch
```

Next, add a small red block for the cameras to observe with:

```
roslaunch exmpl_models add_small_red_block.launch
```

Instead of launching `find_red_pixels_left_cam.launch`, use:

```
roslaunch example_opencv find_features_left_cam.launch
```

This starts the node `find_features`, performs the necessary topic re-mapping and starts `image_view`. With these steps, the output appears as in Fig 6.7. As desired, the OpenCV image displays the detected edges, which correspond to the edges of the block.

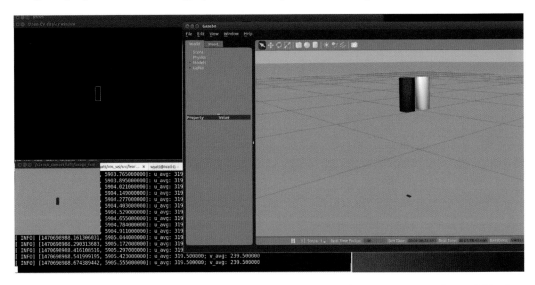

Figure 6.7: Result of running Canny edge detection on left-camera view of a red block

While this is only a simple example of OpenCV capabilities, it illustrates how images can be processed with both low and high-level functions and how OpenCV can be integrated with ROS. Many advanced, open-source ROS packages exist using OpenCV, such as camera calibration, visual odometry, lane finding, facial recognition, and object recognition. Existing non-ROS OpenCV solutions can be ported to ROS with relative ease using `cv_bridge`, and thus much prior effort can be leveraged with ROS.

6.5   WRAP-UP

This chapter introduced several fundamental concepts for using cameras in ROS. Conventions have been presented for assigning coordinate frames to cameras and reconciling these with sensor-plane pixel coordinates. A step toward getting physical coordinates from camera images is intrinsic camera calibration. ROS uses the conventions of OpenCV for representing intrinsic camera calibration. ROS packages support performing camera calibration, *e.g.* using checkerboard-style calibration templates. Results of intrinsic calibration are published on ROS `camera_info` topics.

Camera calibration support in ROS is extended to stereo cameras. The results include individual camera intrinsic parameters, rectification parameters and identification of a baseline between transformed (rectified) images. The value of calibrating stereo cameras is that it becomes possible to infer 3-D coordinates of points of interest using multiple camera views.

Use of OpenCV for low-level image processing was introduced very briefly. OpenCV has an extensive library of image-processing functions, and these can be used within ROS nodes by following the integration steps presented in this chapter. Use of OpenCV functions with ROS will be introduced further in Chapter 8 in the context of computing depth maps from stereo images.

Depth Imaging and Point Clouds

CONTENTS

I NTRODUCTION

Ultimately, a robot needs to understand 3-D information from its sensors. This is essential both for computing navigability (*e.g.* cliffs, ditches, doors, barriers) and manipulation (*e.g.* how to approach and grasp an object of interest). Some depth information can be inferred from 2-D cameras if *a priori* assumptions can be invoked. Preferably, sensors would provide unambiguous 3-D data. There are multiple means in common use for 3-D data acquisition. Three of these are discussed in this chapter: tilting LIDAR, stereo vision, and depth cameras.

7.1   DEPTH FROM SCANNING LIDAR

The most common LIDAR sensors yield measurements in polar coordinates along a single slice plane. Some (expensive) LIDARs contain multiple laser sources tilted at different angles, enabling 3-D data scans in spherical coordinates. A relatively low-cost way to get 3-D data scans from LIDAR is to mount the LIDAR on a movable joint and mechanically scan about an axis perpendicular to the LIDAR's spinning-mirror axis. One approach is to use slip rings for power and signal connections and rotate the LIDAR continuously, as with the Carnegie Robotics Multisense-SL sensor head[1] used on Boston Dynamics Atlas robots in the DARPA Robotics Challenge.[2] A Gazebo model of this device is available open source[3]. An alternative is to mount a single-plane LIDAR on a revolute joint (with joint limits) and oscillate the LIDAR back and forth. Such a tilting LIDAR design was integrated into the Willow Garage PR2 robot.[4]

A minimal model is `lidar_wobbler.urdf` in directory `model` of package `lidar_wobbler` (in Part3 of the accompanying repository). This model combines elements from Section 5.2.1 for LIDAR simulation in Gazebo and elements from Section 3.6 for ROS joint position control. The URDF is given in Listing 7.1.

---

[1] http://carnegierobotics.com/multisense-sl/
[2] http://www.darpa.mil/program/darpa-robotics-challenge
[3] http://gazebosim.org/tutorials?tut=drcsim_multisense&cat=
[4] http://www.willowgarage.com/pages/pr2/specs

Listing 7.1: `lidar_wobbler.urdf`: LIDAR wobbler URDF model

```
1   <?xml version="1.0"?>
2   <robot  name="lidar_wobbler">
3
4   <!-- Used for fixing robot to Gazebo 'base_link' -->
5     <link name="world"/>
6
7     <joint name="glue_frame_to_world" type="fixed">
8       <parent link="world"/>
9       <child link="link1"/>
10    </joint>
11
12   <!-- Base Link; no visual or collision, thus invisible in simulation -->
13     <link name="link1">
14
15       <inertial>
16         <origin xyz="0 0 0.5" rpy="0 0 0"/>
17         <mass value="1"/>
18         <inertia
19           ixx="1.0" ixy="0.0" ixz="0.0"
20           iyy="1.0" iyz="0.0"
21           izz="1.0"/>
22       </inertial>
23     </link>
24
25   <!-- Moveable Link -->
26   <!-- add a simulated lidar, including visual, collision and inertial properties, and ↩
         physics simulation-->
27     <link name="lidar_link">
28         <collision>
29             <origin xyz="0.5 0 0.5" rpy="0 0 0"/>
30             <geometry>
31                 <!-- coarse LIDAR model; a simple box -->
32                 <box size="0.2 0.2 0.2"/>
33             </geometry>
34         </collision>
35
36         <visual>
37             <origin xyz="0 0 0" rpy="0 0 0" />
38             <geometry>
39                 <box size="0.2 0.2 0.2" />
40             </geometry>
41             <material name="sick_grey">
42                 <color rgba="0.7 0.5 0.3 1.0"/>
43             </material>
44         </visual>
45
46         <inertial>
47             <mass value="4.0" />
48             <origin xyz="0 0 0" rpy="0 0 0"/>
49             <inertia ixx="0.01" ixy="0" ixz="0" iyy="0.01" iyz="0" izz="0.01" />
50         </inertial>
51     </link>
52     <!--the above displays a box meant to imply Lidar-->
53
54     <joint name="joint1" type="revolute">
55       <parent link="link1"/>
56       <child link="lidar_link"/>
57       <origin xyz="0 0 1" rpy="0 0 0"/>
58       <axis xyz="0 1 0"/>
59       <limit effort="10.0" lower="0.0" upper="6.28" velocity="0.5"/>
60       <dynamics damping="1.0"/>
61     </joint>
62
63     <transmission name="tran1">
64       <type>transmission_interface/SimpleTransmission</type>
65       <joint name="joint1">
66         <hardwareInterface>EffortJointInterface</hardwareInterface>
67       </joint>
68       <actuator name="motor1">
69         <hardwareInterface>EffortJointInterface</hardwareInterface>
70         <mechanicalReduction>1</mechanicalReduction>
71       </actuator>
72     </transmission>
```

```
73   <gazebo>
74      <plugin name="gazebo_ros_control" filename="libgazebo_ros_control.so">
75         <robotNamespace>/lidar_wobbler</robotNamespace>
76      </plugin>
77   </gazebo>
78
79   <!-- here is the gazebo plug-in to simulate a lidar sensor -->
80   <gazebo reference="lidar_link">
81      <sensor type="gpu_ray" name="sick_lidar_sensor">
82         <pose>0 0 0 0 0 0</pose>
83         <visualize>false</visualize>
84         <update_rate>40</update_rate>
85         <ray>
86            <scan>
87               <horizontal>
88                  <samples>181</samples>
89                  <resolution>1</resolution>
90                  <min_angle>-1.570796</min_angle>
91                  <max_angle>1.570796</max_angle>
92               </horizontal>
93            </scan>
94            <range>
95               <min>0.10</min>
96               <max>80.0</max>
97               <resolution>0.01</resolution>
98            </range>
99            <noise>
100              <type>gaussian</type>
101              <mean>0.0</mean>
102              <stddev>0.01</stddev>
103           </noise>
104        </ray>
105        <plugin name="gazebo_ros_lidar_controller" filename="libgazebo_ros_gpu_laser.so"↩
              >
106           <topicName>/scan</topicName>
107           <frameName>lidar_link</frameName>
108        </plugin>
109     </sensor>
110   </gazebo>
111
112 </robot>
```

In the `lidar_wobbler` model, the chosen model name is `lidar_wobbler`, which is also the chosen `namespace`. This `namespace` applies to joint states, model states and joint commands of the wobbler.

A single movable (revolute) joint is defined, which is capable of tilting the LIDAR about an axis perpendicular to the LIDAR's spinning-mirror axis (*i.e.* the tilt axis lies in the slice plane of the LIDAR). This joint can be controlled by a ROS position controller.

To control the single joint, a YAML file virtually identical to that in Section 3.6 is created in subdirectory `config` of the `lidar_wobbler` package. The contents appear below. Importantly, the `lidar_wobbler` name must match the name in the URDF model file.

```
lidar_wobbler:
  # Publish all joint states -----------------------------------
  joint_state_controller:
    type: joint_state_controller/JointStateController
    publish_rate: 50

  # Position Controllers ---------------------------------------
  joint1_position_controller:
    type: effort_controllers/JointPositionController
    joint: joint1
    pid: {p: 10.0, i: 0.0, d: 10.0, i_clamp_min: -10.0, i_clamp_max: 10.0}
```

Source file `wobbler_sine_commander.cpp` in the `/src` subdirectory of package `lidar_wobbler` commands joint values to topic `/lidar_wobbler/joint1_position_controller/command`.

The joint commands are updated at 100 Hz, computed as sinusoidal motion at an amplitude and frequency specified by the user upon start-up.

To run the wobbler example, first start Gazebo. Since the LIDAR plug-in library assumes use of a GPU, this may require using `optirun` (depending on your hardware).

```
(optirun) roslaunch gazebo_ros empty_world.launch
```

Next, run:

```
roslaunch lidar_wobbler lidar_wobbler.launch
```

The contents of this launch file appear in Listing 7.2

Listing 7.2: `lidar_wobbler.launch`: LIDAR wobbler launch file

```
1  <launch>
2    <!-- Load joint controller configurations from YAML file to parameter server -->
3    <rosparam file="$(find lidar_wobbler)/config/one_dof_ctl_params.yaml" command="load"←
         />
4    <param name="robot_description"
5       textfile="$(find lidar_wobbler)/model/lidar_wobbler.urdf"/>
6
7    <!-- Spawn a robot into Gazebo -->
8    <node name="spawn_urdf" pkg="gazebo_ros" type="spawn_model"
9       args="-param robot_description -urdf -model lidar_wobbler" />
10
11   <!--start up the controller plug-ins via the controller manager -->
12   <node name="controller_spawner" pkg="controller_manager" type="spawner" respawn="←
         false"
13      output="screen" ns="/lidar_wobbler" args="joint_state_controller  ←
            joint1_position_controller"/>
14
15  <!-- start a robot_state_publisher -->
16  <node name="robot_state_publisher" pkg="robot_state_publisher" type="←
         robot_state_publisher" >
17     <remap from="joint_states" to="/lidar_wobbler/joint_states" />
18  </node>
19
20  <!-- start a joint_state_publisher -->
21  <node name="joint_state_publisher" pkg="joint_state_publisher" type="←
         joint_state_publisher" >
22     <remap from="joint_states" to="/lidar_wobbler/joint_states" />
23  </node>
24  </launch>
```

This launch file performs six tasks. It loads the controller parameters onto the parameter server, then loads the robot model onto the parameter server. It spawns the robot model into Gazebo from the parameter server. The ROS joint controller for `joint1` is then loaded. In order for `rviz` to get robot states, a `joint_state_publisher` and a `robot_state_publisher` are started. In this case, both have their topics re-mapped to use the chosen name space `lidar_wobbler`.

With these nodes running, the LIDAR simulator in the robot model in Gazebo obtains range data from the environment. In an empty world, there are no interesting points to sample. From Gazebo, one can insert models to be scanned. In the present example, the starting pen and the Beer model are inserted (and located arbitrarily, interactively).

Beer is among the models on line at `http://gazebosim.org/models`. This model (authored by Maurice Fallon, per the configuration file) has a visual model specified as:

```
      <visual name='visual'>
        <geometry>
          <cylinder>
```

```
            <radius>0.055000</radius>
            <length>0.230000</length>
          </cylinder>
        </geometry>

        <material>
          <script>
            <uri>model://beer/materials/scripts</uri>
            <uri>model://beer/materials/textures</uri>
            <name>Beer/Diffuse</name>
          </script>
        </material>
      </visual>
      </material>
```

Note that the cylindrical model of the can specifies a radius of 0.055 m. In the example shown, the can is positioned to lie on its side, presenting a curved surface to the wobbler. The top-most surface of the can is thus at a height of 0.11 m.

To start the LIDAR oscillating, the sine commander is started with:

```
rosrun lidar_wobbler wobbler_sine_commander
```

For the present example, responses to the user prompts were at a frequency of 0.1 Hz and an amplitude of 1.57 rad.

Finally, `rviz` is started with:

```
rosrun rviz rviz
```

Within `rviz`, set the fixed frame to `world`, and add a `LaserScan` display with topic set to `/scan`. Set the Decay Time parameter within the LaserScan item to some value greater than 0 (e.g. 10 sec for the present example). The result of these operations is shown in Fig 7.1. Figure 7.1 shows that the tilting LIDAR is able to detect the ground plane and the

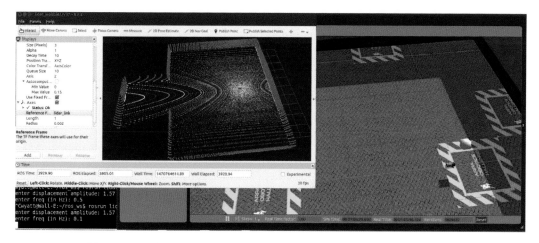

Figure 7.1: Gazebo and rviz views of LIDAR wobbler–wide view

walls of the starting pen (at least those parts with line-of-sight view from the LIDAR).

Figure 7.2 shows a view of the LIDAR data zoomed on the sideways-can model. The resolution of samples of the can's surface is relatively coarse. Nonetheless, one face of the can is discernible.

Using `rviz`, one can select a patch of points and publish their 3-D coordinates using

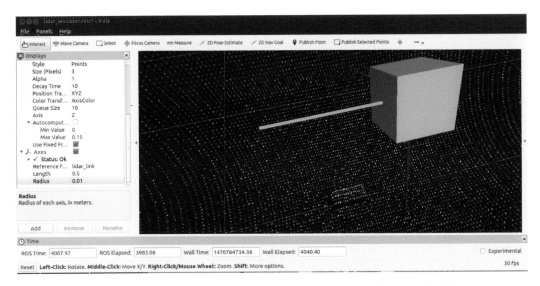

Figure 7.2: Gazebo and rviz views of LIDAR wobbler: zoomed view of sideways can on ground plane

the `PublishSelectedPoints` tool. The result of clicking and dragging on a patch in `rviz` is a publication of points of type `sensor_msgs/PointCloud2` on topic `selected_points`. A separate node to interpret these points can be run with:

```
rosrun pcl_utils compute_selected_points_centroid
```

Details of this node are deferred until Chapter 8, but the result of this node is to compute the centroid of points published on topic `selected_points`. With this node running, selecting a patch of points near the uppermost surface of the sideways can results in `compute_selected_points_centroid` displaying:

```
centroid of selected points is: (0.018509, 0.152455, 0.115234)
```

The z height computed by this node is consistent with the expected value of 0.110 m above the ground plane.

In this example, `rviz` was used to transform and accumulate the sample points. The LIDAR points, originating in polar coordinates with respect to a moving slice plane, are converted into Cartesian coordinates in the world frame, represented as a point cloud. By setting the persistence value in `rviz`, one can display the points acquired over many LIDAR scans. The result produces panoramic 3-D scans.

`rviz` is useful for visualization and for direct interaction with data. However, it is more common to operate on the data with perceptual-processing algorithms. To transform the LIDAR data into Cartesian-space point-cloud data, the `laser_assembler` node can be used (see `http://wiki.ros.org/laser_assembler`).

7.2 DEPTH FROM STEREO CAMERAS

Although a video camera does not provide depth information, depth can be inferred via triangulation from known calibration properties and presumed correspondence of pixels in multiple (typically, left and right) cameras.

The result of intrinsic stereo-camera calibration is a pair of `camera_info` publications, both left and right. Messages on these topics include parameters for de-warping images, rotation matrices for rectifying images for simple left–right image comparisons, and key intrinsic properties, including focal lengths, optical center, and baseline between the rectified images. With this information, incoming raw images (left and right) can be de-warped, rectified, and analyzed to perform triangulation on identified correspondences in the left and right images, from which 3-D coordinates can be computed.

Section 6.4.1 introduced OpenCV code for identifying red pixels and computing their centroid. This code can be used to illustrate triangulation. First, start the multi-camera model in Gazebo with:

```
roslaunch simple_camera_model multicam_simu.launch
```

Add the red-block target using:

```
roslaunch exmpl_models add_small_red_block.launch
```

The find-red-block code can be run separately for the left and right cameras by launching:

```
roslaunch example_opencv find_red_pixels_left_cam.launch
```

and:

```
roslaunch example_opencv find_red_pixels_right_cam.launch
```

These nodes display results for the left camera:

```
u_avg: 319.500000; v_avg: 239.500000
```

and for the right camera:

```
u_avg: 112.481611; v_avg: 239.500000
```

Note that the v coordinates are identical in the left and right cameras. This is expected for rectified cameras (and these cameras are rectified by construction in the multicamera model). The u coordinates differ by approximately 207 pixels. The difference in u values is called the disparity.

As presented in Section 6.3, each of our simulated stereo cameras has a focal length of 1034.47 pixels, and the baseline between the cameras (distance between the parallel optical axes) is 0.02 m. From the respective projection matrices, we can derive the z distance of the target centroid along the optical axes to be: $z = f_x * b/(\Delta u) = 1034.47 * 0.02/207 = 0.100$. From Gazebo, we can see that the block model is located at height 0.4 m above the ground plane. Since the cameras are located 0.5 m above the ground plane, the distance from the camera frame to the block is 0.1 m, which agrees with the computation.

Given the depth coordinate, z, the corresponding x and y values follow, thus yielding 3-D Cartesian coordinates from dual-camera triangulation. In the present case, $(u_{left}, v_{left}) = (319.5, 239.5)$, which is centered in the left-camera image, and thus the centroid coordinates are $(x, y, z) = (0, 0, 0.1)$ with respect to the left-camera optical frame.

This example illustrates how 3-D Cartesian coordinates can be inferred from dual camera views. The (u, v) values in this example in the left and right images were unambiguous, based on the centroids of distinctively red pixels. The same triangulation computation applies to more complex cases, provided one can solve the correspondence problem associating a pixel in the right image with a pixel in the left image.

A ROS node for this purpose is `stereo_image_proc`. This is a sophisticated node that performs a variety of functions, as documented at `http://wiki.ros.org/stereo_image_proc`. To illustrate use of this node, we can use our multicamera Gazebo model, starting it with:

```
roslaunch simple_camera_model multicam_simu.launch
```

At this point, `rostopic list` shows 26 `stereo_camera` topics. We will be interested in the subtopics `/left/camera_info`, `/left/image_raw`, `/right/camera_info` and `/right/image_raw`.

We can then start the `stereo_image_proc` node with:

```
ROS_NAMESPACE=stereo_camera rosrun stereo_image_proc stereo_image_proc
```

In the above command, specifying the name space `stereo_camera` simplifies topic remapping. The inputs to this node are the right and left `image_raw` and `camera_info` topics under the name space `stereo_camera`. The `stereo_image_proc` performs dewarping, rectification and disparity computations to produce a point-cloud output. Actually, with `stereo_image_proc` running, there are over 100 topics within the name space `stereo_image_proc`. Of the topics published by `stereo_image_proc`, we will be concerned with only four of them (all within the name space `stereo_camera`): `/left/image_rect_color`, `/right/image_rect_color`, `/disparity` and `points2`.

The node `stereo_image_proc` presents a large number of `stereo_camera` topics, and image topics can consume substantial communications bandwidth. Since this is a common problem, an underlying transport mechanism, `image_transport` (see `http://wiki.ros.org/image_transport`), manages bandwidth by restricting publication only to topics that have active subscribers. Since we will be using only four of the topics available from `stereo_image_proc`, only these topics will get active publications. Similarly, of the 26 image topics available from our Gazebo simulation of stereo cameras, only the four topics subscribed to by `image_transport` will be active. All unused topics will remain idle, not consuming bandwidth.

With the multicamera simulation running in Gazebo and the `stereo_image_proc` node running, we can bring in models for the stereo system to view. One of the constraints of stereo vision, though, is that objects to be viewed must have sufficient texture to infer depth. Inferring correspondence of pixels in right and left camera views can be performed by recognizing distinctive corresponding patterns in right and left views. However, if the object viewed is bland (*e.g.* a flat-gray ground plane), it is not possible to identify corresponding pixels in right and left views. If the object has sufficient texture, stereo can work well.

For an object with sufficient texture, we can again use the Beer model that refers to a texture file in `<uri>model://beer/materials/textures</uri>`, which contains an image (in PNG format) that gets wrapped onto the surface of the cylinder thus creating interesting and useful high-resolution texture, suitable for stereo imaging.

After inserting the model, then rotating and translating it to present a view of a curved face, the Gazebo view appears as in Fig 7.3. (Note: some $-z$ gravity was imposed to assure the can would lie on the ground plane.)

To see the effects of the node `stereo_image_proc`, we can launch the node `stereo_view` with:

```
rosrun image_view stereo_view stereo:=/stereo_camera image:=image_rect_color
```

Figure 7.3: Gazebo view of stereo-camera model viewing can

This brings up three image display windows, labelled left, right, and disparity, as shown in Fig 7.4. The left and right (rectified) image displays are lined up to visualize the disparity between them. Most of the pixels in the left image display have recognizable corresponding pixels in the right image display. The right-image pixel correspondences are on the same row (v) value as in the left image, although pixels in the right image are shifted to the left (negative u offsets) relative to the left image. For any pair of corresponding left and right pixels, the u-shift is the disparity, which is inversely related to the depth.

The disparity image shows colorization of pixels (with respect to the left-camera reference) for which the `stereo_image_proc` determines a credible correspondence to pixels in the right image. The coloration ranges over values from minimum disparity to maximum disparity, which are tunable parameters.

The lower-right area of Fig 7.4 shows a display that is brought up by running:

```
rosrun rqt_reconfigure rqt_reconfigure
```

This interface allows the user to adjust parameters of the stereo block-matching algorithm interactively. To find pixel correspondences, a block-matching algorithm attempts to find good matches between small patches of pixels (of `correlation_window_size`) in the left image relative to the right image. Since these images are rectified, it is only necessary to search in the x (u) direction in the right image to identify matches with corresponding blocks in the left image. Finding good parameters for block matching can be challenging. Details are described at `http://wiki.ros.org/stereo_image_proc/Tutorials/ ChoosingGoodStereoParameters`. For our case, we can compute some expectations.

The cameras are known to be 0.5 m above the ground plane, so there should be no viewable objects further away than this. The distance between cameras is 0.02 m, there are 640 horizontal pixels, and each focal length is 1034 pixels. For a point $M(x, y, z) = (X, Y, Z)$ with respect to the left-camera optical frame, the corresponding pixel onto which this point projects is $(u, v)_{left} = (u_c + f_x X/Z, v_c + f_y Y/Z)$. This same point from the right-camera optical frame is at $M = (X - b, Y, Z)$, where $b = 0.02$ is the baseline (or inter-ocular distance) between the left and right cameras. Correspondingly, the example point would project onto

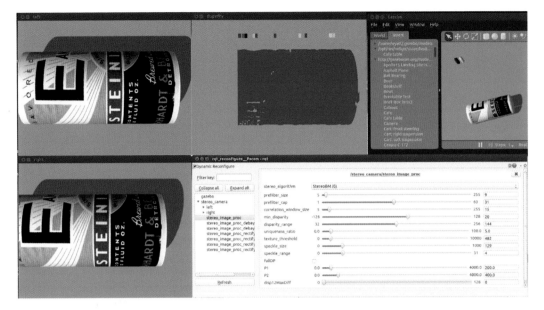

Figure 7.4: Display of can images: right, left and disparity

$(u, v)_{right} = (u_c + f_x(X - b)/Z, v_c + f_y Y/Z)$. The difference between these two projections is $(\Delta u, \Delta v) = (f_x b/Z, 0)$. We see that that $v_{left} = v_{right}$, which is the intent of rectification. For corresponding points in the left and right images, the disparity is $\Delta u = f_x b/Z$, and thus if we know the distance to the point of interest, we can compute the corresponding disparity. Inversely, if we know the disparity of a point in left versus right images, we can compute the distance to that point: $Z = bf_x/(\Delta u)$, *i.e.* the distance (the Z coordinate of the point of interest measured in the left-camera optical frame) is inversely proportional to the disparity.

Since our stereo cameras stare down at the ground plane a distance 0.5 m away, the maximum distance of any points of interest is 0.5 m, which corresponds to a disparity of $\Delta u = f_x b/Z = (1034)(0.02)/(0.5) = 41.4$ pixels. We can thus set the minimum disparity to this value.

The maximum possible disparity would be 480 pixels–a case for which a point would project to maximum u in the left camera and minimum u in the right camera. However, this is the extreme case and likely not of much use. If the maximum disparity is set to a lower value, the block-matching algorithm will not have to evaluate as many cases, and it can run faster. It is thus beneficial to set this value to the lowest practical value, corresponding to the minimum viewing distance anticipated.

For each point that has a (presumed) valid disparity value, one can compute the 3-D coordinates of the corresponding light source (point of interest), which is done by convention with respect to the left-camera optical frame. Three-D points thus computed can be visualized in `rviz`. This is accomplished by publishing the 3-D points in a `pointCloud2` message. The node `stereo_image_proc` performs this computation and publishes the resulting pointcloud on topic `ROS_NAMESPACE/points2` (which is `stereo_camera`, in the present example).

Starting `rviz` with:

```
rosrun rviz rviz
```

one can visualize the 3-D data from stereo vision by adding a `PointCloud2` display item and setting the topic to `/stereo_camera/points2`. A screenshot of `rviz` displaying these points is shown in Fig 7.5. In Fig 7.5, the left-camera optical frame axes are shown, and

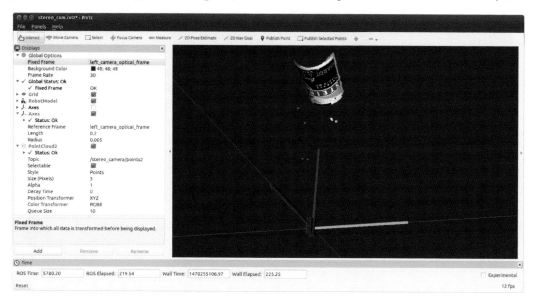

Figure 7.5: `rviz` view of three-dimensional points computed from stereo vision

the display's reference frame is set to left-camera optical frame. Using `rviz`, one can select a patch of points and publish their 3-D coordinates using the `PublishSelectedPoints` tool. Figure 7.6 illustrates an example, zooming on the image approximately along the left-camera's optical axis. The result of clicking and dragging on a patch in `rviz` is a publication of points of type `sensor_msgs/PointCloud2` on topic `selected_points`. To interpret these points, we can again run:

```
rosrun pcl_utils compute_selected_points_centroid
```

With this node running, selecting the patch shown in Fig 7.6 results in the output:

`centroid of selected points is: (0.020989, -0.001343, 0.389879)`

The z component is 0.390. Since the can has a radius of 0.055 m, and it is lying on its side on the ground plane, we would expect the upper surface to be at $z = 0.11m$ in world coordinates. Since the left camera-frame origin is defined to be at height 0.5 m above the ground plane, we would thus expect the surface of the can nearest the camera to be at $z = 0.5 - 0.11 = 0.39m$ from the camera-frame origin. From this simple experiment, we see that the distance computation is consistent. (Note that the x and y values are both small, since a patch was selected close to where the camera optical axis intersected the can.)

The point-cloud view can be rotated in `rviz`, giving the operator a good sense of 3-D. Note, though, that some points in Fig 7.5 are clearly incorrect, apparently floating in space. These correspond to errors in the block-matching algorithm and are the equivalent of optical illusions. While stereo cameras can be very useful, one must be wary of inevitable errors in inferring correspondence of pixels, including no matches due to inadequate texture as well as false matches leading to false points in 3-D. Alternative depth-imaging techniques, *e.g.* using structured light, can be more reliable, but often at the expense of resolution or the inability to include color.

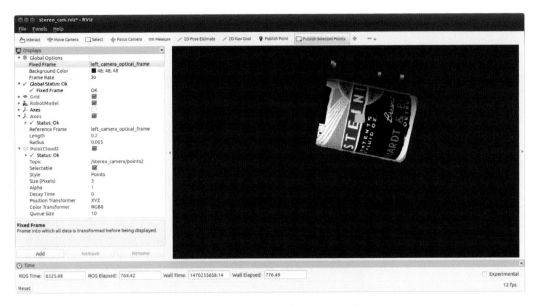

Figure 7.6: Selection of points in `rviz` (cyan patch): stereo vision view

7.3 DEPTH CAMERAS

The Microsoft Kinect camera provides depth and color information. Depth is inferred from distortion of a projected infra-red speckle pattern. This leads to a surprisingly good depth image, although there are limitations. Depth accuracy is fairly coarse; there is significant depth noise and reflections and absorption of the projected light result in both drop-outs (inability to interpret the speckle image) and false 3-D points. Nonetheless, these cameras are surprisingly effective for their cost.

Simulating the Kinect camera was introduced in Section 5.2.3. A similar model is constructed in the `simple_camera_model` package in file `simple_kinect_model.xacro`. The Kinect camera is defined to be oriented looking down from a height 0.5 m above the ground plane. In this model, the parameter `<pointCloudCutoff>0.3</pointCloudCutoff>` is set to 0.3 instead of the original 0.4 m, since our example beer can would come too close to the camera to be perceived by the Kinect.

The simple Kinect model can be launched with:

```
roslaunch simple_camera_model kinect_simu.launch
```

In this launch file, an additional transform publisher is started with:

```
<node pkg="tf" type="static_transform_publisher" name="rcamera_frame_bdcst" args="0 0 ↵
    0 -1.5708 0 -1.5708   camera_link kinect_depth_frame 50"/>
```

This publisher node publishes the transform relating the `kinect_depth_frame` (associated with the Kinect plug-in) with a frame on the robot model `camera_link`. The transform is chosen to align the Kinect optical axis anti-parallel to the world z axis, and to align the Kinect image-plane x axis parallel to the world x axis. Publishing such a transform allows associating Kinect point-cloud coordinates with the world frame.

The Beer model is inserted in Gazebo, then rotated and translated to present a curved face to the camera; `rviz` is then started,

```
rosrun rviz rviz
```

and the **pointCloud2** item topic is set to **kinect/depth/points**. The Kinect emulation also outputs a 2-D image, which can be viewed by running:

```
rosrun image_view image_view image:=/kinect/rgb/image_raw
```

With the Gazebo simulation of the kinect model running along with the static transform publisher node, **rviz** and the **image_view** node, the screen appears as in Fig 7.7. The

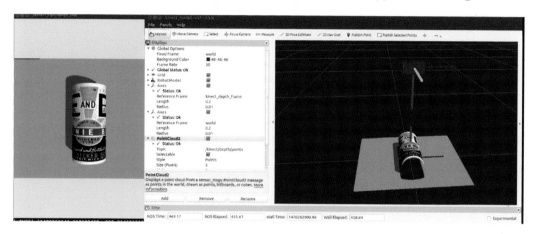

Figure 7.7: Selection of points in **rviz** (cyan patch): Kinect view

point-cloud view appears similar to that of the stereo-camera display in **rviz**. However, the Kinect camera has a wider field of view than the stereo-camera model. Further, the Kinect camera is capable of computing point-cloud points for the ground plane, since it does not rely on texture to infer 3-D coordinates (using instead its own light projection pattern).

The Kinect view, as emulated by the Gazebo plug-in, appears overall superior to the stereo-camera view, with greater field of view, ability to image bland objects, and absence of optical illusions. With the physical Kinect camera, however, the image data contains artifacts, including drop-outs, optical illusions, noise and miscalibration between depth points and associated colors. Nonetheless, once cast into ROS, the Kinect camera data can be treated similarly to any other source of 3-D data, making perceptual processing applicable across multiple sensor types.

7.4 WRAP-UP

This chapter presented three common means for obtaining 3-D non-contact sensory data: tilting LIDAR, stereo vision and depth cameras (specifically, the Kinect sensor). Depth images in ROS are treated as point clouds, which can be published with point-cloud message types. Point-cloud data must be interpreted in context, *e.g.* to identify regions suitable for navigation, to recognize landmarks or fiducials, or to recognize and localize objects of interest to be manipulated. Interpretation of point-cloud data is assisted with the use of the Point Cloud Library, introduced next.

Point Cloud Processing

CONTENTS

I NTRODUCTION
Interpreting 3-D sensory data is enabled with the use of the Point Cloud Library (see
http://pointclouds.org/). This open-source effort is independent of but compatible with
ROS. The Point Cloud Library (PCL) offers an array of functions for interpreting 3-D data.
The present treatment is not intended to be a comprehensive tutorial of PCL. Rather, a
few simple capabilities are introduced that provide useful functionality. Greater expertise
can be gained by consulting the on-line tutorials and code examples. (At the time of this
writing, there appears to be no textbook for teaching PCL.) It is hoped that the incremental
examples discussed here will help to make the on-line resources more accessible.

8.1   SIMPLE POINT-CLOUD DISPLAY NODE

The `pcl_utils` package contains the source program `display_ellipse.cpp` and the associated module `make_clouds.cpp`. This program introduces use of some basic PCL
datatypes and conversions to ROS-compatible messages. This package was created using
`cs_create_pkg` with options of `roscpp`, `sensor_msgs`, `pcl_ros` and `pcl_conversions`.
(Additional dependencies of `std_msgs` and `tf` are used in later examples.)

Within the `display_ellipse` node, a point-cloud object is populated with computed
values that describe an ellipse extruded in the z direction that has coloration that varies in
the z direction.

The source code of the function that populates a point cloud is shown in Listing 8.1.

Listing 8.1: `make_clouds.cpp`: C++ code illustrating populating a point cloud

```
1  //make_clouds.cpp
2  //a function to populate two point clouds with computed points
3  // modified from: from: http://docs.ros.org/hydro/api/pcl/html/←
        pcl__visualizer__demo_8cpp_source.html
4  #include<ros/ros.h>
5  #include <stdlib.h>
6  #include <math.h>
7
8  #include <sensor_msgs/PointCloud2.h> //ROS message type to publish a pointCloud
9  #include <pcl_ros/point_cloud.h> //use these to convert between PCL and ROS datatypes
```

```
10   #include <pcl/ros/conversions.h>
11
12   #include <pcl-1.7/pcl/point_cloud.h>
13   #include <pcl-1.7/pcl/PCLHeader.h>
14
15
16   using namespace std;
17
18   //a function to populate a pointCloud and a colored pointCloud;
19   // provide pointers to these, and this function will fill them with data
20   void make_clouds(pcl::PointCloud<pcl::PointXYZ>::Ptr basic_cloud_ptr,
21           pcl::PointCloud<pcl::PointXYZRGB>::Ptr point_cloud_ptr) {
22       // make an ellipse extruded along the z-axis. The color for
23       // the XYZRGB cloud will gradually go from red to green to blue.
24
25       uint8_t r(255), g(15), b(15); //declare and initialize red, green, blue component ↩
               values
26
27       //here are "point" objects that are compatible as building-blocks of point clouds
28       pcl::PointXYZ basic_point; // simple points have x,y,z, but no color
29       pcl::PointXYZRGB point; //colored point clouds also have RGB values
30
31       for (float z = -1.0; z <= 1.0; z += 0.05) //build cloud in z direction
32       {
33           // color is encoded strangely, but efficiently.  Stored as a 4-byte "float", ↩
                   but
34           // interpreted as individual byte values for 3 colors
35           // bits 0-7 are blue value, bits 8-15 are green, bits 16-23 are red;
36           // Can build the rgb encoding with bit-level operations:
37           uint32_t rgb = (static_cast<uint32_t> (r) << 16 |
38                   static_cast<uint32_t> (g) << 8 | static_cast<uint32_t> (b));
39
40           // and encode these bits as a single-precision (4-byte) float:
41           float rgb_float = *reinterpret_cast<float*> (&rgb);
42
43           //using fixed color and fixed z, compute coords of an ellipse in x-y plane
44           for (float ang = 0.0; ang <= 2.0 * M_PI; ang += 2.0 * M_PI / 72.0) {
45               //choose minor axis length= 0.5, major axis length = 1.0
46               // compute and fill in components of point
47               basic_point.x = 0.5 * cosf(ang); //cosf is cosine, operates on and returns↩
                       single-precision floats
48               basic_point.y = sinf(ang);
49               basic_point.z = z;
50               basic_cloud_ptr->points.push_back(basic_point); //append this point to the↩
                       vector of points
51
52               //use the same point coordinates for our colored pointcloud
53               point.x = basic_point.x;
54               point.y = basic_point.y;
55               point.z = basic_point.z;
56               //but also add rgb information
57               point.rgb = rgb_float; //*reinterpret_cast<float*> (&rgb);
58               point_cloud_ptr->points.push_back(point);
59           }
60           if (z < 0.0) //alter the color smoothly in the z direction
61           {
62               r -= 12; //less red
63               g += 12; //more green
64           } else {
65               g -= 12; // for positive z, lower the green
66               b += 12; // and increase the blue
67           }
68       }
69
70       //these will be unordered point clouds, i.e. a random bucket of points
71       basic_cloud_ptr->width = (int) basic_cloud_ptr->points.size();
72       basic_cloud_ptr->height = 1; //height=1 implies this is not an "ordered" point ↩
               cloud
73       basic_cloud_ptr->header.frame_id = "camera"; // need to assign a frame id
74
75       point_cloud_ptr->width = (int) point_cloud_ptr->points.size();
76       point_cloud_ptr->height = 1;
77       point_cloud_ptr->header.frame_id = "camera";
78
79   }
```

The function `make_clouds()` accepts pointers to PCL `PointCloud` objects and fills these objects with computed data. The object `pcl::PointCloud <pcl::PointXYZ >` is templated to accommodate different types of point clouds. We specifically consider basic point clouds (type `pcl::PointXYZ`), which have no color associated with individual points, and colored point clouds (type `pcl::PointXYZRGB`).

A PCL point-cloud object contains fields for a header, which includes a field for `frame_id`, and components that define the height and width of the point cloud data. Point clouds can be unordered or ordered. In the former case, the point-cloud width will be the number of points, and the point-cloud height will be 1. An unordered point cloud is a "bucket of points" in no particular order.

The first argument of this function is a pointer to a simple point cloud, consisting of (x, y, z) points with no associated intensity or color. The second argument is a pointer to an object of type `pcl::PointXYZRGB`, which will be populated with points that have color as well as 3-D coordinates.

Lines 28 and 29 instantiate variables of type `pcl::PointXYZ` and `pcl::PointXYZRGB`, which are consistent with elements within corresponding point clouds. An outer loop, starting on line 31, steps through values of z coordinates. At each elevation of z, coordinates x and y are computed corresponding to an ellipse. Lines 47 through 49 compute these points and assign them to elements of the **basic_point** object:

```
basic_point.x = 0.5*cosf(ang);
basic_point.y = sinf(ang);
basic_point.z = z;
```

On line 50, this point is appended to the vector of points within the uncolored point cloud:

```
basic_cloud_ptr->points.push_back(basic_point);
```

In lines 53 through 55, these same coordinates are assigned to the colored **point** object. The associated color is added on line 57:

```
point.rgb = rgb_float;
```

and the colored point is added to the colored point cloud (line 58):

```
point_cloud_ptr->points.push_back(point);
```

Encoding of color is somewhat complex. The red, green and blue intensities are represented as values from 0 to 255 in unsigned short (eight-bit) integers (line 25). Somewhat awkwardly, these are encoded as three bytes within a four-byte, single-precision float (lines 37 through 41).

At each increment of z value, the R, G and B values are altered (lines 60 through 67) to smoothly change the color of points in each z plane of the extruded ellipse.

Lines 71 through 77 set some meta-data of the populated point clouds. The lines:

```
    basic_cloud_ptr->width = (int) basic_cloud_ptr->points.size();
    basic_cloud_ptr->height = 1; //height=1 implies this is not an "ordered" point ↵
        cloud
```

specify a height of 1, which implies that the point cloud does not have an associated mapping to a 2-D array and is considered an unordered point cloud. Correspondingly, the width of the point cloud is equal to the total number of points. When the `make_clouds()` function concludes, the pointer arguments **basic_cloud_ptr** and **point_cloud_ptr** point to cloud objects that are populated with data and header information and are suitable for analysis and display.

The `display_ellipse.cpp` file, displayed in Listing 8.2 shows how save a point-cloud image to disk, as well as publish a point cloud to a topic consistent with visualization via `rviz`. This function instantiates point-cloud pointers (lines 24 through 25), which are used as arguments to the `make_clouds()` function (line 31).

Listing 8.2: `display_ellipse.cpp`: C++ code illustrating publishing a point cloud

```
1  //display_ellipse.cpp
2  //example of creating a point cloud and publishing it for rviz display
3
4  #include<ros/ros.h>
5  #include <stdlib.h>
6  #include <math.h>
7  #include <sensor_msgs/PointCloud2.h> //ROS message type to publish a pointCloud
8  #include <pcl_ros/point_cloud.h> //use these to convert between PCL and ROS datatypes
9  #include <pcl/ros/conversions.h>
10 #include <pcl-1.7/pcl/point_cloud.h>
11 #include <pcl-1.7/pcl/PCLHeader.h>
12
13 using namespace std;
14
15 //this function is defined in: make_clouds.cpp
16 extern void make_clouds(pcl::PointCloud<pcl::PointXYZ>::Ptr basic_cloud_ptr,
17         pcl::PointCloud<pcl::PointXYZRGB>::Ptr point_cloud_ptr);
18
19 int main(int argc, char** argv) {
20     ros::init(argc, argv, "ellipse"); //node name
21     ros::NodeHandle nh;
22
23     // create some point-cloud objects to hold data
24     pcl::PointCloud<pcl::PointXYZ>::Ptr basic_cloud_ptr(new pcl::PointCloud<pcl:: ↩
            PointXYZ>); //no color
25     pcl::PointCloud<pcl::PointXYZRGB>::Ptr point_cloud_clr_ptr(new pcl::PointCloud<pcl↩
            ::PointXYZRGB>); //colored
26
27     cout << "Generating example point-cloud ellipse.\n\n";
28     cout << "view in rviz; choose: topic= ellipse; and fixed frame= camera" << endl;
29
30     // -----use fnc to create example point clouds: basic and colored-----
31     make_clouds(basic_cloud_ptr, point_cloud_clr_ptr);
32
33     // we now have "interesting" point clouds in basic_cloud_ptr and ↩
            point_cloud_clr_ptr
34     pcl::io::savePCDFileASCII ("ellipse.pcd", *point_cloud_clr_ptr); //save image to ↩
            disk
35
36     sensor_msgs::PointCloud2 ros_cloud; //a ROS-compatible pointCloud message
37
38     pcl::toROSMsg(*point_cloud_clr_ptr, ros_cloud); //convert from PCL to ROS type ↩
            this way
39
40     //publish the colored point cloud in a ROS-compatible message on topic "ellipse"
41     ros::Publisher pubCloud = nh.advertise<sensor_msgs::PointCloud2> ("/ellipse", 1);
42
43     //publish the ROS-type message; can view this in rviz on topic "/ellipse"
44     //need to set the Rviz fixed frame to "camera"
45     while (ros::ok()) {
46         pubCloud.publish(ros_cloud);
47         ros::Duration(0.5).sleep(); //keep refreshing the publication periodically
48     }
49     return 0;
50 }
```

On line 34, a PCL function is called to save the generated point cloud to disk:

```
pcl::io::savePCDFileASCII ("ellipse.pcd", *point_cloud_clr_ptr); //save image
```

This instruction causes the point cloud to be saved to a file named `ellipse.pcd` (in the current directory, *i.e.* the directory from which this node is run).

For ROS compatibility, point-cloud data is published using the ROS message type `sensor_msgs/PointCloud2`. This message type is suitable for serialization and publications, but it is not compatible with PCL operations. The packages `pcl_ros` and `pcl_conversions` offer means to convert between ROS point-cloud message types and PCL point-cloud objects.

The PCL-style point cloud is converted to a ROS-style point-cloud message suitable for publication (lines 36 and 37). The ROS message `ros_cloud` is then published (line 46) on topic `ellipse` (as specified on line 41, upon construction of the publisher object).

To compile the example program, the `CMakeLists.txt` file must be edited to reference the point-cloud library. Specifically, the lines:

```
find_package(PCL 1.7 REQUIRED)
include_directories(${PCL_INCLUDE_DIRS})
```

are uncommented in the provided `CMakeLists.txt` auto-generated file. This change is sufficient to bring in the PCL library and link this library with the main program.

With a `roscore` running, the example program can be started with

```
rosrun pcl_utils display_ellipse
```

This node has little output, except to remind the user to select topic `ellipse` and frame `camera` to view the point-cloud publication in `rviz`. Starting `rviz` and selecting `camera` as the fixed frame and selecting `ellipse` for the topic in a `PointCloud2` display yields the result in Fig 8.1.

As expected, the image consists of sample points of an extruded ellipse with color variation in the extrusion (z) direction. This image can be rotated, translated and zoomed in `rviz`, which helps one to perceive its 3-D spatial properties.

The function `savePCDFileASCII` invoked on line 34 causes the file to be saved in ASCII format, whereas an alternative command `savePCDFile` would save the point cloud in binary format. The ASCII representation should be used sparingly. ASCII files are typically more than twice the size of binary files. Further, there is (currently) a bug in the ASCII storage that results in some color corruption. Nonetheless, the ASCII option can be useful for understanding PCD file formats.

The ASCII file `ellipse.pcd` can be viewed with a simple text editor. The beginning of this file is:

```
# .PCD v0.7 - Point Cloud Data file format
VERSION 0.7
FIELDS x y z rgb
SIZE 4 4 4 4
TYPE F F F F
COUNT 1 1 1 1
WIDTH 2920
HEIGHT 1
VIEWPOINT 0 0 0 1 0 0 0
POINTS 2920
DATA ascii
0.5 0 -1 2.3423454e-38
0.49809736 0.087155737 -1 2.3423454e-38
0.49240386 0.17364818 -1 2.3423454e-38
0.48296291 0.25881904 -1 2.3423454e-38
```

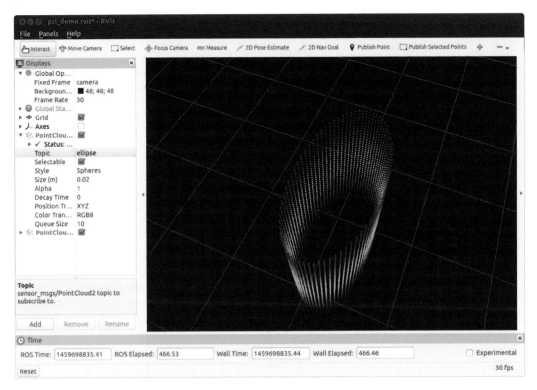

Figure 8.1: Rviz view of point cloud generated and published by `display_ellipse`

```
0.46984631 0.34202015 -1 2.3423454e-38
0.45315388 0.4226183 -1 2.3423454e-38
```

The file format contains metadata describing how the point cloud is encoded, followed by the data itself. (Details of the file format can be found at `http://pointclouds.org/documentation/tutorials/pcd_file_format.php`.) The field names, in order, are `x y z rgb`. The WIDTH of the data is 2920, which is identical to the POINTS. The HEIGHT is simply 1, and WIDTH × HEIGHT = POINTS. This description means that the point cloud is stored merely as a list of points–an unordered point cloud–without any inherent structure (or at least none that is encoded in the file).

Following the description of how the data is encoded, the actual data is listed. Only six lines of this data are shown; the entire data includes 2920 points. As can be seen from the first six lines, all values of z are -1, which is consistent with the outer z-height loop of `make_clouds()` (which starts at $z = -1$). The x and y values are also consistent with the sin and cos values expected. The fourth value is harder to interpret, since the RGB data is encoded as three of four bytes in the single-precision floating-point value.

In the next section, we will see the alternative of an ordered point-cloud encoding.

8.2   LOADING AND DISPLAYING POINT-CLOUD IMAGES FROM DISK

In the previous section, a point cloud was populated with computed data. More typically, one would be working with point-cloud data acquired by some form of depth camera, *e.g.* a depth camera such as the Kinect™, a scanning LIDAR such as the Carnegie Robotics sensor head, or point clouds generated by stereo vision. Such data is typically acquired and processed online during robot operation, but it is also convenient to save snapshots of such data to disk, *e.g.* for use in code development or creation of an image database.

A simple example of reading a PCD file from disk and displaying it to `rviz` is provided in `display_pcd_file.cpp`, which is shown in Listing 8.3.

Listing 8.3: `display_pcd_file.cpp` C++ code illustrating reading PCD file from disk and publishing it as a point cloud

```cpp
1   ///display_pcd_file.cpp
2   // prompts for a pcd file name, reads the file, and displays to rviz on topic "pcd"
3
4   #include<ros/ros.h> //generic C++ stuff
5   #include <stdlib.h>
6   #include <math.h>
7   #include <sensor_msgs/PointCloud2.h> //useful ROS message types
8   #include <pcl_ros/point_cloud.h> //to convert between PCL a nd ROS
9   #include <pcl/ros/conversions.h>
10  #include <pcl/point_types.h>
11  #include <pcl/point_cloud.h>
12  #include <pcl/common/common_headers.h>
13  #include <pcl-1.7/pcl/point_cloud.h>
14  #include <pcl-1.7/pcl/PCLHeader.h>
15
16  using namespace std;
17
18  int main(int argc, char** argv) {
19      ros::init(argc, argv, "pcd_publisher"); //node name
20      ros::NodeHandle nh;
21      pcl::PointCloud<pcl::PointXYZRGB>::Ptr pcl_clr_ptr(new pcl::PointCloud<pcl::↩
            PointXYZRGB>); //pointer for color version of pointcloud
22
23      cout<<"enter pcd file name: ";
24      string fname;
25      cin>>fname;
26
27      if (pcl::io::loadPCDFile<pcl::PointXYZRGB> (fname, *pcl_clr_ptr) == -1) //* load ↩
            the file
28  {
29      ROS_ERROR ("Couldn't read file \n");
30      return (-1);
31  }
32  std::cout << "Loaded "
33              << pcl_clr_ptr->width * pcl_clr_ptr->height
34              << " data points from file "<<fname<<std::endl;
35
36      //publish the point cloud in a ROS-compatible message; here's a publisher:
37      ros::Publisher pubCloud = nh.advertise<sensor_msgs::PointCloud2> ("/pcd", 1);
38      sensor_msgs::PointCloud2 ros_cloud;  //here is the ROS-compatible message
39      pcl::toROSMsg(*pcl_clr_ptr, ros_cloud); //convert from PCL to ROS type this way
40      ros_cloud.header.frame_id = "camera_depth_optical_frame";
41
42      //publish the ROS-type message on topic "/ellipse"; can view this in rviz
43      while (ros::ok()) {
44
45          pubCloud.publish(ros_cloud);
46          ros::spinOnce();
47          ros::Duration(0.1).sleep();
48      }
49      return 0;
50  }
```

Lines 37 through 48 of Listing 8.3 are equivalent to corresponding lines in the previous `display_ellipse` program. The topic is changed to `pcd` (line 37) and the `frame_id` is set to `camera_depth_optical_frame` (line 40), and these must be set correspondingly in `rviz` to view the output.

The primary difference with this example code is that the user is prompted to enter a file name (lines 23 through 25), and this file name is used to load a PCD file from disk and populate the pointer `pcl_clr_ptr` (line 27):

```
if (pcl::io::loadPCDFile<pcl::PointXYZRGB> (fname, *pcl_clr_ptr) == -1)
```

As a diagnostic, lines 32 through 34 inspect the width and height fields of the opened point cloud, multiply these together, and print the result as the number of points in the point cloud.

Example PCD files reside in the accompanying repository within the ROS package `pcd_images`. A snippet of an example PCD file from a Kinect camera image, stored as ASCII, is:

```
# .PCD v0.7 - Point Cloud Data file format
VERSION 0.7
FIELDS x y z rgb
SIZE 4 4 4 4
TYPE F F F F
COUNT 1 1 1 1
WIDTH 640
HEIGHT 480
VIEWPOINT 0 0 0 1 0 0 0
POINTS 307200
DATA ascii
nan nan nan -1.9552709e+38
nan nan nan -1.9552709e+38
nan nan nan -1.9552746e+38
nan nan nan -1.9553241e+38
```

This example is similar to the ellipse PCD file, with the notable exception that the HEIGHT parameter is not unity. The HEIGHT × WIDTH is 640 × 480, which is the sensor resolution of the Kinect color camera. The total number of points is 307,200, which is 640 × 480. In this case, the data is an ordered point cloud, since the listing of point data corresponds to a rectangular array.

An ordered point cloud preserves potentially valuable spatial relationships among the individual 3-D points. A Kinect camera, for example, has each 3-D point associated with a corresponding pixel in a 2-D (color) image array. Ideally, an inverse mapping also associates each pixel in the 2-D array with a corresponding 3-D point. In practice, there are many "holes" in this mapping for which no valid 3-D point is associated with a 2-D pixel. Such missing data is represented with NaN (not-a-number) codes, as is the case with the initial point values in the PCD file snippet shown above.

Coordinate data within a point-cloud object typically is stored as four-byte, single-precision floating-point values. Single-point precision is more than adequate for the level of precision of current 3-D sensors, and thus this lower-precision representation helps reduce file size without compromising precision.

Running the display node with

```
rosrun pcl_utils display_pcd_file
```

prompts the user to enter a file name. If the node is run from the `pcd_images` directory, the file name entered will not require a directory path. Entering the file name `coke_can.pcd` results in the response:

```
Loaded 307200 data points from file coke_can.pcd
```

which is the number of points expected for a Kinect-1 image. In `rviz`, setting the topic to `pcd` and setting the fixed frame to `camera_depth_optical_frame` produces the display shown in Fig 8.2.

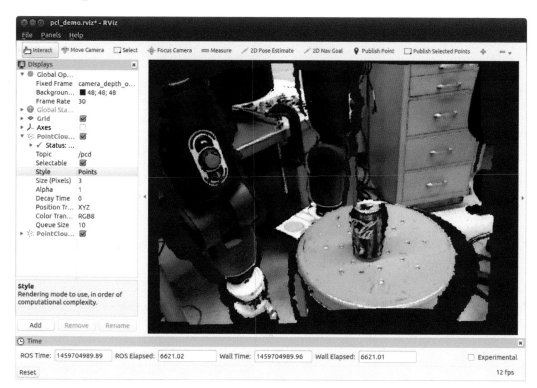

Figure 8.2: Rviz view of image read from disk and published by `display_pcd_file`

8.3 SAVING PUBLISHED POINT-CLOUD IMAGES TO DISK

A convenient utility program in the `pcl_utils` package is `pcd_snapshot.cpp`. The contents of this program are shown in Listing 8.4.

Listing 8.4: `pcd_snapshot.cpp` C++ code illustrating subscribing to point-cloud topic and saving data to disk as PCD file

```
1  //pcd_snapshot.cpp
2  // example of saving a kinect snapshot to a pcd file
3  // need to connect "Kinect" and start it with: roslaunch freenect_launch freenect.←
        launch
4
5  #include <iostream>
6  #include <ros/ros.h>
7  #include <sensor_msgs/PointCloud2.h>
8  #include <pcl_ros/point_cloud.h> //use these to convert between PCL and ROS datatypes
9  #include <pcl/ros/conversions.h>
10 #include <pcl-1.7/pcl/point_cloud.h>
11 #include <pcl-1.7/pcl/PCLHeader.h>
12
13 using namespace std;
14
15 bool got_kinect_image = false; //snapshot indicator
```

```
16  pcl::PointCloud<pcl::PointXYZRGB>::Ptr pclKinect_clr_ptr(new pcl::PointCloud<pcl::↵
        PointXYZRGB>); //pointer for color version of pointcloud
17
18  void kinectCB(const sensor_msgs::PointCloud2ConstPtr& cloud) {
19      if (!got_kinect_image) { // once only, to keep the data stable
20          ROS_INFO("got new selected kinect image");
21          pcl::fromROSMsg(*cloud, *pclKinect_clr_ptr);
22          ROS_INFO("image has %d * %d points", pclKinect_clr_ptr->width, ↵
                pclKinect_clr_ptr->height);
23          got_kinect_image = true;
24      }
25  }
26
27  int main(int argc, char** argv) {
28      ros::init(argc, argv, "pcl_snapshot_main"); //node name
29      ros::NodeHandle nh;
30      ros::Subscriber pointcloud_subscriber = nh.subscribe("/camera/depth_registered/↵
            points", 1, kinectCB);
31
32      //spin until obtain a snapshot
33      ROS_INFO("waiting for kinect data");
34      while (!got_kinect_image) {
35          ROS_INFO("waiting...");
36          ros::spinOnce();
37          ros::Duration(0.5).sleep();
38      }
39      ROS_INFO("got snapshot; saving to file kinect_snapshot.pcd");
40      pcl::io::savePCDFile("kinect_snapshot.pcd", *pclKinect_clr_ptr, true);
41
42      return 0;
43  }
```

In this program, a ROS subscriber subscribes to topic `camera/depth_registered/points` (line 30) with a callback function, `kinectCB()`, listening for messages of type `sensor_msgs/PointCloud2` (lines 18 through 25). When a message is received by the callback function, it converts the message to a `pcl::PointCloud` type (line 21). The callback function signals that it has received a new image by setting `got_kinect_image` to "true."

The main program waits for the `got_kinect_image` signal (line 34), and upon confirmation, the main program saves the acquired image as filename `kinect_snapshot.pcd` within the current directory (line 40):

```
pcl::io::savePCDFile("kinect_snapshot.pcd", *pclKinect_clr_ptr, true);
```

To use this program with a Kinect camera, plug in the camera, and with a roscore running, start the Kinect driver:

```
roslaunch freenect_launch freenect.launch
```

(If the freenect driver is not already installed on your system, it will need to be installed. This installation is included with the provided setup scripts.)

Start `rviz`, and set the fixed frame to `camera_depth_optical_frame`, add a `PointCloud2` item and set its topic to `camera/depth_registered/points`. This will provide a real-time view of the Kinect data, published as a ROS point cloud.

To prepare to capture an image, open a terminal and navigate to a directory relevant for storing the image. When the `rviz` display shows a scene to be captured, run the node:

```
rosrun pcl_utils pcd_snapshot
```

This will capture a single point-cloud transmission and save the captured image to disk (in the current directory) as a PCD file called `kinect_snapshot.pcd`. The saved image will be in binary format, and it will store the colored, ordered point cloud. To take more snapshots, re-name `kinect_snapshot.pcd` to something mnemonic, and re-run `pcd_snapshot`. (If you do not re-name the previous snapshot, the file `kinect_snapshot.pcd` will be overwritten.)

The image in Fig 8.2 was acquired in this manner. The resulting PCD file can be read from disk and displayed, using `display_pcd_file`, or it can be read into another application to perform interpretation of the point-cloud data.

8.4   INTERPRETING POINT-CLOUD IMAGES WITH PCL METHODS

Within the `pcl_utils` package, the program `find_plane_pcd_file.cpp` (displayed in Listing 8.5) illustrates use of a few PCL methods. The functions are illustrated operating on a PCD file. Lines 56 through 65 are similar to `display_pcd_file`, in which the user is prompted for a PCD file name, which is then read from disk. The resulting point cloud is published on topic "pcd" (lines 68, 73 and 108), making it viewable via `rviz`.

Listing 8.5: `find_plane_pcd_file.cpp` C++ code illustrating use of PCL methods

```
1   //find_plane_pcd_file.cpp
2   // prompts for a pcd file name, reads the file, and displays to rviz on topic "pcd"
3   // can select a patch; then computes a plane containing that patch, which is published↵
        on topic "planar_pts"
4   // illustrates use of PCL methods: computePointNormal(), transformPointCloud(),
5   // pcl::PassThrough methods setInputCloud(), setFilterFieldName(), setFilterLimits, ↵
        filter()
6   // pcl::io::loadPCDFile()
7   // pcl::toROSMsg() for converting PCL pointcloud to ROS message
8   // voxel-grid filtering: pcl::VoxelGrid,  setInputCloud(), setLeafSize(), filter()
9   //wsn March 2016
10
11  #include<ros/ros.h>
12  #include <stdlib.h>
13  #include <math.h>
14
15  #include <sensor_msgs/PointCloud2.h>
16  #include <pcl_ros/point_cloud.h> //to convert between PCL and ROS
17  #include <pcl/ros/conversions.h>
18
19  #include <pcl/point_types.h>
20  #include <pcl/point_cloud.h>
21  //#include <pcl/PCLPointCloud2.h> //PCL is migrating to PointCloud2
22
23  #include <pcl/common/common_headers.h>
24  #include <pcl-1.7/pcl/point_cloud.h>
25  #include <pcl-1.7/pcl/PCLHeader.h>
26
27  //will use filter objects "passthrough" and "voxel_grid" in this example
28  #include <pcl/filters/passthrough.h>
29  #include <pcl/filters/voxel_grid.h>
30
31  #include <pcl_utils/pcl_utils.h>   //a local library with some utility fncs
32
33
34  using namespace std;
35  extern PclUtils *g_pcl_utils_ptr;
36
37  //this fnc is defined in a separate module, find_indices_of_plane_from_patch.cpp
38  extern void find_indices_of_plane_from_patch(pcl::PointCloud<pcl::PointXYZRGB>::Ptr ↵
        input_cloud_ptr,
39       pcl::PointCloud<pcl::PointXYZ>::Ptr patch_cloud_ptr, vector<int> &indices);
40
41  int main(int argc, char** argv) {
42      ros::init(argc, argv, "plane_finder"); //node name
43      ros::NodeHandle nh;
44      //pointer for color version of pointcloud
45      pcl::PointCloud<pcl::PointXYZRGB>::Ptr pclKinect_clr_ptr(new pcl::PointCloud<pcl::↵
        PointXYZRGB>);
46      //pointer for pointcloud of planar points found
```

```
47    pcl::PointCloud<pcl::PointXYZRGB>::Ptr plane_pts_ptr(new pcl::PointCloud<pcl::↩
          PointXYZRGB>);
48    //ptr to selected pts from Rvis tool
49    pcl::PointCloud<pcl::PointXYZ>::Ptr selected_pts_cloud_ptr(new pcl::PointCloud<pcl↩
          ::PointXYZ>);
50    //ptr to hold filtered Kinect image
51    pcl::PointCloud<pcl::PointXYZRGB>::Ptr downsampled_kinect_ptr(new pcl::PointCloud<↩
          pcl::PointXYZRGB>);
52
53    vector<int> indices;
54
55    //load a PCD file using pcl::io function; alternatively, could subscribe to Kinect↩
           messages
56    string fname;
57    cout << "enter pcd file name: "; //prompt to enter file name
58    cin >> fname;
59    if (pcl::io::loadPCDFile<pcl::PointXYZRGB> (fname, *pclKinect_clr_ptr) == -1) //* ↩
          load the file
60    {
61        ROS_ERROR("Couldn't read file \n");
62        return (-1);
63    }
64    //PCD file does not seem to record the reference frame;  set frame_id manually
65    pclKinect_clr_ptr->header.frame_id = "camera_depth_optical_frame";
66
67    //will publish  pointClouds as ROS-compatible messages; create publishers; note ↩
          topics for rviz viewing
68    ros::Publisher pubCloud = nh.advertise<sensor_msgs::PointCloud2> ("/pcd", 1);
69    ros::Publisher pubPlane = nh.advertise<sensor_msgs::PointCloud2> ("planar_pts", 1)↩
          ;
70    ros::Publisher pubDnSamp = nh.advertise<sensor_msgs::PointCloud2> ("↩
          downsampled_pcd", 1);
71
72    sensor_msgs::PointCloud2 ros_cloud, ros_planar_cloud, downsampled_cloud; //here ↩
          are ROS-compatible messages
73    pcl::toROSMsg(*pclKinect_clr_ptr, ros_cloud); //convert from PCL cloud to ROS ↩
          message this way
74
75    //use voxel filtering to downsample the original cloud:
76    cout << "starting voxel filtering" << endl;
77    pcl::VoxelGrid<pcl::PointXYZRGB> vox;
78    vox.setInputCloud(pclKinect_clr_ptr);
79
80    vox.setLeafSize(0.02f, 0.02f, 0.02f);
81    vox.filter(*downsampled_kinect_ptr);
82    cout << "done voxel filtering" << endl;
83
84    cout << "num bytes in original cloud data = " << pclKinect_clr_ptr->points.size() ↩
          << endl;
85    cout << "num bytes in filtered cloud data = " << downsampled_kinect_ptr->points.↩
          size() << endl; //
86    pcl::toROSMsg(*downsampled_kinect_ptr, downsampled_cloud); //convert to ros ↩
          message for publication and display
87
88    PclUtils pclUtils(&nh); //instantiate a PclUtils object--a local library w/ some ↩
          handy fncs
89    g_pcl_utils_ptr = &pclUtils; // make this object shared globally, so above fnc can↩
           use it too
90
91    cout << " select a patch of points to find corresponding plane..." << endl; //↩
          prompt user action
92    //loop to test for new selected-points inputs and compute and display ↩
          corresponding planar fits
93    while (ros::ok()) {
94        if (pclUtils.got_selected_points()) { //here if user selected a new patch of ↩
              points
95            pclUtils.reset_got_selected_points(); // reset for a future trigger
96            pclUtils.get_copy_selected_points(selected_pts_cloud_ptr); //get a copy of↩
                  the selected points
97            cout << "got new patch with number of selected pts = " << ↩
                  selected_pts_cloud_ptr->points.size() << endl;
98
99            //find pts coplanar w/ selected patch, using PCL methods in above-defined ↩
                  function
100           //"indices" will get filled with indices of points that are approx co-↩
                  planar with the selected patch
```

```
101        // can extract indices from original cloud, or from voxel-filtered (down-↩
              sampled) cloud
102        //find_indices_of_plane_from_patch(pclKinect_clr_ptr, ↩
              selected_pts_cloud_ptr, indices);
103        find_indices_of_plane_from_patch(downsampled_kinect_ptr, ↩
              selected_pts_cloud_ptr, indices);
104        pcl::copyPointCloud(*downsampled_kinect_ptr, indices, *plane_pts_ptr); //↩
              extract these pts into new cloud
105        //the new cloud is a set of points from original cloud, coplanar with ↩
              selected patch; display the result
106        pcl::toROSMsg(*plane_pts_ptr, ros_planar_cloud); //convert to ros message ↩
              for publication and display
107        }
108        pubCloud.publish(ros_cloud); // will not need to keep republishing if display ↩
              setting is persistent
109        pubPlane.publish(ros_planar_cloud); // display the set of points computed to ↩
              be coplanar w/ selection
110        pubDnSamp.publish(downsampled_cloud); //can directly publish a pcl::↩
              PointCloud2!!
111        ros::spinOnce(); //pclUtils needs some spin cycles to invoke callbacks for new↩
              selected points
112        ros::Duration(0.1).sleep();
113      }
114
115   return 0;
116 }
```

A new feature introduced in this program, voxel filtering, appears in lines 78 through 81:

```
pcl::VoxelGrid<pcl::PointXYZRGB> vox;
vox.setInputCloud(pclKinect_clr_ptr);
vox.setLeafSize(0.02f, 0.02f, 0.02f);
vox.filter(*downsampled_kinect_ptr);
```

These instructions instantiate a PCL-library object of type `VoxelGrid` that operates on data of type `pcl::PointXYZRGB` (colored point clouds). The `VoxelGrid` object, vox, has member functions that perform spatial filtering for down-sampling. (For greater detail, see http://pointclouds.org/documentation/tutorials/voxel_grid.php#voxelgrid.) Typical of PCL methods, the first operation with vox is to associate point-cloud data with it (on which the object will operate). PCL attempts to avoid making copies of data, since point-cloud data can be voluminous. Pointers or reference variables are typically used to direct the methods to the data already in memory. After affiliating vox with the point cloud read from disk, parameters are set for leaf size. The x, y and z values have all been set to 0.02 (corresponding to a 2 cm cube).

With the instruction `filter`, the vox object filters the input data and produces a down-sampled output (which is populated in a cloud via the pointer `downsampled_kinect_ptr`). Conceptually, down-sampling by this method is equivalent to subdividing a volume into small cubes and assigning each point of the original point cloud to a single cube. After all points are assigned, each non-empty cube is represented by a single point, which is the average of all points assigned to that cube. Depending on the size of the cubes, the data reduction can be dramatic, while the loss of resolution may be acceptable.

In the example code, lines 86 and 110 convert the down-sampled cloud to a ROS message and publish the new cloud to the topic `downsampled_pcd`. The sizes of the original and down-sampled clouds are printed out (lines 84 and 85).

Running `rosrun pcl_utils find_plane_pcd_file` prompts for a file name. Using the same file as in Fig 8.2 (`coke_can.pcd`), downsampling at the chosen resolution (2cm cubes) results in printed output:

```
done voxel filtering
num bytes in original cloud data = 307200
num bytes in filtered cloud data = 6334
```

showing that down-sampling reduced the number of points by nearly a factor of 50 (in this case). Bringing up `rviz`, and setting the fixed frame to `camera_depth_optical_frame` and choosing the topic `downsampled_pcd` in a `PointCloud2` display item produces the display in Fig 8.3. In this example, the severity of the downsampling is obvious, yet the image is still recognizable as corresponding to Fig 8.2. With the data reduction, algorithms applied to point-cloud interpretation can run significantly faster. (The choice of downsampling resolution would need to be tuned for any specific application). In addition to demonstrat-

Figure 8.3: Rviz view of image read from disk, down-sampled and published by `find_plane_pcd_file`

ing downsampling, the `find_plane_pcd_file` illustrates a means of interacting with the point-cloud data to assist interpretation of the data. The example code is linked with an example library that contains additional point-cloud processing examples, `pcl_utils.cpp` (in the `pcl_utils` package). This library defines a class, `PclUtils`, an object of which is instantiated on line 88. In addition, this object is shared with external functions via a global pointer, which is defined on line 35 and assigned to point to the PclUtils object on line 89.

One of the features of `PclUtils` is a subscriber that subscribes to the topic `selected_points`. The publisher on this topic is an `rviz` tool, `publish_selected_points`, which allows a user to select points from an `rviz` display via click and drag of a mouse. The (selectable) points lying within the rectangle thus defined are published to the topic `selected_points` as a message type `sensor_msgs::PointCloud2`. When the user selects a region with this tool, the corresponding (underlying) 3-D data points are published, which awakens a callback function that receives and stores the corresponding message.

Within `find_plane_pcd_file`, the value `pclUtils.got_selected_points()` is tested (line 94) to see whether the user selected a new patch of points. If so, a copy of the corresponding selected-points data is obtained (line 96) using the `PclUtils` method:

```
pclUtils.get_copy_selected_points(selected_pts_cloud_ptr); //get a copy of the ↩
    selected points
```

When a new copy of selected points is obtained, a function is called to analyze the down-sampled image in terms of a cue from the selected points (line 103):

```
find_indices_of_plane_from_patch(downsampled_kinect_ptr, selected_pts_cloud_ptr, ↩
    indices);
```

The function `find_indices_of_plane_from_patch()` is implemented in a separate module, `find_indices_of_plane_from_patch.cpp`, which resides in package `pcl_utils`. The contents of this module are shown in Listing 8.6.

Listing 8.6: `find_indices_of_plane_from_patch.cpp` C++ module showing PCL functions for interpretation of point-cloud data

```
1   //find_indices_of_plane_from_patch.cpp
2
3   #include <ros/ros.h>
4   #include <stdlib.h>
5   #include <math.h>
6
7   #include <sensor_msgs/PointCloud2.h>
8   #include <pcl_ros/point_cloud.h> //to convert between PCL and ROS
9   #include <pcl/ros/conversions.h>
10
11  #include <pcl/point_types.h>
12  #include <pcl/point_cloud.h>
13  //#include <pcl/PCLPointCloud2.h> //PCL is migrating to PointCloud2
14
15  #include <pcl/common/common_headers.h>
16  #include <pcl-1.7/pcl/point_cloud.h>
17  #include <pcl-1.7/pcl/PCLHeader.h>
18
19  //will use filter objects "passthrough" and "voxel_grid" in this example
20  #include <pcl/filters/passthrough.h>
21  #include <pcl/filters/voxel_grid.h>
22
23  #include <pcl_utils/pcl_utils.h>  //a local library with some utility fncs
24
25
26  using namespace std;
27  PclUtils *g_pcl_utils_ptr;
28
29  void find_indices_of_plane_from_patch(pcl::PointCloud<pcl::PointXYZRGB>::Ptr ↩
        input_cloud_ptr,
30          pcl::PointCloud<pcl::PointXYZ>::Ptr patch_cloud_ptr, vector<int> &indices) {
31
32      float curvature;
33      Eigen::Vector4f plane_parameters;
34
35      pcl::PointCloud<pcl::PointXYZRGB>::Ptr transformed_cloud_ptr(new pcl::PointCloud<↩
            pcl::PointXYZRGB>); //pointer for color version of pointcloud
36
37      pcl::computePointNormal(*patch_cloud_ptr, plane_parameters, curvature); //pcl fnc ↩
            to compute plane fit to point cloud
38      cout << "PCL: plane params of patch: " << plane_parameters.transpose() << endl;
39
40      //next, define a coordinate frame on the plane fitted to the patch.
41      // choose the z-axis of this frame to be the plane normal--but enforce that the ↩
            normal
42      // must point towards the camera
43      Eigen::Affine3f A_plane_wrt_camera;
44      // here, use a utility function in pclUtils to construct a frame on the computed ↩
            plane
45      A_plane_wrt_camera = g_pcl_utils_ptr->make_affine_from_plane_params(↩
            plane_parameters);
46      cout << "A_plane_wrt_camera rotation:" << endl;
```

```
47    cout << A_plane_wrt_camera.linear() << endl;
48    cout << "origin: " << A_plane_wrt_camera.translation().transpose() << endl;
49
50    //next, transform all points in input_cloud into the plane frame.
51    //the result of this is, points that are members of the plane of interest should ←
          have z-coordinates
52    // nearly 0, and thus these points will be easy to find
53    cout << "transforming all points to plane coordinates..." << endl;
54    //Transform each point in the given point cloud according to the given ←
          transformation.
55    //pcl fnc: pass in ptrs to input cloud, holder for transformed cloud, and desired ←
          transform
56    //note that if A contains a description of the frame on the plane, we want to ←
          xform with inverse(A)
57    pcl::transformPointCloud(*input_cloud_ptr, *transformed_cloud_ptr, ←
          A_plane_wrt_camera.inverse());
58
59    //now we'll use some functions from the pcl filter library;
60    pcl::PassThrough<pcl::PointXYZRGB> pass; //create a pass-through object
61    pass.setInputCloud(transformed_cloud_ptr); //set the cloud we want to operate on--←
          pass via a pointer
62    pass.setFilterFieldName("z"); // we will "filter" based on points that lie within ←
          some range of z-value
63    pass.setFilterLimits(-0.02, 0.02); //here is the range: z value near zero, -0.02<z←
          <0.02
64    pass.filter(indices); //  this will return the indices of the points in   ←
          transformed_cloud_ptr that pass our test
65    cout << "number of points passing the filter = " << indices.size() << endl;
66    //This fnc populates the reference arg "indices", so the calling fnc gets the list←
          of interesting points
67 }
```

The objective of this function is, given `patch_cloud` points, which are presumably nearly coplanar, finding all other points in `input_cloud` that are nearly coplanar with the input patch. The result is returned in a vector of point indices, specifying which points in input cloud qualify as coplanar. This allows the operator to interactively select points displayed in `rviz` to induce the node to find all points coplanar with the selection. This may be useful *e.g.* for defining walls, doors, floor and table surfaces corresponding to the operator's focus of attention. The example function illustrates use of PCL methods `computePointNormal()`, `transformPointCloud()`, and `pcl::PassThrough` filter methods `setInputCloud()`, `setFilterFieldName()`, `setFilterLimits()`, and `filter()`.

On line 37 of `find_indices_of_plane_from_patch()`, the input cloud (a presumed nearly coplanar patch of points) is analyzed to find a least-squares fit of a plane to the points.

```
pcl::computePointNormal(*patch_cloud_ptr, plane_parameters, curvature); //pcl fnc to ←
    compute plane fit to point cloud
```

The fitted plane is described in terms of four parameters: the components (nx, ny, nz) of the plane's surface normal, and the minimum distance of the plane from the camera's origin (focal point). These parameters are returned in an Eigen-type 4×1 vector via a the reference argument `plane_parameters`. Additionally, the (scalar) curvature is returned in the `curvature` argument. The function `computePointNormal()` uses an eigenvalue-eigenvector approach to provide a robust best-estimate to a plane fit through the provided points. (See `http://pointclouds.org/documentation/tutorials/normal_estimation.php` for an explanation of the mathematical theory behind this implementation.)

When fitting a plane to points, there is an ambiguity regarding direction of the plane normal. The data alone cannot distinguish positive versus negative normal. Physically, solid objects have surfaces that distinguish inside from outside, and by convention the surface normal is defined to point toward the exterior of the object.

In the current example, the point-cloud data is presumed to be expressed with respect to the camera's viewpoint. Consequently, all points observable by the camera must correspond to surfaces that have a surface normal pointing (at least partially) toward the camera. When fitting a plane to points, if the computed surface normal has a positive z component, the normal vector must be negated to be logically consistent.

Additionally, the distance of a plane from a point (the focal point, in this case) is unambiguously defined. However, a signed distance is more useful, where the distance is defined as the displacement of the plane from the reference point along the plane's normal. Since all surface normals must have a negative z component, all surfaces also must have a negative displacement from the camera-frame origin.

These corrections are applied as necessary within the function `make_affine_from_plane_params()` (invoked on line 45). This function is a method within the `pcl_utils` library. The implementation of this method is displayed in Listing 8.7.

Listing 8.7: Method `make_affine_from_plane_params()` from `Pcl_utils` library

```
1  // given plane parameters of normal vec and distance to plane, construct and return an↵
        Eigen Affine object
2  // suitable for transforming points to a frame defined on the plane
3
4  Eigen::Affine3f PclUtils::make_affine_from_plane_params(Eigen::Vector4f ↵
        plane_parameters) {
5      Eigen::Vector3f plane_normal;
6      double plane_dist;
7      plane_normal(0) = plane_parameters(0);
8      plane_normal(1) = plane_parameters(1);
9      plane_normal(2) = plane_parameters(2);
10     plane_dist = plane_parameters(3);
11     return (make_affine_from_plane_params(plane_normal, plane_dist));
12  }
13  //this version takes separate args for plane normal and plane distance
14  Eigen::Affine3f PclUtils::make_affine_from_plane_params(Eigen::Vector3f plane_normal, ↵
        double plane_dist) {
15     Eigen::Vector3f xvec,yvec,zvec;
16     Eigen::Matrix3f R_transform;
17     Eigen::Affine3f A_transform;
18     Eigen::Vector3f plane_origin;
19     // define a frame on the plane, with zvec parallel to the plane normal
20     zvec = plane_normal;
21     if (zvec(2)>0) zvec*= -1.0; //insist that plane normal points towards camera
22     // this assumes that reference frame of points corresponds to camera w/ z axis ↵
            pointing out from camera
23     xvec<< 1,0,0; // this is arbitrary, but should be valid for images taken w/ zvec= ↵
            optical axis
24     xvec = xvec - zvec * (zvec.dot(xvec)); // force definition of xvec to be ↵
            orthogonal to plane normal
25     xvec /= xvec.norm(); // want this to be unit length as well
26     yvec = zvec.cross(xvec);
27     R_transform.col(0) = xvec;
28     R_transform.col(1) = yvec;
29     R_transform.col(2) = zvec;
30     //cout<<"R_transform = :"<<endl;
31     //cout<<R_transform<<endl;
32     if (plane_dist>0) plane_dist*=-1.0; // all planes are a negative distance from the↵
            camera, to be consistent w/ normal
33     A_transform.linear() = R_transform; // directions of the x,y,z axes of the plane's↵
            frame, expressed w/rt camera frame
34     plane_origin = zvec*plane_dist; //define the plane-frame origin here
35     A_transform.translation() = plane_origin;
36     return A_transform;
37  }
```

Lines 20 and 21 test and correct (as necessary) the surface normal of the provided patch of points. Similarly, line 32 assures that the signed distance of the plane from the camera's origin is negative.

This function constructs a coordinate system on the plane defined by the plane parameters. The coordinate system has its z axis equal to the plane normal. The x and y axes are constructed (lines 23 through 26) to be mutually orthogonal and to form a right-hand coordinate system together with the z axis. These axes are used to populate the columns of a 3×3 orientation matrix (lines 27 through 29), which constitutes the `linear` field of an `Eigen::Affine` object (line 33). The origin of the constructed coordinate system is defined to be the point on the plane closest to the camera-frame origin. This is computed as a displacement from the camera origin, in the direction opposite the plane normal, by (signed) distance equal to the plane-distance parameter (line 34). The resulting vector is stored as the translation() field of the `Eigen::Affine` object. The resulting populated affine object is returned.

After function `find_indices_of_plane_from_patch()` calls function `make_affine_from_plane_params()` (line 45 of Listing 8.6), the resulting affine is used to transform all of the input-cloud data. Noting that object `A_plane_wrt_camera` specifies the origin and orientation of a coordinate frame defined on the plane, the input data can be transformed into the plane frame using the inverse of this affine object. A PCL function, `transformPointCloud()` can be used to transform a complete point cloud. This is invoked on line 57 of `find_indices_of_plane_from_patch()`, specifying the input cloud (`*input_cloud_ptr`), an object to hold the transformed cloud (`*transformed_cloud_ptr`) and the desired transformation to apply to the input (`A_plane_wrt_camera.inverse()`).

```
pcl::transformPointCloud(*input_cloud_ptr, *transformed_cloud_ptr, A_plane_wrt_camera.↵
    inverse());
```

After being transformed into the coordinate frame defined on the plane, the data becomes easier to interpret. Ideally, all points coplanar with the plane's coordinate frame will have a z component of zero. In dealing with noisy data, one must specify a tolerance to accept points that have nearly-zero z components. This is done using PCL filters in lines 60 through 64:

```
pcl::PassThrough<pcl::PointXYZRGB> pass; //create a pass-through object
pass.setInputCloud(transformed_cloud_ptr); //set the cloud we want to operate on-
pass.setFilterFieldName("z"); // "filter" based on some range of z-value
pass.setFilterLimits(-0.02, 0.02); //set range: z value near zero, -0.02<z<0.02
pass.filter(indices); //  returns indices of the points that pass our test
```

A `PassThrough` filter object is instantiated from the PCL library that operates on colored point clouds. This object is directed to operate on the point cloud specifed using the `setInputCloud()` method. The desired filter operation is specified by the `setFilterFieldName()` method, which in this case is set to accept (pass through) points based on inspection of their z coordinates. The `setFilterLimits()` method is called to specify the tolerance; here, points with z values between -2 cm and 2 cm will be considered sufficiently close to zero. Finally, the method `filter(indices)` is called. This method performs the filtering operation. All points within the specified input cloud that have z values between -0.02 and 0.02 are accepted, and these accepted points are identified in terms of their indices in the point data of the input cloud. These indices are returned within the vector of integer indices. The result is identification of points within the original point cloud that are (nearly) coplanar with the user's selected patch.

The calling function, the main program within `find_plane_pcd_file` given in Listing 8.5, invokes the function `find_indices_of_plane_from_patch()` (line 103) whenever the user selects a new set of points in `rviz`. This function returns with a list of points (in `indices`) that qualify as coplanar with the selected patch. Although these indices were selected from the transformed input cloud (expressed in the constructed plane's frame), the same indices apply to the points in the original cloud (expressed with respect to the camera frame). The corresponding points are extracted from the original point cloud in line 104 using the PCL method `copyPointCloud()`:

```
pcl::copyPointCloud(*downsampled_kinect_ptr, indices, *plane_pts_ptr); //extract these↵
    pts into new cloud
```

With this operation, the points referenced by index values in `indices` are copied from the down-sampled point cloud into a new cloud called `plane_pts`. This cloud is converted to a ROS message (line 106) and published (line 109) using a publisher object (line 69) defined to publish to topic `planar_pts`.

By running the node `find_plane_pcd_file` on the same input file as before, points were selected from the `rviz` display as shown in Fig 8.4. A rectangle of points on the stool

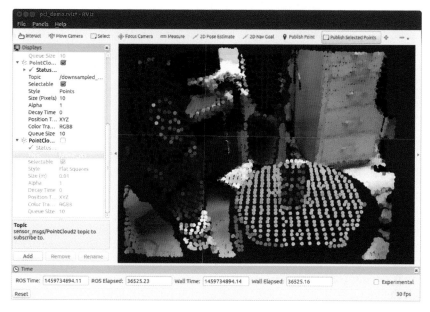

Figure 8.4: Scene with down-sampled point cloud and patch of selected points (in cyan)

has been selected, as indicated by light cyan coloring of these points. The act of selecting points causes the `find_plane_pcd_file` node to compute a plane through the points, then find all points from the (down-sampled) point cloud that are (nearly) coplanar, and publish the corresponding cloud on topic `planar_pts`.

Figure 8.5 shows the points chosen from the down-sampled point cloud that were determined to be coplanar with the selected patch. The points on the surface of the stool are correctly identified. However, additional points lie in the same plane, including points on the furniture and the robot in the background, that are not part of the surface of the stool. The selected points could be further filtered to limit consideration to x and y values as well (*e.g.*, based on the computed centroid of the selected patch), thus limiting point extraction to only points on the surface of the stool. Points above the stool may also be extracted, yielding identification of points corresponding to objects on a surface of interest.

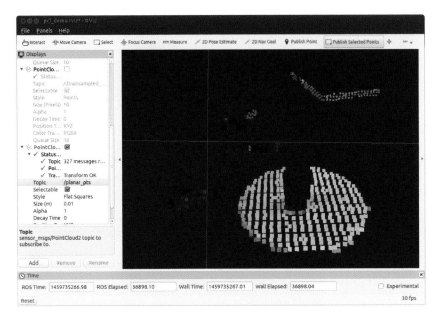

Figure 8.5: Points computed to be coplanar with selected points

8.5 OBJECT FINDER

The need for a robot to use vision for manipulation leads to introduction of another Kinect model. The `simple_kinect_model2.xacro` in package `simple_camera_model` contains a Kinect simulation with the camera pose similar to that which might be used by a robot, in this case elevated approximately 2 m from the floor and tilted looking downward, 1.2 rad from horizontal.

To provide an interesting scene for the camera, a world file has been defined in the `worlds` package, along with a complementary launch file. Running:

```
roslaunch worlds table_w_block.launch
```

brings up Gazebo with a world model containing the DARPA Robotics Challenge starting pen, a cafe table, and a toy block.

After starting Gazebo with this world model, one can add the Kinect model with the launch file:

```
roslaunch simple_camera_model kinect_simu2.launch
```

This launch file also brings up `rviz` with a saved settings file. The result of these two operations is shown in Fig 8.6. The colors in `rviz` are not consistent with the colors in Gazebo, due to an inconsistency of how Gazebo and `rviz` encode colors. However, emulated cameras in Gazebo use the model colors as seen in Gazebo, so this is not an issue for development of vision code incorporating color. The block can be seen on the table in both the Gazebo model and in the Kinect data displayed in `rviz`. The camera model in the Gazebo scene is simply a white rectangular prism, apparently floating in space (since there is no visual model for a link supporting the camera).

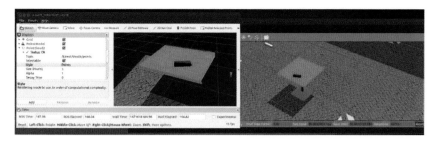

Figure 8.6: Scene of object on table viewed by Kinect camera

From the Gazebo display, selecting the model `toy_block`, one can see that the block coordinates are $(x, y, z) = (0.5, -0.35, 0.792)$, $(R, P, Y) = (0, 0, 0.43)$. The block model has a thickness of 0.035 m, and the block's frame is in the middle of the prism.

A PCL processing node, `object_finder_as`, offers an action server that accepts an object code and attempts to find the coordinates of that object in the scene. Highlights of this action server are described in the following.

At the start of the main function, an object of the class `ObjectFinder` is instantiated:

```
ObjectFinder object_finder_as; // create an instance of the class "ObjectFinder"
```

This class is defined in the same file. It instantiates `objectFinderActionServer`. It also instantiates an object of the class `PclUtils` (from the package `pcl_utils`), which contains a collection of functions that operate on point clouds, as well as a subscriber to `/kinect/depth/points`, which receives publications of point clouds from the (simulated) Kinect camera.

Also within `main()`, the transform from the Kinect sensor frame to the world frame is found using a transform listener (line 233):

```
tfListener.lookupTransform("base_link", "kinect_pc_frame", ros::Time(0), ↵
    stf_kinect_wrt_base);
```

This transform is tested until it is successfully received, then the resulting transform is converted to an Eigen type:

```
g_affine_kinect_wrt_base = object_finder_as.xformUtils_.transformTFToAffine3f(↵
    tf_kinect_wrt_base);
```

This transform is subsequently used to transform all Kinect point-cloud points to express them with respect to the world frame. (It should be noted that achieving sufficient precision of this transform requires extrinsic calibration; see Section 16.1. In the present case with the simulator, the transform is known unrealistically precisely.)

Once `main()` has completed these initialization steps, the action-server waits for goal messages, and work in response to receipt of a goal message is performed within the callback function `ObjectFinder::executeCB()`.

At the start of this callback, a `snapshot` command is sent to the `pclUtils` object (line 139 of `object_finder_as`):

```
pclUtils_.reset_got_kinect_cloud();
```

Next, (line 147, within executeCB()), the point cloud obtained is transformed to world coordinates using the Kinect camera transform obtained at start-up:

```
pclUtils_.transform_kinect_cloud(g_affine_kinect_wrt_base);
```

After this transformation, all point cloud points are expressed in coordinates with respect to the world frame.

With the points transformed into world coordinates, it is easier to find horizontal surfaces, since these correspond to points with (nearly) identical z coordinates. On line 153, a pclUtils function. find_table_height(), is called to find the height of a table. This function filters the point-cloud data to retain only points that lie within a specified range: (xmin, xmax, ymin, ymax, zmin, zmax, dz_resolution). Within this function, points are extracted in thin, horizontal slabs, each of thickness dz_resolution. Stepping through these filters from zmin to zmax, the number of points in each layer is tested, and the layer with the most points is assumed to be the height of the table surface. This function is invoked in executeCB as:

```
table_ht = pclUtils_.find_table_height(0.0, 1, -0.5, 0.5, 0.6, 1.2, 0.005);
```

After finding the table height, executeCB switches to a case corresponding to an object code specified in the goal message. (For example, to find the TOY_BLOCK model, a client of this server should set the object code in the goal message to ObjectIdCodes::TOY_BLOCK_ID defined in the header file <object_manipulation_properties/object_ID_codes.h>.) The approach taken in this simple object finder action server example is to have a special-case recognition function for an explicit list of objects. More generally, an object finder server should consult a network-based source of objects, with object codes and associated properties useful for recognition, grasp and usage.

In the present construction, ObjectIdCodes::TOY_BLOCK_ID switches to a case that invokes (line 177):

```
found_object = find_toy_block(surface_height, object_pose); //special case for toy ↩
    block
```

The special-purpose function find_toy_block() (lines 72-107) invokes the pclUtils function find_plane_fit():

```
pclUtils_.find_plane_fit(0, 1, -0.5, 0.5, surface_height + 0.035, surface_height +↩
    0.06, 0.001,
        plane_normal, plane_dist, major_axis, centroid);
```

This function filters points to within a box bounded by parameters (xmin, xmax, ymin, ymax, zmin, zmax) and finds the horizontal slab of thickness dz within this range that contains the most points (thus identifying points nominally on the top surface of the block). This function then fits a plane to these points. Knowledge of the table-surface height and block dimensions helps to identify points on the top surface of the block. These points are nearly all coplanar. A plane is fit to this data using an eigenvector-eigenvalue approach, from which we obtain the centroid, the surface normal, and the major axis of the block.

The object finder action server returns a best estimate of the coordinate frame of the requested object. In the case of the block, the top surface is used to help establish a local coordinate system. However, knowing that the block thickness is 0.035, this offset is applied to the z component of the frame origin in order to return a coordinate frame consistent with the model frame (which is in the center of the block).

An example action client to test this server is `example_object_finder_action_client`. The example code can be run as follows. With Gazebo initialized as per Fig 8.6, start the object finder server with:

```
rosrun object_finder object_finder_as
```

Additionally, we can use the triad display node introduced in Section 5.1.2 to display computed frame results:

```
rosrun example_rviz_marker triad_display
```

The test action client node can be started with:

```
rosrun object_finder example_object_finder_action_client
```

This node sends a goal request to the object finder action server to find a toy-block model in the current Kinect scene. Screen output from this node displays:

```
got pose x,y,z = 0.497095, -0.347294, 0.791365
got quaternion x,y,z, w = -0.027704, 0.017787, -0.540053, 0.840936
```

The centroid is very close to the expected values of $(0.5, -0.35, 0.792)$. The quaternion has small x and y terms, indicating that the object rotation corresponds to a rotation about vertical, as expected.

To visualize the result, the perceived object pose is illustrated in `rviz` with a marker (by publishing the identified pose to the `display_triad` node). The result is shown in Fig 8.7. The origin of the displayed frame appears to be correctly located in the middle of the block. Further, the z axis points upward, normal to the top face, and the x axis of the estimated block frame points along the major axis of the block, as desired.

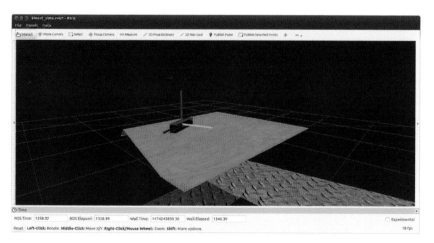

Figure 8.7: Object-finder estimation of toy-block model frame using PCL processing. Origin is in the middle of the block, z axis is normal to the top face, and x axis is along the major axis.

A notable limitation of this simple object finder is that it cannot handle multiple objects on a table. In contrast, the function to identify the table surface will be immune to multiple objects on the surface, provided enough of the table surface is exposed. A natural extension of this simple example would be to cluster points that lie above the table surface, then separately apply object recognition and localization algorithms to the separate clusters of points. Another limitation of this example is its focus on recognizing a single object type. Means to recognize alternative object types would need to be implemented, and the algorithms used might be very different (*e.g.* an application of deep learning is a candidate).

Despite the limitations of this simple example, it nonetheless illustrates conceptually how one may go about designing PCL-based code to recognize and localize objects. The intent is that an action server should allow clients to request poses of specific objects.

This example will be revisited in Part VI in the context of integrating perception and manipulation.

8.6 WRAP-UP

Cameras are the most popular sensors in robotics, since they provide high-resolution information about objects in the environment. Non-contact sensing is valuable for planning navigation and manipulation. Increasingly, 3-D sensing is more common, including stereo vision, panoramic LIDARs, and use of structured light for depth perception.

The OpenCV and Point-Cloud Library projects created tools for developers to interpret camera and depth-camera data. Higher-level interpretation of scenes, including object identification and localization, will certainly continue to evolve, as such capability is essential for robots operating autonomously in unstructured environments. Excellent progress toward this end is the Object Recognition Kitchen (`http://wiki.ros.org/object_recognition`), which integrates a variety of pattern matching and localization techniques to recognize and localize objects.

IV

Mobile Robots in ROS

INTRODUCTION

Mobile-robot navigation can be subdivided roughly into path planning and driving. A path planner is responsible for proposing a collision-free, efficient and navigable path to be followed. A driving subsystem is responsible for controlling the robot to execute such a plan safely and precisely. Achieving good navigation requires development of multiple, inter-related subsystems.

Global path planning requires a model (map) of the environment, and thus mapmaking and representation are important topics.

Global path planning typically is performed with simplifying assumptions on dynamics, resulting in coarse plans. These plans are converted to efficient, achievable trajectories through trajectory planning.

Steering algorithms are used to achieve precise following of plans, but this requires knowledge of path-following errors to be corrected. Localization is thus a crucial module used to obtain robot coordinates from which one can compute path-following errors.

The subject of path planning branches into a variety of cases, including *a priori* specification of a path to be followed, autonomous path planning based on knowledge of a map, planning with uncertainty or partial map information, local path planning in unknown environments with global positioning available, or exploration behaviors without benefit of maps or global positioning information.

These topics are broad and are subjects of ongoing research. It is not presumed that this treatment will cover the field of mobile robotics. Rather, use of ROS for mobile platforms will be described with respect to illustrative algorithms. Further, the examples here will be limited to differential-drive ground vehicles, although the concepts are extensible to more complex systems, including holonomic vehicles, walking and crawling robots, aerial vehicles, and underwater vehicles.

CHAPTER 9

Mobile-Robot Motion Control

CONTENTS

INTRODUCTION
The driving subsystem of a mobile robot is responsible for making it follow a desired path with sufficient precision and safety. This typically involves three subsystems: a desired-state generator that specifies a viable sequence of states to lead the robot through a desired path; a robot state estimator that can be used to estimate deviations of the robot from the desired trajectory; and a steering algorithm that evaluates estimated states relative to desired states to compute and execute corrective actions. These three topics will be introduced in this chapter. For the purposes of this chapter, it will be assumed that a desired path has been specified. The subject of path planning (means to compute a desired path) will be deferred to Chapter 10.

The topics of desired-state generation and steering are integrated within the nav-stack in ROS (introduced in Chapter 10). However, one would want to substitute the default nav-stack steering algorithm for a custom-designed steering algorithm optimized for a specific platform to achieve more reliable and more precise motion control.

9.1 DESIRED STATE GENERATION

The purpose of a desired-state generator is to compute and stream a sequence of robot states that corresponds to following a specified path while conforming to dynamic constraints (limitations on angular and translational velocities and accelerations). The resulting desired states can be used either for open-loop control or closed-loop control.

9.1.1 From paths to trajectories

The state of a mobile platform corresponds to its pose and twist. To be explicit, one must declare a reference frame on the vehicle as well as some reference frame of interest (*e.g.* coordinates of a map, absolute GPS coordinates, or simply coordinates relative to wherever the robot starts up). The vehicle frame's pose can be specified as its position (x, y, z coordinates of the origin of the vehicle frame with respect to the chosen reference frame) and its orientation (*e.g.* a quaternion expressing the orientation of the vehicle frame with respect to the reference frame). Additionally, the vehicle's twist is expressed as x, y and z velocities, as well as x, y and z rotational velocities with respect to the reference frame.

For a wheeled vehicle navigating on a plane, the pose can be expressed more compactly in terms of x, y and heading (ψ) coordinates. In the present treatment, orientation sometimes will be referred to informally as "heading" and will use the variable ψ. Use of "heading" in place of "orientation" for vehicles confined to a plane implies a scalar angle, which is simpler and more easily visualized than a full quaternion. The term "heading" can be ambiguous, since traditional navigation defines heading relative to north with angle measured clockwise. Here, heading will be defined in conventional engineering terms: the angle of the vehicle's x axis relative to the reference frame x axis, measured from reference frame to vehicle frame as a positive rotation about the reference frame z axis. In the plane (where the z axis points up, normal to the plane), this is a counter-clockwise rotation from the reference-frame x axis to the vehicle-frame x axis.

For navigation in the plane, one can convert orientation to heading as follows:

Listing 9.1: Function to convert orientation quaternion to scalar heading angle in plane

```
double   convertPlanarQuat2Psi(geometry_msgs::Quaternion quaternion) {
    double quat_z = quaternion.z;
    double quat_w = quaternion.w;
    double psi = 2.0 * atan2(quat_z, quat_w); // conversion from quaternion to heading↩
        for planar motion
    return psi;
}
```

This function takes a quaternion (expressed as a ROS message type, `geometry_msgs::Quaternion`) and converts this to a simple rotation in the x-y plane. The x and y components of a quaternion corresponding to a pure rotation about z are simply zero.

The inverse function, converting a heading angle into a quaternion, can be performed as:

Listing 9.2: Function to convert heading of planar-motion vehicle to quaternion

```
geometry_msgs::Quaternion convertPlanarPsi2Quaternion(double psi) {
    geometry_msgs::Quaternion quaternion;
    quaternion.x = 0.0;
    quaternion.y = 0.0;
    quaternion.z = sin(psi / 2.0);
    quaternion.w = cos(psi / 2.0);
    return (quaternion);
}
```

This function accepts a scalar heading angle and converts it to a ROS quaternion. Note that these conversions apply only to motion confined to a plane, for which only one degree of freedom of rotation is unconstrained. In this context, "heading" and "orientation" can be used interchangeably.

In addition to pose, a robot state requires specification of the twist. A twist is a vector of six velocities: three translational velocities and three rotational velocities. For motion in a plane, only three of these six components are relevant: v_x, v_y and ω_z. The remaining components should be set to zero.

A ROS message type that is conventionally used to express a robot's state is `nav_msgs/Odometry`, which is comprised of the following components:

```
std_msgs/Header header
  uint32 seq
  time stamp
  string frame_id
string child_frame_id
geometry_msgs/PoseWithCovariance pose
  geometry_msgs/Pose pose
    geometry_msgs/Point position
      float64 x
      float64 y
      float64 z
    geometry_msgs/Quaternion orientation
      float64 x
      float64 y
      float64 z
      float64 w
  float64[36] covariance
geometry_msgs/TwistWithCovariance twist
  geometry_msgs/Twist twist
    geometry_msgs/Vector3 linear
      float64 x
      float64 y
      float64 z
    geometry_msgs/Vector3 angular
      float64 x
      float64 y
      float64 z
  float64[36] covariance
```

This message type can be used to publish desired states since it contains fields for both pose and twist. It is also useful to specify the reference frame in which the states are specified, which can be done via the `frame_id` component of the header. The covariance fields (which are useful in the context of state estimation) will not be needed for specifying desired states. Their existence makes the odometry data structure more cumbersome than necessary, but use of the odometry message is conventional in ROS, and using this datatype for desired states will make these messages consistent with state estimation publications.

It is also conventional to send a path request to a mobile robot using the `nav_msgs/Path` message type, which is defined as:

```
std_msgs/Header header
  uint32 seq
  time stamp
  string frame_id
geometry_msgs/PoseStamped[] poses
```

```
std_msgs/Header header
  uint32 seq
  time stamp
  string frame_id
geometry_msgs/Pose pose
  geometry_msgs/Point position
    float64 x
    float64 y
    float64 z
  geometry_msgs/Quaternion orientation
    float64 x
    float64 y
    float64 z
    float64 w
```

Note that specification of a path contains a vector (variable-length array) of (time-stamped) poses. This specification is exclusively geometric. It does not contain any velocity nor timing information. It may also be the case that a path is defined very coarsely, *e.g.* in terms of via points. In order to follow the path with consideration of speed and acceleration limits, the path should be augmented to create a virtually continuous stream of states (poses and twists). Such translation converts a path into a trajectory. The resulting stream of outputs constitutes desired states, and the corresponding node is a desired-state generator.

Desired state messages should be updated and published at a relatively high frequency (*e.g.* 50 Hz would be suitable for slowly moving mobile robots). The sequence of states should be smooth and should conform to constraints on achievable speeds and accelerations.

To illustrate the design of a desired-state generator, we will consider a simple path description: a polyline, which is a simple sequence of line segments connected at vertices. A polyline path is easily specified in a `nav_msgs/Path` message. Here, a `Path` message will be interpreted literally to mean the robot should follow the corresponding polyline, which requires a spin-in-place rotation at each vertex. Given non-zero mass and rotational inertia of the robot and non-infinite translational and rotational acceleration limits, it follows that a polyline path requires that the robot come to a full halt at each vertex of the path. Following the path corresponds to pure forward translation along each line segment followed by pure rotation at each vertex. As a consequence, polyline paths are inefficient.

A more sophisticated trajectory generator would consider fitting higher-order path segments through the specified via points, *e.g.* including circular arcs or splines. Execution of motion along such paths would coordinate simultaneous translational and rotational velocities to sweep through a continuous trajectory without halting. While polyline paths are assumed here for simplicity, the approach presented here is extensible to more sophisticated trajectories.

The process of converting the crude `Path` specification to fine-grained sequence of states is sometimes referred to as filtering. This process can be performed in real time while the robot is running, or it can be performed in advance as a pre-computation. In order to assign viable desired states, some degree of look-ahead is required, where the look-ahead distance must exceed the braking distance required to come to a full halt. Polyline paths are the simplest case, since the robot must come to a full halt at each vertex of the path. Thus the required look-ahead distance corresponds to the next subgoal (vertex) in the vector of path poses.

With the polyline simplification, there are only two behaviors to consider: rotate from an initial heading to a desired heading, and move forward for some desired distance. In both cases, the robot should come to a complete halt at the end of the motion. Doing so requires that the velocity (linear or angular) ramp up from zero to some peak velocity, then ramp back down to zero velocity just as the desired goal pose is reached. A common technique for profiling the velocity is the trapezoidal velocity profile.

A trapezoidal velocity profile provides a simple means to construct dynamically feasible trajectories. If a robot's trajectory must induce a net displacement of d_{travel} and must conform to a maximum velocity and a maximum acceleration (translational or rotational), the trapezoidal velocity profile approach involves three phases:

1. Ramp up the velocity from initial velocity (zero, in our case) to some peak, v_{cruise}, at constant ramp-up acceleration, a_{RU}.

2. Maintain constant velocity, v_{cruise}, for some distance, d_{cruise} (equivalently, some time, $t_{cruise} = d_{cruise}/v_{cruise}$).

3. Ramp down the velocity from v_{cruise} to zero at constant ramp-down deceleration, a_{RD}.

The values of v_{cruise} and d_{cruise} must be chosen to satisfy the velocity constraint $|v_{cruise}| < v_{speedLimit}$ and to achieve the desired total travel distance, d_{travel}. In the above, the speed $v_{speedLimit}$ could correspond to the maximum achievable velocity of the robot, or it could be defined to be a more conservative speed limit (*e.g.* for navigating cautiously through constricted or risky areas). Similarly, the values of ramp-up and ramp-down accelerations, a_{RU} and a_{RD}, could be set to correspond to the physical acceleration and deceleration limits of the robot, or these accelerations could be set to lower values for gentler motion (although a_{RD} might be set to physical limits, if an emergency braking condition is called for).

The trapezoidal velocity profile constraints can be satisfied as follows. Define a maximum ramp-up distance as $d_{RUmax} = \frac{1}{2}v_{speedLimit}^2/a_{RU}$. Define a maximum ramp-down distance as $d_{RDmax} = \frac{1}{2}v_{speedLimit}^2/a_{RD}$. These values define two cases to consider for computing the velocity profile: short moves and long moves. A short move corresponds to $d_{travel} \leq d_{RUmax} + d_{RDmax}$, and long moves are $d_{travel} > d_{RUmax} + d_{RDmax}$.

For short moves, the feasible velocity profile will not be able to reach $v_{speedLimit} = v_{cruise}$. Rather, the peak velocity will be $v_{peak} < v_{speedLimit}$, and the velocity profile will be triangular ($d_{cruise} = 0$, *i.e.* a degenerate trapezoid) instead of trapezoidal. The triangular profile will consist of a ramp-up phase covering distance $d_{RU} = \frac{1}{2}v_{peak}^2/a_{RU}$, and a ramp-down phase that covers an additional ramp-down distance, $d_{RD} = \frac{1}{2}v_{peak}^2/a_{RD}$. These two phases must cover the desired distance, so $d_{travel} = d_{RU} + d_{RD}$. Given the desired travel distance and the ramp-up and ramp-down accelerations, one can solve for the corresponding v_{peak} as:

$$v_{peak} = \sqrt{2d_{travel}a_{RD}a_{RU}/(a_{RD} + a_{RU})} \qquad (9.1)$$

For the common case of $a_{RD} = a_{RU} = a_{ramp}$, this simplifies to $v_{peak} = \sqrt{d_{travel}a_{ramp}}$. Given v_{peak}, the ramp-up and ramp-down times are $\Delta t_{RU} = v_{peak}/a_{RU}$ and $\Delta t_{RD} = v_{peak}/a_{RD}$. The total move time is $\Delta t_{move} = \Delta t_{RU} + \Delta t_{RD}$.

For the short-move case (*i.e.* triangular velocity profile case), the corresponding distance covered as a function of time, starting from time t_0, is:

$$d(t) = \begin{cases} \frac{1}{2}a_{RU}(t - t_0)^2 & \text{for } 0 \leq t - t_0 \leq \Delta t_{RU} \\ d_{travel} - \frac{1}{2}|a_{RD}|(\Delta t_{move} - (t - t_0))^2 & \text{for } t_{RU} \leq t - t_0 \leq \Delta t_{move} \end{cases} \qquad (9.2)$$

For the long-move case, $d_{travel} > d_{RUmax} + d_{RDmax}$, the velocity profile will be a trapezoid consisting of three phases. The first phase is velocity ramp-up, during which the acceleration will be constant at a_{RU}. The velocity will ramp up to v_{cruise} (which may be set to $v_{speedLimit}$, or to some more conservative value). This phase will cover a distance d_{RUmax} in ramp-up time $\Delta t_{RU} = v_{cruise}/a_{RU}$. In the second phase, the velocity will be held constant at

v_{cruise}. This phase will cover a distance $d_{cruise} = d_{travel} - d_{RUmax} - d_{RDmax}$ in time $\Delta t_{cruise} = d_{cruise}/v_{cruise}$. The third phase will be a velocity ramp-down, which will cover distance d_{RDmax} in time $\Delta t_{RD} = v_{cruise}/a_{RD}$. The total move time is $t_{move} = \Delta t_{RU} + \Delta t_{cruise} + \Delta t_{RD}$.

For the trapezoidal velocity profile case, the corresponding distance covered as a function of time, starting from time t_0, is:

$$d(t) = \begin{cases} \frac{1}{2}a_{RU}(t-t_0)^2 & \text{for } 0 \leq t - t_0 \leq \Delta t_{RU} \\ d_{RUmax} + v_{cruise}(t - \Delta t_{RU}) & \text{for } \Delta t_{RU} \leq t - t_0 < \Delta t_{RU} + \Delta t_{cruise} \\ d_{travel} - \frac{1}{2}|a_{RD}|(\Delta t_{move} - (t-t_0))^2 & \text{for } \Delta t_{RU} + \Delta t_{cruise} \leq t - t_0 \leq \Delta t_{move} \end{cases}$$

$$(9.3)$$

Implementation of these equations for generating feasible trajectories from path segments is illustrated in example software described next.

9.1.2 A trajectory builder library

The package `traj_builder` (in `Part_4` of the `learning_ros` repository) contains a library defining a class for constructing triangular and trapezoidal velocity profiles. The code is lengthy, and thus it is not detailed here in entirety.

The class definition for the `traj_builder` library is in `traj_builder.h`. The header includes default values for maximum angular and linear accelerations (assumed to be identical for ramp-up and ramp-down), for maximum angular and linear velocities, and for the time-step resolution of the trajectories to be constructed. A default minimum-distance path-segment tolerance is also defined. Polyline line segments shorter than this tolerance are ignored, in which case the trajectory only defines reorienting to a desired heading. All of these parameters can be changed by respective `set` functions defined in-line in the header file.

Several utility functions are defined as member functions. The function `min_dang(psi)` evaluates periodic alternatives of a desired delta angle and returns the option of smallest magnitude. The function `sat(x)` returns the saturated value of x, saturated at $+1$ or -1. The function `sgn(x)` returns the sign of x ($+1$, -1 or 0). Conversion from a quaternion to heading is performed by the function `convertPlanarQuat2Psi()`, and the reverse direction conversion is computed by `convertPlanarPsi2Quaternion()`. The function `xyPsi2PoseStamped(x,y,psi)` accepts a pose defined within a plane (described by coordinates x and y and heading ψ) and returns a `geometry_msgs::PoseStamped` object populated with the corresponding 6-D components. The remaining member functions perform computations to convert a path segment into a corresponding trajectory.

The main function in the `traj_builder` library is `build_point_and_go_traj()`. The contents of this function are provided in Listing 9.3.

Listing 9.3: Top-level function to convert path segment to trajectory (from `traj_builder.cpp`)

```
1  void TrajBuilder::build_point_and_go_traj(geometry_msgs::PoseStamped start_pose,
2          geometry_msgs::PoseStamped end_pose,
3          std::vector<nav_msgs::Odometry> &vec_of_states) {
4      ROS_INFO("building point-and-go trajectory");
5      nav_msgs::Odometry bridge_state;
6      geometry_msgs::PoseStamped bridge_pose; //bridge end of prev traj to start of new ↩
           traj
7      vec_of_states.clear(); //get ready to build a new trajectory of desired states
8      ROS_INFO("building rotational trajectory");
9      double x_start = start_pose.pose.position.x;
10     double y_start = start_pose.pose.position.y;
```

```
11      double x_end = end_pose.pose.position.x;
12      double y_end = end_pose.pose.position.y;
13      double dx = x_end - x_start;
14      double dy = y_end - y_start;
15      double des_psi = atan2(dy, dx); //heading to point towards goal pose
16      ROS_INFO("desired heading to subgoal = %f", des_psi);
17      //bridge pose: state of robot with start_x, start_y, but pointing at next subgoal
18      //  achieve this pose with a spin move before proceeding to subgoal with ↩
            translational
19      //  motion
20      bridge_pose = start_pose;
21      bridge_pose.pose.orientation = convertPlanarPsi2Quaternion(des_psi);
22      ROS_INFO("building reorientation trajectory");
23      build_spin_traj(start_pose, bridge_pose, vec_of_states); //build trajectory to ↩
            reorient
24      //start next segment where previous segment left off
25      ROS_INFO("building translational trajectory");
26      build_travel_traj(bridge_pose, end_pose, vec_of_states);
27  }
```

This function constructs trajectories one path segment at a time, which is useful for polyline paths. It takes arguments of a start pose, a goal pose, and a reference to a vector of nav_msgs::Odom objects, which it will fill with a sequence of states that correspond to a smooth and executable trajectory from start to goal. This function builds a trajectory that does the following:

- Finds heading from start to goal coordinates

- Computes a spin-in-place trajectory to point the robot toward the goal pose

- Computes a forward-motion trajectory to move in a straight line, stopping at the origin of the goal pose

The computed trajectories are stored in a vector of Odometry objects, sampled every time step dt_ (set by default to 20 ms, but changeable via the function set_dt(double dt)). Each Odometry object contains a state description corresponding to incremental subgoals. The pose components are populated based on x, y, and orientation ψ values for planar motion. The twist components are populated from corresponding forward velocity and rotational velocity. For polyline paths, the sequence of desired states starts at rest and ends at rest, first performing reorientation, then straight-line motion. Whether rotating or translating, the computed trajectories correspond to either triangular velocity profiles or trapezoidal velocity profiles, depending on the move distance and the acceleration and speed-limit parameters.

For the initial spin-in-place behavior, the robot must reorient to point toward the origin of the goal pose. Typically, this means the orientation of the goal pose will be ignored. However, if the start and goal poses have nearly identical (x, y) values, (as determined by the parameter path_move_tol_), the goal pose is interpreted to contain the desired goal heading, and only reorientation (no translation) is computed. Thus, if one wants to travel to a pose and reach both the specified coordinates and specified orientation, one may repeat the last pose value as an additional subgoal, and this will lead to a computed trajectory that concludes with re-orienting to the desired final heading.

In Listing 9.3, the vector of states to be computed is initially cleared (line 7). Lines 9 through 15 compute the heading required to point from the origin of the start pose to the origin of the goal pose. This heading, des_psi, becomes the goal of the initial part of the trajectory–a spin-in-place move. For this reorientation move, a bridge pose, (an intermediate goal pose), is constructed by copying the (x, y) coordinates of the start pose but changing the bridge pose heading to point toward the destination coordinates. Line 21 accomplishes this by setting the orientation of bridge_pose to a quaternion corresponding to the computed desired heading, des_psi.

On line 23, the function `build_spin_traj()` is called with arguments of the start pose, the bridge pose, and a reference to the vector of desired states. This function computes a profiled trajectory with ramp-up, possible constant velocity, and ramp-down of the angular velocity that is consistent with ending up at rest with heading `des_psi`.

On line 26, the function `build_travel_traj()` is called with arguments to specify starting at the bridge pose and terminating at the (x, y) coordinates of the goal pose, as well as a reference to the partially constructed vector of desired states. The `build_travel_traj()` function appends additional states to this vector to specify a triangular or trapezoidal velocity profile corresponding to straight-line motion from the initial (x, y) coordinates to the goal (x, y) coordinates. As noted, the robot will not necessarily end up oriented consistent with the orientation of the goal pose. To enforce re-orienting to a specified pose, the goal pose should be repeated as a goal, resulting in a spin-in-place motion with no translation motion.

The functions `build_spin_traj()` and `build_travel_traj()` evaluate the move distance corresponding to the arguments for start and goal pose, and these functions will call appropriate helper functions to build either a triangular or trapezoidal trajectory. One of these four functions, `build_triangular_travel_traj()`, is shown in Listing 9.4.

Listing 9.4: Low-level function to compute triangular velocity profile trajectory for straight-line path (from `traj_builder.cpp`)

```cpp
// constructs straight-line trajectory with triangular velocity profile,
// respective limits of velocity and accel
void TrajBuilder::build_triangular_travel_traj(geometry_msgs::PoseStamped start_pose,
        geometry_msgs::PoseStamped end_pose,
        std::vector<nav_msgs::Odometry> &vec_of_states) {
    double x_start = start_pose.pose.position.x;
    double y_start = start_pose.pose.position.y;
    double x_end = end_pose.pose.position.x;
    double y_end = end_pose.pose.position.y;
    double dx = x_end - x_start;
    double dy = y_end - y_start;
    double psi_des = atan2(dy, dx);
    nav_msgs::Odometry des_state;
    des_state.header = start_pose.header; //really, want to copy the frame_id
    des_state.pose.pose = start_pose.pose; //start from here
    des_state.twist.twist = halt_twist_; // insist on starting from rest
    double trip_len = sqrt(dx * dx + dy * dy);
    double t_ramp = sqrt(trip_len / accel_max_);
    int npts_ramp = round(t_ramp / dt_);
    double v_peak = accel_max_*t_ramp; // could consider special cases for reverse ↩
        motion
    double d_vel = alpha_max_*dt_; // incremental velocity changes for ramp-up

    double x_des = x_start; //start from here
    double y_des = y_start;
    double speed_des = 0.0;
    des_state.twist.twist.angular.z = 0.0; //omega_des; will not change
    des_state.pose.pose.orientation = convertPlanarPsi2Quaternion(psi_des); //constant
    // orientation of des_state will not change; only position and twist
    double t = 0.0;
    //ramp up;
    for (int i = 0; i < npts_ramp; i++) {
        t += dt_;
        speed_des = accel_max_*t;
        des_state.twist.twist.linear.x = speed_des; //update speed
        //update positions
        x_des = x_start + 0.5 * accel_max_ * t * t * cos(psi_des);
        y_des = y_start + 0.5 * accel_max_ * t * t * sin(psi_des);
        des_state.pose.pose.position.x = x_des;
        des_state.pose.pose.position.y = y_des;
        vec_of_states.push_back(des_state);
    }
    //ramp down:
```

```
43        for (int i = 0; i < npts_ramp; i++) {
44            speed_des -= accel_max_*dt_; //Euler one-step integration
45            des_state.twist.twist.linear.x = speed_des;
46            x_des += speed_des * dt_ * cos(psi_des); //Euler one-step integration
47            y_des += speed_des * dt_ * sin(psi_des); //Euler one-step integration
48            des_state.pose.pose.position.x = x_des;
49            des_state.pose.pose.position.y = y_des;
50            vec_of_states.push_back(des_state);
51        }
52        //make sure the last state is precisely where requested, and at rest:
53        des_state.pose.pose = end_pose.pose;
54        //but final orientation will follow from point-and-go direction
55        des_state.pose.pose.orientation = convertPlanarPsi2Quaternion(psi_des);
56        des_state.twist.twist = halt_twist_; // insist on starting from rest
57        vec_of_states.push_back(des_state);
58 }
```

The function `build_triangular_travel_traj()` takes three arguments: a start pose, an end pose and a reference to a vector of desired states. This function assumes that the robot is already pointing from the start pose toward the goal pose. Lines 6 through 12 extract the (x, y) coordinates from the start and end poses and compute the heading angle from start to pose, `psi_des`. An object of `nav_msgs::Odometry`, called `des_state`, is instantiated, and its header is populated with a copy of the start pose header (line 15). Primarily, the value of this is to retain the `frame_id` of the input poses. On line 16, all components of twist are set to zero in the `des_state`. While building the trajectory, only the value of x velocity will be altered; all of the other five components will remain 0.

The trip length (in meters) and the ramp-up time (in seconds) are computed on lines 17 and 18. The trajectory will be computed as a sequence of samples of state. The number of samples for the ramp-up phase is computed on line 19. The peak velocity that will be achieved in the triangular trajectory is computed on line 20.

The robot's initial state is prescribed by the (x, y) coordinates of the start pose, the required heading, `des_psi`, and initial velocities of 0 (lines 23 through 27). A loop to compute the ramp-up trajectory begins on line 31. In lines 33 through 39, the desired state is updated corresponding to time t (which is incremented by `dt_` each iteration of the loop). The updates are consistent with the equations presented in Section 9.1.1. These updates are populated in `des_state` (lines 38 and 39) and each such state is appended to the variable length array `vec_of_states` (line 40).

Lines 43 through 51 repeat the computations of state samples, but for the ramp-down phase. Instead of using an analytic expression in this loop, an alternative computation is performed using numerical integration (lines 44, 46 and 47). Because numerical integration can result in rounding errors, the final state of the trajectory is coerced to match the desired goal coordinates (lines 53 through 57).

When this function returns, the vector of states will contain the newly computed trajectory, appended to whatever contents the vector originally contained. If this vector is to be cleared to start computing a new trajectory, it must be cleared by the parent function that calls `build_triangular_travel_traj()`. (This is true of all four lower-level trajectory builder functions.)

The function `build_triangular_spin_traj()` is virtually identical to `build_triangular_travel_traj()`. The triangular spin trajectory builder ramps up angular velocity, then ramps down again. One important difference is that one must take into account the sign of the spin direction, which may be positive or negative. Thus, ramping up may consist of ramping to more negative angular velocities. The translational trajectory builder function could be augmented to consider negative motions as well, which would be necessary to perform backing up (reverse) motions.

The functions `build_trapezoidal_travel_traj()` and `build_trapezoidal_spin_traj()` perform equivalent logic, but for long moves that include a middle phase of travel at constant velocity. These functions behave similarly, taking arguments of start pose, end pose and reference to a vector of states and returning a computed trajectory in the vector of states.

A function `build_braking_traj()` is also declared. However, its implementation is empty. The intent of this function is to provide computation of a trajectory for a graceful halt (a ramp-down trajectory with desirable deceleration). Such a trajectory would be valuable for a robot to perform unanticipated braking, *e.g.* due to map errors or unexpected obstacles (*e.g.* debris or pedestrians). Planning a halt trajectory on the fly in an emergency situation might seem like unnecessary delay. However, such planning requires only about 1 ms to compute on an unexceptional computer, and thus this presents no significant delay. Implementation of this function is left as an exercise.

Illustration of use of the trajectory-builder library is provided by the node `traj_builder_example_main` in the `traj_builder` package. This node hard-codes start and goal poses, and it iterates computing trajectories from start to goal, then back from goal to start. Each computed trajectory is published, in sequence, every 20 ms to the topic `desState`. By running this function, the desired state values can be plotted with `rqt_plot`. An example result is shown in Fig 9.1, where the start coordinates are $(0,0)$ and the goal coordinates are $(2,-4)$. The velocity limits were set to large values to produce triangular velocity profiles. It can be seen from Fig 9.1 that the angular (blue) and translational (brown) velocities alternately ramp up and down triangularly. The corresponding heading (magenta), x position (green) and y position (orange) transition smoothly between the goal values as blended quadratics with an inflection point. A second example, shown in Fig 9.2, has specified start

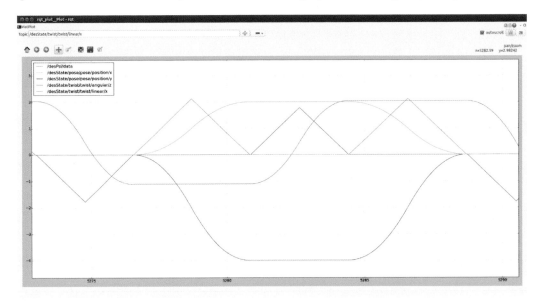

Figure 9.1: Example computed triangular velocity profile trajectories

coordinates of $(0,0)$ and goal coordinates of $(2,-4)$ with speed limits of 1.0 m/sec and 1.0 rad/sec. The resulting angular (blue) and translational (magenta) velocities have trapezoidal profiles, as expected. The corresponding x, y and ψ values are smooth, reach the desired goals, and satisfy the acceleration and velocity constraints.

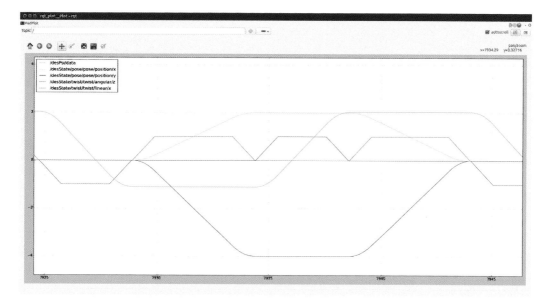

Figure 9.2: Example computed trapezoidal velocity profile trajectories

9.1.3 Open-loop control

A simple approach to using computed trajectories is open-loop control. To use it, merely copy the twist values of speed and spin rate from each desired state to command motion of a mobile robot via the `cmd_vel` topic. From `traj_builder_example_main`, the following lines build a trajectory, step through the trajectory (at 20 ms intervals), strip the twist component from each desired state, and publish that twist to the `cmd_vel` topic.

```
trajBuilder.build_point_and_go_traj(g_start_pose, g_end_pose, vec_of_states);
ROS_INFO("publishing desired states and open-loop cmd_vel");
for (int i = 0; i < vec_of_states.size(); i++) {
    des_state = vec_of_states[i];
    des_state.header.stamp = ros::Time::now();
    des_state_publisher.publish(des_state);
    des_psi = trajBuilder.convertPlanarQuat2Psi(des_state.pose.pose.↩
        orientation);
    psi_msg.data = des_psi;
    des_psi_publisher.publish(psi_msg);
    twist_commander.publish(des_state.twist.twist); //FOR OPEN-LOOP CTL ONLY!
    looprate.sleep(); //sleep for defined sample period, then do loop again
}
```

Note that publication of the twist values to `cmd_vel` is a special case. Commonly, the desired states should be published at the sample rate of the computed trajectories (as is done in the above code snippet), and these published states may be used for control. However, interpreting the desired states as open-loop commands and publishing to `cmd_vel` (via `twist_commander.publish(des_state.twist.twist);`) invokes open-loop control, which is seldom adequate.

Figure 9.3 shows an example result of commanding our model mobile robot (mobot) with open-loop commands. In this case, `traj_builder_example_main` was used to command the robot to go 5 m back and forth along the x axis. The angular acceleration was set to a low value ($0.1\ rad/sec^2$) to help improve pointing accuracy with slow turns. The computed trajectories include a trapezoidal velocity profile for forward motion (green) and triangular

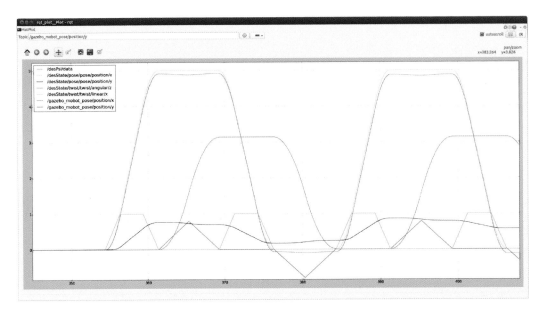

Figure 9.3: Example open-loop control of mobot. Intended motion is along x axis.

angular-velocity profiles for turning (blue). The corresponding desired y value remains 0 throughout the desired trajectory, and the desired x value has a smooth trajectory from 0 to 5 m, then (after a 180-degree turn) back again from 5 m to 0. An important feature of Fig 9.3 is the black trace, corresponding to the actual x value of the robot. Although the robot starts with perfect initial conditions ($x = 0$, $y = 0$, $\psi = 0$), it drifts off course fairly quickly. Initially, the robot is commanded to move straight ahead, but it drifts to its left by approximately 0.8 m. Although the robot model has no asymmetry, it still tends to drift to its left in simulation, presumably due to numerical imprecision in the Gazebo physics simulation. Although such asymmetry was not intentionally modeled, such behavior is commonplace. A mobile robot will not steer perfectly straight with open-loop motion commands. The degree of drift will vary with robot properties (*e.g.* the robot mechanics, its electronic motor driver, its speed controller, uneven wear of the drive wheels and influence of the uncontrolled casters) as well as terrain properties (*e.g.* irregular or slippery surfaces).

Figure 9.4 shows the result of open-loop control under the same conditions, except that the maximum angular acceleration has been increased to 1.0 rad/sec^2. In Fig 9.4, the blue trace shows that the robot has drifted off course by roughly 3 m. This emphasizes that use of open-loop control should be restricted to short travel distances and low accelerations, unless large path deviations are permissible (*e.g.* for random exploration).

9.1.4 Desired state publishing

The example node `traj_builder_example_main` illustrated use of the `traj_builder` library and publishing desired states at a fixed rate. This example, however, was restrictive in that goal poses were hard coded. Further, there was no provision for responding to sensor or emergency-stop (E-stop) commands.

A more flexible node for publishing desired states is contained in the package `mobot_pub_des_state`. This package includes a class definition for `DesStatePublisher`, which is described in the header file `pub_des_state.h` and implemented in the file `pub_des_state.cpp`. In this example, `DesStatePublisher` is not a library (although it

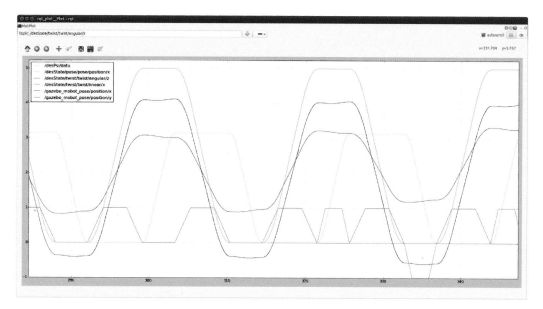

Figure 9.4: Example open-loop control of mobot with higher angular acceleration limit. Intended motion is along x axis.

could be). Instead, the CMakeLists.txt file compiles pub_des_state.cpp as a module together with pub_des_state_main.cpp with the instruction:

```
cs_add_executable(mobot_pub_des_state src/pub_des_state_main.cpp src/pub_des_state.cpp↩
    )
```

The main program, pub_des_state_main.cpp, shown in Listing 9.5, is very brief.

Listing 9.5: pub_des_state_main.cpp: desired state publisher, main program

```
1  #include "pub_des_state.h"
2
3  int main(int argc, char **argv) {
4      ros::init(argc, argv, "des_state_publisher");
5      ros::NodeHandle nh;
6      //instantiate a desired-state publisher object
7      DesStatePublisher desStatePublisher(nh);
8      //dt is set in header file pub_des_state.h
9      ros::Rate looprate(1 / dt); //timer for fixed publication rate
10     desStatePublisher.set_init_pose(0,0,0); //x=0, y=0, psi=0
11     //put some points in the path queue--hard coded here
12     desStatePublisher.append_path_queue(5.0,0.0,0.0);
13     desStatePublisher.append_path_queue(0.0,0.0,0.0);
14
15     // main loop; publish a desired state every iteration
16     while (ros::ok()) {
17         desStatePublisher.pub_next_state();
18         ros::spinOnce();
19         looprate.sleep(); //sleep for defined sample period, then do loop again
20     }
21 }
```

This program includes the header file for the DesStatePublisher class, which contains all other necessary header files. This header also sets a value for "dt", which is the sample period that will be used to generate trajectory arrays of desired states. Line 10 sets the initial pose of the robot. In practice, this should be based on sensors that estimate the robot's initial pose in some reference frame (*e.g.* a map frame).

Lines 12 and 13 show how one can append goal poses to a path queue. The function `append_path_queue()` should be used rarely in practice. Rather, points should be added to the path queue via a service client of `append_path_queue_service`.

Lines 16 through 20 comprise the main loop of this node. At each iteration, the member method `pub_next_state()` of the `desStatePublisher` object is called, which causes this object to advance one step within its state machine. Typically, this results in accessing the next state from a computed trajectory and publishing this state to the topic `/desState`, but several other behaviors can be invoked.

An object of class `DesStatePublisher` has four services. The `append_path_queue_service` expects a service request message as defined in the `mobot_pub_des_state` package. This service message, `path.srv`, contains a field `path` of type `nav_msgs/Path`. This message type contains a vector of poses. A service client can populate this vector of poses with any desired sequence of subgoals, then send it to the service `append_path_queue_service` to add the listed points to the end of the current path queue. An example program to do this is `pub_des_state_path_client.cpp`. This node creates a client of `append_path_queue_service`, defines five stamped poses, appends them one at a time to the vector of poses within the `nav_msgs/Path` message, then sends this message as a request to the append-path service.

With the `mobot_pub_des_state` node running, invoking `pub_des_state_path_client` sends five stamped poses to the desired-state publisher, which adds these poses as new subgoals to a C++ `queue` object of subgoals. An example result is shown in Fig 9.5. Starting

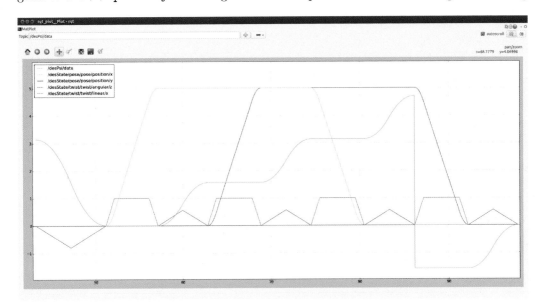

Figure 9.5: Trajectory generation for 5 m × 5 m square path

from $(x, y) = (0, 0)$ and with orientation $\psi = 0$, the desired x velocity has a trapezoidal profile (magenta trace) resulting in a smooth trajectory of x values advancing from 0 to 5 m (cyan trace). Once this position goal is reached, the desired states correspond to re-orientation via a triangular angular-velocity profile (blue trace), resulting in the heading changing from 0 to $\pi/2$ (green trace). At this heading, the desired y coordinate is increased smoothly to 5 m using a trapezoidal velocity profile. After reaching this subgoal, the heading is smoothly changed to π. The next advance brings the desired state to $(x, y) = (0, 5)$. This is followed by smooth motion to $(x, y) = (0, 0)$, then re-orientation to $\psi = 0$. This shows how the desired-state publisher can receive pose subgoals and sequence through them with smooth trajectories that conform to dynamic limitations.

When the path queue within the desired-state publisher is empty, the desired-state publisher continues to publish the last command sent (which should be a zero-twist command). With the addition of new subgoals (poses) in the queue, the `DesStatePublisher` object will process each new pose as a vertex in a polyline using the `traj_builder` library. While this behavior is convenient, a few additional features are needed.

An important addition is the ability to respond to an emergency-stop signal. A service is provided for this: `estop_service`. One can send a trigger (message type `std_srvs::Trigger`) to this service, which will invoke an E-stop. Emergency-stop logic can be complex, as memory is required. Upon initiating an E-stop, the desired-state node should compute a feasible deceleration profile; publish the states of this trajectory to completion; then hold the final state until the E-stop is reset. For E-stop reset, the desired-state publisher should plan a trajectory from the halted state to the next unattained subgoal, then invoke publications of this recovery trajectory.

Another important capability is the ability to clear the current path queue. If the robot is halted, *e.g.* due to an E-stop, it is often the case that the original plan is no longer valid. Subgoals in the path queue may need to be flushed before new subgoals can be specified. The service `flush_path_queue_service` accepts a trigger (message type `std_srvs::Trigger`) that causes the current path queue to be cleared.

The main function of the `DesStatePublisher` class is `pub_next_state()`, the contents of which are shown in Listing 9.6 (which is extracted from `pub_des_state.cpp`).

Listing 9.6: key function, `pub_next_state()` from `pub_des_state.cpp`

```
void DesStatePublisher::pub_next_state() {
    // first test if an e-stop has been triggered
    if (e_stop_trigger_) {
        e_stop_trigger_ = false; //reset trigger
        //compute a halt trajectory
        trajBuilder_.build_braking_traj(current_pose_, des_state_vec_);
        motion_mode_ = HALTING;
        traj_pt_i_ = 0;
        npts_traj_ = des_state_vec_.size();
    }
    //or if an e-stop has been cleared
    if (e_stop_reset_) {
        e_stop_reset_ = false; //reset trigger
        if (motion_mode_ != E_STOPPED) {
            ROS_WARN("e-stop reset while not in e-stop mode");
        }
        //OK...want to resume motion from e-stopped mode;
        else {
            motion_mode_ = DONE_W_SUBGOAL; //this will pick up where left off
        }
    }

    //state machine; results in publishing a new desired state
    switch (motion_mode_) {
        case E_STOPPED: //this state must be reset by a service
            desired_state_publisher_.publish(halt_state_);
            break;

        case HALTING: //e-stop service callback sets this mode
            //if need to brake from e-stop, service will have computed
            // new des_state_vec_, set indices and set motion mode;
            current_des_state_ = des_state_vec_[traj_pt_i_];
            current_des_state_.header.stamp = ros::Time::now();
            desired_state_publisher_.publish(current_des_state_);
```

```
35          current_pose_.pose = current_des_state_.pose.pose;
36          current_pose_.header = current_des_state_.header;
37          des_psi_ = trajBuilder_.convertPlanarQuat2Psi(current_pose_.pose.←┘
                orientation);
38          float_msg_.data = des_psi_;
39          des_psi_publisher_.publish(float_msg_);
40
41          traj_pt_i_++;
42          //segue from braking to halted e-stop state;
43          if (traj_pt_i_ >= npts_traj_) { //here if completed all pts of braking ←┘
                traj
44              halt_state_ = des_state_vec_.back(); //last point of halting traj
45              // make sure it has 0 twist
46              halt_state_.twist.twist = halt_twist_;
47              seg_end_state_ = halt_state_;
48              current_des_state_ = seg_end_state_;
49              motion_mode_ = E_STOPPED; //change state to remain halted
50          }
51          break;
52
53      case PURSUING_SUBGOAL: //if have remaining pts in computed traj, send them
54          //extract the i'th point of our plan:
55          current_des_state_ = des_state_vec_[traj_pt_i_];
56          current_pose_.pose = current_des_state_.pose.pose;
57          current_des_state_.header.stamp = ros::Time::now();
58          desired_state_publisher_.publish(current_des_state_);
59          //next three lines just for convenience--convert to heading and publish
60          // for rqt_plot visualization
61          des_psi_ = trajBuilder_.convertPlanarQuat2Psi(current_pose_.pose.←┘
                orientation);
62          float_msg_.data = des_psi_;
63          des_psi_publisher_.publish(float_msg_);
64          traj_pt_i_++; // increment counter to prep for next point of plan
65          //check if we have clocked out all of our planned states:
66          if (traj_pt_i_ >= npts_traj_) {
67              motion_mode_ = DONE_W_SUBGOAL; //if so, indicate we are done
68              seg_end_state_ = des_state_vec_.back(); // last state of traj
69              path_queue_.pop(); // done w/ this subgoal; remove from the queue
70              ROS_INFO("reached a subgoal: x = %f, y= %f",current_pose_.pose.←┘
                    position.x,
71                          current_pose_.pose.position.y);
72          }
73          break;
74
75      case DONE_W_SUBGOAL: //suspended, pending a new subgoal
76          //see if there is another subgoal is in queue; if so, use
77          //it to compute a new trajectory and change motion mode
78
79          if (!path_queue_.empty()) {
80              int n_path_pts = path_queue_.size();
81              ROS_INFO("%d points in path queue",n_path_pts);
82              start_pose_ = current_pose_;
83              end_pose_ = path_queue_.front();
84              trajBuilder_.build_point_and_go_traj(start_pose_, end_pose_,←┘
                    des_state_vec_);
85              traj_pt_i_ = 0;
86              npts_traj_ = des_state_vec_.size();
87              motion_mode_ = PURSUING_SUBGOAL; // got a new plan; change mode to ←┘
                    pursue it
88              ROS_INFO("computed new trajectory to pursue");
89          } else { //no new goal? stay halted in this mode
90              // by simply reiterating the last state sent (should have zero vel)
91              desired_state_publisher_.publish(seg_end_state_);
92          }
93          break;
94
95      default: //this should not happen
96          ROS_WARN("motion mode not recognized!");
97          desired_state_publisher_.publish(current_des_state_);
98          break;
99  }
```

The state-machine logic of function **pub_next_state()** is based on setting the **motion_mode_** member variable to one of the states: E_STOPPED, DONE_W_SUBGOAL,

PURSUING_SUBGOAL or HALTING. An illustration of this state-machine logic is shown in Fig 9.6. The state transition possibilities are as follows.

If an E-stop trigger is received (via the service estop_service), the flag e_stop_trigger_ is set to true. As a result, lines 3 through 10 of pub_next_state() reset the E-stop trigger; use the trajectory-builder object to compute a braking trajectory and put this trajectory in des_state_vec_; set motion_mode_ to HALTING; set the member variable npts_traj_ to the number of states in the braking trajectory; and initialize the state index traj_pt_i_ to 0. The state machine will then be in HALTING mode, prepared to perform planned braking.

A complement to triggering an E-stop is recovering from an E-stop. An E-stop is cleared via clear_estop_service, which sets the value of e_stop_reset_ to true. Lines 12 through 21 deal with this case. The e_stop_reset_ trigger is reset, and the motion mode is set to DONE_W_SUBGOAL. The effect of this state change is that the pub_next_state() function will treat this condition as though the robot had been halted because it was out of goals. Subsequently, if there is at least one goal in the queue, it will be processed according to the DONE_W_SUBGOAL logic. Note that an E-stop does not remove the current subgoal from the queue, so re-setting an E-stop can resume with a plan to reach this unattained pose. However, it might also be the case that flush_path_queue_service has emptied the queue during the E-stop condition (if higher-level code finds this appropriate), in which case the most recent subgoal will be forgotten.

After checking the status of reset triggers, pub_next_state() enters a switch-case block to handle the various motion mode cases.

The case (motion-mode state) HALTING (lines 29 through 51) provides the logic for ramping down velocities to a halt. The member variables current_des_state_ and current_pose_ are updated by using index traj_pt_i_ into the array (vector) des_state_vec_. Note that the trajectory in this vector will have been computed for braking due to an E-stop trigger. The extracted desired state is published to /desState. In addition, for debug and visualization purposes, lines 37 through 39 compute the desired scalar heading and publish this to topic desPsi. The index into the vector of desired states is then incremented and tested. When this index reaches the last value in the vector of desired states, lines 43 through 50 set current_des_state_ to the last state in the braking trajectory (for use in E-stop recovery), set an equivalent halt_state_ (with zero twist), and set the motion mode to E_STOPPED.

For case E_STOPPED, the halt_state_ is repeatedly published. The E_STOPPED mode does not advance the state machine. This is a terminal state, until or unless an E-stop reset condition is received (which results in the motion mode changing to DONE_W_SUBGOAL).

The case PURSUING_SUBGOAL is similar to HALTING. This mode steps through states from des_state_vec_, publishing them to the desState topic, updating current_des_state_, computing and publishing the scalar heading, and incrementing the index into current_des_state_ (lines 53 through 64). When the last point in the plan is reached, lines 66 through 71 perform the following. The motion mode is changed to DONE_W_SUBGOAL, the last state commanded is saved in seg_end_state_, and the current (achieved) subgoal is removed (popped) from the path queue.

Finally, the motion mode state DONE_W_SUBGOAL prepares a plan for achieving the next subgoal (lines 75 through 93). If there is at least one subgoal in the path queue (line 79), a new trajectory plan is computed using the trajectory builder library (lines 82 through 84), and the trajectory index is reset to zero, number of trajectory points in the plan is set, and the motion mode is changed to PURSUING_SUBGOAL for execution of the plan.

If there are no points in the path queue, DONE_W_SUBGOAL republishes the end state from the previously achieved subgoal. (This is initialized to the start state within the constructor of DesStatePublisher.)

This code provides a minimal set of capabilities for executing plans and responding to unexpected conditions. It is up to higher-level code to construct intelligent plans and invoke halts as appropriate (*e.g.* based on sensation of potential collision or unsafe terrain). At present, this code is limited to polyline paths. A useful extension would be to handle curved or spline paths to achieve more graceful and more efficient navigation.

An additional node, `open_loop_controller`, invokes open-loop control from publications of desired states. This node simply subscribes to the `desState` topic, strips off the twist term, and re-publishes to `cmd_vel`.

These nodes can be run as follows. First, start Gazebo with:

```
roslaunch gazebo_ros empty_world.launch
```

Load the mobile-robot model onto the parameter server and spawn into Gazebo with:

```
roslaunch mobot_urdf mobot.launch
```

Start the open-loop controller:

```
rosrun mobot_pub_des_state open_loop_controller
```

Start the desired-state publisher:

```
rosrun mobot_pub_des_state mobot_pub_des_state
```

Send a path request via a client node (a square path, in this case):

```
rosrun mobot_pub_des_state pub_des_state_path_client
```

In addition to running prescribed paths, the following services can be invoked:

- `rosservice call estop_service`

- `rosservice call clear_estop_service`

Note, though, that the E-stop behavior is not fully functional, since the braking-trajectory function within TrajBuilder is not implemented (left as an exercise for the reader).

The code presented here can be used for open-loop control to execute prescribed paths. However, open-loop control is seldom adequate. Often, a robot must navigate with precision, *e.g.* to pass through doorways, dock with a charger, or approach a workstation within a pose tolerance, or stay within a lane on a highway. Open-loop control uses only twist commands and makes no reference to the pose within each desired state. To take advantage of the pose component of desired state, the robot needs to know its actual pose in space. By comparing actual pose to desired pose, a steering algorithm can improve navigation precision. The subject of state estimation is introduced next.

pub_des_state state-machine logic

Inputs: e_stop_trigger;
e_stop_reset;
append_path;
flush_path

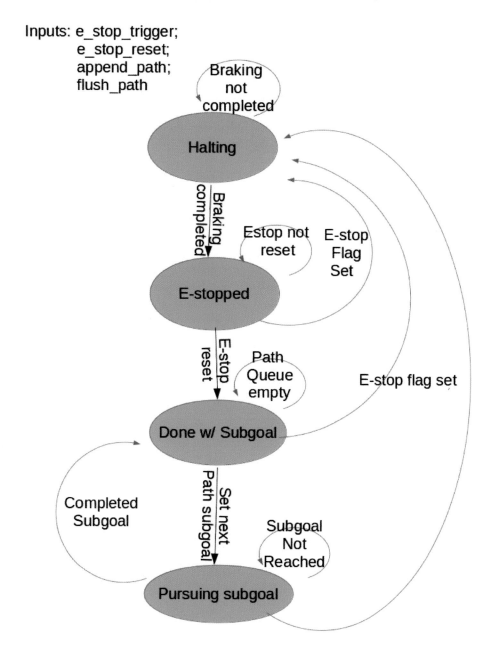

Figure 9.6: Logic of state machine for publishing desired states

9.2   ROBOT STATE ESTIMATION

Use of open-loop control results in drift from the intended path, which accumulates with distance travelled. Open-loop control is thus unsuitable for travelling significant distances or when there are demanding tolerances on path following (*e.g.* navigating through a narrow doorway). For more precise and reliable path following, a feedback steering algorithm that requires reference to both desired states and to estimated actual states of the system is used. State estimation is introduced in this section, starting with simple odometry and extending to use of GPS and LIDAR.

9.2.1   Getting model state from Gazebo

For development and analysis purposes only, it can be useful to get model states from Gazebo. One should be careful, however, not to rely on this information for code to be deployed on real robots. Gazebo has omniscient awareness of system states because it computes these states. On a physical system, however, the state with respect to the world is unknown. Considerable effort is often required to estimate the state of a system in the world.

For the purpose of evaluating system performance, one can consult the Gazeb published system state and use this information to compute the difference between estimated system state and actual system state (as computed by Gazebo). One can use Gazebo's published model states for developing steering algorithms, although it is important that the algorithms be tolerant of the non-idealities of state estimators. The ideal system state from Gazebo can also be sampled, modified and re-published to emulate realistic absolute sensors, such as Global Positioning Systems. By adding appropriate levels of noise to the Gazebo states, one can create a virtual sensor that is useful for development of localization algorithms.

Starting our mobot simulation with:

```
roslaunch gazebo_ros empty_world.launch
```

```
roslaunch mobot_urdf mobot.launch
```

we can see the topics published, including **gazebo/model_states**. This topic carries messages of type **gazebo_msgs/ModelStates**. Running:

```
rosmsg show gazebo_msgs/ModelStates
```

displays the format of this message type:

```
string[] name
geometry_msgs/Pose[] pose
  geometry_msgs/Point position
    float64 x
    float64 y
    float64 z
  geometry_msgs/Quaternion orientation
    float64 x
    float64 y
    float64 z
    float64 w
geometry_msgs/Twist[] twist
  geometry_msgs/Vector3 linear
```

```
        float64 x
        float64 y
        float64 z
   geometry_msgs/Vector3 angular
        float64 x
        float64 y
        float64 z
```

This message type contains a vector of names, a vector of poses and a vector of twists. The message is interpreted as follows: the order in which a model name appears in the `name` array is the same as the order in which the respective pose and twist appear in the pose and twist arrays. A `rostopic echo gazebo/model_states` example output is:

```
name: ['ground_plane', 'mobot']
pose:
  -
    position:
      x: 0.0
      y: 0.0
      z: 0.0
    orientation:
      x: 0.0
      y: 0.0
      z: 0.0
      w: 1.0
  -
    position:
      x: 0.00080270286594
      y: 0.170554714089
      z: -1.95963022377e-05
    orientation:
      x: -1.33139494634e-05
      y: 0.0020008448071
      z: 0.0033807500955
      w: 0.999992283456
twist:
  -
    linear:
      x: 0.0
      y: 0.0
      z: 0.0
    angular:
      x: 0.0
      y: 0.0
      z: 0.0
  -
    linear:
      x: 2.66486083024e-05
      y: 0.000452450930556
      z: -0.00510658997482
    angular:
      x: 5.29981190091e-05
      y: -0.00334270636901
      z: -3.15214058467e-06
```

From the echo, we see that Gazebo is publishing the states of only two models:

ground_plane and mobot. The ground plane simply has (identically) zero pose and zero velocity, and these properties remain constant, since the ground plane is statically joined to the world frame. The mobot is the model we care about. By accessing the second pose and the second twist in the respective arrays, we can obtain Gazebo's claim of the robot's state.

Inconveniently, one cannot assume that the mobot will always be the second model within the arrays of states published by Gazebo. One must identify the index location of the desired model to access the corresponding state information.

Listing 9.7 shows the code in mobot_gazebo_state.cpp, which is contained in a package of the same name within the accompanying code repository. The code can be used to emulate GPS. Lines 21 through 29 search through the list of names for a match to mobot. If a match is found, the corresponding array index is contained in imodel. Lines 31 and 32 access the corresponding pose and re-publish this on the topic gazebo_mobot_pose. Lines 33 through 37 copy the ideal pose to the object g_noisy_mobot_pose. The x and y components of the pose are corrupted with Gaussian random noise, where the mean (0) and standard deviation (1.0) are set on line 14. The heading of g_noisy_mobot_pose is suppressed (line 34), and the resulting corrupted (partial) pose is published to topic gazebo_mobot_noisy_pose. By adding noise to the ideal pose information, one can make a more realistic emulation of a GPS source. Such sources do not have drift, but position information does contain noise. Further, heading information from GPS is generally not trustworthy. For the emulated GPS, the heading information has been suppressed, enforcing that code developed based on this simulated signal cannot accidentally rely on information that is expected to be poor in practice. The noisy GPS emulator signal from this node will be used later in this chapter to show how to compute localization that uses GPS.

The topic gazebo_mobot_pose contains ideal position and orientation data. It can be used to develop steering algorithms and to compare localization results with the correct answer (which is typically unknown in practice).

Listing 9.7: mobot_gazebo_state.cpp: code to get mobot model states from Gazebo and re-publish both as ideal states and as noisy states

```
1  #include <ros/ros.h>
2  #include <gazebo_msgs/ModelStates.h>
3  #include <geometry_msgs/Pose.h>
4  #include <string.h>
5  #include <stdio.h>
6  #include <math.h>
7  #include <random>
8
9  geometry_msgs::Pose g_mobot_pose; //this is the pose of the robot in the world, ↩
        according to Gazebo
10 geometry_msgs::Pose g_noisy_mobot_pose; //added noise to x,y, and suppress orientation
11 geometry_msgs::Quaternion g_quat;
12 ros::Publisher g_pose_publisher;
13 ros::Publisher g_gps_publisher;
14 std::normal_distribution<double> distribution(0.0,1.0); //args: mean, std_dev
15 std::default_random_engine generator;
16 void model_state_CB(const gazebo_msgs::ModelStates& model_states)
17 {
18    int n_models = model_states.name.size();
19    int imodel;
20    //ROS_INFO("there are %d models in the transmission",n_models);
21    bool found_name=false;
22    for (imodel=0;imodel<n_models;imodel++) {
23       std::string model_name(model_states.name[imodel]);
24       if (model_name.compare("mobot")==0) {
25          //ROS_INFO("found match: mobot is model %d",imodel);
26          found_name=true;
27          break;
```

```
28        }
29    }
30    if(found_name) {
31        g_mobot_pose= model_states.pose[imodel];
32        g_pose_publisher.publish(g_mobot_pose);
33        g_noisy_mobot_pose = g_mobot_pose;
34        g_noisy_mobot_pose.orientation = g_quat;
35        g_noisy_mobot_pose.position.x += distribution(generator);
36        g_noisy_mobot_pose.position.y += distribution(generator);
37        g_gps_publisher.publish(g_noisy_mobot_pose); //publish noisy values
38        //double randval = distribution(generator);
39        //ROS_INFO("randval =%f",randval);
40        }
41    else
42        {
43            ROS_WARN("state of mobot model not found");
44        }
45  }
46
47  int main(int argc, char **argv) {
48      ros::init(argc, argv, "gazebo_model_publisher");
49      ros::NodeHandle nh;
50
51      g_pose_publisher= nh.advertise<geometry_msgs::Pose>("gazebo_mobot_pose", 1);
52      g_gps_publisher = nh.advertise<geometry_msgs::Pose>("gazebo_mobot_noisy_pose", 1);
53      ros::Subscriber state_sub = nh.subscribe("gazebo/model_states",1,model_state_CB);
54      //suppress the orientation output for noisy state; fill out a legal, constant ↵
                quaternion
55      g_quat.x=0;
56      g_quat.y=0;
57      g_quat.z=0;
58      g_quat.w=1;
59      ros::spin();
60  }
```

The Gazebo-state publisher node can be started with:

```
rosrun mobot_gazebo_state mobot_gazebo_state
```

For the mobot following a 5×5 m desired trajectory under open-loop control, the resulting republished Gazebo-model state is shown in Fig 9.7. Published noisy GPS will be used in Section 9.2.3 to show how to merge sensor sources for localization.

9.2.2 Odometry

When creating a ROS interface to a vehicle, there are two essential needs: a means to command motion (*e.g.* via `cmd_vel`) and a means to report back estimated state (typically via the topic `odom`). For our differential-drive vehicle, we assume that the two wheels are independently controllable via left and right-wheel velocity servos, and that each wheel has some type of encoder capable of providing incremental wheel angles (or, alternatively, wheel velocities). From speed and spin commands, we need to derive corresponding left and right-wheel joint commands, and from measured wheel motions, we need to derive and publish updates of estimated pose and twist of the robot.

A differential mapping between incremental wheel motions, $d\theta_l$ and $d\theta_r$, and corresponding incremental pose updates can be derived with the help of Fig 9.8.

With reference to the figure, we specify a coordinate frame on the vehicle with origin between the drive wheels, x axis pointing forward and y axis pointing to the left. With incremental motions of the left and right wheels, the robot will advance its frame origin forward by amount Δs and will change its heading by amount $\Delta \psi$. The relation depends on the wheel diameter, D, and the track (distance between the wheels), T. (It is implicitly assumed that the left and right wheel diameters are equal, although this will not be precisely

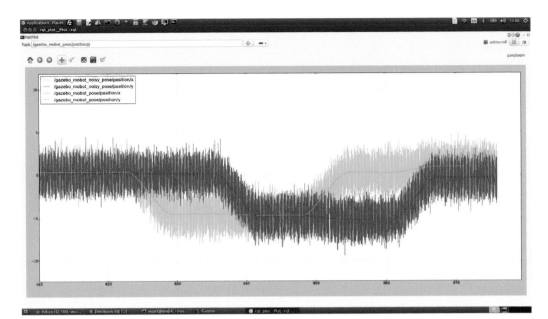

Figure 9.7: Re-published mobot Gazebo state and noisy state with robot executing square trajectory under open-loop control

---

true.) The relations are:

$$\Delta s = \frac{1}{4}D(d\theta_l + d\theta_r) = K_s(d\theta_l + d\theta_r) \tag{9.4}$$

where $d\theta_l$ and $d\theta_r$ are in radians, and s and D are in meters. The change in heading is

$$\Delta\psi = \frac{1}{2}D(d\theta_r - d\theta_l)/T = K_\psi(d\theta_r - d\theta_l) \tag{9.5}$$

If at time t the estimate of pose is $(\tilde{x}(t), \tilde{y}(t), \tilde{\psi}(t))$, and if at time $t + dt$ incremental motions of the left and right wheels are $d\theta_l$ and $d\theta_r$, then the updated pose at time $t + dt$ is approximately:

$$\begin{bmatrix} \tilde{x}(t+dt) \\ \tilde{y}(t+dt) \\ \tilde{\psi}(t+dt) \end{bmatrix} = \begin{bmatrix} \tilde{x}(t) \\ \tilde{y}(t) \\ \tilde{\psi}(t) \end{bmatrix} + \begin{bmatrix} K_s(d\theta_l + d\theta_r)cos(\tilde{\psi}) \\ K_s(d\theta_l + d\theta_r)sin(\tilde{\psi}) \\ K_\psi(d\theta_r - d\theta_l) \end{bmatrix} \tag{9.6}$$

The robot's twist can be computed as well, which is:

$$\begin{bmatrix} \tilde{v}_x \\ \tilde{v}_y \\ \tilde{\omega}_z \end{bmatrix} = \begin{bmatrix} K_s(d\theta_l + d\theta_r)cos(\tilde{\psi})/dt \\ K_s(d\theta_l + d\theta_r)sin(\tilde{\psi})/dt \\ K_\psi(d\theta_r - d\theta_l)/dt \end{bmatrix} \tag{9.7}$$

An implementation of these equations is in `mobot_drifty_odom.cpp`, within a package of the same name. The source code appears in Listings 9.8 through 9.10 (as well as in the accompanying code repository).

This node uses the parameters TRACK (line 13) and two wheel radii (lines 10 and 11) in the computation of state updates. These values agree with the mobot URDF specifications,

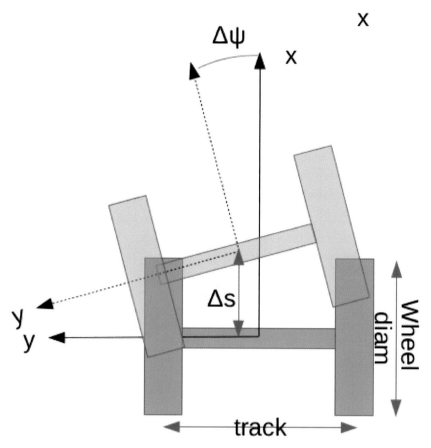

Figure 9.8: Differential-drive kinematics: incremental wheel rotations yield incremental pose changes

and they can be altered to simulate the effects of imperfect odometry computations. Wheel-joint increments used in pose and twist updates are based on subscription to the topic `joint_states` (line 125). The joint-state data (specifically, the rotation angles of the left and right wheels) is checked every 10 ms (as set by the loop timer on line 128).

Lines 53 through 71 of the joint-state subscriber callback test that the data is good and that both wheel-joint topics are present. If the wheel-joint data is determined to be valid, the pose and twist estimates are updated via lines 73 through 86.

The `mobot_drifty_odom` node can be started with:

```
rosrun mobot_drifty_odom mobot_drifty_odom
```

By also running the `mobot_gazebo_state` node, one can compare odometry to the actual pose, as reported by Gazebo. If the robot is started from the home pose, (0,0,0), it keeps track of robot state quite well, as shown in Fig 9.9. The odometry values for x and y closely follow the Gazebo (ideal) states for x and y.

Note that although state estimation from integration of odometry is quite good (in this case), that does not mean that the robot follows the desired trajectory well. In fact, path tracking shown in Fig 9.9 is not very good. The trapezoids of x and y should have reached and tracked horizontal values identically equal to 0 m or 5 m. The problem with the path fol-

Listing 9.8: `mobot_drifty_odom.cpp`: code to compute and publish odom from incremental wheel-joint rotations, preamble

```cpp
#include <ros/ros.h>
#include <geometry_msgs/Pose.h>
#include <geometry_msgs/TransformStamped.h>
#include <sensor_msgs/JointState.h>
#include <nav_msgs/Odometry.h>
#include <string.h>
#include <stdio.h>
#include <math.h>
#include <tf/transform_broadcaster.h>

//from URDF:  <xacro:property name="tirediam" value="0.3302" />
const double R_LEFT_WHEEL = 0.3302 / 2.0;
const double R_RIGHT_WHEEL = R_LEFT_WHEEL + 0.005; //introduce error--tire diam diff
//from URDF: <xacro:property name="track" value=".56515" />
const double TRACK = 0.560515; //0.56515; //0.560; // track error

const double wheel_ang_sham_init = -1000000.0;
bool joints_states_good = false;

nav_msgs::Odometry g_drifty_odom;
sensor_msgs::JointState g_joint_state;
ros::Publisher g_drifty_odom_pub;
ros::Subscriber g_joint_state_subscriber;
//tf::TransformBroadcaster* g_odom_broadcaster_ptr;
geometry_msgs::TransformStamped g_odom_trans;

double g_new_left_wheel_ang, g_old_left_wheel_ang;
double g_new_right_wheel_ang, g_old_right_wheel_ang;
double g_t_new, g_t_old, g_dt;
ros::Time g_cur_time;
double g_odom_psi;

geometry_msgs::Quaternion convertPlanarPsi2Quaternion(double psi) {
    geometry_msgs::Quaternion quaternion;
    quaternion.x = 0.0;
    quaternion.y = 0.0;
    quaternion.z = sin(psi / 2.0);
    quaternion.w = cos(psi / 2.0);
    return (quaternion);
}
```

Listing 9.9: `mobot_drifty_odom.cpp`: code to compute and publish odom from incremental wheel-joint rotations, callback function

```cpp
42  void joint_state_CB(const sensor_msgs::JointState& joint_states) {
43      double dtheta_right, dtheta_left, ds, dpsi;
44      int n_joints = joint_states.name.size();
45      int ijnt;
46      int njnts_found = 0;
47      bool found_name = false;
48
49      g_old_left_wheel_ang = g_new_left_wheel_ang;
50      g_old_right_wheel_ang = g_new_right_wheel_ang;
51      g_t_old = g_t_new;
52      g_cur_time = ros::Time::now();
53      g_t_new = g_cur_time.toSec();
54      g_dt = g_t_new - g_t_old;
55
56      for (ijnt = 0; ijnt < n_joints; ijnt++) {
57          std::string joint_name(joint_states.name[ijnt]);
58          if (joint_name.compare("left_wheel_joint") == 0) {
59              g_new_left_wheel_ang = joint_states.position[ijnt];
60              njnts_found++;
61          }
62          if (joint_name.compare("right_wheel_joint") == 0) {
63              g_new_right_wheel_ang = joint_states.position[ijnt];
64              njnts_found++;
65          }
66      }
67      if (njnts_found < 2) {
68          //ROS_WARN("did not find both wheel joint angles!");
69          //for (ijnt = 0; ijnt < n_joints; ijnt++) {
70          //    std::cout<<joint_states.name[ijnt]<<std::endl;
71          //}
72      }
73      else {
74          //ROS_INFO("found both wheel joint names");
75      }
76      if (!joints_states_good) {
77          if (g_new_left_wheel_ang > wheel_ang_sham_init / 2.0) {
78              joints_states_good = true; //passed the test
79              g_old_left_wheel_ang = g_new_left_wheel_ang;
80              g_old_right_wheel_ang = g_new_right_wheel_ang; //assume right is good now ←┘
                   as well
81          }
82      }
83      if (joints_states_good) { //only compute odom if wheel angles are valid
84          dtheta_left = g_new_left_wheel_ang - g_old_left_wheel_ang;
85          dtheta_right = g_new_right_wheel_ang - g_old_right_wheel_ang;
86          ds = 0.5 * (dtheta_left * R_LEFT_WHEEL + dtheta_right * R_RIGHT_WHEEL);
87          dpsi = dtheta_right * R_RIGHT_WHEEL / TRACK - dtheta_left * R_LEFT_WHEEL / ←┘
                   TRACK;
88
89          g_drifty_odom.pose.pose.position.x += ds * cos(g_odom_psi);
90          g_drifty_odom.pose.pose.position.y += ds * sin(g_odom_psi);
91          g_odom_psi += dpsi;
92          //ROS_INFO("dthetal, dthetar, dpsi, odom_psi, dx, dy= %f, %f %f, %f %f %f", ←┘
                   dtheta_left, dtheta_right, dpsi, g_odom_psi,
93          //         ds * cos(g_odom_psi), ds * sin(g_odom_psi));
94          g_drifty_odom.pose.pose.orientation = convertPlanarPsi2Quaternion(g_odom_psi);
95
96          g_drifty_odom.twist.twist.linear.x = ds / g_dt;
97          g_drifty_odom.twist.twist.angular.z = dpsi / g_dt;
98          g_drifty_odom.header.stamp = g_cur_time;
99          g_drifty_odom_pub.publish(g_drifty_odom);
100     }
101 }
```

Listing 9.10: `mobot_drifty_odom.cpp`: code to compute and publish odom from incremental wheel-joint rotations, main program

```
103  int main(int argc, char **argv) {
104      ros::init(argc, argv, "drifty_odom_publisher");
105      ros::NodeHandle nh;
106      //inits:
107      g_new_left_wheel_ang = wheel_ang_sham_init;
108      g_old_left_wheel_ang = wheel_ang_sham_init;
109      g_new_right_wheel_ang = wheel_ang_sham_init;
110      g_old_right_wheel_ang = wheel_ang_sham_init;
111      g_cur_time = ros::Time::now();
112      g_t_new = g_cur_time.toSec();
113      g_t_old = g_t_new;
114
115      //initialize odom with pose and twist defined as zero at start-up location
116      g_drifty_odom.child_frame_id = "base_link";
117      g_drifty_odom.header.frame_id = "drifty_odom";
118      g_drifty_odom.header.stamp = g_cur_time;
119      g_drifty_odom.pose.pose.position.x = 0.0;
120      g_drifty_odom.pose.pose.position.y = 0.0;
121      g_drifty_odom.pose.pose.position.z = 0.0;
122      g_drifty_odom.pose.pose.orientation.x = 0.0;
123      g_drifty_odom.pose.pose.orientation.y = 0.0;
124      g_drifty_odom.pose.pose.orientation.z = 0.0;
125      g_drifty_odom.pose.pose.orientation.w = 1.0;
126
127      g_drifty_odom.twist.twist.linear.x = 0.0;
128      g_drifty_odom.twist.twist.linear.y = 0.0;
129      g_drifty_odom.twist.twist.linear.z = 0.0;
130      g_drifty_odom.twist.twist.angular.x = 0.0;
131      g_drifty_odom.twist.twist.angular.y = 0.0;
132      g_drifty_odom.twist.twist.angular.z = 0.0;
133
134      ros::Rate timer(100.0); // a 100Hz timer
135
136      g_drifty_odom_pub = nh.advertise<nav_msgs::Odometry>("drifty_odom", 1);
137      g_joint_state_subscriber = nh.subscribe("joint_states", 1, joint_state_CB);
138      while (ros::ok()) {
139          ros::spin();
140      }
141  }
```

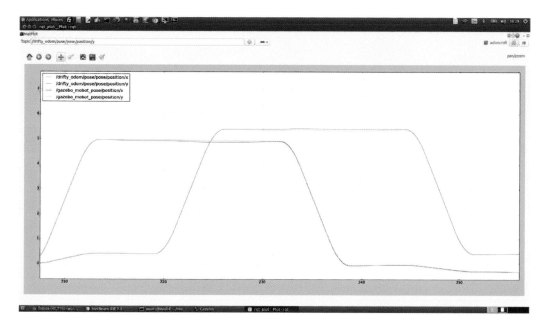

Figure 9.9: Comparison of Gazebo ideal state and `odom` estimated state as simulated robot follows 5 m × 5 m square path

lowing is that the differential-drive controller (subject to friction, torque saturation, inertial effects and controller bandwidth limitations) is imperfect, and the commanded speeds are not achieved identically. To improve on the path following, it is necessary to use a feedback steering controller, which will be introduced in Section 9.3.

The quality of state estimation from integration of odometry as shown in Fig 9.9 is unrealistically precise. More realistic odometry can be simulated by altering parameters of the kinematic model. As an example, line 13 of Listing 9.8 was changed to:

```
const double R_RIGHT_WHEEL = R_LEFT_WHEEL+0.005;
```

Adding 5 mm to the assumed wheel diameter is equivalent to having a modeling error that fails to recognize a tire wear differential of 5 mm. When `mobot_drifty_odom` is run with the same 5 m × 5 m square path command, the odometry estimates of pose diverge badly from the Gazebo-reported poses, as shown in Fig 9.10. Odometry drift is a significant problem that can be corrected only with reference to some additional sensing, *e.g.* LIDAR sensing of the environment or GPS signals. Nonetheless, the odometry signal is very useful, particularly when combined with additional sensors.

Often, the robot will have an Inertial Measurement Unit (IMU), and this additional sensory information can be folded into the odometry estimate. An IMU is insensitive to some sources of error in estimating pose from wheel kinematics alone, including effects of slipping or skidding, uneven terrain, imprecision of kinematic parameters (*e.g.* wheel diameters and track), and unmeasured backlash in the drive train. On the other hand, one must be careful with bias offsets in IMU signals, which can cause odometry estimates to integrate to infinity, even when the robot is standing still.

When creating a ROS interface to a new mobile platform, code similar to `mobot_drifty_odom` should be written (unless such a driver already exists, which is increasingly common). The result of this node is publication of estimated state to the topic

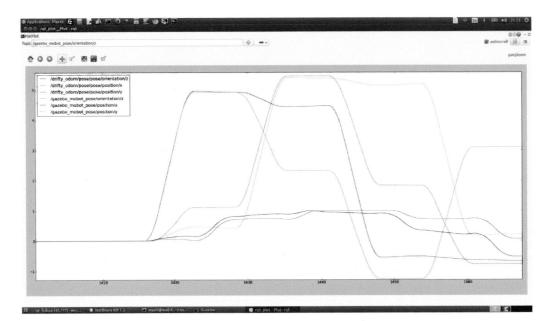

Figure 9.10: Ideal state and odom state estimates for 5 m × 5 m path diverge. Note 5 mm right wheel diameter error with odom estimate

odom. For the STDR robot, odom messages are published, although they are merely identical to the equivalent `cmd_vel` inputs, without consideration of realistic dynamics. The mobot simulation in Gazebo uses a differential-drive plug-in, which includes the equivalent of Listing 9.8 within the plug-in.

A second node needed to interface to a mobile platform acts on input commands on the `vel_cmd` topic. This node should convert speed and spin commands into right-wheel and left-wheel velocity commands. Conversion from speed and spin to wheel joint commands can be computed as follows. For a known sample period, dt, Equations 9.4 and 9.5 can be re-written as:

$$\begin{bmatrix} ds/dt \\ d\psi/dt \end{bmatrix} = \begin{bmatrix} K_s & K_s \\ K_\psi & -K_\psi \end{bmatrix} \begin{bmatrix} d\theta_r/dt \\ d\theta_l/dt \end{bmatrix} \tag{9.8}$$

which can be inverted to yield:

$$\begin{bmatrix} d\theta_r/dt \\ d\theta_l/dt \end{bmatrix} = \frac{2}{K_s K_\psi} \begin{bmatrix} K_\psi & K_s \\ K_\psi & -K_s \end{bmatrix} \begin{bmatrix} ds/dt \\ d\psi/dt \end{bmatrix} \tag{9.9}$$

Equation 9.9 may be used to convert incoming speed (ds/dt) and spin ($d\psi/dt$) commands from the `cmd_vel` topic into corresponding right and left-wheel speed commands, $d\theta_r/dt$ and $d\theta_l/dt$. A local velocity controller should enforce that the commanded wheel speeds are achieved. Inevitably, there will be errors between the commanded wheel velocities and the realized wheel velocities. Odometry thus should be based on measured wheel increments rather than the assumption that the wheel-speed commands are achieved precisely, although the latter approach can be taken if necessary, as is done with the STDR simulator.

As noted, computation of odometry suffers from modeling imperfections, leading to accumulation of errors that can be arbitrarily large. Addressing this issue requires additional sensors. A second problem with odometry is its choice of reference frame.

For both the pose and the twist, the estimated values correspond to the robot's base frame expressed with respect to some ground reference frame. The reference frame, however is ambiguous. By default, for odometry computations, the reference frame pose of $(0, 0, 0)$ corresponds to the robot pose when the odometry node begins its computations. This pose (which is different every time the robot's odometry node starts) is known as the odometry frame.

The approach in ROS to addressing odometry drift is somewhat odd but pragmatic. Two common absolute reference frames are `world` (*e.g.* latitude, longitude and heading) and `map`, for which map coordinates can be stated with respect to defined reference frame on the map (a definition of $(x, y) = (0, 0)$). Often, the robot base that is computing and publishing its odometry information does not have access to other sensors that could help establish the robot's absolute pose. Instead, the base continues to publish its odometry signals that are increasingly less accurate over time and distance travelled. Soon, the pose information from this topic is seemingly worthless. However, the odom signal is still valuable, if combined properly with additional sensors. In ROS, this is done with the computation of a transform from the `odom` frame to the `world` (or `map`) frame.

Consider a world frame. There is a correct value of state, (x, y, ψ), in the world (whether or not these values are known). The pose of the robot with respect to the world can be represented as a transform, as introduced in Chapter 4. A transform $T_{odom/world}$ can be used to express the position and orientation of the `odom` frame with respect to the `world` frame. At start-up, the odom publisher node assumes that its initial pose is $(x, y, \psi) = (0, 0, 0)$, and this will be the reference origin for the odometry frame. In world coordinates, these values may be referred to as $(x_{world}, y_{world}, \psi_{world})$. Equivalently, these values may be expressed as components within $T_{odom/world}$. As the robot moves and odometry information is updated, the state estimates may be expressed as $T_{robot/odom}$. The robot's pose in world coordinates can be computed as $T_{robot/world} = T_{odom/world} T_{robot/odom}$. As noted, the odometry-based state estimates accumulate drift errors as the robot moves. The ROS approach for accounting for this drift is to express the error using variable transformation $T_{odom/world}$. That is, one could say the odometry state estimate is correct, but the odom reference frame has drifted from its original pose in the world. To reconcile the `odom` state with the `world` frame, one must update $T_{odom/world}$.

This odd perspective has some benefits. As noted, the mobile base responsible for publishing odom may be unaware of additional sensors, and thus all it may be able to do is to continue publishing its increasingly poor state estimates. However, if the transform $T_{odom/world}$ that reconciles the odometry values with world-frame states can be found, the odometry signal is still useful. The odometry signal is quite smooth and is updated frequently. These are important virtues for use in feedback for steering. In contrast, additional sensors, such as GPS, may be noisy and update at a much lower rate, making them unsuitable for steering feedback. World-frame sensors (*e.g.* GPS) do have the benefit of having zero drift. Such signals thus can be used to help update $T_{odom/world}$, reconciling odometry publications with world-frame coordinates. Although odometry state estimates drift to arbitrarily large errors, such drift is typically slow. Therefore, absolute (world-frame) sensors do not require fast updates nor must they have low noise in order to provide ongoing corrections to $T_{odom/world}$.

9.2.3 Combining odometry, GPS and inertial sensing

Odometry offers fast and smooth incremental state estimation. However, it suffers from cumulative drift, particularly if wheel slip occurs. Robots with treads perform skid steer, that can cause estimation of heading to have large uncertainty. Even without skid steer, heading estimation can be affected significantly by wheel slip. If heading estimation is flawed, translation estimates from odometry can suffer dramatically.

An improvement over odometry for rotation estimation can be obtained using an Inertial Measurement Unit (IMU). IMUs vary widely in cost and performance, from low-cost MEMS-based chips to high-cost laser-gyro units. IMUs typically offer six outputs, three translational accelerations, and three components of rotation rate.

To emulate an IMU in Gazebo, the plug-in `libgazebo_ros_imu.so` can be used, as illustrated in the model file `mobot_w_imu.xacro` in package `mobot_urdf`, which follows.

```
<?xml version="1.0"?>
<robot
    xmlns:xacro="http://www.ros.org/wiki/xacro" name="mobot">
    <xacro:include filename="$(find mobot_urdf)/urdf/mobot.xacro" />

<link name="imu_link">
</link>

<joint name="imu_joint" type="fixed">
    <parent link="base_link"/>
    <child link="imu_link" />
</joint>

 <!--add IMU plug-in-->
<gazebo>
  <plugin name="imu_plugin" filename="libgazebo_ros_imu.so">
    <alwaysOn>true</alwaysOn>
    <updateRate>100.0</updateRate>
    <bodyName>imu_link</bodyName>
    <topicName>imu_data</topicName>
    <gaussianNoise>1e-06</gaussianNoise>
  </plugin>
</gazebo>
</robot>
```

This model file imports the previously introduced mobot model, and it additionally incorporates the IMU plug-in. When this model file is spawned in Gazebo, results of the IMU emulation will be published on `imu_data`, and this data will be published at the specified rate of 100 Hz.

For navigation on a plane, the most important component of the IMU is the rotation about the z axis, which is the rate of change of heading. This measurement will be more reliable than inferring heading from odometry. Its chief limitation is a bias offset which, when integrated, can accumulate to arbitrarily large heading errors.

Integration of odometry or IMU signals both suffer from accumulation of errors. In contrast, a Global Positioning System (GPS) signal is absolute. A GPS signal, however, can be noisy, can have limited precision (*e.g.* a few meters), and its estimation of heading is untrustworthy. However, one can combine the virtues of odometry, IMU and GPS to infer a pose estimate that has the benefits of smoothness and good precision without drift. An example of this is the file `localization_w_gps.cpp` in the package `localization_w_gps`. This node assumes use of odometry (which may suffer from drift, notably in yaw), an IMU and GPS (which may have significant noise).

To run this node, first start Gazebo:

```
roslaunch gazebo_ros empty_world.launch
```

Then launch the mobot model with IMU:

```
roslaunch mobot_urdf mobot_w_imu.launch
```

Run a (realistically) flawed odometry publisher:

```
rosrun mobot_drifty_odom mobot_drifty_odom
```

Start emulation of a noisy GPS sensor with:

```
rosrun localization_w_gps localization_w_gps
```

Then start the example localization node:

```
rosrun localization_w_gps localization_w_gps
```

The robot can be moved around conveniently using keyboard teleoperation by running:

```
rosrun teleop_twist_keyboard teleop_twist_keyboard.py
```

The localization node will integrate information from the imperfect odometry publications, the z rotation rate from the IMU, and noisy GPS data.

Within the node `localization_w_gps`, the main loop contains the following lines (lines 142 and 143):

```
x_est = (1-K_GPS)*x_est + K_GPS*g_x_gps; //incorporate gps feedback
y_est = (1-K_GPS)*y_est + K_GPS*g_y_gps; //ditto
```

These lines cause the estimated x and y positions to converge on the corresponding GPS values. However, with a low value of `K_GPS`, this convergence will be relatively immune to GPS noise.

Lines 142 and 143 integrate odometry based on the reported forward speed and best estimate of heading:

```
dl_odom_est = MAIN_DT*g_odom_speed; //moved this far in 1 DT
move_dist+= dl_odom_est; //keep track of cumulative move distance
```

Heading is updated based on integrating IMU-based z rotation rate (lines 146 through 148):

```
yaw_est+= MAIN_DT*g_omega_z_imu; //integrate the IMU's yaw to estimate heading
        if (yaw_est<-M_PI) yaw_est+= 2.0*M_PI; //remap periodically
        if (yaw_est>M_PI) yaw_est-= 2.0*M_PI;
```

Incremental x and y motion is estimated from best estimate of heading and incremental motion from odometry-based speed, and this model-based incremental motion is incorporated in the x and y position estimates (lines 150 through 153):

```
dx_odom = dl_odom_est*cos(yaw_est); //incremental x and y motions, as
dy_odom = dl_odom_est*sin(yaw_est); //inferred from speed and heading est
x_est+= dx_odom; //cumulative x and y estimates updated from odometry
y_est+= dy_odom;
```

Although GPS can help correct drift of x and y coordinates, GPS does not provide credible heading information. However, heading drift can be corrected by considering the influence of translational motion. If the robot is estimated to advance by at least some minimal distance, *i.e.* if (fabs(move_dist) > L_MOVE), one can compare the incremental progress in x and y based on integration of odometry versus incremental progress in x and y based on absolute (albeit noisy) GPS. This is done in lines 156 through 174:

```
if (fabs(move_dist) > L_MOVE) { //if moved this far since last yaw update,
  //time to do another yaw update based on GPS
  //since last update, express motion in polar coords based on gps
  dang_gps = atan2((y_est-y_est_old),(x_est-x_est_old));
  //similarly, update in polar coords based on odometry
  dang_odom = atan2(delta_odom_y,delta_odom_x);
  //if gps and odom disagree on avg heading, make a correction
  yaw_err = dang_gps - dang_odom;
  if (yaw_err>M_PI) yaw_err-=2.0*M_PI;
  if (yaw_err<-M_PI) yaw_err+=2.0*M_PI;
  //K_YAW should not be lareger than unity; smaller --> less noise sensitivity
  yaw_est+=K_YAW*yaw_err; //here's the yaw update due to gps
  y_est_old=y_est; //save this state as a checkpoint for next yaw update
  x_est_old=x_est;
  move_dist=0;
  delta_odom_y=0;
  delta_odom_x=0;
}
```

With this logic, the robot is inferred to have advanced at least by distance L_MOVE (a tunable parameter). During this motion, the x and y values have changed by amounts computed based on GPS in contrast to odometry. In polar coordinates, the motion can be expressed as a change in angle of dang_gps according to GPS, and dang_odom according to odometry. If odometry started with the correct heading, these values should agree. If the discrepancy is attributable to an error in initial heading of odometry, this heading error may be inferred to equal yaw_err = dang_gps - dang_odom. Since the GPS signals are noisy, one may elect to incorporate only a fraction of this computed correction, as modulated by the gain K_YAW, which should have a value between 0 and 1.

With a heading correction based on travel distance of L_MOVE, the travel distance counter is reset to distance = 0.

With feedback on x, y and heading, computed localization can have relatively low noise and low drift. This is illustrated in the examples of Figures 9.11 through 9.14.

Figure 9.11 shows the pose estimate of the robot at rest, after the pose estimate has converged toward the actual pose. There remains some noise on the pose estimate, but this is dramatically smaller than the noise present in the GPS signal (gray, noisy trace after time 4342).

Figure 9.12 shows the rate of convergence of position estimates toward GPS-based position. With the chosen gains, the x and y values converge on the correct x and y values, although this convergence can take about 10 sec. During this time, the heading estimate persists at a large value, with no correction toward the actual heading. Until the robot moves, the heading error is not apparent.

Figure 9.13 shows how heading is corrected once the robot starts moving. In this example, the robot's motion is dominantly in the $-y$ direction, since the heading is roughly -90 degrees. Since the estimate of the heading is grossly wrong (initially estimated at approximately 1 rad), odometry predicts that the robot's forward velocity results in increasing y (whereas GPS determines the motion has decreasing y). This discrepancy is computed by the localization algorithm and is applied to correct the heading. The heading estimate (red trace) decreases, converging on the true heading (over about 10 sec).

Figures 9.11 through 9.13 illustrate convergence from large initial-condition errors, and this convergence is relatively slow (at the example gains). However, once approximate convergence is achieved, relatively precise tracking can be achieved. Figure 9.14 illustrates the estimated pose versus the true pose while the robot is teleoperated through a variety of

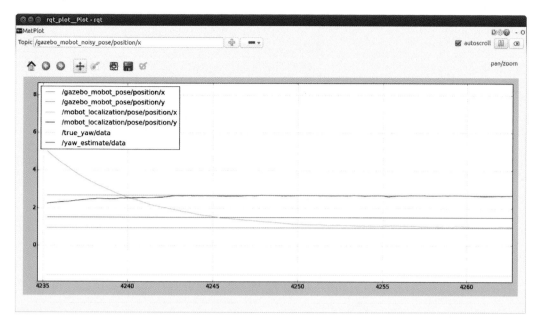

Figure 9.11: Pose estimate and ideal pose (time 4330 to 4342) and noisy pose per GPS (time 4342 to 4355)

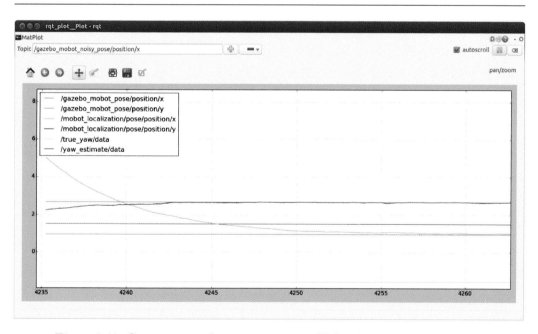

Figure 9.12: Convergence of pose estimate to GPS values with robot at rest

translation and spin moves. The x, y and heading track the actual values well, in spite of the fact that there is no absolute heading sensor and the GPS signals have large modeled noise.

In practice, the GPS values would have much lower noise than this example suggests. The

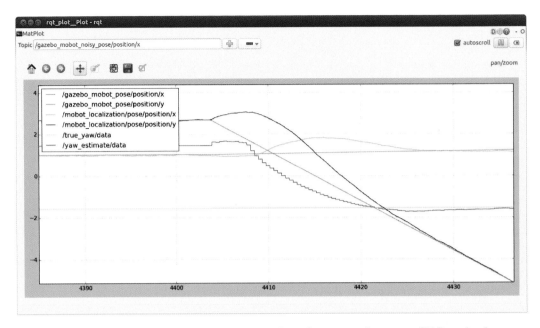

Figure 9.13: Convergence of heading estimate based on error between GPS and odometry estimates of translation

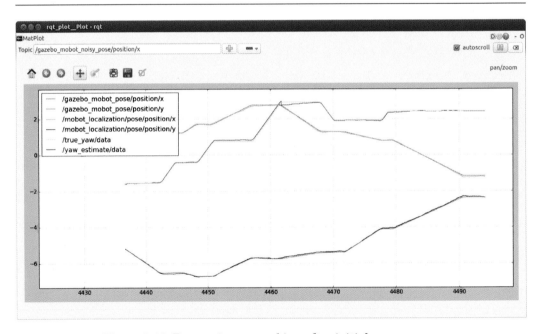

Figure 9.14: Pose estimate tracking after initial convergence

tunable parameters of `K_YAW`, `K_GPS` and `L_MOVE` should be tuned to appropriate weightings based on levels of odometry uncertainty, GPS noise and IMU drift.

ROS packages exist for localization merging multiple sensors, *e.g.* using extended Kalman filtering. (See *e.g.* `http://wiki.ros.org/robot_pose_ekf` and `http://wiki.ros.org/`

robot_pose_ekf/Tutorials/AddingGpsSensor.) The simple localization algorithm presented in this section can be generalized to six-DOF, *e.g.* for submersible or aerial vehicles, or for wheeled vehicles traversing very uneven terrain. It is unclear, though, whether the sensor and model behaviors satisfy the assumptions of Kalman filtering, including Gaussian noise with zero mean. Inevitably, tuning of parameters will be required for good performance in the context of a specific mobile platform with specific sensors and specific environmental conditions.

9.2.4 Combining odometry and LIDAR

One of the types of direct environmental sensing particularly applicable to indoor environments is LIDAR. Use of LIDAR for localization is most convenient when one has a pre-determined map of the environment. Fortunately, creating maps with LIDAR is relatively simple and convenient using existing ROS packages, as will be described in Section 10.1. Here, we assume that a map is available, and we wish to establish the robot's pose within the map with the help of LIDAR signals.

A useful package for LIDAR-based localization is Adaptive Monte Carlo Localization (AMCL). From the package wiki at `http://wiki.ros.org/amcl`:

> AMCL is a probabilistic localization system for a robot moving in 2D. It implements the adaptive (or KLD-sampling) Monte Carlo localization approach (as described by Dieter Fox), which uses a particle filter to track the pose of a robot against a known map.

To illustrate use of this package, start Gazebo with the mobot model, add a model environment (the OSRF starting pen, in this example), along with supporting nodes to publish desired states and perform open-loop control based on desired states. These steps are combined in the launch file `mobot_startup_open_loop.launch`. The system can be started with:

```
roslaunch gazebo_ros empty_world.launch
```

```
roslaunch mobot_urdf mobot_startup_open_loop.launch
```

(The above steps could also be combined within a single launch file, as well as the steps to follow.)

A map is loaded using the ROS `map_server` package, described at `http://wiki.ros.org/map_server`. An example (partial) map of the starting pen environment has been created and saved in the directory `maps/starting_pen`. The `maps` directory is created as a ROS package, by virtue of containing a `package.xml` file and a `CMakeLists.txt` file. However, this directory does not contain any actual code. By creating this directory as a ROS package, however, one can conveniently navigate to this directory with `roscd`, or refer to this directory in launch files with the expression `$(find maps)`.

The starting pen map can be loaded by running:

```
roscd maps
```

followed by:

```
rosrun map_server map_server starting_pen/starting_pen_map.yaml
```

This node publishes the specified map on topic `map` with message type `nav_msgs/OccupancyGrid`. The example map is comprised of 4000 × 4000 cells, with each cell 5 cm × 5 cm. Each cell contains a gray-scale value indicating whether that cell is occupied (black), empty (white) or uncertain (implied by shade of gray between white and black). This would be a fairly large message to send repeatedly, when it is, in fact, updated only rarely. The map server thus uses a special publication option "latched" that avoids republications except when the message (map) value is changed or when a new subscriber starts.

With the LIDAR signal available (on topic "scan" for our mobot model) and with a map loaded (the starting-pen map, appropriate for the mobot placed within the starting pen), the AMCL localization algorithm can be run. The AMCL node is started with:

```
rosrun amcl amcl
```

The AMCL node expects to find a map on topic `map` and expects to find LIDAR data on topic `scan`. If the LIDAR data is published to a different topic, this can be accommodated with topic re-mapping. For example, if the LIDAR data is published to topic `/laser/scan`, AMCL should be started with topic re-mapping as follows:

```
rosrun amcl amcl scan:=/laser/scan
```

With AMCL running, this node attempts to discover the pose of the robot with respect to the map. As a particle-filter based method, it presents a large collection of candidate poses with varying probabilities. The distribution of particles under consideration can be visualized by displaying `PoseArray` in `rviz`, and having it subscribe to the topic `particlecloud` (which is published by AMCL). For our example, this display on start-up of AMCL appears as in Fig 9.15.

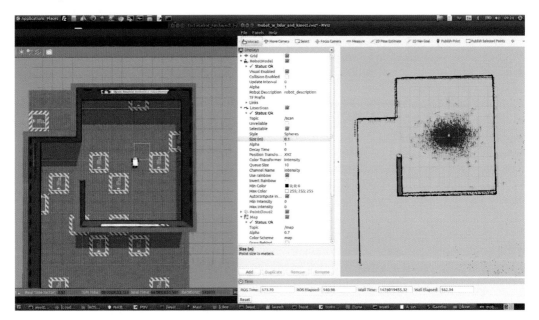

Figure 9.15: Distribution of candidate poses for mobot within map upon start-up. The candidate poses have large initial variance.

Note that the set of candidate poses under consideration (as illustrated by the arrows in Fig 9.15) has a fairly large variance.

In the example shown, the initial pose estimate is actually quite close to the correct pose (although the initial uncertainty is high). This can be seen by estimating the pose in `rviz` relative to the starting-pen map and comparing this to the ground truth shown in Gazebo. Further, the LIDAR pings in `rviz` (red dots) line up convincingly with the walls in the starting-pen map.

(Please note: Gazebo emulation of LIDAR uses a plug-in library that runs on graphics hardware. Computers that do not have compatible graphics chips will falsely display LIDAR points at minimum radius. If running within a virtual machine, it may be necessary to toggle 3-D acceleration in the settings to get LIDAR emulation to work. `optirun` may be helpful in some cases. See `https://wiki.ubuntu.com/Bumblebee`.)

The AMCL node will attempt to deduce a logical initial pose, which is quite successful in the example of Fig 9.15. In other cases, it may be necessary to provide manual assistance to establish an approximate initial pose. The initial pose estimate can be provided interactively with the `rviz` tool "2D pose estimate", which enables the user to click and drag a vector, implying the initial position and heading AMCL should assume. Using this tool, the robot pose shown in the map will move to the user's suggested pose, and AMCL will attempt localization from there.

Assuming at least a coarse initial estimate of pose, AMCL's estimate of pose will improve as the robot acquires multiple clues from LIDAR with changing viewpoints. To acquire such evidence for localization, the robot can be commanded to move in a 3 m × 3 m square (open loop) using a client of the desired-state publisher by running:

```
rosrun mobot_pub_des_state pub_des_state_path_client_3x3
```

Alternatively, the robot can be commanded to move under keyboard teleoperation by running:

```
rosrun teleop_twist_keyboard teleop_twist_keyboard.py
```

After running the 3 × 3 square motion, the result is shown in Fig 9.16, where the candidate poses are more tightly bundled. (In fact, the candidate poses are hardly visible, concentrated primarily under the robot between the drive wheels.)

With AMCL running, pose estimates (with respect to the map frame) are published to the topic `amcl_pose` with message type `geometry_msgs/PoseWithCovarianceStamped`. These updates are relatively infrequent (on the order of seconds), and thus they are unsuitable for steering feedback. Further, these are only poses; twist information is not included.

At the same time, a pose estimate (along with twist) is being published to the `odom` topic that is smooth and has much higher bandwidth. The relation between the `odom` frame and the `map` frame can be expressed as a transform. Transforms between `odom` and `map` are recomputed and published (to the `tf` topic) by AMCL (with updates of map pose estimates from AMCL). The resulting transforms can be observed using a tool within the `tf` package. Running:

```
rosrun tf tf_echo odom map
```

requests that the transform from `odom` frame to the `map` frame be displayed. An example display from this command is:

```
At time 516.100
```

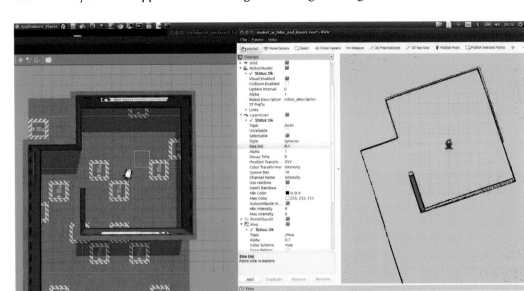

Figure 9.16: Distribution of candidate poses for mobot within map after motion. The candidate poses are concentrated in a small bundle near the true robot pose.

```
- Translation: [0.002, 0.024, 0.002]
- Rotation: in Quaternion [0.002, 0.001, -0.003, 1.000]
            in RPY (radian) [0.004, 0.003, -0.006]
            in RPY (degree) [0.214, 0.145, -0.332]
At time 517.058
- Translation: [0.002, 0.024, 0.002]
- Rotation: in Quaternion [0.002, 0.001, -0.003, 1.000]
            in RPY (radian) [0.004, 0.003, -0.006]
            in RPY (degree) [0.214, 0.145, -0.332]
At time 518.059
- Translation: [0.002, 0.024, 0.002]
- Rotation: in Quaternion [0.002, 0.001, -0.003, 1.000]
            in RPY (radian) [0.004, 0.003, -0.006]
            in RPY (degree) [0.214, 0.145, -0.332]
```

This transform can be obtained programmatically within a node and used to correct for odometry drift. Note, though, that sudden (typically small) jumps in this transform can result in corresponding jerky behavior of a vehicle that uses steering feedback based on such localization. It may be desirable to low-pass filter the map-to-odom transforms before using transformed values in feedback. In addition, updates of AMCL are much too slow (*e.g.* 0.5 Hz).

In brief, AMCL has the virtue of zero drift, but its update rate is too slow for steering feedback. In contrast, odometry has a high rate of update, but it suffers from cumulative drift. It is possible to combine these alternative state measures to realize the benefits of both. This is illustrated in the package odom_tf.

The odom_tf package has a library (of the same name) that defines a class OdomTf. The implementation file, OdomTf.cpp, has two callback functions:

```
void OdomTf::odomCallback(const nav_msgs::Odometry& odom_rcvd)
```

and

```
void OdomTf::amclCallback(const geometry_msgs::PoseWithCovarianceStamped& amcl_rcvd)
```

The amclCallback function responds to messages on the topic /amcl_pose. The AMCL node publishes (stamped) poses to this topic corresponding to the pose of the base_link with respect to the map frame. These signals are imperfect, and (in the present example) they are updated only at 0.5 Hz. However, they are absolute and do not suffer from cumulative drift. When each such AMCL-based pose is received, the callback function amclCallback() repackages this pose as a transform, specifying frame_id = map and child_frame = base_link. The result is populated in the member variable stfAmclBaseLinkWrtMap_.

The odomCallback responds to messages on the topic drifty_odom. (More typically, one should subscribe to the odom topic, but for illustrative purposes, the odom topic is deliberately corrupted and republished as drifty_odom). In the odomCallback, the pose of the robot with respect to the odom frame is received as a message, and this pose is converted to a stamped transform, stfBaseLinkWrtDriftyOdom_.

The two alternative localization transforms, stfBaseLinkWrtDriftyOdom_ and stfAmclBaseLinkWrtMap_, can be reconciled through the following interpretation. Recognizing that the pose of the robot expressed in the odom frame is subject to drift, one can think of this, equivalently, as the odom reference frame drifting with respect to the map frame. If odometry is perfect, then the odom frame will remain stationary in the map frame. However, to the extent that odometry has cumulative drift, the odom frame can be observed to move relative to the map frame. It should be appreciated that, although the robot may be moving quickly, the odom reference frame would move only slowly with respect to the map frame (corresponding to the rate of error accumulation). The pose of the odom frame with respect to the map frame is expressed as a stamped transform in the member variable stfDriftyOdomWrtBase_.

The transform stfDriftyOdomWrtBase_ is updated every time a new AMCL pose is received by amclCallback(). This is achieved by multiplying transforms:

```
stfDriftyOdomWrtBase_ = xform_utils.stamped_transform_inverse(←↩
    stfBaseLinkWrtDriftyOdom_);
xform_utils.multiply_stamped_tfs(stfAmclBaseLinkWrtMap_,stfDriftyOdomWrtBase_,←↩
    stfDriftyOdomWrtMap_)
```

Each time AMCL publishes a new estimate of the robot's pose with respect to the map, the corresponding transform stfAmclBaseLinkWrtMap_ is updated. Within amclCallback(), the most recent value of the transform corresponding to robot pose with respect to odom, stfBaseLinkWrtDriftyOdom_, is used to deduce an update for the odom base frame with respect to the map frame, stfDriftyOdomWrtBase_. This is done by inverting the transform stfBaseLinkWrtDriftyOdom to obtain stfDriftyOdomWrtBase, which can be used consistently in transform multiplication with stfAmclBaseLinkWrtMap to obtain stfDriftyOdomWrtMap_.

The value of stfDriftyOdomWrtMap_ is updated at the frequency of AMCL updates, which can be a low frequency. However, if the rate of drift of odometry is low, then the transform between the map frame and the odom frame remains (approximately) valid for a relatively long period of time (presumably, for a duration longer than the period between AMCL updates). Under this assumption, odom localization can be corrected with the transform stfDriftyOdomWrtMap_ at a high rate, using the full frequency of odom updates, and re-using the `stfDriftyOdomWrtMap_` potentially hundreds of times between AMCL up-dates. In this fashion, one can achieve fast and smooth updates of localization with zero cumulative drift. This is accomplished each time `odom` is updated, via the callback function `odomCallback`. Specifically, this is accomplished with the operation:

```
xform_utils.multiply_stamped_tfs(stfDriftyOdomWrtMap_,stfBaseLinkWrtDriftyOdom_,↩
    stfEstBaseWrtMap_);
```

The stamped transform `stfEstBaseWrtMap_` is the best estimate of the robot base frame with respect to the `map` frame. This estimate is updated at the full frequency of `odom`, and it is corrected with the most current transform of the `odom` frame with respect to the `map` frame.

The transform `stfEstBaseWrtMap_` is published on the `tf` topic. Additionally, the (stamped) pose of the robot with respect to the `map` frame (as inferred from this fusion of odometry and AMCL) is extracted from `stfEstBaseWrtMap_` to populate the member variable `estBasePoseWrtMap_`. The position and orientation from this pose can be used in steering algorithms, since it is both updated frequently and not subject to drift. Use of this estimate of localization for steering feedback is described in Section 9.4.

9.3 DIFFERENTIAL-DRIVE STEERING ALGORITHMS

Vehicle steering algorithms provide feedback that attempts to make a mobile robot follow a desired path. Many variations exist, depending on how the desired trajectories are spec-ified, what signals are available for feedback and the kinematics of the vehicle's steering mechanism.

Some vehicles have omnidirectional steering, such as the PR2 robot.[1] In this case, one can command three degrees of freedom of twist (velocity): a forward speed, a sideways speed and a spin rate. Steering algorithms for omnidirectional motion capability are relatively straightforward.

A more difficult and more common variation involves a steering mechanism as used in road vehicles, called Ackermann steering ([17]). In this case, one has control over forward speed and over a steering angle (heading of the front wheels, *e.g.* as imposed by a steering wheel). The angular velocity (yaw rate) of the vehicle depends on both the steering angle and the vehicle's forward speed. The vehicle cannot spin independent of forward motion. Further, the vehicle is kinematically constrained such that it cannot (or should not) slip sideways. A simple approach to steering for such vehicles is the wagon-handle algorithm, a form of "pure pursuit" steering (see *e.g.* [29]).

A third case, which will be the focus of this chapter, is differential-drive kinematics. Examples include skid-steered vehicles, such as iRobot's tracked PackBot,[2] and Clearpath's four-wheeled Husky ground vehicle.[3] Alternatively (and more gracefully), some differential-drive robots have two drive wheels and passive casters. This design type includes powered wheelchairs, the Roomba[4] or Turtlebot,[5] and the Pioneer 3.[6] The simple mobot model introduced in Section 3.7 is an example of this class of mobile robots. For such designs, one can control two degrees of freedom independently: speed and yaw rate (spin). We have seen control of this type of robot in simulation, both for the simple two-dimensional robot

---

[1]http://www.willowgarage.com/pages/pr2/overview
[2]http://www.irobot.com/For-Defense-and-Security/Robots/510-PackBot.aspx#Military
[3]http://www.clearpathrobotics.com/husky-unmanned-ground-vehicle-robot/
[4]http://www.irobot.com/For-the-Home/Vacuuming/Roomba.aspx
[5]http://www.turtlebot.com/
[6]http://www.mobilerobots.com/ResearchRobots/PioneerP3DX.aspx

(STDR) and the mobot model, using speed and spin commands via the `cmd_vel` topic. We begin with a kinematic analysis of differential-drive vehicles.

9.3.1 Robot motion model

Consider a simple vehicle that travels in the x–y plane and can be commanded with a speed, v, and an angular-rotation rate, ω. The robot has a 3-D pose that can be specified as its x and y coordinates plus its heading, ψ. The heading will be defined as measured counter-clockwise from the world frame (or chosen reference frame) x axis. The motion of the robot is described by the following differential equations:

$$\frac{d}{dt}\begin{bmatrix} x \\ y \\ \psi \end{bmatrix} = \begin{bmatrix} v\cos(\psi) \\ v\sin(\psi) \\ \omega \end{bmatrix} \tag{9.10}$$

It will be useful to define path following error coordinates d_{err}, a lateral offset error, and ψ_{err}, a heading error, relative to a desired path. An example desired path segment is a directed line segment starting from (x_0, y_0), and with tangent $\mathbf{t} = [\cos(\psi_{des}), \sin(\psi_{des})]^T$ and with edge normal $\mathbf{n} = [-\sin(\psi_{des}), \cos(\psi_{des})]^T$, where ψ_{des} is the angle of the segment relative to the x axis of the reference frame (measured counter-clockwise). The edge normal is a positive 90-degree rotation of $\mathbf{t}$, where positive rotation is defined as up (rotation about z, where the z axis is defined normal to the x–y plane, consistent with forming a right-hand coordinate system).

With respect to this directed line segment, we can compute a lateral offset error of the robot at position (x, y) as:

$$d_{err} = \left(\begin{bmatrix} x \\ y \end{bmatrix} - \begin{bmatrix} x_0 \\ y_0 \end{bmatrix} \right) \cdot \mathbf{n} \tag{9.11}$$

The heading error, ψ_{err}, is computed as the actual heading versus the desired heading: $\psi_{err} = \psi - \psi_{des}$. With rotational variables, it is important to consider periodicity. For example, consider a desired heading of $\psi_{des} = 0$ (*i.e.* pointing parallel to the x axis) and an actual heading of $\psi = 0.1$. This corresponds to a heading error of 0.1 rad, which is close to the desired heading. However, for a heading of $\psi = 6.0$, the formula $\psi_{err} = \psi - \psi_{des}$ would seem to imply a heading error of 6.0 rad, which is a large heading error. Instead, the periodic solutions of this result should be checked to find the smallest error interpretation. The value of $\psi_{err} - 2\pi$ corresponds to a negative heading error of approximately -0.28 rad. That is, the heading is more easily corrected by rotating a positive 0.28 rad than by rotating negative 6 rad. This periodic condition must be checked within iterations of a control algorithm or it can lead to gross instability. We will choose to define the heading error as the smallest magnitude option among periodic alternatives.

We will define a 2-D path following error vector in terms of our two error components as:

$$\mathbf{e} = \begin{bmatrix} d_{err} \\ \psi_{err} \end{bmatrix} \tag{9.12}$$

The components may be interpreted as a lateral offset, which is positive to the left of the path, and a heading error, which is measured positive as a counter-clockwise rotation of heading away from the path tangent.

In the following, we will assume that the robot is moving forward at speed v and that we can command a superimposed rotation rate, ω, with the objective of driving both components of the error vector to zero. If the error vector is zero, the robot will have zero offset (*i.e.* will be positioned on top of the path segment) and zero heading error (*i.e.* will be pointing consistent with the desired heading). For a differential-drive robot, an objective of a steering algorithm is to compute an appropriate spin command, ω, to drive the error components to zero, and to do so with desirable dynamics (*e.g.* rapidly and stably).

9.3.2 Linear steering of a linear robot

We first consider a simplified system: a linear approximation of the robot dynamics controlled by a linear controller. For simplicity, assume that the desired path is the positive x axis, for which the error vector is simply $\mathbf{e} = [y, \psi]^T$ (assuming ψ is expressed as the smallest absolute-value option among periodic alternatives).

Assuming a linear controller, we may command $\omega = \mathbf{Ke}$, where $\mathbf{K} = [K_d, K_\psi]$. The two components of $\mathbf{K}$ are control gains, which should be chosen to result in desirable dynamics of the robot approaching the desired path.

Consider simplified (linearized) vehicle dynamics of:

$$\frac{d}{dt}\begin{bmatrix} x \\ y \\ \psi \end{bmatrix} = \begin{bmatrix} v \\ v\psi \\ \omega \end{bmatrix} \tag{9.13}$$

corresponding to a small-angle approximation for small values of ψ.

Equivalently, the error dynamics can be expressed as:

$$\frac{d}{dt}\mathbf{e} = \begin{bmatrix} 0 & v \\ 0 & 0 \end{bmatrix}\mathbf{e} + \begin{bmatrix} 0 \\ 1 \end{bmatrix}\omega \tag{9.14}$$

Substituting in our control policy for ω yields:

$$\frac{d}{dt}\mathbf{e} = \begin{bmatrix} 0 & v \\ 0 & 0 \end{bmatrix}\mathbf{e} + \begin{bmatrix} 0 \\ 1 \end{bmatrix}\mathbf{Ke} = \begin{bmatrix} 0 & v \\ K_d & K_\psi \end{bmatrix}\mathbf{e} \tag{9.15}$$

In the LaPlace domain, this corresponds to:

$$\begin{bmatrix} 0 \\ 0 \end{bmatrix} = \begin{bmatrix} s & -v \\ -K_d & s - K_\psi \end{bmatrix}\mathbf{e} \tag{9.16}$$

which has a characteristic equation of:

$$0 = s^2 - K_\psi s - vK_d \tag{9.17}$$

Any combination of negative values for the control gains theoretically results in a stably controlled system (for this linear approximation). We can choose gains intelligently by choosing values interpreted in terms of the generic second-order system response: $0 = s^2 + 2\zeta w_n s + w_n^2$. If we choose $w_n = 6$ (roughly 2π, or about 1 Hz) and $\zeta = 1$ (*i.e.* critical damping), we would expect convergence to the desired path with a time constant of approximately 1 sec with zero overshoot.

Figures 9.17, 9.18, and 9.19 show a simulation of the linearized system with the chosen controller. The initial offset error and the heading error both converge to zero within about 1 sec, as expected, and the offset error does not overshoot, consistent with a critically-damped system. The control effort, *i.e.* the ω command, is shown in Fig 9.19. The spin command starts with a strongly negative ω, which changes sign before converging on zero as the robot converges on zero offset error and zero heading error.

9.3.3 Linear steering of a non-linear robot

In the preceding section, we considered a linear controller acting on a hypothetical linear system (a small-angle approximation of our robot dynamics). We should evaluate the consequences of attempting use of a linear controller on a more realistic model of the robot. With

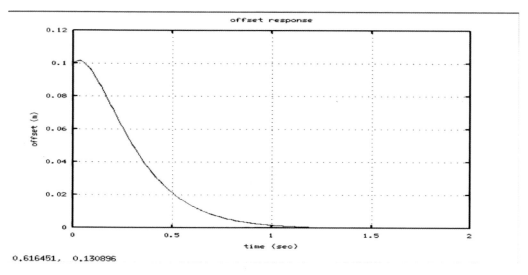

0.616451, 0.130896

Figure 9.17: Offset response versus time, linear model, linear controller, 1 Hz controller, critically damped

---

the same gains as in Section 9.3.2 and the dynamic model of Section 9.3.1 (not linearized), we consider a directed line-segment desired path, specified as follows. The start of the line segment is at $(x, y) = (1, 0)$, and the slope of the desired path is 45 degrees. Path heading error and offset error are computed as described in Section 9.3.1. The angular-rotation command is computed based on the linear control algorithm and control gains of Section 9.3.2.

Figures 9.20 and 9.21 show the response to relatively small initial errors, displayed as lateral offset error, Fig 9.20, and path versus desired path in the x–y plane, Fig 9.21. The response is good, similar to that of the linear analysis.

Figure 9.22 shows the response of the linear controller on the non-linear robot for a variety of initial conditions for which convergence is well behaved. In this view, the heading error is plotted as a function of displacement, which is a phase space plot. It can be observed that as the robot approaches convergence to the desired path segment, the heading error (robot heading minus specified path heading) is approximately linearly proportional to the lateral offset error. This observation will be used to help design a non-linear controller that shares characteristics with the linear controller.

Unlike linear systems, the stability of a non-linear system can depend on initial conditions. Tuning the controller for a desirable response may give the impression that it is well behaved, but different initial conditions can result in a wildly unstable response. Figures 9.23 and 9.24 show an example with a large initial offset error. The robot spins in circles, potentially dangerously, and it fails to converge on the desired path.

In short, a linear controller cannot be trusted on an actual robot. A non-linear controller is considered next, which blends strategies for large offset errors with the behavior of a linear controller for small path-following errors.

9.3.4 Non-linear steering of a non-linear robot

In the previous sections, it was shown that a linear controller works well as a steering algorithm, provided the path following errors are sufficiently small. The linear control algorithm computes a spin rate command as a weighted sum of heading and displacement errors. How-

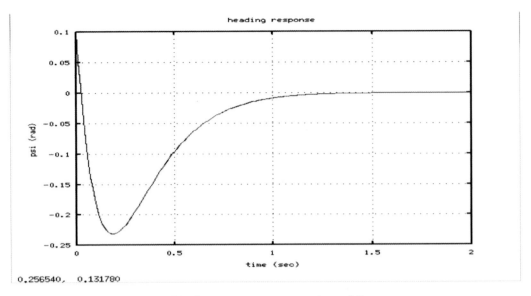

Figure 9.18: Heading response versus time of linear system

ever, when the robot is far from the goal (or has a large heading error), linear mapping is inappropriate.

If the lateral displacement is large, the robot should first attempt to head directly toward the desired path. To do so, the best heading is to orient the robot toward the path, with robot heading orthogonal to path heading. In this case, a heading error of $\pm\pi/2$ is desirable and should not be penalized. This may be considered an approach phase or a reaching phase.

Once the lateral displacement between the path segment and the robot is sufficiently small, the reaching phase should transition to a path following phase, for which a linear controller is well suited. In the path following phase, the heading should tend to align with the path tangent (*i.e.* the heading error should tend toward zero) as the lateral offset decreases.

The above observations can be integrated in a formula that combines the approach behavior and the path following behavior as a function of lateral offset. To construct this controller, we define four different heading variables: ψ_{path}, the ideal heading along a specified path; ψ_{state}, the current heading of the robot; $\psi_{strategy}$, a computed strategy of desirable heading to achieve convergence to the specified path; and ψ_{cmd}, a heading command to be sent to a low-level heading controller.

The non-linear steering controller described here depends on a low-level heading controller. The purpose of the heading controller is to cause the robot to change its orientation, ψ_{state}, to point in a commanded direction, ψ_{cmd}. A simple heading controller may be of the form:

$$\omega = K_\psi(\psi_{cmd} - \psi_{state}) \tag{9.18}$$

although more sophisticated controllers may be used for better performance.

In implementation, the value of $(\psi_{cmd} - \psi_{state})$ should be checked for periodicity, choosing the direction of rotation that is fastest (*i.e.* using the smallest rotation angle to reach the commanded heading). The gain value, K_ψ, should be tuned to achieve good response from the system.

It is also desirable to have the controller explicitly respect saturation limits. If the speed

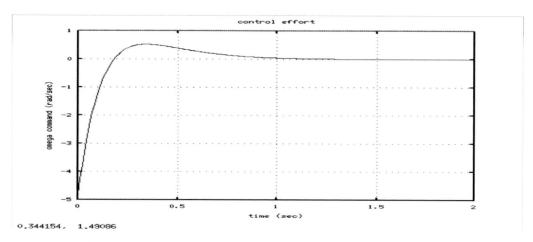

0.344154, 1.49086

Figure 9.19: Control effort history of linear system versus time. Note that spin command may exceed actual physical limitations

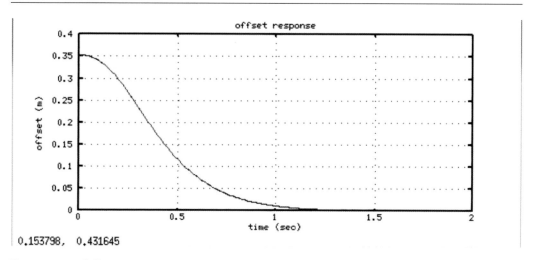

0.153798, 0.431645

Figure 9.20: Offset versus time, non-linear model, 1 Hz linear controller. Response to small initial error is good, similar to linear model response.

limit imposed on rotation rate (whether a hard or soft limit) is ω_{max}, this can be respected by saturating the feedback algorithm to put limits on the spin command, ω_{cmd}:

$$\omega_{cmd} = \omega_{max} f_{sat}(K_\psi(\psi_{cmd} - \psi_{state})) \tag{9.19}$$

The saturation function, f_{sat}, is defined as:

$$f_{sat} = \begin{cases} -1 & \text{if } x < -1 \\ 1 & \text{if } x > 1 \\ x & \text{otherwise} \end{cases} \tag{9.20}$$

Assuming the heading controller works well, it remains to compute a strategy for heading commands that will lead the robot to converge on the specified path. This mapping should result in values of $\psi_{strategy}$ that are continuous and smooth and achieve the desired result of good convergence to the specified path regardless of initial conditions. Mapping should

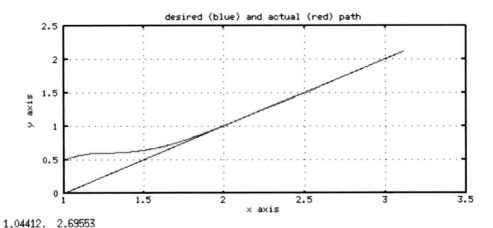

1.04412. 2.69553

Figure 9.21: Path following of non-linear robot with linear controller and small initial error. Convergence to precise path following is well behaved.

---

also blend approach strategies for large d_{err} (point toward the path, orthogonal to the path) and small d_{err} (behave like a linear controller for path following).

Initially, consider a desired path corresponding to the x axis (in the chosen reference frame). The path heading for this case is $\psi_{path}=0$, the lateral offset error is $d_{err} = y$, and the heading error is $\psi_{err} = \psi_{state}$. A non-linear approach strategy can be constructed that specifies $\psi_{strategy}$ as a function of lateral offset, d_{err}, as follows:

$$\psi_{strategy}(d_{err}) = -(\pi/2)f_{sat}(d_{err}/d_{thresh}) \tag{9.21}$$

In this formula, the value d_{thresh} (with units of meters) is a parameter to be tuned. When the lateral offset error approaches d_{thresh}, the heading strategy approaches $\pm\pi/2$, causing the robot to orient perpendicular to (and pointing toward) the current path segment. For values of offset error, d_{err}, that are small compared to d_{thresh}, the heading strategy is proportional to the offset error, which conforms to the behavior of the convergence region of a linear steering algorithm, as seen in Fig 9.22.

To generalize the approach strategy to path segments that are oriented arbitrarily, the specified path heading, ψ_{path}, should be added to the computed-strategy heading, yielding:

$$\psi_{cmd} = \psi_{path} + \psi_{strategy} \tag{9.22}$$

The resulting ψ_{cmd} becomes the input to the lower-level heading controller.

9.3.5 Simulating non-linear steering algorithm

The package `mobot_nl_steering` contains an implementation of the non-linear steering algorithm as a class, `SteeringController`. The code is lengthy, and thus is not displayed here in full. Some key excerpts are interpreted here.

A member function that accounts for periodicity to find an angular error is:

```
double SteeringController::min_dang(double dang) {
    while (dang > M_PI) dang -= 2.0 * M_PI;
    while (dang < -M_PI) dang += 2.0 * M_PI;
    return dang;
}
```

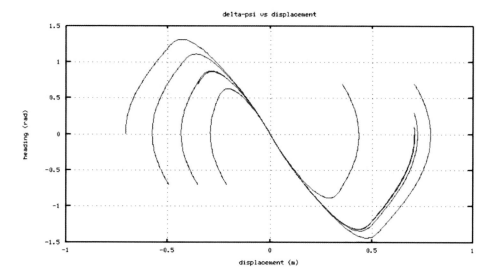

Figure 9.22: Phase space of linear control of non-linear robot. Note linear relation between displacement and heading error near the convergence region.

This function is useful for finding the direction and magnitude of the minimum angular distance motion from some initial angle to some desired angle.

Implementation of the saturation function is:

```
double SteeringController::sat(double x) {
    if (x>1.0) {
        return 1.0;
    }
    if (x< -1.0) {
        return -1.0;
    }
    return x;
}
```

The main steering function is `SteeringController::mobot_nl_steering()`. The following code segment:

```
double tx = cos(des_state_psi_); // [tx,ty] is tangent of desired path
double ty = sin(des_state_psi_);
double nx = -ty; //components [nx, ny] of normal to path, points to left of ←
    desired heading
double ny = tx;

double dx = state_x_ - des_state_x_;  //x-error relative to desired path
double dy = state_y_ - des_state_y_;  //y-error

lateral_err_ = dx*nx+dy*ny; //lateral error is error vector dotted with path ←
    normal
                            // lateral offset error is positive if robot is ←
                                to the left of the path
double trip_dist_err = dx*tx+dy*ty; // progress error: if positive, then we are ←
    ahead of schedule
//heading error: if positive, should rotate -omega to align with desired heading
double heading_err = min_dang(state_psi_ - des_state_psi_);
double strategy_psi = psi_strategy(lateral_err_); //heading command, based on NL ←
    algorithm
controller_omega = omega_cmd_fnc(strategy_psi, state_psi_, des_state_psi_); //spin←
    command
```

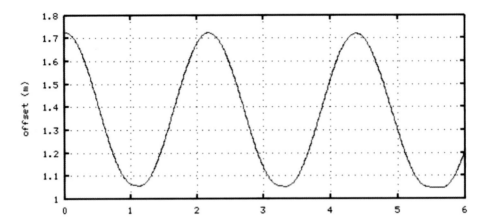

Figure 9.23: Linear controller on non-linear robot with larger initial displacement. Displacement error versus time oscillates.

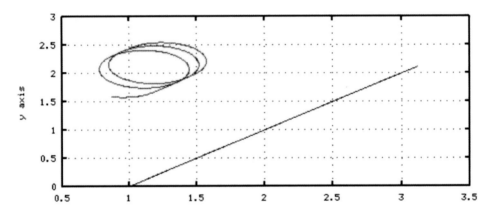

Figure 9.24: Path from linear controller on non-linear robot with large initial error y versus x. Robot spins in circles, failing to converge on desired path.

computes the path tangent, **t**, given a desired path heading `des_state_psi_`, as well as the path normal, **n**, the lateral offset error, `lateral_err_`, and the heading error, `heading_err`, taking periodicity into consideration. The value of `strategy_psi` is computed as a function of the lateral error. This function is given by:

```
double SteeringController::psi_strategy(double offset_err) {
    double psi_strategy = -(M_PI/2)*sat(offset_err/K_LAT_ERR_THRESH);
    return psi_strategy;
}
```

which implements our derived non-linear expression for heading strategy. This is used in the heading controller:

```
double SteeringController::omega_cmd_fnc(double psi_strategy, double psi_state, double↩
    psi_path) {
  psi_cmd_ = psi_strategy+psi_path;
```

```
double omega_cmd = K_PSI*(psi_cmd_ - psi_state);
omega_cmd = MAX_OMEGA*sat(omega_cmd/MAX_OMEGA); //saturate the command at specified ↩
    limit
return omega_cmd;
}
```

The values `K_PSI` and `K_LAT_ERR_THRESH` are parameters to be tuned for desirable response (*e.g.* fast but with low overshoot).

To invoke the controller on our simulated mobile robot, we need an estimate of the robot's state published on topic `gazebo_mobot_pose` as a message of type `geometry_msgs::Pose`. For the purpose of evaluating our controller, this state estimation is taken from the Gazebo publication on topic `gazebo/model_states`, parsed for the state of the model mobot, and republished on topic `gazebo_mobot_pose`. This is performed by the `mobot_gazebo_state` node in the `mobot_gazebo_state` package. Simulation experiments can be run by starting the simulator:

```
roslaunch gazebo_ros empty_world.launch
```

loading the robot model:

```
roslaunch mobot_urdf mobot.launch
```

starting the mobot state publisher:

```
rosrun mobot_gazebo_state mobot_gazebo_state
```

and running the controller:

```
rosrun mobot_nl_steering mobot_nl_steering
```

Desired path states can be published to the topic `/desState` with message type `nav_msgs::Odometry`. However, by default, the desired path is the world x axis (pointing in the positive direction). The robot can be re-positioned and re-oriented in Gazebo, and the non-linear steering controller will attempt to have the robot converge on the current path segment (*e.g.* the x axis).

For chosen parameters of:

```
const double K_PSI= 5.0; // control gains for steering
const double K_LAT_ERR_THRESH = 3.0;
// dynamic limitations:
const double MAX_SPEED = 1.0; // m/sec; tune this
const double MAX_OMEGA = 1.0; // rad/sec; tune this
```

an example response is shown in Fig 9.25. The robot starts out with a lateral displacement error of approximately 4 m (positive, *i.e.* shifted to the left of the desired path). Since this is a large displacement error, the heading strategy is computed to be $-\pi/2$. Since the robot's initial heading is zero, the difference between actual and desired headings is large, resulting in saturating the ω command at -1.0 rad/sec. This command persists while the robot re-orients and moves forward, closer to the path. After about 2 sec, the heading strategy starts to re-orient toward zero (the path heading). The ω command becomes positive to achieve more positive headings (approaching zero heading). After about 4 sec, the remainder of the trajectory looks like a linear controller, with both heading error and offset error smoothly converging to zero.

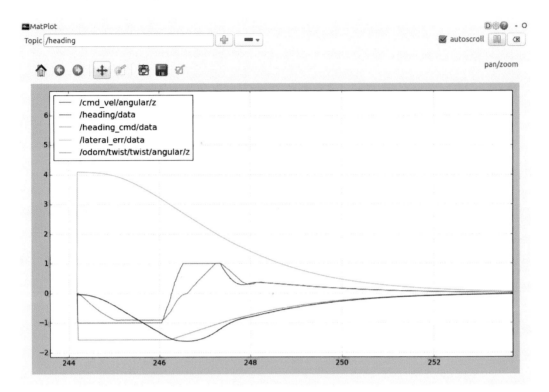

Figure 9.25: State versus time for non-linear control of non-linear robot. Behavior at large initial errors is appropriate and different from linear control, while response blends to linear-control behavior for small errors near convergence.

9.4 STEERING WITH RESPECT TO MAP COORDINATES

The steering algorithms of Section 9.3 rely on the state estimation techniques of Section 9.2. A specific example using integration of AMCL with odometry, as described in Section 9.2.4, is illustrated here. In the package `lin_steering`, the node `lin_steering_wrt_amcl` (with source code of the same name) performs such steering.

The main program within `lin_steering_wrt_amcl.cpp` is shown in Listing 9.11.

Listing 9.11: Main program of `lin_steering_w_amcl.cpp`: code to perform linear feedback steering with respect to localization based on integrated AMCL and odometry

```
int main(int argc, char** argv)
{
    ros::init(argc, argv, "steeringController"); //node name
    ros::NodeHandle nh;
    XformUtils xform_utils;
    geometry_msgs::PoseStamped est_st_pose_base_wrt_map;
    ros::Rate sleep_timer(UPDATE_RATE); //a timer for desired rate, e.g. 50Hz

    OdomTf odomTf(&nh); //instantiate an OdomTf object to fuse amcl and drifty odom
    while (!odomTf.odom_tf_is_ready()) {
        ROS_WARN("waiting on odomTf warm-up");
        ros::Duration(0.5).sleep();
        ros::spinOnce();
    }
```

```
    ROS_INFO("main: instantiating an object of type SteeringController");
    SteeringController steeringController(&nh);

    ROS_INFO("starting steering algorithm");
    while (ros::ok()) {
        ros::spinOnce();
        //get pose from computed stamped transform of base_link w/rt map, combining ↩
            amcl and drifty_odom
        est_st_pose_base_wrt_map = xform_utils.get_pose_from_stamped_tf(odomTf.↩
            stfEstBaseWrtMap_);
        g_odom_tf_x = est_st_pose_base_wrt_map.pose.position.x;
        g_odom_tf_y = est_st_pose_base_wrt_map.pose.position.y;
        g_odom_tf_phi = xform_utils.convertPlanarQuat2Phi(est_st_pose_base_wrt_map.↩
            pose.orientation);
        steeringController.lin_steering_algorithm(); // compute and publish twist ↩
            commands and cmd_vel and cmd_vel_stamped
        sleep_timer.sleep();
    }
    return 0;
}
```

This main function instantiates an object of type `OdomTf` and an object of type `SteeringController`. Both of these depend on a timed loop and on `ros::spinOnce()` for them to receive updates via their callback functions and to perform their respective computations. The `OdomTf` object updates (and publishes) its current best estimate of the stamped transform `stfEstBaseWrtMap_`, which is the current best estimate of the robot's pose in map coordinates. As described in Section 9.2.4, the `OdomTf` object transforms pose of the robot in odometry coordinates into pose of the robot in map coordinates. This is performed at the full update rate of odometry publications, using the most current estimate of odometry-frame drift with respect to map coordinates, and computed based on AMCL publications.

The main function of `lin_steering_wrt_amcl.cpp` contains a timed loop in which `OdomTf` is updated, and from these updates the pose of the robot is described in x, y and heading, expressed in map coordinates. These values are copied to the global variables `g_odom_tf_x`, `g_odom_tf_y` and `g_odom_tf_phi`, respectively, for use by the steering algorithm.

With the call to `ros::spinOnce()` in the main loop, the `ros::spinOnce()` command also enables updates via callback functions within the `SteeringController` object. A callback function in this object receives the most current desired-state publication (as published by a desired-state publisher, as described in Section 9.1.4).

Having updated the estimate of the robot's state (in map coordinates), as well as the desired state of the robot (in map coordinates), the linear steering algorithm is called. The `lin_steering_algorithm()` function within the `SteeringController` object is shown in Listing 9.12.

Listing 9.12: Steering function within `lin_steering_w_amcl.cpp`

```
void SteeringController::lin_steering_algorithm() {
    double controller_speed;
    double controller_omega;

    double tx = cos(des_state_phi_); //components of tangent to desired path
    double ty = sin(des_state_phi_);
    double nx = -ty; //components of normal to desired path
    double ny = tx;

    double heading_err;
    double lateral_err;
    double trip_dist_err; // error is scheduling...are we ahead or behind?
```

```
// have access to: des_state_vel_ , des_state_omega_, des_state_x_, des_state_y_ , ⤶
    des_state_phi_
// and corresponding state estimate values , in map coordinates
double dx = des_state_x_ - g_odom_tf_x; //error between desired and actual x-⤶
    coordinate value , in map coordinates
double dy = des_state_y_ - g_odom_tf_y; //error between desired and actual y-⤶
    coordinate value , in map coords

lateral_err = dx*nx + dy*ny; //compute sideways offset from desired path
trip_dist_err = dx*tx + dy*ty; //compute progress along path
heading_err = min_dang(des_state_phi_ - g_odom_tf_phi); // watch out for ⤶
    periodicity of heading

// use linear feedback to reduce the three errors:
controller_speed = des_state_vel_ + K_TRIP_DIST*trip_dist_err; //speed up/slow ⤶
    down to null out trip dist err

//steering feedback , based on both heading error and lateral offset error
controller_omega = des_state_omega_ + K_PHI*heading_err + K_DISP*lateral_err;
controller_omega = MAX_OMEGA*sat(controller_omega/MAX_OMEGA); // saturate omega ⤶
    command at specified limits

// send out corresponding speed/spin commands:
twist_cmd_.linear.x = controller_speed;
twist_cmd_.angular.z = controller_omega;
twist_cmd2_.twist = twist_cmd_; // copy the twist command into twist2 message
twist_cmd2_.header.stamp = ros::Time::now(); // look up the time and put it in the⤶
    header
cmd_publisher_.publish(twist_cmd_);
//this second publication is simply for debug/visualization; in includes a stamped⤶
    header,
// which makes it convenient to plot with rqt_plot
cmd_publisher2_.publish(twist_cmd2_);
}
```

The linear-steering function uses the most recent desired-state update and the most recent robot state-estimate update to compute errors in x, y and heading. The x and y errors are transformed into alternative values of lateral offset and trip-distance error. Trip distance error is used to modulate forward speed to keep the robot on schedule with forward progress. Lateral offset and heading error are used to control the angular velocity to steer the robot to converge on the desired path with the desired heading. The feedback computations are used to populate a Twist message, which is then published to the cmd_vel topic to control the robot. It is important that this publication be updated periodically at a high enough update rate for the steering to be stable.

This steering algorithm can be observed running with the following commands. First, start up the mobot in the DARPA starting pen with:

```
roslaunch mobot_urdf mobot_in_pen.launch
```

Next, launch multiple nodes with:

```
roslaunch odom_tf mobot_w_odom_tf.launch
```

The above launch performs the following. The map_server node is started, with reference to a map of the starting pen. An amcl node is started, specifically considering non-holonomic robot mobility. The mobot_drifty_odom node is started, providing a deliberately imperfect odometry publication (so as to illustrate the value of integration with amcl). A mobot_pub_des_state node is started, ready to accept path commands from a client. The lin_steering_wrt_amcl node is started, which uses information published by both mobot_drifty_odom and mobot_pub_des_state. For visualizing the result, rviz is started (with a pre-set configuration), and a triad_display node is started to help display internal pose values computed within lin_steering_wrt_amcl.

Figure 9.26 shows the state of the robot in both Gazebo and `rviz` upon start-up. Note that the global (fixed) frame for `rviz` has been set to `map`. Four frames are displayed: the true `base_link` of the robot, the pose of the robot as computed/updated by AMCL, the (drifty) odometry reference frame, and the estimated pose of the robot from integration of AMCL and (drifty) odom. Upon start-up, these four frames all coincide, since there has been no motion yet to induce drift of odometry, and the slow update rate of AMCL cannot be seen with the robot stationary.

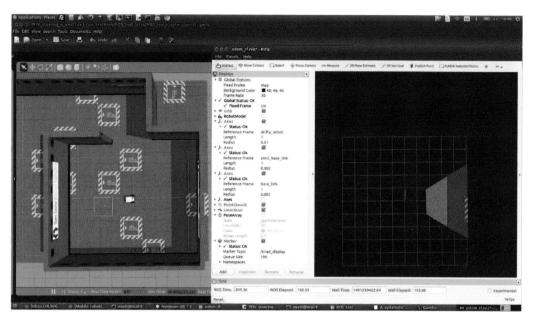

Figure 9.26: Initial state of mobot in starting pen after launch. Note that in the `rviz` view, four frames are displayed, but these frames initially coincide.

To get the robot to move, a path client sends via points to the desired-state publisher node. An example client relevant to the starting pen can be run with:

```
rosrun mobot_pub_des_state starting_pen_pub_des_state_path_client
```

This client sends three via points, directing the robot to exit the starting pen and end at a location outside the pen. Two snapshots of progress after sending this path service request are shown in Figures 9.27 and 9.28. In Figure 9.27, the four frames displayed in `rviz` are no longer coincident. The odometry frame, near the bottom of the `rviz` view, shows significant drift in both translation and rotation. With ideal odometry, this frame would have remained identically at its start-up pose. AMCL updates help to identify the cumulative odometry-frame drift, and this is the result displayed.

Near the top of the `rviz` view, the other three frames are shown, and these are relatively close to each other. The top-most frame is the true pose of the `base_link`. The lowest of the three clustered frames is the robot's estimated pose per AMCL. This frame is lagging behind, and it updates in discrete jumps at a relatively low update rate. Between these two frames is the estimated pose of the robot, based on integration of AMCL and odometry. Ideally, this frame would be coincident with the true `base_link` frame, although there is some lag (attributed to delayed updates of AMCL). In the present case, the odometry drift is fairly severe. With better-tuned odometry, the odometry frame would drift more slowly

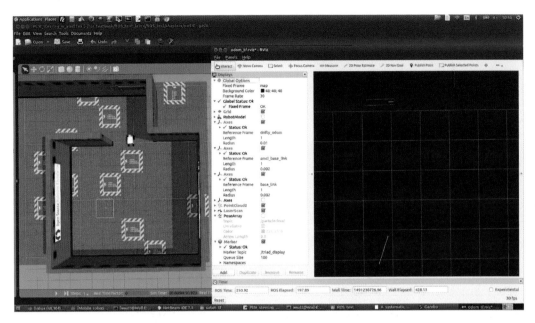

Figure 9.27: Snapshot of progress of mobot steering with respect to state estimation from integrated AMCL and odometry. Note that the odometry frame, near the bottom of the `rviz` view, has translated and rotated significantly since startup, illustrating the amount of cumulative drift of odometry to this point.

and AMCL transforms would remain valid for longer durations. Correspondingly, integrated AMCL and odometry would more closely track the true pose of the robot. Even with this large odometry drift, the estimated pose is still suitable for steering, since it is updated frequently and it does not drift with accumulating errors.

Figure 9.28 shows the status of the robot near its goal destination. Having traveled a longer distance, the odometry-based pose estimate has become worse. This is visualized by drift of the odometry reference frame (which, ideally, would remain coincident with its initial pose). At this point, the odometry reference frame, shown in the lower right of the `rviz` view of Fig 9.28, has drifted nearly outside the pen and has rotated by nearly 90 degrees. In spite of this, the pose estimation from integrated AMCL and odometry remains reasonably precise. In the `rviz` view in Fig 9.28, there are three frames that are close to each other. The bottom-most of these three frames is the true robot pose, the top-most frame is the most recent pose per AMCL, and the middle frame is the estimated robot pose from integration of AMCL and odometry. All three frames agree on the robot's orientation. The integrated AMCL and odometry frame is close to the robot's true pose, and this frame is updated with each publication of `odom`, making it suitable for steering.

Running the above code illustrates success of steering using integrated AMCL and odometry. Using this estimate for steering corrections, the robot succeeds in driving accurately along the intended path. In contrast, steering with respect to uncorrected odometry would result in collisions early in the robot's trajectory.

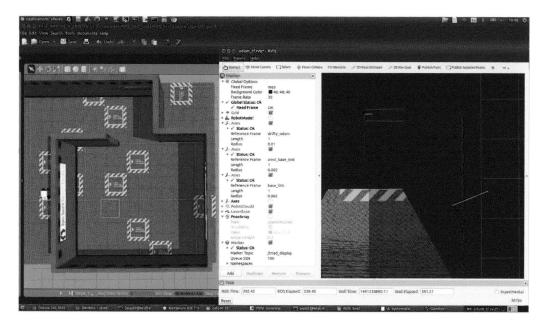

Figure 9.28: Result of mobot steering with respect to state estimation from integrated AMCL and odometry. The odometry frame, lower right in the `rviz` view, has translated and rotated dramatically during navigation. Nonetheless, the robot's true pose, the AMCL pose estimate and the integrated AMCL and odometry pose estimate are shown to be in approximate agreement.

9.5   WRAP-UP

This chapter presented an introduction to mobile-robot motion control. Essential components include sufficiently detailed specification of desired states (trajectories); estimation of actual robot states (localization); and feedback for driving a robot's trajectory towards a desired trajectory (steering algorithms).

The presentation to this point has assumed the existence of a viable path. Generating a viable path that takes into consideration obstacle avoidance in a non-trivial environment is the subject of path planning, which is introduced next.

Mobile-Robot Navigation

CONTENTS

INTRODUCTION
To this point, consideration of planning has been restricted to point-and-move along some specified polyline. More generally, a vehicle should be able to plan and execute its own motion toward a goal destination. In doing so, it should consider all *a priori* information available, such as a pre-recorded map, as well as unexpected obstacles encountered en route. In ROS, this is addressed with what is often referred to as the navigation stack. The navigation stack assumes use of maps, localization, and obstacle sensing. With specifications of vehicle dimensions, costmaps representing poses in the environment as having corresponding penalties are computed. A global planner computes collision-free paths from the robot's current pose to a specified goal destination. While following a global path plan, the robot senses and responds to obstacles, typically resulting in path deviations that bypass the obstacle and re-join the global plan.

In this chapter, use of the navigation stack via the package `move_base` is introduced. We start with describing a process for map making, then show how maps can be used with the nav-stack and how action clients can invoke the functionality of the nav-stack.

10.1 MAP MAKING

The `gmapping` package can be used to create maps using a mobile robot and a LIDAR sensor. Per the `gmapping` wiki, `http://wiki.ros.org/gmapping`:

> The `gmapping` package provides laser-based SLAM (Simultaneous Localization and Mapping), as a ROS node called `slam_gmapping`. Using `slam_gmapping`, you can create a 2-D occupancy grid map (like a building floorplan) from laser and pose data collected by a mobile robot.

To use this package, the robot's LIDAR must be mounted horizontally with a fixed mount, and the robot must publish odometry. This package post-processes LIDAR and `tf` data and attempts to build a map that explains the LIDAR data relative to the robot's movements.

Mapping can be performed online, or as a post-process. For the post-processing approach, the first step in map making is to acquire (bag) data. As an example, start Gazebo:

```
roslaunch gazebo_ros empty_world.launch
```

The `mobot_startup_open_loop.launch` file in the package `mobot_urdf` starts the mobot model, brings in the `starting_pen` model, starts the desired-state publisher, and starts an open-loop controller. The robot model used in this launch contains LIDAR, a Kinect sensor and a color camera. `rviz` is also launched. Perform these operations with:

```
roslaunch mobot_urdf mobot_startup_open_loop.launch
```

With the robot running, `rviz` should show that the LIDAR is active and that it can sense the walls of the starting pen. At this point, the robot should wander (slowly) through some path, exploring its environment while bagging data on the topics `tf` and `scan` (where `scan` is the LIDAR topic for the mobot model).

To record the data, navigate to a desired directory. The current example uses the `maps` directory within the `learning_ros` repository. From the chosen directory, start recording data with:

```
rosbag record -O mapData /scan /tf
```

While data is being recorded, have the robot move. An example pre-scripted motion for a 3 m × 3 m square path may be invoked with:

```
rosrun mobot_pub_des_state pub_des_state_path_client_3x3
```

If desired, this can be re-run to have the robot perform a second (or third) repeat of the path. Due to open-loop control, the robot will drift from the desired path, so repeats of this path request will yield novel viewpoints, which are useful for map making.

When using a real robot for map making, one can start the bagging process, then use a joystick to manually drive the robot through the environment of interest. When doing so, one should take care to move the robot slowly, particularly in turning motions. Such data will be easier to post-process to reconcile scans from different poses.

Once enough data has been collected, kill the `rosbag` process (with control-C). Gazebo can be killed as well. The bagged data will be contained in the `mapData` file within the chosen directory.

Given a `rosbag` recording that contains LIDAR scan data and transform (`tf`) data, the `rosbag` can be post-processed to compute a corresponding map. To do so, start a roscore, then, from the directory in which the `rosbag` data resides, run:

```
rosrun gmapping slam_gmapping scan:=/scan
```

The remapping option `scan:=/scan` is not necessary in the present example, since the LIDAR scan topic for the mobot model is already called `scan`. However, if the robot's scan topic were, *e.g.* `/robot0/scan`, one should use topic re-mapping with the option: `scan:=/robot0/scan`. At this point, the `gmapping` map builder is subscribed to the `tf` topic and to the LIDAR topic. To publish the acquired data to the respective topics, playback the data with:

```
rosbag play mapData.bag
```

The above command assumes that the recorded data is called `mapData.bag` (an option chosen during `rosbag` recording), and it is assumed that this command is run from the directory in which `mapData.bag` resides. With both `rosbag play` and `gmapping slam_mapping` running, the map making node will receive publications of LIDAR and `tf` data and will attempt to build a map from this data.

The map making process can be watched dynamically, if desired. To do so, bring up `rviz`:

```
rosrun rviz rviz
```

Add a `map` display item. In the display panel, expand this new item and choose the topic `/map`. While the recording is played back, one can watch a display of the map being built.

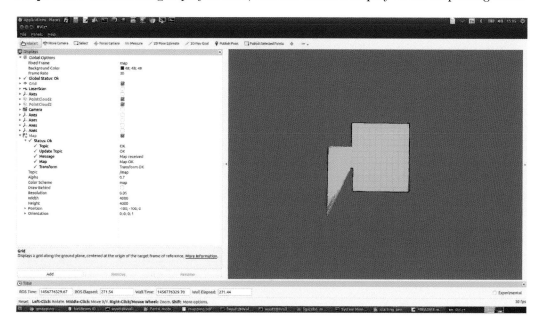

Figure 10.1: `rviz` view of map constructed from recorded data of mobot moving within starting pen

Figure 10.1 shows an example 2-D occupancy map in which shades of gray imply the degree of certainty of a cell being occupied. Black cells correspond to high probability of occupancy, and white cells imply a high probability of being vacant. Gray cells imply no information is available. The display in Fig 10.1 is a credible fit to the starting pen. The exit hallway is incompletely mapped, since the robot did not travel to poses from which a line-of-sight LIDAR acquisition could detect the occluded hallway.

When the bag-file playback is complete, the created map needs to be saved to disk. This is done with the command:

```
rosrun map_server map_saver -f newMap
```

where `newMap` is the file name, which should be chosen mnemonically. This will create two files: `newMap.yaml` and `newMap.pgm`. For the example map created, the `newMap.yaml` file contains:

```
image: newMap.pgm
resolution: 0.050000
origin: [-100.000000, -100.000000, 0.000000]
negate: 0
```

```
occupied_thresh: 0.65
free_thresh: 0.196
```

These parameters declare that the map data can be found in the file named `newMap.pgm`, and that the resolution of cells within this map is 5 cm. The threshold values are used by planning algorithms to decide whether a cell is to be called occupied or vacant.

The map data is contained in `newMap.pgm`. This is in an image format, which can be viewed directly just by double-clicking the file within Linux. If desired, the map may be touched up using an image editor program.

As an alternative to logging (bagging) data and post-processing to create a map, the map making can be performed interactively. To do so, one should start a robot in an environment of interest, and make sure that the LIDAR sensor is active, that its transform to the robot's base frame is published, that odometry is published, and that the robot is prepared to be moved under teleoperation. For our simulated mobile robot, this can be accomplished using:

```
roslaunch mobot_urdf mobot_in_pen.launch
```

This launch file starts Gazebo, spawns a model of the starting pen, loads the mobot model onto the parameter server, spawns the robot model into Gazebo, and starts a robot-state publisher. In the present case, the LIDAR link is already part of the robot model, and thus its transform to the robot-base frame will be published by the robot-state publisher. When running a physical robot, a static transform publisher would be needed to establish the pose of the LIDAR relative to the robot's base frame.

With the robot started in the environment of interest, three nodes should be run: `gmapping`, `rviz` and `teleop_twist_keyboard` (or some equivalent means to command the robot). These nodes can be started with assistance of a launch file in the package `mobot_mapping`:

```
roslaunch mobot_mapping mobot_startup_gmapping.launch
```

This launch file launches `rviz` using a stored configuration file that includes a `map` display (on topic `map`). When these nodes are launched, the display appears as in Fig 10.2. As the robot moves through its environment, more of the map is constructed. Figures 10.3 through 10.5 show the evolution of the map as the robot turns slightly counter-clockwise, after a full revolution, then after beginning to enter the exit hallway.

The starting pen map in `maps/starting_pen`, as previously invoked in Fig 9.15, was constructed by this process, including having the robot exit the starting pen and encircle the pen to establish both interior and exterior map representations.

It was shown in Section 9.2.4 that a map can be used to help establish a robot's pose relative to that map. In addition, having a map supports making motion plans, as will be described next.

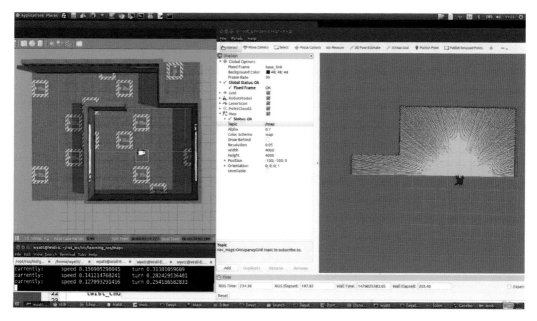

Figure 10.2: Initial view of `gmapping` process with mobot in starting pen

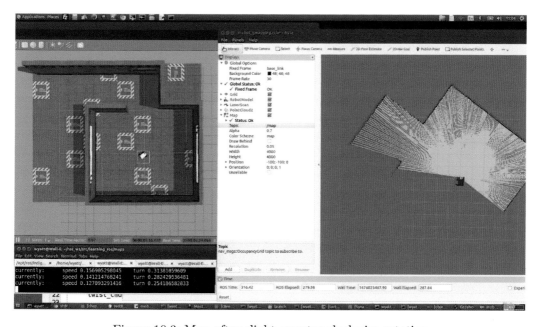

Figure 10.3: Map after slight counter-clockwise rotation

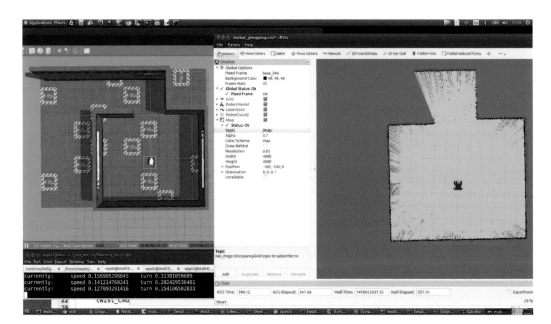

Figure 10.4: Map after full rotation

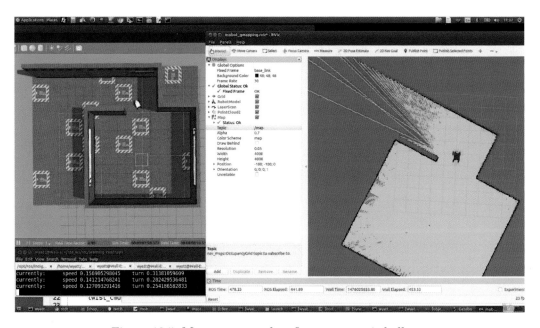

Figure 10.5: Map state as robot first enters exit hallway

10.2 PATH PLANNING

One of the most commonly used packages in ROS is `move_base`, which incorporates many features. The `move_base` node incorporates global planning, local planning and steering. Global and local planning are based on the notion of costmaps, described at `http://wiki.ros.org/costmap_2d`.

An example costmap can be visualized by starting up the mobot model in the starting pen:

```
roslaunch mobot_urdf mobot_in_pen.launch
```

then running `mobot_startup_navstack.launch`:

```
roslaunch mobot_nav_config mobot_startup_navstack.launch
```

This launch file loads the starting pen map (with `map_server`), starts `AMCL`, starts `rviz` with a preset configuration, and starts `move_base` with multiple configuration settings (which will be discussed in detail).

The above launch produces the `rviz` view of Fig 10.6. This display shows the map of the starting pen, construction of which was described in Section 10.1. The overlay of red dots corresponding to LIDAR pings shows that `AMCL` successfully localized the robot with respect to the map, since the LIDAR pings line up with walls within the map.

Additionally, the map's walls have a fattened, colorized border. These borders correspond to cost penalties, encoded as color. To use the costmap, one defines a footprint for the robot (as part of the configuration). For planning purposes, a candidate path solution is penalized for allowing non-zero cost cells to be contained within the robot's footprint. Moving close to a barrier may be acceptable, but at some modest cost. However, allowing a forbidden (lethal) cell within the footprint would score a candidate path as non-viable. The gray areas of the map carry no penalty, and path solutions that keep the robot's footprint entirely within the non-colored regions have the lowest penalty assigned.

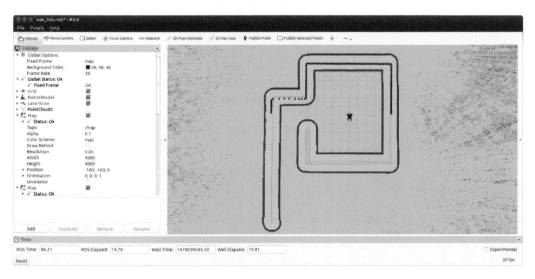

Figure 10.6: Global costmap of starting pen

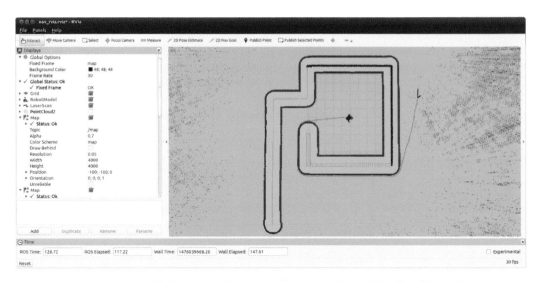

Figure 10.7: Global plan (blue trace) to specified 2dNavGoal (triad)

Figure 10.7 shows an example of global path planning with respect to the fattened map. The blue trace displays the plan computed by the global planner. The goal for this example was set interactively through use of the `2dNavGoal` tool in `rviz`. By clicking this tool, then clicking and dragging the mouse somewhere within an unoccupied region of the map, one can set both the origin and heading of a navigation goal. In Fig 10.7, the goal thus set is displayed as a frame (colored triad). The path displayed keeps a buffer distance from all walls while trying to reach the goal in minimum distance. A local planner (incorporating a steering algorithm) is responsible for driving the robot along the proposed path solution.

Global path solutions are computed with respect to a known map. However, unexpected barriers can invalidate the plan. For this reason, a local costmap is also computed and interpreted. Figure 10.8 shows a local costmap overlaid on the global costmap. In this case, a barrier (a construction barrel) was added to Gazebo in front of the robot. This obstacle was not present during the original mapping process, and thus it is unknown to the global costmap. Such cases are commonplace, as furniture may be rearranged, pallets may be moved about a factory, or people, vehicles or other robots may appear unexpectedly as obstacles. As shown in Fig 10.8, the local costmap is myopic; it is only aware of obstacles sensed within some defined window relative to the robot. Although the sensor may have longer range, the local costmap is deliberately represented as a short-range interpretation. In Fig 10.8, the barrel has only just entered the robot's local awareness.

As the barrier is detected, the path plan is re-generated to go around the obstacle, as shown by the blue trace in Fig 10.8.

From this brief introduction, it is apparent that much computation is occurring behind the scenes. Further, considerable tuning is required to obtain good performance. Tuning is prescribed using four configuration files: `costmap_common_params.yaml`, `local_costmap_params.yaml`, `global_costmap_params.yaml` and `base_local_planner_params.yaml`. Setting configuration parameters for the nav-stack is described in `http://wiki.ros.org/navigation/Tutorials/RobotSetup`, and in greater detail in `http://wiki.ros.org/costmap_2d` for costmap parameters and `http://wiki.ros.org/base_local_planner` for planning parameters. For our example mobot, these files reside in the package `mobot_nav_config`. The file `costmap_common_params.yaml` contains:

```
obstacle_range: 3.0
raytrace_range: 3.5
```

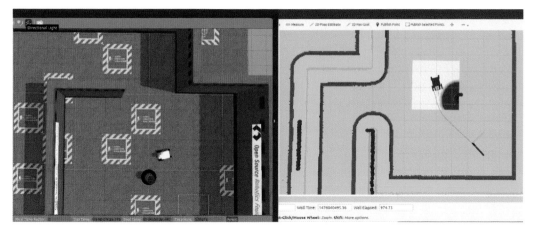

Figure 10.8: Global and local costmaps with unexpected obstacle (construction barrel)

```
#footprint: [[x0, y0], [x1, y1], ... [xn, yn]]
#footprint: [[0.2,0.4], [-0.8,0.4], [-0.8,-0.4], [0.2,-0.4]]
#alt: ir of robot:
robot_radius: 0.8
inflation_radius: 1.0

observation_sources: laser_scan_sensor

laser_scan_sensor: {sensor_frame: lidar_link, data_type: LaserScan, topic:
  /scan, marking: true, clearing: true}
```

These parameters specify that the LIDAR sensor data will have a limit of awareness of obstacles only out to 3.0 m, and that clearing of obstacles may occur out to 3.5 m. (Obstacles are cleared by recognizing that a LIDAR ray extends beyond some threshold radius, and thus there are no obstacles along that ray out to the threshold radius.) A footprint may be defined either by specifying coordinates of a polygon (expressed in the robot's base frame), or by specifying a simple radius about the robot's base frame origin. In the present case, the commented-out polygon is a reasonable approximation to the outline of the mobot. However, planning with respect to a polygon is more complex than planning with respect to a circle—particularly since the mobot's origin (between the drive wheels) is significantly offset from the center of the bounding rectangle. The alternative of setting a radius must be conservative in order to enclose the full outline of the robot (particularly since the robot's origin is not in the center of the robot's footprint). Assuming a conservative circular boundary simplifies planning and leads to fewer deadlock conditions. However, such a conservative boundary also prevents finding viable plans through narrow (but feasible) passageways.

An inflation radius of 1 m is also quite conservative. Plans will be penalized for coming closer than this to walls and other obstacles.

The `observation_sources` field should contain all sensors that will be used for obstacle detection. Most commonly (as in this case), a LIDAR will be used. Various depth camera types are also options. Such sensors have the benefit of being aware of obstacles that do not appear in a 2-D LIDAR's slice plane. For each sensor type listed, one must also specify the sensor frame and the sensor's message type. Dynamic local costmaps are marked by sensing obstacles and cleared by sensing empty space. Marking and clearing will ordinarily be enabled by setting these fields to `true`.

The parameters set within `costmap_common_params.yaml` are applied to both global and local planning. Additional parameters are specific to the global and local planners. The file `global_costmap_params.yaml` is shown below:

```
global_costmap:
  global_frame: /map
  robot_base_frame: /base_link
  update_frequency: 1.0
  static_map: true
```

This file specifies that global planning should be performed in the `map` frame, and the robot's path will be specified with respect to its `base_link` frame. For the global planner to succeed, a transform must be published between the `base_link` and the `map` frame. If a map is loaded and if `AMCL` is running, this transform will be published by `AMCL`. If these frames have been assigned other names, the names must be reflected in the configuration file. The `static_map` parameter has been set to `true` to recognize that a full map has already been obtained and loaded. In contrast, a map that is being constructed dynamically, *e.g.* using SLAM, would not be static.

Costmap parameters specific to the local planner are specified in `local_costmap_params.yaml`, shown here:

```
local_costmap:
  global_frame: /odom
  robot_base_frame: base_link
  update_frequency: 1
  publish_frequency: 1
  static_map: false
  rolling_window: true
  width: 3
  height: 3
  resolution: 0.05
```

In this case, the local planner uses the `odom` frame, since odometry is updated faster and more smoothly than `map` coordinates, which is important for steering control. The reference frame on the robot is, again, the `base_link`. The local costmap is constructed and updated dynamically based on LIDAR sensing, and thus the `static_map` parameter is `false`. The local costmap is computed only within a relatively small window about the robot, defined as a `rolling_window` with specified dimensions and resolution.

The `base_local_planner_params.yaml` file is given below:

```
TrajectoryPlannerROS:
  max_vel_x: 1
  min_vel_x: 0.1
  max_vel_theta: 1.0
  min_in_place_vel_theta: 0.2

  acc_lim_theta: 1
  acc_lim_x: 1
  acc_lim_y: 0

  holonomic_robot: false
```

This file sets parameters relevant for steering, including minimum and maximum forward and spin velocities and accelerations. Importantly, since our robot model uses differential drive, it is not holonomic, *i.e.* it cannot slip or "crab" sideways. To inform the planner of this, the parameter `holonomic_robot` is set to `false`.

Having defined values for the parameters required for `move_base`, these parameters should be loaded onto the parameter server for use by `move_base`. This is accomplished through a launch file. The `mobot_startup_navstack.launch` file in package `mobot_nav_config` performs these functions for our example mobile robot. The contents of this file are:

```
<launch>
  <!-- Run the map server and load the starting-pen map -->
  <node name="map_server" pkg="map_server" type="map_server"
    args="$(find maps)/starting_pen/starting_pen_map.yaml"/>

   <!-- use amcl for localization-->
   <!--node pkg="amcl" type="amcl" name="amcl" output="screen"-->
  <!-- alt: try differential-drive amcl config params -->
  <include file="$(find amcl)/examples/amcl_diff.launch"/>

   <!--alt: instead of loading a map and localizing on it, could run slam dynamically-->
   <!--node name="slam_gmapping" pkg="gmapping" type="slam_gmapping"/-->


<!-- launch rviz using a specific config file -->
 <node pkg="rviz" type="rviz" name="rviz" args="-d
   $(find mobot_nav_config)/nav_rviz.rviz"/>

<!--move_base w/ navstack config files-->
 <node pkg="move_base" type="move_base" respawn="false"
   name="move_base" output="screen">
   <rosparam file="$(find mobot_nav_config)/costmap_common_params.yaml" command="load"
     ns="global_costmap" />
    <rosparam file="$(find mobot_nav_config)/costmap_common_params.yaml" command="load"
      ns="local_costmap" />
    <rosparam file="$(find mobot_nav_config)/base_local_planner_params.yaml"
      command="load" />

    <rosparam file="$(find mobot_nav_config)/local_costmap_params.yaml"
      command="load" />
    <rosparam file="$(find mobot_nav_config)/global_costmap_params.yaml"
      command="load" />
  </node>

</launch>
```

This launch file first loads the starting pen map, then it starts `AMCL` running. In this case, a version of `AMCL` for differential-drive robots is used. (Knowing that the robot cannot slip sideways is helpful in constraining sensor-versus-map localization fits.) `rviz` is also launched, with reference to a pre-set configuration file. Finally, the `move_base` node is started, which includes loading the configuration files onto the parameter server for use by `move_base`. In loading these files, the parameters in `costmap_common_params.yaml` are needed by both the global and local planners, and these parameters should be consistent for both. To enforce this consistency, the shared parameters are loaded into the name spaces of the `global_costmap` and `local_costmap` separately, drawing on the same source file.

Getting good performance from `move_base` can require considerable tuning of the associated parameters. If these parameters are set inappropriately for a given robot, it can fail to find viable solutions, get stuck at obstacle boundaries, steer erratically, and oscillate at the goal pose.

In addition to tuning the configuration parameters, one has the option of writing a global planner and/or local planner (and steering algorithm) and having these alternative planners operate within `move_base` (see `http://wiki.ros.org/nav_core`). One can also use `move_base` to get to approximate goal poses, then use custom controls to publish to `cmd_vel` for local precision behaviors (*e.g.* to approach a charging station approximately with `move_base`, then perform precision navigation, presumably with sensor feedback, to complete a docking maneuver).

10.3 EXAMPLE MOVE-BASE CLIENT

The package `example_move_base_client` (with source code `example_move_base_client.cpp`) contains an example of how to construct an action client of the move-base action server. The example can be run by first starting Gazebo with the mobot in the starting pen:

```
roslaunch mobot_urdf mobot_in_pen.launch
```

then loading configuration files and starting move-base using `mobot_startup_navstack.launch`:

```
roslaunch mobot_nav_config mobot_startup_navstack.launch
```

The example move-base client has a hard-coded goal destination. This is communicated to the move-base action server by running:

```
rosrun example_move_base_client example_move_base_client
```

This results in generation of a global plan and start of navigation along that plan. Figure 10.9 shows a screenshot of the mobot en route to the goal destination.

Figure 10.9: Example move-base client interaction. Destination is sent as goal message to move-base action server that plans a route (blue trace) to the destination (triad goal marker).

The example code that sends a navigation goal is shown in Listing 10.1.

Listing 10.1: `example_move_base_client.cpp`: example action client of move-base

```cpp
#include<ros/ros.h>
#include <actionlib/client/simple_action_client.h>
#include <actionlib/client/terminal_state.h>
#include <move_base_msgs/MoveBaseAction.h>
#include <Eigen/Eigen>
#include <Eigen/Dense>
#include <Eigen/Geometry>
#include <geometry_msgs/PoseStamped.h>
#include <tf/transform_listener.h>
#include <xform_utils/xform_utils.h>


geometry_msgs::PoseStamped g_destination_pose;

void set_des_pose() {
    g_destination_pose.header.frame_id="/map";
    g_destination_pose.header.stamp = ros::Time::now();
    g_destination_pose.pose.position.z=0;
    g_destination_pose.pose.position.x = -8.8;
    g_destination_pose.pose.position.y = 0.18;
    g_destination_pose.pose.orientation.z= -0.707;
    g_destination_pose.pose.orientation.w= 0.707;
}

void navigatorDoneCb(const actionlib::SimpleClientGoalState& state,
        const move_base_msgs::MoveBaseResultConstPtr& result) {
    ROS_INFO(" navigatorDoneCb: server responded with state [%s]", state.toString().
        c_str());
}

int main(int argc, char** argv) {
    ros::init(argc, argv, "example_navigator_action_client"); // name this node
    ros::NodeHandle nh; //standard ros node handle
    set_des_pose(); //define values for via points
    tf::TransformListener tfListener;
    geometry_msgs::PoseStamped current_pose;
    move_base_msgs::MoveBaseGoal move_base_goal;
    XformUtils xform_utils; //instantiate an object of XformUtils

    bool tferr=true;
    ROS_INFO("waiting for tf between map and base_link...");
    tf::StampedTransform tfBaseLinkWrtMap;
    while (tferr) {
        tferr=false;
        try {
                //try to lookup transform, link2-frame w/rt base_link frame; this will
                    test if
            // a valid transform chain has been published from base_frame to link2
                tfListener.lookupTransform("map","base_link", ros::Time(0),
                    tfBaseLinkWrtMap);
            } catch(tf::TransformException &exception) {
                ROS_WARN("%s; retrying...", exception.what());
                tferr=true;
                ros::Duration(0.5).sleep(); // sleep for half a second
                ros::spinOnce();
            }
    }
    ROS_INFO("tf is good; current pose is:");
    current_pose = xform_utils.get_pose_from_stamped_tf(tfBaseLinkWrtMap);
    xform_utils.printStampedPose(current_pose);

    actionlib::SimpleActionClient<move_base_msgs::MoveBaseAction> navigator_ac("
        move_base", true);

    // attempt to connect to the server:
    ROS_INFO("waiting for move_base server: ");
    bool server_exists = false;
    while ((!server_exists)&&(ros::ok())) {
        server_exists = navigator_ac.waitForServer(ros::Duration(0.5)); //
```

```
66        ros::spinOnce();
67        ros::Duration(0.5).sleep();
68        ROS_INFO("retrying...");
69    }
70    ROS_INFO("connected to move_base action server"); // if here, then we connected to↩
          the server;
71    //geometry_msgs/PoseStamped target_pose
72    move_base_goal.target_pose = g_destination_pose;
73
74    ROS_INFO("sending goal: ");
75    xform_utils.printStampedPose(g_destination_pose);
76    navigator_ac.sendGoal(move_base_goal,&navigatorDoneCb);
77
78
79    bool finished_before_timeout = navigator_ac.waitForResult(ros::Duration(120.0));
80        //bool finished_before_timeout = action_client.waitForResult(); // wait ↩
              forever...
81    if (!finished_before_timeout) {
82        ROS_WARN("giving up waiting on result ");
83        return 1;
84    }
85
86    return 0;
87 }
```

The move-base action client depends on package `move_base_msgs`, and it includes the corresponding header file (line 4: `#include <move_base_msgs/MoveBaseAction.h>`). For illustration, a transform listener and the `xform_utils` class are also included.

The function `set_des_pose()` merely assigns a hard-coded destination to a `geometry_msgs::PoseStamped` object.

Lines 42 through 54 invoke repeated attempts to find the current pose of the robot in the map frame. Once this is successful, the transform is converted to a pose (line 56) and this pose is displayed (line 57).

An action client of `move_base` is instantiated (line 59), using the action message defined in `move_base_msgs::MoveBaseAction`. The action client attempts to connect to the action server (lines 61 through 70).

A goal message type is instantiated (line 36) and populated with a desired pose (line 72). This goal message is sent to the action server (line 76), and the client then waits for the `move_base` action server to respond.

By virtue of sending this goal message, the `move_base` action server plans a global plan from start pose to destination pose, attempting to minimize a cost function that includes penalties for moving too close to obstacles (walls) and a path-length penalty. The resulting global solution is the blue trace in Fig 10.9. The local planner sends velocity commands to the robot to follow local plans that attempt to converge on the global plan while avoiding unexpected obstacles that are perceived along the way.

10.4 MODIFYING NAVIGATION STACK

The navigation stack incorporates multiple components, including static and dynamic costmaps, a global planner, and a local planner. The purpose of the local planner is to respond to unexpected obstacles encountered while following the global plan. The default local planner uses a dynamic window approach (see `http://wiki.ros.org/dwa_local_planner`). This reactive planner considers a set of parameterized circular arcs as alternatives to following an invalid global plan (with the intent of re-joining the global plan once the unexpected obstacle is cleared). In this approach, `cmd_vel` is re-computed during each local-planner controller cycle, which can result in a different arc choice for each iteration. This approach can behave well if the mobile robot does a good job of following each local plan, since iterations of re-planning would then result in reiterating the former (still valid) plan.

However, if the robot does a poor job of following local plans using open-loop velocity commands, the local planner can behave poorly, rejecting its former local plan and substituting a new local plan during each control cycle. In the case of the mobot Gazebo simulation, the robot tends to list to its left. Although this is an artifact of Gazebo's physics simulation, it is nonetheless a behavior that may be expected in real robots. With a bias error in attempting to execute `cmd_vel`, the local planner fails to bring the robot back to its original global plan. Consequently, the robot may veer off of a path, fail to successfully drive through navigable narrow passages, or demonstrate poor convergence to goal coordinates.

The local planner can be improved by incorporating a feedback steering algorithm, as discussed in Section 9.3. One way to do this would be to remap the `cmd_vel` topic of `move_base` to some other topic, and run a new node that publishes to the robot's true `cmd_vel` topic. However, it is convenient to integrate with the considerable resources of `move_base`, including exploiting the global planner and accessing the costmaps. ROS provides a means to do this by writing and using a plug-in. A minimal example plug-in that replaces the local planner in `move_base` is given in the package `example_nav_plugin`. This plug-in contains no useful logic; it simply commands the robot to move in a circular arc for 5 sec each time a new navigation goal is specified. The purpose of this example is merely to illustrate the steps to creating a plug-in. (For more details on plug-ins in ROS, see `http://wiki.ros.org/pluginlib`.)

Creating a plug-in requires minor variations on the files `package.xml` and `CMakeLists.txt`, creation of an additional, small XML file, a specific parameter setting in a corresponding launch file, and application-specific composition of the source files (cpp and header files). These steps are detailed below; composition of the plug-in source code is treated last.

For the minimal example provided, the `package.xml` file follows:

Listing 10.2: `package.xml` file for `example_nav_plugin` package

```
1  <?xml version="1.0"?>
2  <package>
3    <name>example_nav_plugin</name>
4    <version>0.0.0</version>
5    <description>The example_nav_plugin package</description>
6
7    <buildtool_depend>catkin</buildtool_depend>
8    <buildtool_depend>catkin_simple</buildtool_depend>
9    <build_depend>roscpp</build_depend>
10 <build_depend>nav_core</build_depend>
11 <build_depend>xform_utils</build_depend>
12   <run_depend>roscpp</run_depend>
13 <run_depend>nav_core</run_depend>
14 <run_depend>xform_utils</run_depend>
15
16   <export>
17     <!-- the following line refers to another xml file, in which some
18     info re/ the new plugin library is defined  -->
19     <nav_core plugin="${prefix}/nav_planner_plugin.xml" />
20   </export>
21 </package>
```

Note the statement on line 19, which specifies the name of an additional XML file (called `nav_planner_plugin.xml`). The contents of this additional XML file are:

Listing 10.3: `nav_planner_plugin.xml`

```
1  <library path="lib/libminimal_nav_plugin">
2      <class name="example_nav_plugin/MinimalPlanner" type="MinimalPlanner" ↩
          base_class_type="nav_core::BaseLocalPlanner">
3          <description>Example local planner plugin- causes the robot to move in a CCW ↩
              circle.</description>
4      </class>
5  </library>
```

In this additional XML file, one specifies the new library name, the new class name, and the base class from which the new class is derived. In the present case, the name of the library specified in this package's `CMakeLists.txt` file is `minimal_nav_plugin`. The compiler will mangle this name to prepend `lib` and append `.so` and will place the resulting library (`libminimal_nav_plugin.so`) in the /devel/lib directory of the ROS workspace.

Within the source code of the new plug-in library, a new class is defined. The source code for this minimal example defines the class `MinimalPlanner`. The XML file notes that the definition of `MinimalPlanner` is within the package `example_nav_plugin`. The XML file also notes that the new class, `MinimalPlanner`, is derived from the base class `nav_core::BaseLocalPlanner`. This is the name of the base class that must be used to replace the local planner. (To replace a global planner, use the base class `nav_core::BaseGlobalPlanner` instead.)

The `CMakeLists.txt` file within the package `example_nav_plugin` contains the line:

```
add_library(minimal_nav_plugin src/minimal_nav_plugin.cpp)
```

This specifies that we are creating a library called `minimal_nav_plugin`. This library will behave as a plug-in, but no further variations are required in the `CMakeLists.txt` file.

The source code for the plug-in will define a class that, in this example, is called `MinimalPlanner` (consistent with the references in `nav_planner_plugin.xml`). Assuming the corresponding code has been compiled into the named library (`minimal_nav_plugin`, in this case), the new plug-in can be used within `move_base` by setting a corresponding parameter in the launch file. Within the package `example_nav_plugin`, the `mobot_w_minimal_plugin.launch` file, specifies how to bring up the navigation stack with use of the new local planner. Only one line of this launch file is different from the default bring-up:

```
<param name="base_local_planner" value="example_nav_plugin/MinimalPlanner"/>
```

This line specifies that `move_base` should replace the default `base_local_planner` with the new class, `MinimalPlanner`, from the package `example_nav_plugin`.

Finally, we consider the code source files that compose the new plug-in. The header file for our new class, `minimal_nav_plugin.h` appears in Listing 10.4.

Listing 10.4: `minimal_nav_plugin.h`

```
1  #ifndef MINIMAL_NAV_PLUGIN_H
2  #define MINIMAL_NAV_PLUGIN_H
3
4  #include <nav_core/base_local_planner.h>
5  #include <nav_msgs/Path.h>
6
7      class MinimalPlanner : public nav_core::BaseLocalPlanner {
8      public:
9          MinimalPlanner(); //optionally, pass in args to access components of move_base
10         //MinimalPlanner(std::string name, costmap_2d::Costmap2DROS* costmap_ros);
11
```

```
12          /** overridden classes from interface nav_core::BaseGlobalPlanner **/
13          void initialize(std::string name, tf::TransformListener * tf, costmap_2d::↩
                  Costmap2DROS * costmap_ros);
14          bool isGoalReached();
15          bool setPlan(const std::vector< geometry_msgs::PoseStamped > &plan);
16          bool computeVelocityCommands(geometry_msgs::Twist &cmd_vel);
17
18      private:
19          ros::Time tg;
20          unsigned int old_size;
21          tf::TransformListener * handed_tf;
22      };
23
24  #endif
```

This header file defines the class `MinimalPlanner`, which is derived from `nav_core::BaseLocalPlanner`. The choice of class name must be consistent with the names used in launch file and in the plug-in's XML file. Four `BaseLocalPlanner` functions are overridden by the new class. The function `initialize()` will get called by the constructor. The function `isGoalReached()` should specify a stopping condition. The function `setPlan()` will return `true` if a new global plan has been received, and a reference variable for that plan will provide access to the new plan. Most importantly, the function `computeVelocityCommands` is responsible for setting the relevant paramters of `cmd_vel` (forward velocity and yaw rate, in the present case). By setting these values, the `move_base` node will publish these values as commands to the robot.

The function `computeVelocityCommands` will be called by `move_base` at the controller-frequency rate. The rate default is 20 Hz, but this can be changed within the code or by setting a configuration parameter. The current value on the parameter server can be found programmatically with:

```
nh_.getParam("/move_base/controller_frequency", controller_rate_);
```

The example minimal local planner plug-in source code is relatively short; it is shown in Listing 10.5.

Listing 10.5: `minimal_nav_plugin.cpp`

```
1   //package name, header name for new plugin library
2   #include <example_nav_plugin/minimal_nav_plugin.h>
3
4   #include <pluginlib/class_list_macros.h>
5
6   PLUGINLIB_EXPORT_CLASS(MinimalPlanner, nav_core::BaseLocalPlanner);
7
8   MinimalPlanner::MinimalPlanner(){
9       //nothing ot fill in here; "initialize" will do the initializations
10  }
11
12  //put inits here:
13  void MinimalPlanner::initialize(std::string name, tf::TransformListener * tf, ↩
            costmap_2d::Costmap2DROS * costmap_ros){
14      ros::NodeHandle nh(name);
15
16      old_size = 0;
17      handed_tf = tf;
18  }
19
20  bool MinimalPlanner::isGoalReached(){
21      //For demonstration purposes, sending a single navpoint will cause five seconds of↩
            activity before exiting.
22      return ros::Time::now() > tg;
23  }
24
```

```
25   bool MinimalPlanner::setPlan(const std::vector< geometry_msgs::PoseStamped > &plan){
26       //The "plan" that comes in here is a bunch of poses of varaible length, calculated↩
             by the global planner(?).
27       //We're just ignoring it entirely, but an actual planner would probably take this ↩
             opportunity
28       //to store it somewhere and maybe update components that refrerence it.
29       ROS_INFO("GOT A PLAN OF SIZE %lu", plan.size());
30       //If we wait long enough, the global planner will ask us to follow the same plan ↩
             again. This would reset the five-
31       //second timer if I just had it refresh every time this function was called, so I ↩
             check to see if the new plan is
32       //"the same" as the old one first.
33       if(plan.size() != old_size){
34           old_size = plan.size();
35           tg = ros::Time::now() + ros::Duration(5.0);
36       }
37       return true;
38   }
39
40   bool MinimalPlanner::computeVelocityCommands(geometry_msgs::Twist &cmd_vel){
41       //This is the meat-and-potatoes of the plugin, where velocities are actually ↩
             generated.
42       //in this minimal case, simply specify constants; more generally, choose vx and ↩
             omega_z
43       // intelligently based on the goal and the environment
44       // When isGoalReached() is false, computeVelocityCommands will be called each ↩
             iteration
45       // of the controller--which is a settable parameter.  On each iteration, values in
46       // cmd_vel should be computed and set, and these values will be published by ↩
             move_base
47       // to command robot motion
48       cmd_vel.linear.x = 0.2;
49       cmd_vel.linear.y = 0.0;
50       cmd_vel.linear.z = 0.0;
51       cmd_vel.angular.z = 0.2;
52       return true;
53   }
```

A more extensive plug-in example is given in the package `test_plugin` of the accompanying repository. This package defines a class `test_planner::TestPlanner` with more capability. This alternative plug-in implements a linear steering algorithm, as introduced in Section 9.3.2. It uses the global plan generated by the global planner, from which it computes a time series of desired states. The robot's pose is inferred from odometry and from AMCL, which localizes the robot relative to a map (of the starting pen, in this example). Linear feedback steering enables the robot to stay on course better, since it compensates for the robot's tendency to side-slip. The example code in this package also shows how a plug-in can access the nav-stack's `CostMaps`. This functionality can be used to construct a local planner, such as a "bug" planner that can follow walls, to enable re-connecting with the planned global path.

In addition to overriding the default local planner with a plug-in, one can also write plug-ins for alternative global planners. Additionally, one can create plug-ins for costMaps, which can be useful for annotating features that are not directly accessible to the robot's sensors. Additional costMap layers can be used, *e.g.* to impose "no fly" zones for known dangerous areas, such as cliffs or stairwells, glass walls, or terrain that is unnavigable.

10.5   WRAP-UP

The navigation stack is one of the most popular capabilities in ROS. It incorporates consideration of *a priori* maps, dimensions of the robot, mobility of the robot, and obstacles perceived en route. Subject to constraints on robot speeds and accelerations, a global path plan is constructed from an initial pose to a specified goal pose. Execution of the plan involves a local planner, which takes into consideration perception of nearby obstacles and plans and executes local motions to avoid the obstacles and reconnect with the global plan.

Tuning the parameters of the navigation stack for a specific robot can be time consuming and involves trial and error. Even with these parameters tuned, however, the performance of `move_base` may be inadequate, particularly if the robot must navigate precisely through narrow passageways (including doorways). It thus may be necessary to interleave `move_base` goals with custom desired-state generation and steering algorithms to achieve a good balance between generality and precision motion control.

Navigation of the mobot will be revisited in Section VI, in which a combined mobile base and manipulator are integrated to perform mobile manipulation.

V

Robot Arms in ROS

I NTRODUCTION
This section introduces concepts in planning and control of robot arms. The presentation approach is ground-up. At the lowest level, joint controllers are specified, whether position controlled, velocity controlled or force controlled. The next higher layer of abstraction is specification and execution of joint-space trajectories, which include coordination of joint displacements at planned arrival times. Through consideration of forward and inverse kinematics, one can compute joint-space solutions that are appropriate for performing specific tasks. To construct a viable joint-space trajectory, one must consider Cartesian motion plans and choose among inverse kinematic options to construct corresponding feasible joint-space trajectory plans.

These concepts are considered in greater detail in the context of a specific robot, the Baxter robot (from Rethink Robotics), culminating in an object-grabber action server for manipulation.

Low-Level Control

CONTENTS

I NTRODUCTION
This chapter will explore variations on low-level joint control in ROS. In Section 3.5, robot-arm control concepts in ROS were introduced using a separate feedback node that interacted with Gazebo via Gazebo topics. This approach, while suitable for illustrative purposes, is not preferred, since sample rates and latencies due to communication via topics limits the achievable performance of such controllers. Instead, as introduced in Section 3.6, it is preferable to interact with Gazebo using plug-ins. Design of Gazebo plug-ins will not be covered here, but the interested designer can learn about plug-ins from online tutorials (see `http://gazebosim.org/tutorials?tut=ros_gzplugins`). Existing plug-ins for position controllers and velocity controllers will be discussed.

Using a Gazebo plug-in for joint control requires a fair amount of detailed specification (see `http://gazebosim.org/tutorials/?tut=ros_control`). One specifies joint range-of-motion limits, velocity limits, effort limits and damping. The URDF must specify a transmission and actuator for each controlled joint. The `gazebo_ros_control.so` library is included within a Gazebo tag in the robot model file. Control gains are specified in an associated YAML file. Finally, one specifies in a launch file to spawn individual joint controllers (position or velocity). Having specified the control type to use, one needs to tune feedback parameters, for which ROS offers some helpful tools.

This section will illustrate use of three simple controllers: position control, velocity control and force control (which will require modelling a force sensor). These examples will use a single degree-of-freedom robot with a prismatic joint.

11.1 A ONE-DOF PRISMATIC-JOINT ROBOT MODEL

The associated package `example_controllers` includes the file `prismatic_1dof_robot_description_w_jnt_ctl.xacro`. This model file is largely similar to the `minimal_robot_description.urdf` in package `minimal_robot_description`, described in Section 3.2, except that its single degree of freedom is a prismatic joint instead of a revolute joint. Key components of this file include the following lines:

```
<joint name="joint1" type="prismatic">
  <parent link="link1"/>
  <child link="link2"/>
  <origin xyz="0 0 1" rpy="0 0 0"/>
  <axis xyz="0 0 1"/>
  <limit effort="1000.0" lower="-1.0" upper="0.0" velocity="100.0"/>
  <dynamics damping="10.0"/>
</joint>
```

which defines joint1 and specifies upper and lower range of motion, maximum control effort (1000 N force), and a maximum velocity (100 m/sec).

The following lines, declaring a transmission and actuator, are also required to interface with ROS controllers:

```
<transmission name="tran1">
  <type>transmission_interface/SimpleTransmission</type>
  <joint name="joint1">
    <hardwareInterface>EffortJointInterface</hardwareInterface>
  </joint>
  <actuator name="motor1">
    <hardwareInterface>EffortJointInterface</hardwareInterface>
    <mechanicalReduction>1</mechanicalReduction>
  </actuator>
</transmission>
```

The controller plug-in library is included in the URDF with the following lines:

```
<gazebo>
  <plugin name="gazebo_ros_control" filename="libgazebo_ros_control.so">
    <robotNamespace>/one_DOF_robot</robotNamespace>
  </plugin>
</gazebo>
```

(The model also includes a simulated force sensor, but discussion of this is deferred for now.)

The ROS control plug-in includes options for both position and velocity controllers. The specific controller to be used is established by launching appropriate nodes.

11.2   EXAMPLE POSITION CONTROLLER

The launch file prismatic_1dof_robot_w_jnt_pos_ctl.launch contains:

```
<launch>
  <!-- Load joint controller configurations from YAML file to parameter server -->
  <rosparam file="$(find example_controllers)/control_config/one_dof_pos_ctl_params.yaml"
    command="load"/>

<!-- Convert xacro model file and put on parameter server -->
<param name="robot_description" command="$(find xacro)/xacro.py
  '$(find example_controllers)/prismatic_1dof_robot_description_w_jnt_ctl.xacro'" />

  <!-- Spawn a robot into Gazebo -->
  <node name="spawn_urdf" pkg="gazebo_ros" type="spawn_model"
    args="-param robot_description -urdf -model one_DOF_robot" />

  <!--start up the controller plug-ins via the controller manager -->
  <node name="controller_spawner" pkg="controller_manager" type="spawner" respawn="false"
    output="screen" ns="/one_DOF_robot" args="joint_state_controller
    joint1_position_controller"/>

</launch>
```

This launch file loads control parameters from file **one_dof_pos_ctl_params.yaml**, converts the model xacro file into a URDF file and loads it on the parameter server, spawns the robot model into Gazebo for simulation, and loads the controller **joint1_position_controller**. The controller parameters must be consistent with the chosen type (position controller). Further, the joint name associated with the controller (**joint1**) must be part of the controller name loaded.

The YAML file with control parameters, **one_dof_pos_ctl_params.yaml**, is listed below:

```
one_DOF_robot:
  # Publish all joint states ------------------------------------
  joint_state_controller:
    type: joint_state_controller/JointStateController
    publish_rate: 50

  # Position Controllers ----------------------------------------
  joint1_position_controller:
    type: effort_controllers/JointPositionController
    joint: joint1
    pid: {p: 400.0, i: 0.0, d: 0.0}
```

The control parameter file must reference the name of the robot (**one_DOF_robot**), and for each joint (only **joint1** in this case), the position-control gains must be listed. In the example, the proportional gain is 400 N/m, and all other gains are suppressed to zero. Damping is provided implicit to the joint definition (which contains the specification: **<dynamics damping="10.0"/>**). In fact, the derivative gain in the default position controller is quite noisy, presumably due to use of backwards-difference velocity estimates. Implicit joint damping in the robot model is better behaved than derivative feedback.

Choosing good control gains can be challenging, but ROS offers tools that can help. A detailed tutorial can be found at **http://gazebosim.org/tutorials/?tut=ros_control**. In choosing gains, one must keep in mind that the physics engine runs with a default time step of 1 ms, and thus control bandwidths must be designed to accommodate this limitation. For 1 kHz update rate, controllers should be limited (roughly) to less than 100 Hz bandwidth. In practice, numerical artifacts appear well below this frequency. For the one-DOF example, the mass of **link2** was set to 1 kg, and the proportional gain was set to 400. The undamped natural frequency of the system is thus $\sqrt{(K_p/m)} = 20$ rad/sec, or about 3 Hz.

The system response to sinusoidal command inputs can be observed using **rqt_plot**, and control gains can be adjusted interactively using **rqt_gui**. Sinusoidal command inputs can also be specified using **rqt_gui**. In the **example_controllers** package, an alternative excitation node called **one_dof_sine_commander**, prompts for frequency and amplitude and publishes sinusoidal commands to the **joint1** command topic. The system can be tested as follows. First, start Gazebo with:

```
roslaunch gazebo_ros empty_world.launch
```

Then spawn the robot and its controller with the associated control gains using:

```
roslaunch example_controllers prismatic_1dof_robot_w_jnt_pos_ctl.launch
```

Start **rqt_gui** and **rqt_plot**:

```
rqt_plot
rosrun rqt_gui rqt_gui
```

Choose topics to plot in `rqt_plot` to be `one_DOF_robot/joint1_position_controller/command/data`, `one_DOF_robot/joint_states/position[0]`, and `one_DOF_robot/joint_states/effort[0]`.

Using `rqt_gui` is somewhat involved, as it presents many options. One can add a `MessagePublisher` and select the topic `one_DOF_robot/joint1_position_controller/command`. Expanding this topic, the sole item is `data`. One can edit the `expression` value on this row to define the command to be published. This can be as simple as a constant value, or it can be an expression to be evaluated dynamically (typically, a sinusoidal function).

In a separate panel of `rqt_gui`, one can perform dynamic reconfigure to vary control gains interactively. Expanding the `joint1` control topic down to the level of `PID` exposes sliders for the control gains. These will be initialized to the values in the control-parameter YAML file. They can be adjusted while the robot is running, so one can immediately observe their effects in `rqt_plot`.

Figure 11.1 shows a snapshot of this process. The Gazebo window shows the robot; the `rqt_gui` window shows the selection of joint command publisher and PID values; and the `rqt_plot` window shows the transient response of position command, actual position, and control effort due to a step in command from −0.2 m to −0.5 m. The control effort in Fig

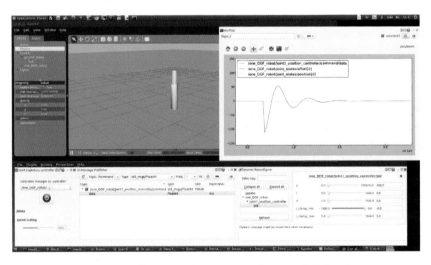

Figure 11.1: Servo tuning process for one-DOF robot

11.1 is large, squeezing the displacement response to a small range. Figure 11.2 shows a zoom of this transient to reveal the step command (in red) and the position response (in cyan). The response shows several overshoots and convergence over roughly half a second. Alternatively, the published stimulus can be a sinusoid, as shown in Fig 11.3. For this case, the published command is the function `0.1*sin(i/5)-0.5`, which is a sinusoid of amplitude 0.1 m, offset by −0.5 m (which is the middle of the joint's range of motion). The frequency follows from the publication rate, which is set to 100 Hz. The value of "i" increments at this rate, and thus the frequency of this sinusoid is 20 rad/sec. Since our undamped natural frequency is 20 rad/sec, we expect to see poor tracking at this frequency. As expected, the position response lags the command by a phase of 90 degrees at this frequency of excitation.

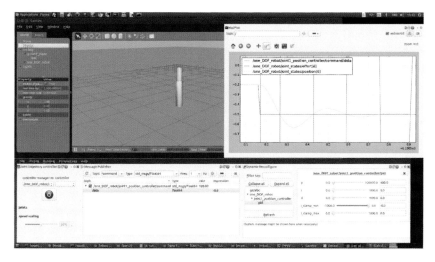

Figure 11.2: Servo tuning process for one-DOF robot: zoom on position response

The PID sliders can be varied to test alternative gains and their influence on response to step inputs or sinusoids.

A limitation of the position controller is that the derivative gain does not accept velocity feed-forward commands. Consequently, dynamic tracking performance is limited. An alternative is to use a ROS velocity controller instead of a position controller, and to perform position feedback via an external node (a successive loop closure technique).

11.3 EXAMPLE VELOCITY CONTROLLER

An alternative launch file that references the same robot model is `prismatic_1dof_robot_w_jnt_vel_ctl.launch`, which contains:

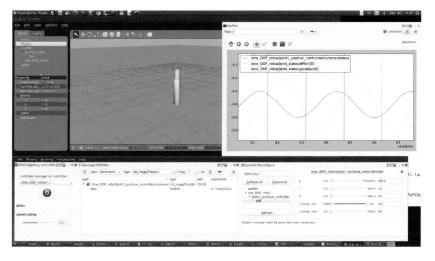

Figure 11.3: Servo tuning process for one-DOF robot: 20 rad/sec sinusoidal command

```
<launch>
  <!-- Load joint controller configurations from YAML file to parameter server -->
  <rosparam file="$(find example_controllers)/control_config/one_dof_vel_ctl_params.yaml"
    command="load"/>

<!-- Convert xacro model file and put on parameter server -->
<param name="robot_description" command="$(find xacro)/xacro.py
  '$(find example_controllers)/prismatic_1dof_robot_description_w_jnt_ctl.xacro'" />

  <!-- Spawn a robot into Gazebo -->
  <node name="spawn_urdf" pkg="gazebo_ros" type="spawn_model"
    args="-param robot_description -urdf -model one_DOF_robot -J joint1 -0.5" />
  <!--start up the controller plug-ins via the controller manager -->
  <node name="controller_spawner" pkg="controller_manager" type="spawner" respawn="false"
    output="screen" ns="/one_DOF_robot" args="joint_state_controller
    joint1_velocity_controller"/>

</launch>
```

As with the position controlled launch, this launch file loads control parameters from a file (one_dof_vel_ctl_params.yaml), converts the model xacro file into a URDF file and loads it on the parameter server, spawns the robot model into Gazebo for simulation, and loads a controller (joint1_velocity_controller instead of joint1_position_controller). Again, the joint name associated with the controller (joint1) must be part of the controller name loaded (joint1_velocity_controller). In this launch file, an additional option used in spawning the model is -J joint1 -0.5. This initializes the robot's joint1 position to the middle of its range of motion.

The YAML file with control parameters, one_dof_vel_ctl_params.yaml, is listed below:

```
one_DOF_robot:
  # Publish all joint states -----------------------------------
  joint_state_controller:
    type: joint_state_controller/JointStateController
    publish_rate: 50

  # Velocity Controllers ---------------------------------------
  joint1_velocity_controller:
    type: effort_controllers/JointVelocityController
    joint: joint1
    pid: {p: 200.0, i: 0.0, d: 0.0}
```

This file references the robot name (one_DOF_robot) and the joint name(s) (joint1). The same process is used to analyze the velocity controller. First, start Gazebo:

```
roslaunch gazebo_ros empty_world.launch
```

The robot is then spawned using the alternative launch file and gains:

```
roslaunch example_controllers prismatic_1dof_robot_w_jnt_vel_ctl.launch
```

The rqt_gui and rqt_plot tools are started:

```
rqt_plot
rosrun rqt_gui rqt_gui
```

Within `rqt_gui`, the `joint1_velocity_controller` is adjusted to change the velocity gain "P". The commanded and actual joint velocities are plotted, as shown in Fig 11.4, and the responses to commands can be observed as the gain is varied.

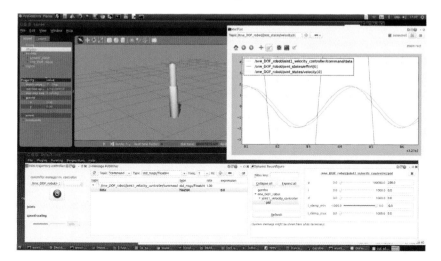

Figure 11.4: Velocity controller tuning process for one-DOF robot

The commanded velocity for the example shown in Fig 11.4 came from a separate node:

```
rosrun example_controllers one_dof_sine_commander
```

This node prompts the user for an amplitude and frequency. The responses for the case of Fig 11.4 were 0.1 m and 3 Hz.

Analysis of the velocity controller is somewhat more difficult, since `link2` can drift from its center to hit its displacement limits, which introduces distortions when attempting to tune gains. To address this, two adjustments were made. First, gravity was set to zero in Gazebo, so `link2` would not collapse to its lowest level. Second, the robot was started with `joint1` in the the middle of its range of motion. Although it may drift from this average position over time, this placement allows some time to obtain performance data before interacting with joint limits. An outer loop that considers position can correct for the drift of velocity control. An example is presented next in the context of force feedback.

11.4 EXAMPLE FORCE CONTROLLER

A third style of control incorporates consideration of interaction forces to achieve programmable compliance. The one-DOF prismatic robot model includes the following lines to emulate a force–torque sensor:

```
<!--model a f/t sensor; set up a link and a joint -->
  <link name="ft_sensor_link">
     <collision>
         <origin xyz="0 0 0" rpy="0 0 0"/>
         <geometry>
           <cylinder length="0.005" radius="0.05"/>
         </geometry>
     </collision>
```

```
        <visual>
            <origin xyz="0 0 0" rpy="0 0 0" />
            <geometry>
                <cylinder length="0.005" radius="0.05"/>
            </geometry>
        </visual>

        <inertial>
            <mass value="0.01" />
            <origin xyz="0 0 0" rpy="0 0 0"/>
            <inertia ixx="0.01" ixy="0" ixz="0" iyy="0.01" iyz="0" izz="0.01" />
        </inertial>
    </link>

    <!--ideally, this would be a static joint; until fixed in gazebo, must-->
        <!--have dynamic jnt-->
    <joint name="ft_sensor_jnt" type="prismatic">
        <parent link="link2"/>
        <child link="ft_sensor_link"/>
        <origin xyz="0 0 1" rpy="0 0 0"/>
        <axis xyz="0 0 1"/>
        <limit effort="1000.0" lower="0" upper="0.0" velocity="0.01"/>
        <dynamics damping="1.0"/>
    </joint>


<!-- Enable the Joint Feedback -->
<gazebo reference="ft_sensor_jnt">
    <provideFeedback>true</provideFeedback>
</gazebo>
<!-- The ft_sensor plugin -->
<gazebo>
    <plugin name="ft_sensor" filename="libgazebo_ros_ft_sensor.so">
        <updateRate>1000.0</updateRate>
        <topicName>ft_sensor_topic</topicName>
        <jointName>ft_sensor_jnt</jointName>
    </plugin>
</gazebo>
```

An additional link, `ft_sensor_link`, is defined with visual, collision and inertial properties. This link is connected to parent `link2` via a joint defined as `ft_sensor_jnt`. Ideally, this would be a static joint, but Gazebo (at present) requires that the joint be movable. It is thus defined as a prismatic joint, but with equal lower and upper joint limits.

A Gazebo tag brings in the plug-in `libgazebo_ros_ft_sensor.so`, which is associated with the `ft_sensor_jnt` and configured to publish its data to topic `ft_sensor_topic`.

When this model is spawned, the topic `ft_sensor_topic` shows up, and:

```
rostopic info ft_sensor_topic
```

shows that this topic carries messages of type `geometry_msgs/WrenchStamped`.

If we start the position-controlled one-DOF robot:

```
roslaunch example_controllers prismatic_1dof_robot_w_jnt_pos_ctl.launch
```

then command a position of -0.2 for `joint1` (*i.e.* nearly fully extended), we can observe transients as we drop a weight on the robot, using:

```
roslaunch exmpl_models add_cylinder_weight.launch
```

This introduces a large cylinder model into Gazebo at an initial height of 4 m. The cylinder has a mass of 10 kg and falls towards the ground plane under the influence of gravity (which was reduced to $-5.0 m/s^2$ for this example). Figure 11.5 shows the response. Due to the additional mass supported by the robot, the resonant frequency and the damping are reduced, resulting in many oscillations before settling down. The force sensor sees a spike of impact when the weight makes contact. As `link2` rebounds, the weight is tossed back in the air, and it subsequently makes three more impacts (and loses contact three more times) before the robot and weight maintain steady contact. From this point on, the robot and payload dynamics is linear, producing a sinusoidal motion due to the robot's position controller acting like a spring.

The actuator effort has the opposite sign as the force-sensor z value, and the actuator effort magnitude is 5 N greater than that of the sensor value. This is because the actuator is supporting the weight of the payload (10 kg) plus the weight of `link2` (1 kg), whereas the force sensor is influenced only by the weight of the payload.

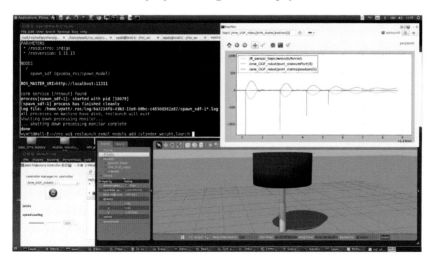

Figure 11.5: Force sensor and actuator effort due to dropping 10 kg weight on one-DOF robot

An anomaly of this example is that the joint effort and force sensor equilibrate to the wrong values. In steady state, the actuator effort should balance the weight of `link2` plus payload, which should be the weight of the 1 kg `link2` plus 10 kg cylinder. Instead, the system equilibrates to the equivalent of supporting slightly more than 7 kg. This error may be related to the physics engine's difficulty handling sustained contact.

The behavior of the position controller responding to the dropped weight is a mass-spring-damper system. However, the response is somewhat unrealistic, since the model has zero Coulomb friction and the transmission is perfectly back-driveable. A similar response can be obtained from more realistic robots by commanding velocities that would correspond to an ideal mass-spring-damper system. An implementation is given within the `example_controllers` package as `nac_controller.cpp`. This node subscribes to the joint state and to the force sensor topic. Within a tight loop it deduces the acceleration that would occur if the system were an ideal mass-spring-damper being acted on by forces exerted at the force sensor location. The values of the virtual mass, stiffness and damping to be emulated are contained within the variables `M_virt`, `K_virt` and `B_virt`, respectively. The key lines within the main loop of `nac_controller.cpp` are:

```
f_virt = K_virt*(x_attractor-g_link2_pos) + B_virt*(0-g_link2_vel);
f_net =    g_force_z + f_virt;
acc_ideal = f_net/M_virt;
v_ideal+= acc_ideal*dt;
```

In the above, `f_virt` is a virtual force comprised of two terms. The first term is the force exerted by a virtual spring of stiffness `K_virt` stretched between an attractor position, `x_attractor`, and the actual position of `link2`, `g_link2_pos`, as reported by the joint state publisher. The second term is due to a virtual damper of damping `B_virt` acting between a reference at zero velocity and `link2` moving at velocity `g_link2_vel`, also as reported by the joint-state publisher.

In addition to the virtual forces, a physical force is exerted at the robot's endpoint, which is sensed by the force sensor. The value of sensed force in the z direction, `g_force_z`, is obtained by subscribing to the force sensor topic. The sum of the sensed physical force and the virtual forces is the net force, `f_net`. A virtual mass, `M_virt`, acted on by this net force would have an acceleration, `acc_ideal`, of `f_net/M_virt`. Correspondingly, the velocity of this mass, `v_ideal`, is the time integral of the ideal acceleration. This model-based velocity is then sent as a command to a stiff velocity controller. Preferably, the value of `M_virt` should be close to the actual mass of `link2` (thus making this controller a natural admittance controller).

To run this controller, again start Gazebo:

```
roslaunch gazebo_ros empty_world.launch
```

Then spawn the robot using velocity control:

```
roslaunch example_controllers prismatic_1dof_robot_w_jnt_vel_ctl.launch
```

Then start the NAC node:

```
rosrun example_controllers nac_controller
```

The example NAC controller was assigned values of `M_virt` = 1.0 kg, `K_virt` = 1000 N/m and `B_virt` = 50 N/(m/sec). With this controller running and gravity set to $-9.8\ m/s^2$, the 10 kg weight was dropped on the robot, resulting in the transient shown in Figure 11.6. By tuning the values of `K_virt` and `B_virt`, the transient behavior can be adjusted to a desirable response.

Although the controllers here have been illustrated only in one degree of freedom, the approach generalizes to complex arms. Figure 11.7 shows a seven-DOF arm holding a weight using natural admittance control. Source code for this example is in package `arm7dof_nac_controller`. Conceptually, the approach is the same as the one-DOF controller. Sensed forces and virtual forces are combined (as vectors rather than scalars), and robot inertia (a matrix, rather than a constant) is considered to compute accelerations (in joint space) expected to result from the effort interactions. The joint accelerations are integrated to compute equivalent joint velocities, and these velocities are commanded to joint velocity servos of the robot.

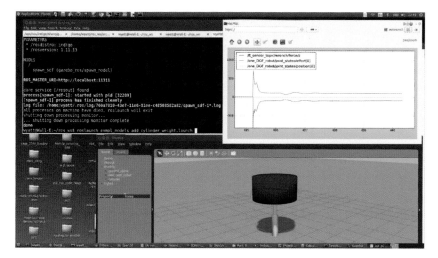

Figure 11.6: NAC controller response to dropping 10 kg weight on one-DOF robot

As a result, the seven-DOF arm behaves like a mass-spring-damper system. It catches a dropped weight gracefully, without excessive impact forces or oscillations.

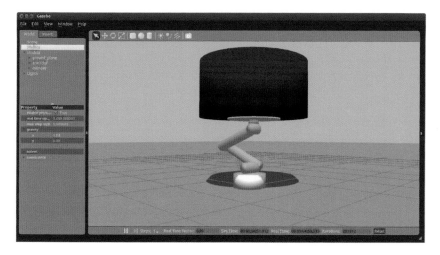

Figure 11.7: Seven-DOF arm catching and holding weight using NAC

11.5 TRAJECTORY MESSAGES FOR ROBOT ARMS

In ROS, the conventional means to execute a joint space motion plan is to populate a trajectory message and send it to a trajectory execution action server.

As described within a URDF file, a robot model consists of a collection (a tree) of robot links, connected pair-wise via joints. Each joint allows a single degree of freedom of motion: a rotation or a translation. Either type of motion may be referred to generically as a displacement. Displacement commands are input to low-level servo controllers, which exert torques or forces (generically referred to as efforts) to attempt to achieve a desired state. The desired state is comprised of desired displacements, possibly augmented by desired velocities, possibly further augmented by desired accelerations. The controller compares the desired state to the actual (measured) state to derive a control effort, to be imposed by the respective actuator.

If the desired state is too far from the actual state, the controller will try to exert unreasonable efforts, resulting in effort saturation and leading to poor following of or convergence to desired states. The resulting motion can be unpredictable and dangerous.

In order for the robot to achieve a desirable motion, motion commands should be evaluated in advance to make sure they are achievable within the following constraints:

- The robot should not hit objects in its environment.

- The path in joint space should conform to the constraints of minimum and maximum range of motion of the joints.

- The joint velocities should remain within the velocity limits of the respective actuators.

- The required joint efforts should remain within the effort limits of the actuators.

The first two requirements are imposed on path planning. The path defines a sequence of poses to be realized. The second two requirements are imposed on trajectory planning, which augments a path with timing. To convert a path to a trajectory, one must augment each pose with an arrival time (time from start of motion) computed such that velocity and effort constraints are satisfied.

One condition to avoid is a step command to any joint. If a joint is commanded to move instantaneously from A to B, the motion will be physically impossible to achieve, and the resulting behavior will likely be undesirable.

A necessary (but not sufficient) requirement of a safe robot command is that the command should be updated frequently (*e.g.* 100 Hz or faster is typical), and the commands should form an approximation of a smooth, continuous stream for all joints.

Since this is a common requirement, ROS includes a message type for this style of command: `trajectory_msgs/JointTrajectory`.

Invoking:

```
rosmsg show trajectory_msgs/JointTrajectory
```

shows that this message consists of the following fields:

```
std_msgs/Header header
  uint32 seq
  time stamp
  string frame_id
string[] joint_names
trajectory_msgs/JointTrajectoryPoint[] points
  float64[] positions
  float64[] velocities
  float64[] accelerations
  float64[] effort
  duration time_from_start
```

In the header, the `frame_id` is not meaningful, since the commands are in joint space (desired state of each joint).

The vector of strings `joint_names` should be populated with text names assigned to the joints (consistent with naming in the URDF file). For a serial chain robot, joints are conventionally known by integers, starting from joint 1, the joint closest to ground (most proximal joint), and progressing sequentially to the most distal joint. However, robots with multiple arms and/or legs are not so easily described, and thus names are introduced.

In specifying a vector of desired joint displacements, one must associate the joint commands with the corresponding joint names. Generally, there is no requirement to specify the joint commands in any specific order or requirement to specify all joint commands on every iteration. For example, one might command a neck rotation only in one instance, then follow that by a separate command to a subset of joints of the right arm, etc. However, some packages require that all joint states be specified in every command (whether or not it is desired to move all joints). Other packages receiving trajectory messages might implicitly depend on specifying joint commands in a specific, fixed order, ignoring the `joint_names` field (although this is not preferred).

The bulk of the trajectory message is a vector of type `trajectory_msgs/JointTrajectoryPoint`. This type contains four variable-length vectors and a `duration`. A trajectory command can use as few as one of these fields and as many as all four. One common minimal usage is to specify only the joint displacements in the `positions` vector. This can be adequate, particularly for low-speed motions controlled by joint position feedback. Alternatively, the trajectory command might communicate with a velocity controller, *e.g.* for speed control of wheels. A more sophisticated motion plan communicating with a more sophisticated joint controller would include multiple fields (specifying both positions and velocities is common).

To command coordinated motion of all seven joints of a seven-DOF arm, for example, one would populate (at least) the `positions` vector for each of N points to visit, starting from the current pose and ending at some desired pose. Preferably, these points would be relatively close in space (*i.e.* with relatively small changes in any one joint displacement command between sequential point specifications). Alternatively, coarser trajectories may be communicated to a trajectory interpolator, which breaks up motions between coarse subgoals into streams of fine motion commands.

It is desirable that each point include specification of joint velocities that are consistent with the specified displacements and arrival times (although it is legal to specify a trajectory without specifying joint velocities). Alternatively, generation of consistent velocity commands might be left to a lower-level trajectory interpolator node.

Each joint-space point to be visited must specify a `time_from_start`. These time specifications must be monotonically increasing for sequential points and should be consistent with the velocity specifications. It is the user's responsibility to evaluate whether the specified joint displacements, joint velocities and point arrival times are all self-consistent and achievable within the robot's limitations.

The `time_from_start` value, specified for each joint-space point to be visited, distinguishes a path from a trajectory. If one were to specify only the sequence of poses to be realized, the result would be a path description. By augmenting space (path) information with time, the result is a trajectory (in joint space).

While a trajectory message may be fine-grained and lengthy, it is more common (and practical) to send messages that consist of a sequence of subgoals (with adequate resolution). Fine-grained commands can be generated by interpolation via an action server. An illustrative example is provided in the package `example_trajectory`.

In this package, an action message is defined: `TrajAction.action`. The `goal` field of this action message contains `trajectory_msgs::JointTrajectory`. This action message is used by a trajectory client to send goals to a trajectory action server.

Two illustrative nodes are the `example_trajectory_action_client` and `example_trajectory_action_server`. The client computes a desired trajectory, in this case consisting of samples of a sinusoidal motion of `joint1`. (This could easily be extended to N joints, but this is sufficient for testing on the `minimal_robot`.) The samples are deliberately taken at irregular and fairly coarse time intervals. An example output is shown in Fig 11.8. The values of position command (radians) show that the originating sinusoidal function is

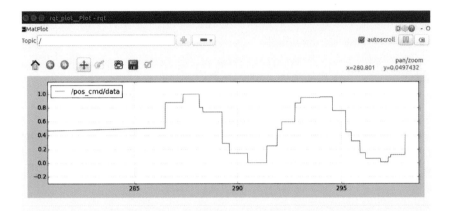

Figure 11.8: Coarse trajectory published at irregularly sampled points of sine wave

sampled coarsely and irregularly. This was done to illustrate the generality of the trajectory message, and show that it need not be fine grained.

The joint command samples are packaged into a trajectory message along with associated arrival times, and this message is transmitted to the trajectory action server within the action goal.

The `example_trajectory_action_server` receives the goal message and interpolates linearly between specified points, resulting in the profile shown in Fig 11.9. The result-

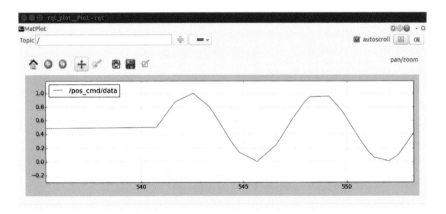

Figure 11.9: Action-server linear interpolation of coarse trajectory

ing profile is piece-wise linear, but sufficiently smooth; the `minimal_robot` can follow this command reasonably well, yielding a smooth motion.

The example is run with:

```
roslaunch minimal_robot_description minimal_robot.launch
```

which brings up the one-DOF robot and its controller. The trajectory interpolation action server is started with:

```
rosrun example_trajectory example_trajectory_action_server
```

and the corresponding trajectory client is started with:

```
rosrun example_trajectory example_trajectory_action_client
```

This client generates a coarse trajectory and sends it to the action server within a goal message. In turn, the action server interpolates this trajectory and sends a fast, smooth stream of commands to the robot.

For the example trajectory action server, the velocities specified by the client in the goal message are ignored. In the one-DOF robot case, the minimal joint controller does not accept velocity feed-forward commands. More generally, one could do better with a servo controller that accepts both position and velocity feed-forward commands. In this case, including consistent velocity commands sent to the servo controller would improve tracking performance. These velocity commands could be generated on the fly by the trajectory interpolator or included in a longer, fine-grained trajectory message.

The example trajectory interpolator is piece-wise linear. It would also be desirable to have a smoother interpolator, *e.g.* cubic splines. (See *e.g.* http://sdk.rethinkrobotics. com/wiki/Joint_Trajectory_Action_Server for a joint trajectory action server for the Baxter robot, written in Python, which uses cubic spline interpolation.)

The intelligence of the trajectory action server is a design decision. Optimizing a trajectory with considerations of speed, precision and collision avoidance is a computationally difficult problem. Further, the optimization criteria may change from one instance to another. The trajectory specification, even if it is coarse, must still take into consideration issues of collisions and speed and effort constraints. Rather than perform this optimization twice (for construction of a viable trajectory, further smoothing and optimization of that trajectory by the action server), one could perform trajectory optimization as part of the planning process that generates a trajectory specification. This could yield a dense sampling of points in the trajectory specification, along with compatible velocities, accelerations and gravity load compensation. If this approach is taken, then linear interpolation of the dense trajectory commands should be adequate in general, although the trajectory action server should also pass along the corresponding velocity and effort values contained within these optimized trajectories.

Trying to make the trajectory action server more intelligent than mere linear interpolation is thus not warranted and may, in fact, interfere with optimization of pre-computed trajectories. Thus, this simple example may be considered adequate in general.

This discussion of trajectory messages applies directly to ROS-Industrial (see http: //rosindustrial.org). To bridge ROS to existing industrial robots, one can write the equivalent of the example_trajectory_action_server in the native language of the target robot (and run it on the native robot controller). This (non-ROS) program also requires custom communications to receive packets containing the equivalent of trajectory messages. A complementary ROS node would receive trajectory messages via a ROS topic or ROS goal message, translate them into the format expected by the robot's communication program, and transmit the corresponding packets to the robot controller for interpolation and execution.

Correspondingly, the robot controller would also run a program that samples and transmits the robot's joint state (at least the joint angles). A ROS node would receive these values in some custom format, translate them into ROS sensor_msgs::JointState messages and publish these on the topic `joint_states` for use by other ROS nodes (including rviz). Using this approach, the ROS-Industrial consortium has retrofit ROS interfaces to a growing number of industrial robots.

11.6 TRAJECTORY INTERPOLATION ACTION SERVER FOR A SEVEN-DOF ARM

The package `arm7dof_traj_as` contains a helpful library, `arm7dof_trajectory_streamer`, and an action server, `arm7dof_action_server` extends the minimal-robot example to seven degrees of freedom. This action server responds to goal messages that contain joint-space trajectories. The incoming trajectories may be coarse, since they are interpolated at 50 Hz (a parameter defined in the header file `arm7dof_trajectory_streamer.h`).

An example client program, `arm7dof_traj_action_client_prompter`, illustrates use of the trajectory action server. This example client first sends the robot to a hard-coded pose, then prompts the user to enter joint numbers and joint values, which it then commands to the robot via the trajectory action server. To run this example, first start Gazebo:

```
roslaunch gazebo_ros empty_world.launch
```

Next, bring up the seven-DOF robot arm with its position controllers, using:

```
roslaunch arm7dof_model arm7dof_w_pos_controller.launch
```

Note that these two steps are required only for robot simulation. When interacting with a real robot with a ROS interface, the actual robot dynamics takes the place of Gazebo's physics engine, and the real controllers take the places of the ROS controllers. Typically, the physical robot would host the roscore. The robot would expose its topics for publishing robot state and receiving joint commands.

Next, running:

```
rosrun arm7dof_traj_as arm7dof_traj_as
```

brings up the trajectory action server, which provides a trajectory interface between low-level joint angle commands and higher-level trajectory plans. This node should be run during both simulation and physical operation.

An example interactive trajectory client can be started with:

```
rosrun arm7dof_traj_as arm7dof_traj_action_client_prompter
```

This will pre-position the robot, then accept commands from the user (one joint at a time). More generally, a trajectory client would be instantiated within a higher-level application that performs trajectory planning, presumably based on sensory information.

The seven-DOF trajectory-interpolation action server here will be modified slightly to apply it to the left and right arms of a Baxter robot, which will be introduced in Chapter 14.

11.7 WRAP-UP

This chapter has presented low-level joint-control options in ROS, including position control, velocity control, and extensions to force control with respect to a force–torque sensor (available as a plug-in in Gazebo). At the joint level, smooth and coordinated motion is achieved using a joint-space interpolator that operates on trajectory messages. Higher-level controls, *e.g.* for Cartesian motion control, ultimately depend on lower-level control in joint space.

Building on joint-level control, we next consider how to compute desirable joint-space trajectories with consideration of desirable end-effector motion. This is the subject of forward and inverse kinematics, which is discussed next.

Robot Arm Kinematics

CONTENTS

I NTRODUCTION

Robot arm kinematics is a starting point for most textbooks in robotics. A fundamental question is, given a set of joint displacements, where is the end effector in space? To be explicit, one must define reference frames. A frame of interest on a robot may be its tool flange, or some point on a tool tip (*e.g.* welder, laser, glue dispenser, grinder, ...) or a frame defined with respect to a gripper. It is useful to compute the pose of the defined frame of interest with respect to the robot's base frame. In a separate transform, one can express the robot's base frame with respect to a defined world frame. The task of computing the pose of a robot's end effector with respect to its base frame given a set of joint displacements constitutes forward kinematics. For open kinematic chains (*i.e.* most robot arms), forward kinematics is always simple to compute, yielding a unique, unambiguous pose as a function of joint displacements.

While forward kinematics is relatively simple, the more relevant question is typically the inverse: given a desired pose of the end effector frame, what set of joint angles will achieve this objective? Solving this inverse kinematics relationship is plagued with problems. There may be no viable solutions (goal is not reachable), there may be a finite number of equally valid solutions at very different sets of joint angles (typical for six-DOF robots), or there may be an infinity of solutions (typical for robots with more than six joints). Further, while forward kinematics solutions can be obtained using a standard procedure, solving for inverse kinematic solutions may require *ad hoc* algorithms exploiting creative insights.

The kinematics-and-dynamics package KDL (see, *e.g.* `http://wiki.ros.org/orocos_kinematics_dynamics`) is a popular tool for computing forward kinematics and robot dynamics. One of its capabilities is that it can reference a robot model on the parameter server to obtain the information necessary to compute forward kinematics. Thus, any open-chain robot in ROS already has a corresponding forward kinematics solver (via KDL). KDL also computes Jacobians (between any two named frames).

Unfortunately, the more useful and more difficult problem of inverse kinematics is not handled well in general. If the robot of interest has specific properties (*e.g.* six DOFs with a spherical wrist), inverse kinematic solutions can be computed analytically, yielding exact expressions for all possible solutions. However, robots with redundant kinematics (more than six DOFs) are increasingly popular, due to their enhanced dexterous workspace, and this makes computing inverse kinematics ambiguous. Further, many robot designs do not conform to the equivalent of a spherical wrist, and for such designs, there may be no known analytic inverse kinematics solution.

Numerical techniques (gradient search using a kinematic Jacobian) are often employed when analytic inverse kinematic solutions are not available. As is typical with numerical searches, this approach can be slow, fail to converge, get stuck in local minima, and (when successful) typically returns a single solution when multiple solutions or an infinity of solutions exist.

When working with a particular robot, if an inverse kinematics package is available, the available solution may be sufficient. If a suitable package is not already available, it may be necessary to design one.

This chapter will introduce examples of forward and inverse kinematics libraries and use these to illustrate common issues in robot kinematics.

12.1 FORWARD KINEMATICS

A simple example to illustrate kinematics is in the package **rrbot**. This robot model borrows from the on-line tutorial at `http://gazebosim.org/tutorials?tut=ros_urdf` for the **rrbot** (a two-DOF robot with two revolute joints). In the subdirectory **model** the file `rrbot.xacro` defines a simple, planar manipulator with three links and two joints. Figure 12.1 shows Gazebo and `rviz` views of this robot, which was launched using:

```
roslaunch gazebo_ros empty_world.launch
roslaunch rrbot rrbot.launch
```

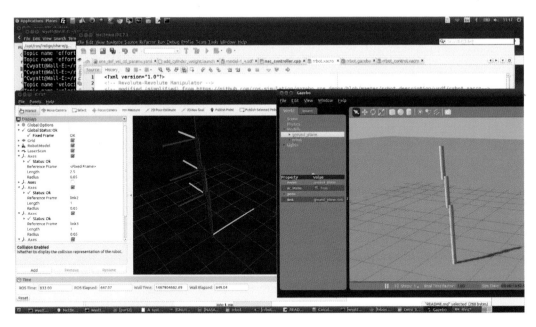

Figure 12.1: Gazebo and `rviz` views of rrbot

The rrbot has frames defined for each of its links. To be consistent with Denavit-Hartenberg convention, the frames were defined such that the z axes pass through revolute joint axes (for joints 1 and 2), as follows:

```
<joint name="joint1" type="continuous">
  <parent link="link1"/>
  <child link="link2"/>
```

```
  <origin xyz="${height1 - axel_offset} 0  ${width} " rpy="0 0 0"/>
  <axis xyz="0 0 1"/>
  <dynamics damping="0.7"/>
</joint>
<joint name="joint2" type="continuous">
  <parent link="link2"/>
  <child link="link3"/>
  <origin xyz="${height2 - axel_offset*2} 0 ${width} " rpy="0 0 0"/>
  <axis xyz="0 0 1"/>
  <dynamics damping="0.7"/>
</joint>
```

It is also convenient to define a frame of interest at the tip of the last link. To do so, a fictitious static joint is defined in `rrbot.xacro` as follows:

```
<joint name="flange_jnt" type="fixed">
  <parent link="link3"/>
  <child link="flange"/>
  <origin xyz="${height3 - axel_offset} 0  0" rpy="0 0 0"/>
</joint>
```

By defining this frame, `tf` will compute transformations out to this frame of interest.
The fictitious link `flange` is defined simply as:

```
<link name="flange"/>
```

Since the link `flange` is connected by a static joint, it is not necessary to define inertial, visual or collision components. The same technique is useful for modeling mounting sensors (rigidly) to robots or defining tool frames on specific end effectors.

With the above joint definitions, the equivalent Denavit-Hartenberg parameters, as defined in the header file `rrbot/include/rrbot/rrbot_kinematics.h`, are:

```
const double DH_a1=0.9; //link length: distance from joint1 to joint2 axes
const double DH_a2=0.95; //link length: distance from joint2 axis to flange-z axis

const double DH_d1 = 0.1;// offset along parent z axis from frame0 to frame1
const double DH_d2 = 0.0; // zero offset along parent z axis from frame1 to flange

const double DH_alpha1 = 0; //joint1 axis is parallel to joint2 axis
const double DH_alpha2 = 0; //joint2 axis is parallel to flange z-axis

//could define robot "home" angles different than DH home pose; reconcile with these ↩
    offsets
const double DH_q_offset1 = 0.0;
const double DH_q_offset2 = 0.0;
```

Converting between a URDF and DH models can be confusing. In the present case, having defined the joint axes consistent with DH convention, the z axes of DH frames 1 and 2 are parallel (as well as the z axis of the `flange` frame), and consequently the `alpha` angles are zero, and the `a` parameters are the link lengths. Since there is an offset from `link1` to `link2` along the z direction, there is also a non-zero `d1` parameter.

In the `rrbot` package, the source file `rrbot_fk_ik.cpp` compiles to a library that computes forward and inverse kinematics for the rrbot. The function `compute_A_of_DH()` returns a 4 × 4 coordinate-transform matrix, as described in Chapter 4. In general, a coordinate transform between successive DH frames follows from specification of the four DH parameters, as shown in `compute_A_of_DH()`.

```cpp
Eigen::Matrix4d compute_A_of_DH(double q, double a, double d, double alpha) {
    Eigen::Matrix4d A;
    Eigen::Matrix3d R;
    Eigen::Vector3d p;

    A = Eigen::Matrix4d::Identity();
    R = Eigen::Matrix3d::Identity();
    //ROS_INFO("compute_A_of_DH: a,d,alpha,q = %f, %f %f %f",a,d,alpha,q);
    double cq = cos(q);
    double sq = sin(q);
    double sa = sin(alpha);
    double ca = cos(alpha);
    R(0, 0) = cq;
    R(0, 1) = -sq*ca; //% - sin(q(i))*cos(alpha);
    R(0, 2) = sq*sa; //%sin(q(i))*sin(alpha);
    R(1, 0) = sq;
    R(1, 1) = cq*ca; //%cos(q(i))*cos(alpha);
    R(1, 2) = -cq*sa; //%
    //%R(3,1)= 0; %already done by default
    R(2, 1) = sa;
    R(2, 2) = ca;
    p(0) = a * cq;
    p(1) = a * sq;
    p(2) = d;
    A.block<3, 3>(0, 0) = R;
    A.col(3).head(3) = p;
    return A;
}
```

A general procedure for computing forward kinematics simply involves computing each successive 4×4 transformation matrix, then multiplying these matrices together. This is implemented in the function `fwd_kin_solve_()` within the `rrbot_fk_ik.cpp` library as follows:

```cpp
Eigen::Matrix4d Rrbot_fwd_solver::fwd_kin_solve_(Eigen::VectorXd q_vec) {
    Eigen::Matrix4d A = Eigen::Matrix4d::Identity();
    //%compute A matrix of frame i wrt frame i-1 for each joint:
    Eigen::Matrix4d A_i_iminusi;
    Eigen::Matrix3d R;
    Eigen::Vector3d p;
    //cout << "A_base_link_wrt_world_:" << endl;
    //cout << A_base_link_wrt_world_ << endl;
    for (int i = 0; i < NJNTS; i++) {
        A_i_iminusi = compute_A_of_DH(q_vec[i] + DH_q_offsets[i], DH_a_params[i],
            DH_d_params[i], DH_alpha_params[i]);
        A_mats_[i] = A_i_iminusi;
    }
    //now, multiply these together
    //A_base_link_wrt_world_ * A_frame1_wrt_base_link = A_frame1_wrt_world
    A_mat_products_[0] = A_base_link_wrt_world_ * A_mats_[0];

    for (int i = 1; i < NJNTS; i++) {
        A_mat_products_[i] = A_mat_products_[i - 1] * A_mats_[i];
    }
    //Eigen::Vector4d test_O_vec; //some test code to get the coordinates of the ↩
        flange frame
    //test_O_vec<<0,0,0,1;
    //cout<<"test Amat prod: "<<A_base_link_wrt_world_*A_mats_[0]*A_mats_[1] *↩
        test_O_vec<<endl;
    return A_mat_products_[NJNTS - 1]; //tool flange frame
}
```

This same forward-kinematics code is applied as well to additional example robots in `learning_ros/Part_5`, including a six-DOF ABB robot (model IRB120), the six-DOF Universal Robots UR10, the Baxter robot (with seven-DOF arms) and a robot model based on the NASA Goddard satellite-servicer arm (called arm7dof). To customize the forward kinematics code for a specific robot, it is necessary only to define the number of joints and values for the DH parameters in the respective header file. If using the KDL package, that step is not even necessary, since KDL will parse the robot model to obtain the parameter values.

Forward and inverse kinematics can be tested using the `test_rrbot_fk.cpp` test function. The rrbot, test function, and supporting tools are brought up with:

```
roslaunch gazebo_ros empty_world.launch
roslaunch rrbot rrbot.launch
rosrun rrbot test_rrbot_fk
rosrun rqt_gui rqt_gui
rosrun tf tf_echo world flange
```

A screenshot with these nodes running is shown in Fig 12.2. The robot can be moved to

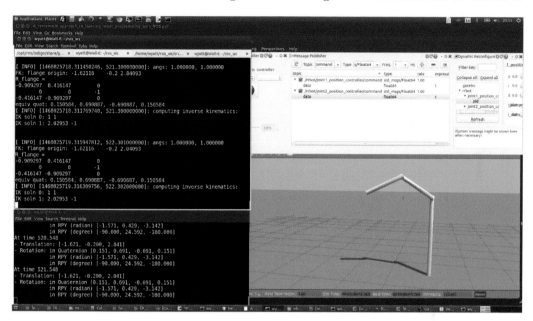

Figure 12.2: Gazebo, `tf_echo` and `fk` test nodes with rrbot

arbitrary poses using `rqt_gui`. In the example shown, the joints are both commanded to angles of 1.0 rad. The `test_rrbot_fk` node subscribes to `joint_states` and prints the published joint values, confirming that the robot is at the commanded joint angles. (In this test, gravity was set to zero, so there is no joint error from gravity droop.) Forward kinematics of the `flange` frame with respect to the base (world) frame is computed and displayed as:

```
FK: flange origin: -1.62116    -0.2 2.04093
```

To test this result, the `tf_echo` node is run to get the `flange` frame with respect to the world frame. This function displays:

```
- Translation: [-1.621, -0.200, 2.041]
```

To the precision displayed, the results are identical.

A second, more realistic test can be run with:

```
roslaunch irb120_description irb120.launch
rosrun irb120_fk_ik irb120_fk_ik_test_main
rosrun rqt_gui rqt_gui
rosrun tf tf_echo world link7
```

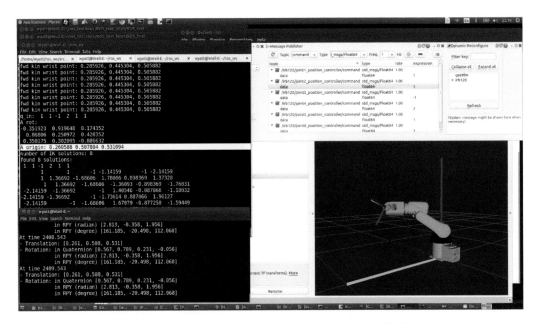

Figure 12.3: `rviz`, `tf_echo` and `fk` test of ABB IRB120

This launches a model of the ABB IRB120 industrial robot. Using `rqt_gui`, the six joints are commanded to the values $[1, 1, -1, 2, 1, 1]$. The resulting pose is shown in Fig 12.3. The `tf_echo` node prints out the pose of the `flange` frame (link7) with respect to the world. It displays:

`- Translation: [0.261, 0.508, 0.531]`

The test node `irb120_fk_ik_test_main` subscribes to the joint states and uses these values to compute forward kinematics. This node prints out:

`A origin: 0.260588 0.507804 0.531094`

Again, to the precision displayed, the results are identical. Additional spot checks yield similar results.

12.2 INVERSE KINEMATICS

The code used to compute forward kinematics for the two-DOF rrbot and the example six-DOF arm is virtually identical, except for the number of joints and the specific numerical values of the DH parameters. The algorithmic approach is valid in general for all open kinematic chains. In contrast, inverse kinematics can require robot-specific algorithms and may be problematic for handling multiple solutions, which is exacerbated with kinematically-redundant robots.

Starting with the two-DOF rrbot, inverse kinematics can be computed using the law of cosines. The `rrbot_fk_ik.cpp` program includes a class for forward kinematics and a class for inverse kinematics. The inverse kinematics function `ik_solve()` takes an argument for the desired end-effector pose (as an `Eigen::Affine3object`), and it populates a `std::vector` of `Eigen::Vector2` objects comprising alternative solutions. There can be 0, 1 or 2 solutions for the rrbot.

With only two degrees of freedom, it is (typically) only possible to satisfy two objectives. For example, one cannot find solutions for an arbitrary desired position (x, y, z) of the tool flange. The rrbot can only reach flange coordinates for which $y = 0.2$. If the requested flange position is $(x, 0.2, z)$, then solutions may be possible.

The function `ik_solve()` first calls `solve_for_elbow_ang()`, which returns 0, 1 or 2 reachable elbow (joint2) angles. These angles follow from the realization that the distance from the shoulder to the flange is a function only of joint2 (not joint1). The value of joint2 angle can be found from the law of cosines: $c^2 = a^2 + b^2 - 2ab\cos(C)$. In this case, a and b are the lengths of the two movable links and c is the desired distance from the robot's shoulder (joint1) to the tool flange (tip of distal link). These three lengths form a triangle, and the angle C is the angle opposite the leg with length c. This relationship can be solved for angle C using an inverse cosine. However, there are zero, one or two solutions for C in this equation. If, for example, a flange pose is requested that is beyond the reach of the robot (*i.e.* $c > a + b$), there are no solutions for C.

When the requested position is within reach of the robot, there are typically two solutions. (A special case occurs when the arm is fully extended, precisely reaching a desired position; in this case there is only one solution.) Most commonly, if there are any solutions, there are two solutions (commonly referred to as elbow up and elbow down).

For each of the elbow solutions, one can find the corresponding shoulder angle (`joint1` angle) that achieves the desired flange position. This is computed using the function `solve_for_shoulder_ang()`, reduces the shoulder-angle problem to solving an equation of the form $r = A\cos(q) + B\sin(q)$ for the value of q. This equation occurs frequently in inverse-kinematics algorithms and is solved within the function `solve_K_eq_Acos_plus_Bsin(K, A, B, q_solns)`. There is some ambiguity, though, since two solutions are obtained, and only one is consistent with a correct robot pose. For a given elbow angle, both candidate shoulder angles are tested using forward kinematics, and the correct shoulder angle is thus identified. Typically, if the desired point is reachable, there will be two valid solutions, (q_{1a}, q_{2a}) and (q_{1b}, q_{2b}).

As an example, the desired flange position of:

```
flange origin: -1.62116    -0.2 2.04093
```

has two solutions, as computed and displayed by `test_rrbot_fk`:

```
IK soln 0: 1 1
IK soln 1: 2.02953 -1
```

These solutions are illustrated in Fig 12.4 Inverse kinematics for the ABB IRB120 is more

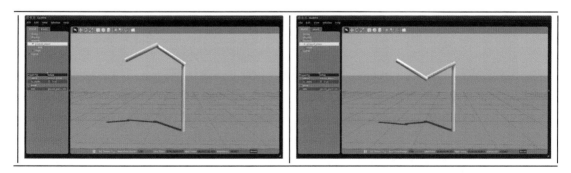

Figure 12.4: Two IK solutions for rrbot: elbow up and elbow down

complex. Fortunately, this robot design incorporates a spherical wrist. (A spherical wrist is a design for which the last three joint axes intersect in a point.) Existence of a spherical wrist simplifies the IK problem, since one can solve for position of the wrist point (the point of intersection of the last three joint axes) in terms of the proximal joints (joints 1, 2 and 3, in the case of the IRB120). Transformation matrices, law of cosines, and solutions of $r = A\cos(q) + B\sin(q)$ are again exploited to solve for joint-angles $q1$, $q2$ and $q3$. In addition to elbow-up and elbow-down solutions, the robot can also rotate 180 degrees about its turret (base) to reach backward to achieve two more elbow-up, elbow-down solutions. There thus can be four distinct solutions that place the wrist point at its desired position. For each of these solutions, the spherical wrist offers two distinct solutions. All together, these comprise eight IK solutions. A specific example for the IRB120 robot was run using the `irb120_fk_ik_test_main` node, starting the robot at joint angles $(1, 1, -1, 2, 1, 1)$. Forward kinematics was computed, then this flange pose was used to generate inverse-kinematic solutions. For this case, there were eight solutions, including re-discovery of the original joint angles, $(1, 1, -1, 2, 1, 1)$. The eight solutions for this case are:

```
found 8 solutions:
1          1            -1        2         1          1
1          1            -1       -1.14159  -1         -2.14159
1          1.36692      -1.68606  1.78066   0.898369   1.37328
1          1.36692      -1.68606 -1.36093  -0.898369  -1.76831
-2.14159  -1.36692      -1        1.40546  -0.887866  -1.18032
-2.14159  -1.36692      -1       -1.73614   0.887866   1.96127
-2.14159  -1            -1.68606  1.67079  -0.877258  -1.59449
-2.14159  -1            -1.68606 -1.4708    0.877258   1.5471
```

These solutions show that the first three joint angles appear twice in separate solutions, which differ only in the last three (wrist) joint angles. All eight of the solutions for this example are shown in Fig 12.5. The two distinct wrist solution for each combination of $(q1, q2, q3)$ may not be easily discernible, but these solutions are quite distinct in $(q4, q5, q6)$.

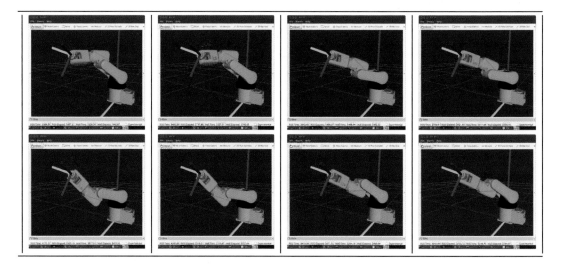

Figure 12.5: Example of eight IK solutions for ABB IRB120

Although the example shown has eight IK solutions, the number of solutions can vary. If a requested point is beyond the reach of the robot (or within the body of the robot) there will be no viable solutions. If IK solutions require moving one or more joints beyond their allowable range of motion, there will be fewer than eight valid solutions. If the desired pose is reachable at a singular pose (when two or more joints align such that their joint axes are colinear), there will be an infinity of IK solutions (describable as a function of the sum or difference of the joint angles of the aligned joints).

A related example is in package `ur_fk_ik`, which is a library of forward and inverse kinematics functions for the Universal Robots UR10 robot. This six-DOF robot does not have a spherical wrist, but an analytic inverse-kinematic method does exist. Typically, there are eight IK solutions for each desired tool flange pose (if reachable) within 0 to 360 degree joint angle ranges.

With an array of IK solutions, one requires an algorithmic means to choose among the solutions to execute one of them. Choosing a best solution can be ambiguous, but one expectation is that the robot will not jump suddenly, *e.g.* between elbow-up and elbow-down solutions, during execution of a trajectory.

How to choose among candidate solutions is even more challenging when the robot has more than six degrees of freedom (or, more generally, when the robot degrees of freedom is larger than the number of task constraints). In such cases, there is a continuum (or multiple, distinct continua) of solutions to choose from. An example is in `arm7dof`.

The robot model in `arm7dof_model` is based on the NASA RESTORE arm design being considered for servicing satellites in space. A Gazebo view of a proposed NASA design is shown in Fig 12.6. Frame assignments for this seven-DOF arm are shown in Fig 12.7. The

Figure 12.6: Proposed NASA satellite-servicer arm, Gazebo view

model in package `arm7dof_model` is a simplified, abstracted version of the satellite servicer arm. This model (with a large disk appended to the tool flange) appears as in Fig 12.8. This design incorporates a spherical wrist, which simplifies the inverse-kinematics algorithm. For each solution of $(q1, q2, q3, q4)$ that places the wrist point in a desired position, there are two solutions $(q5, q6, q7)$ that achieve the desired flange pose (satisfying both desired position and orientation of the flange). However, if the desired wrist point is reachable, there is typically a continuum (or multiple, disconnected continua) of solutions in the space $(q1, q2, q3, q4)$.[1]

---

[1] By DH convention, joint-angle numbering starts at 1 at the most proximal joint and increments sequentially to the last joint. However, indexing in C++ starts from 0, and thus joint numbers are sometimes numbered from 0.

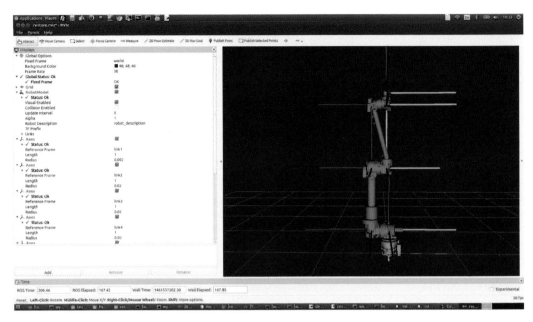

Figure 12.7: Proposed NASA satellite-servicer arm, `rviz` view with frames displayed

An approach to inverse kinematics for this arm is presented in [2], in which solutions are obtained as a function of a free variable, such as the elbow-orbit angle or the base turret angle. The latter approach is taken here. Within the package `arm7dof_fk_ik`, a forward and inverse kinematics library contains a key function `ik_wrist_solns_of_q0()`. This function accepts coordinates of a desired wrist point along with a desired turret angle ($q1$) and it returns all of the viable solutions for joint angles ($q2, q3, q4$). As with the ABB six-DOF robot, once $q1$ is constrained, only (these) three joints imply the wrist-point coordinates, and there can be up to four distinct solutions. Also like the ABB six-DOF robot, each of these wrist-point solutions theoretically has two solutions for ($q5, q6, q7$) (though these may or may not be reachable within the robot's joint range-of-motion constraints). In at least some instances, there thus will be eight distinct, viable IK solutions (at given angle $q1$). For the pose in Fig 12.8, there are six distinct, reachable IK solutions at $q1 = 0$.

The choice of base angle, $q1$, for an optimal IK solution is unclear. In fact, the best choice of $q1$ may only be knowable in a broader context. To defer choosing a value of $q1$, one can compute IK solutions for a range of $q1$ candidates, sampled at some chosen resolution. The function `ik_solns_sampled_qs0()` accepts a specified flange pose and fills in a (reference variable) `std::vector` of seven-DOF IK solutions. The sampling resolution of $q1$ is set by the assigned value of `DQ_YAW`, which is set to 0.1 rad in the header file `arm7dof_kinematics.h`. For the pose shown in Fig 12.8, 228 valid IK solutions were computed at a sampling resolution of 0.1 rad over $q1$ from 0 to 2π. This example was computed using the following code in package `arm7dof_fk_ik`:

```
roslaunch gazebo_ros empty_world.launch
roslaunch arm7dof_model arm7dof_w_pos_controller.launch
rosrun rqt_gui rqt_gui (to command joint angles)
rosrun arm7dof_fk_ik arm7dof_fk_ik_test_main2
```

Since all of these solutions are computed analytically, they present no numerical issues, such as failure to converge. Further, the analytic approach yields a large number of options,

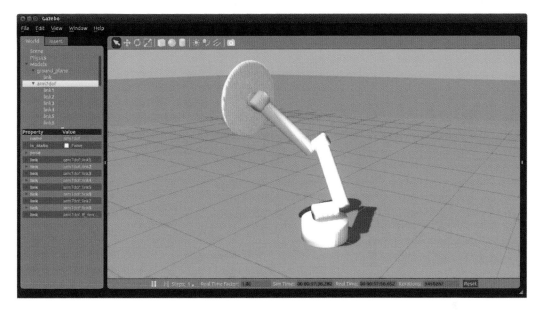

Figure 12.8: Approximation of proposed NASA satellite-servicer arm

not simply a single solution (in contrast to a numerical approach). The solution process is also relatively efficient. For the example shown, computing 228 IK solutions required approximately 30 msec on a laptop using one core of a 2 GHz Intel i7.

The number of solutions expands dramatically in singular poses. Figure 12.7 shows the seven-DOF arm in a singular pose, for which the joint axis of joint 7 is colinear with the joint axis of joint 5. In such poses, given any solution, one can generate a space of solutions by rotating joint 7 by $\Delta\theta$ and simultaneously rotating joint 5 by $-\Delta\theta$, thus generating a continuum of solutions even at fixed $q1$.

For some robot arms, an analytic solution may not be known. In such cases, one must resort to numerical methods. If an approximate analytic solution is known, it can be useful to solve for approximate solutions and use these as starting points for a numerical search. In the numerical search, the manipulator Jacobian is a key component. Jacobians, like forward kinematics, are easy to compute and are well behaved. The example kinematic libraries provided include Jacobian computations.

12.3 WRAP-UP

This chapter has presented a brief overview of robot-arm forward and inverse kinematics. Forward kinematics for open kinematic chains has been solved in general. The `tf` package computes forward kinematics in real time, based on robot model information available in a URDF. The popular kinematics-and-dynamics package KDL also offers forward-kinematics solutions. Inverse kinematics, on the other hand, does not have a general solution, and for kinematically-redundant robots, there is an entire range of IK solution options. If a robot of interest does not already have an inverse-kinematic solution implemented in ROS, the developer may need to create one, as illustrated in the examples of this chapter.

Forward and inverse kinematics is only one step in motion control of robot arms. At a higher level of abstraction, one needs to plan a smooth sequence of robot arm poses that achieves some desirable behavior. The issue of robot arm motion planning is considered next.

Arm Motion Planning

CONTENTS

I NTRODUCTION
Robot arm motion planning is a broad field with a variety of sub-problems. For example, one may be concerned with the optimal trajectory a robot can execute to get from some start pose to some final pose in minimum time, subject to constraints on the actuator efforts and joint ranges of motion. Another variant is planning arm motions that optimize the mechanical advantage of available actuators, *e.g.* as in weight lifting. A common problem is planning how to command a robot's joints to move from an initial pose to a final pose (expressed either in joint space or in task space) while avoiding collisions (with the robot or with entities in the environment). The task to be performed may require a specified end-effector trajectory, *e.g.* as in laser cutting, sealant dispensing, or seam welding. In such cases, the velocity of the endpoint may be constrained by the task (*e.g.* for optimal material removal rates, torch speeds or dispensing rates). In demanding cases, motion planning must be performed in real time, *e.g.* to generate motion to catch a ball. At the other extreme, some motion plans for stereotyped behaviors (*e.g.* folding arms into a compact pose for transportation) may be computed offline with the results saved for execution as a rote skill. In cases of large communication latency, supervisory control may be the best approach. In this case, one may plan a motion and preview the result in simulation before committing the remote system to execute the approved plan.

Arm motion planning packages exist in ROS, and the sophistication and breadth of options may be expected to grow. In the current context, a single example planning approach will be described to illustrate how arm motion planning can integrate within a ROS-based hierarchy.

The example planning problem considered here assumes that one has a specified six-DOF path in Cartesian space that the robot's tool flange must satisfy. For example, a seven-DOF arm may be required to move a knife blade along a specified path (along a workpiece) while maintaining a specified orientation of the knife (with respect to the workpiece), equivalently constraining six-DOF poses of the tool flange. It will be assumed that the Cartesian path can be adequately approximated by N_{path} sequential samples along the continuous path. At each sample point along the path, the seven-DOF arm typically would have infinitely many inverse-kinematic options. Pragmatically, this may be approximated by sampled IK solutions, on the order of hundreds of options per Cartesian knot point. Even in this simplified scenario, the planning problem can be intimidating. For example, if there were 100 samples along the Cartesian trajectory and 200 IK solutions are computed for each such pose, there would be 200^{100} joint-space path options, which is overwhelmingly large.

Fortunately, this planning problem can be simplified dramatically using dynamic programming. The approach described here is composed of two layers: a Cartesian-space planner and an underlying joint-space planner.

13.1 CARTESIAN MOTION PLANNING

The package `cartesian_planner` contains the source files for multiple Cartesian planners, including `arm7dof_cartesian_planner.cpp`, `ur10_cartesian_planner.cpp`, and `baxter_cartesian_planner.cpp`, which define respective planning libraries for the seven-DOF arm, a Universal Robots UR10 arm, and the Baxter robot. The role of these libraries is to sample points along a Cartesian path, compute IK options for each sampled pose along the path, and (with the help of the joint-space planner) find the optimal sequence of joint-space solutions to traverse the desired Cartesian path.

In general, the Cartesian path to be followed is task dependent. The `cartesian_planner` is illustrative of how to sample a path and compute an attractive joint-space trajectory, but it is limited in its generality. The Cartesian planners described here have multiple motion-planning options:

- Specify a start pose, `q_vec`, in joint-space coordinates, and specify a goal pose in Cartesian space. Compute a joint-space trajectory from the start angles to one of the IK solutions for the goal pose.

- Specify a start pose, `q_vec`, in joint space, and specify a goal pose in Cartesian space. Compute a Cartesian trajectory by sampling points along a straight-line path. A trajectory will be generated such that desired final orientation of the gripper will be obtained quickly, then preserved throughout the linear move.

- Specify a start pose, `q_vec`, and a desired delta-p Cartesian motion. Sampled Cartesian poses will be generated starting from the initial pose and translating in a straight line by the specified displacement vector while holding orientation fixed at the initial orientation.

- Specify Cartesian start and end poses with respect to the arm's base. Only the orientation of the end pose will be considered. Orientation of start pose is ignored. A sequence of Cartesian poses is generated that maintains constant orientation while moving in a straight line from the start position to the end position.

The node `example_arm7dof_cart_path_planner_main.cpp` shows how to use the Cartesian planner `arm7dof_cartesian_planner.cpp`, which relies on support from the corresponding kinematics library `arm7dof_fk_ik` and from the generic `joint_space_planner`. Similarly, respective example main programs are provided for the Baxter robot and for the Universal Robots UR10 robot.

Running the node `example_arm7dof_cart_path_planner_main` will produce an output file `arm7dof_poses.path` of joint-space poses that produce Cartesian motion with the tool flange orientation constant. In this example, the desired motion specifications are hard-coded in the main program, corresponding to maintaining orientation of the tool flange pointing up while translating from x_{start} to x_{end} at constant $y_{desired}$ and $z_{desired}$. Samples along this path are computed every 5 cm. These path specification values can be edited to test alternative motions. With a roscore running, the seven-DOF arm planner example can be run with:

```
rosrun   cartesian_planner  example_arm7dof_cart_path_planner_main
```

For this example, 60 samples are computed along the Cartesian path. At each sample point, inverse-kinematic options are computed at increments of 0.1 rad of the base joint. This results in roughly 200 IK solutions per Cartesian sample point (ranging from 130 to 285 IK solutions at each Cartesian pose, for this example).

Key lines of the code `example_arm7dof_cart_path_planner_main.cpp` are:

```
CartTrajPlanner cartTrajPlanner; //instantiate a cartesian planner object

R_gripper_up = cartTrajPlanner.get_R_gripper_up();

//specify start and end poses:
a_tool_start.linear() = R_gripper_up;
a_tool_start.translation() = flange_origin_start;
a_tool_end.linear() = R_gripper_up;
a_tool_end.translation() = flange_origin_end;

//do a Cartesian plan:
found_path = cartTrajPlanner.cartesian_path_planner(a_tool_start, a_tool_end, ↵
    optimal_path);
```

The planner option invoked in this example is specification of both start and end Cartesian poses. After calling `cartesian_path_planner()`, a joint-space plan will be returned in `optimal_path`. This plan is written to an output file called `arm7dof_poses.path`, which consists of 61 lines (one for each Cartesian sample point), the first few lines of which are:

```
2.9, -0.798862, 0.227156, -1.35163, -0.191441, 2.12916, -3.16218
2.9, -0.752214, 0.210785, -1.42965, -0.173242, 2.1637, -3.15245
2.9, -0.707348, 0.196287, -1.50431, -0.156966, 2.19621, -3.1424
```

Each line specifies seven joint angles corresponding to a joint-space pose that is an IK solution of the corresponding desired Cartesian-space pose. Selection of optimal joint-space poses from among the candidate IK solutions is performed by a supporting library, the joint-space planner, which will be discussed next.

The planning libraries can be used as in this example for offline planning. Alternatively, the planning libraries can be used within action servers that perform motion planning and execution on demand. This option will be discussed further, after introducing a generic joint-space planning technique.

13.2   DYNAMIC PROGRAMMING FOR JOINT-SPACE PLANNING

Finding optimal joint-space solutions from among IK options can be challenging. A dynamic-programming approach to joint-space planning described here is applicable to a wide array of robots, at least for low-dimensional null spaces, such as seven-DOF arms.

For example, the motion plan for a seven-DOF arm illustrated by `example_arm7dof_cart_path_planner_main.cpp` results in 61 Cartesian samples, each of which has between 130 and 285 joint-space IK options. For each Cartesian sample point, a single IK solution must be chosen. However, these cannot be selected independently within each Cartesian sample point. Transitions between successive sample points must result in smooth motion, *i.e.* no large jumps in joint angles. Avoiding large jumps is not as simple as looking one step ahead, as invoking a greedy algorithm, since such an approach can lead to trajectories that result in poor options (large jumps) downstream. Instead, the entire path must be considered in context.

An approach to finding optimal sequences of joint-space solutions among a large number of options is dynamic programming. Conceptually, consider the feed-forward graph in Fig

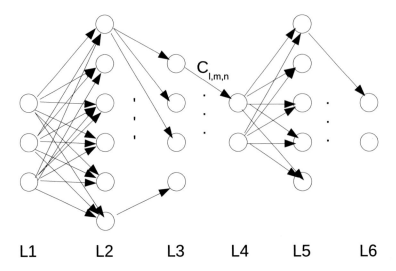

Figure 13.1: Conceptualization of a feed-forward network dynamic-programming problem

13.1. The example network shown consists of six layers (columns), each comprised of some number of nodes (circles within each column). The network is feed-forward in the following sense. The m'th node in layer l, denoted as node $n_{l,m}$, may have a link from to the n'th node in layer $l+1$ (node $n_{(l+1),n}$). However, node $n_{l,m}$ has no link to any node in any layer other than layer $l+1$. (*e.g.* links are never directed backward, nor do they skip over subsequent layers).

A path within this network starts at some node in layer $L1$ and concludes at some node in layer $L6$, visiting a total of six nodes (including start and finish) in the path. Any such path has an associated cost C that is the sum of state costs and transition costs. A state cost is a cost associated with passing through a given node ($c_{l,m}$ for node m in layer l). A transition cost is the cost associated with following a link from node $n_{l,m}$ to node $n_{(l+1),n}$, which may be referred to as transition cost $c_{l,m,n}$ (*e.g.* as labelled in Fig 13.1). The path cost C is a scalar consisting of the sum of all state and transition costs (six state costs and five transition costs, in this example). Thus a candidate path can be scored with a penalty function of path cost. The optimization objective of the proposed problem is to find the path from $L1$ to $L6$ that has the minimum path cost.

With respect to our joint-space planning problem, we can interpret each layer as a six-DOF task-space pose sampled along a specified Cartesian path. The objective is to move the tool flange forward along this path, *e.g.* performing a cutting operation, and we can label (*e.g.*) six points along this path sequentially. For a task such as cutting (or drawing, welding, scribing, glue-dispensing, etc.) it would not make sense to skip over any of these points nor to ever go backward. Rather, each of the Cartesian subgoals must be achieved in sequence. We require that our end effector visit each of the Cartesian poses. However, computing inverse kinematics, we find that there are multiple joint-space solutions that achieve pose l. We may refer to the m'th joint-space solution at the l'th Cartesian pose to be node $n_{l,m}$. (Depending on the chosen sampling resolution, we may consider hundreds of IK alternatives at each task-space sample or layer). Each node within our network describes a unique joint-space pose of the robot. A sequence of nodes, chosen 1 from each layer consecutively, describes a (discretized) joint-space trajectory that is guaranteed to pass through the sequential Cartesian-space samples along the task-space trajectory.

The state cost for a given node in this context may be scored by multiple criteria. For example, a robot pose that results in a self collision or a collision with the environment may be assigned a cost of infinity (or, more simply, removed from the network altogether). A node (pose) that is a near collision may have a high state cost. Nodes that are near singularities might also be penalized. Other pose-dependent criteria might also be considered. Transition costs should penalize large moves. For example, a path that included moving the base joint by π moving from layer l to layer $l + 1$ would result in high accelerations of high-inertia joints, which would be slow, disruptive, and presumably would result in wild tool-flange gyrations while moving only a short distance along the Cartesian path. Similarly, a sudden wrist flip from layer l to layer $l + 1$ would be unacceptable, although the IK solutions may be correct at both of these layers. Thus, large changes in joint angles from layer to layer should be penalized.

Different optimal path solutions will result from different strategies of assigning state and transition penalties. The designer must choose how to assign costs to joint-space poses, how to assign transition costs, the number and locations of the Cartesian-space sample points (layers), and how many IK solutions (nodes) to consider at each sample point. Having cast the planning problem as a feed-forward graph, we can apply graph-solving techniques to our abstracted problem, yielding a sequence of joint-space poses that advance the robot through the chosen Cartesian-space poses.

For this type of graph, dynamic programming can be applied effectively. The process is as follows. Working backward from the final layer, each node will be assigned a `cost to go`. For the final layer, the cost to go is simply the state cost associated with each final-layer node.

Backing up to layer $L - 1$ (layer $L5$, in our example), each node in this layer is assigned a cost to go. For our example, for node 1 in layer $L5$, $n_{5,1}$, there are only two options: to transition to node $n_{6,1}$ at a cost of $c_{5,1} + c_{5,1,1} + c_{6,1}$, or transition to node $n_{6,2}$ at a cost of $c_{5,2} + c_{5,1,2} + c_{6,2}$. The cost to go for node $n_{5,1}$ will be assigned the minimum of these two options, which we may label as $C_{5,1}$. The cost to go for node $n_{5,2}$ is computed similarly, resulting in an assigned value $C_{5,2}$.

Continuing to work backward through the network, when computing cost to go for nodes in layer l, we may assume that the cost to go has already been computed for all nodes in layer $l + 1$. To compute the cost to go for node m in layer l, $C_{l,m}$, consider each node n in layer l+1 and compute the corresponding cost $C_{l,m,n} = c_{l,m} + C_{(l+1),n}$. Find the minimum cost (over all options $n_{(l+1),n}$ in layer $l + 1$) among these options and assign it to $C_{l,m}$. In this process, a cost to go is computed for every node in the network, and computing these costs is comparably efficient at every layer.

Given cost to go assignments for every node, the optimal path through the network can be found by following the steepest gradient of costs $C_{l,m}$ through the network.

A simple numerical example is illustrated with the help of Figs 13.2 through 13.6. This sequence of views shows how the cost to go computation propagates backward from the goal states. As the cost to go is computed, many of the options can be eliminated as provably suboptimal in the context of a globally optimal solution. Typically, a single link will survive the cost to go optimization at each layer of computation. In this simple example with integer transition costs, ties could exist, resulting in multiple surviving links from each such incremental optimization. In practice, however, the cost will be a floating-point value and ties will not occur.

In the illustrated example, it becomes clear early on that node I cannot be part of the optimal solution, and thus the goal node must be node J. Working backward through the layers, links are pruned and incremental transition costs are replaced with the optimal cost to go associated with each node. As the process sweeps back to the starting layer, a single

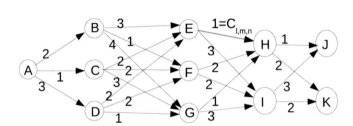

Figure 13.2: Simple numerical example of dynamic programming consisting of five layers with transition costs assigned to each allowed transition

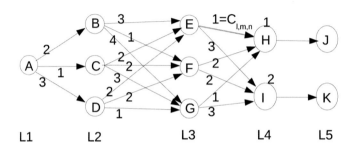

Figure 13.3: First step of solution process. Minimum cost to go is computed for nodes H and I. Two of the four links from layer four to layer five are removed because they are provably suboptimal

path through the network results (Fig 13.6). This solution process is quite efficient, and the resulting path is provably optimal.

A library that performs this optimal-path computation is in the package `joint_space_planner`. This planner assumes all state costs are merely zero (with the presumption that dangerous poses have already been removed from the network). Transition costs are assigned to be a weighted sum of squares of delta joint angles corresponding to a transition. One must provide an entire network and a vector of weights (associated with joints) to the planner constructor, and the planner will return the optimal path through the network. The network passed to the solver is represented as:

```
vector<vector<Eigen::VectorXd> > &path_options
```

The `path_options` argument is a (std) vector of (std) vectors of (Eigen) vectors. The outermost vector contains L elements corresponding to L layers in the network. Each layer within this vector contains a vector of nodes (IK options at this point along the trajectory).

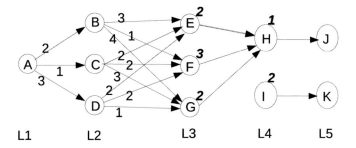

Figure 13.4: Second step of solution process. Minimum cost to go is computed for nodes E, F and G in layer 3 (costs 2, 3 and 2, respectively). Three of the six links have been pruned, leaving only the single optimal option for each node in layer three. It can be seen at this point that node I will not be involved in the optimal solution.

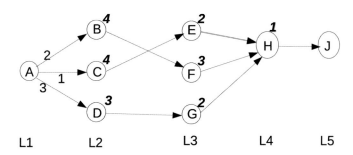

Figure 13.5: Third step of solution process. Minimum cost to go is computed for nodes B, C and D in layer two (costs 4, 4 and 3, respectively). Six of the nine links have been pruned, leaving only the single optimal option for each node in layer two.

The number of nodes within a layer may vary from layer to layer. Each node within a layer is an N-jointed IK solution, expressed as an Eigen-type vector. The network solver is written to accommodate arbitrary dimensions in terms of number of layers, number of nodes in any layer, and number of dimensions of each node. However, care should be taken to avoid excessive numbers of layers and numbers of nodes per layer, as the planning can become slow.

In the example `example_arm7dof_cart_path_planner_main.cpp`, a seven-DOF robot is considered moving along a Cartesian path sampled every 5 cm, resulting in 61 layers (including start and end poses). Each layer has between 130 and 285 IK joint-space options. On one core of a 2 GHz i7 Intel processor, the planning process requires about 8 sec, which includes approximately 3 sec to compute approximately 13,000 IK solutions and about 5 sec to find the minimum-cost path through the resulting network.

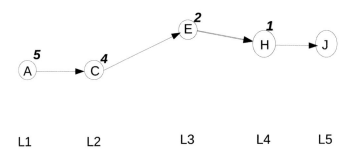

Figure 13.6: Final step of solution process. Minimum cost to go is computed from node A. Only a single path through the network constituting the optimal solution survives.

The result of the example plan is the file `arm7dof_poses.path`, which lists the seven-DOF joint angles recommended for each of the 61 steps along the Cartesian path.

The example Cartesian planners in package `cartesian_planner`, including 7DOFarm, UR10 and Baxter, all use the same joint-space planner library described here. The library is flexible enough to accommodate any number of joints, any number of IK solutions per Cartesian sample point, and any number of Cartesian sample points along a Cartesian path. However, excessively large networks will require correspondingly large computation times.

It should be noted, however, that the simple joint-space planning optimization illustrated here does not take into consideration potential collisions, either with the environment or with the robot's own body. Joint-space solutions that would result in collisions should be removed from the graph before planning a path through the network.

13.3   CARTESIAN-MOTION ACTION SERVERS

In Section 13.1, Cartesian motion planning libraries were introduced. Each such motion library is customized for a specific robot type. However, the objective is the same in each case: a gripper or tool is to be moved in space to a destination pose, often constrained to move along a Cartesian path. Ideally, this can be accomplished with commands from a higher level that are robot agnostic. An example approach to this uses Cartesian motion action servers.

In the `cartesian_planner` package, the nodes `baxter_rt_arm_cart_move_as.cpp` and `ur10_cart_move_as.cpp` use their respective Cartesian planner libraries and they present an action-server interface called `cartMoveActionServer`. Action clients can communicate with these action servers using a generic action message, defined in `cart_move.action` in the `cartesian_planner` package. This action message includes the following mnemonics in the goal field:

```
uint8 GET_TOOL_POSE = 5
uint8 GET_Q_DATA = 7

#requests for motion plans;
uint8 PLAN_PATH_CURRENT_TO_WAITING_POSE=20
#plan a joint-space path from current arm pose to some IK soln of Cartesian goal
uint8 PLAN_JSPACE_PATH_CURRENT_TO_CART_GRIPPER_POSE = 21
```

```
#rectilinear translation w/ fixed orientation
uint8 PLAN_PATH_CURRENT_TO_GOAL_DP_XYZ = 22
#plan cartesian path from current arm pose to goal gripper pose
uint8 PLAN_PATH_CURRENT_TO_GOAL_GRIPPER_POSE=23
```

Each of these is an action code that can be specified by a client.

Components of the goal message are:

```
#goal:
int32 command_code
geometry_msgs/PoseStamped des_pose_gripper
float64[] arm_dp #a 3-D vector displacement relative to current pose
float64[] q_goal
float64 time_scale_stretch_factor
```

A client should fill in the `command_code` field with one of the defined action-code values. For the case of `PLAN_PATH_CURRENT_TO_WAITING_POSE`, no additional fields need to be filled in. For requesting either a joint-space or Cartesian-space plan to a goal pose, the client must specify the corresponding command code in the goal message and must also specify the desired gripper pose in the `des_pose_gripper` field. If requesting a relative motion along a specified vector, the `des_pose_gripper` field is not needed, but the `arm_dp` field must be filled in. (Joint-space goal specifications are also possible, although this mode is less general and should be avoided).

An example client program that uses this action interface is `example_generic_cart_move_ac.cpp` in the `cartesian_planner` package. This node instantiates an action client that connects to the action server `cartMoveActionServer` (through use of the library `cart_motion_commander`) and communicates via the action message `cart_move.action`. This client illustrates invoking various actions offered by the Cartesian-motion action server, including: planning and executing a path to a pre-defined waiting pose; planning and executing a path to an absolute tool pose; and planning and executing a Cartesian motion relative to the current tool pose. The class `ArmMotionCommander` in the `cart_motion_commander` library encapsulates details for invoking these behaviors, including populating a `cart_move` goal message, sending this message to the action server, and returning the response code received by the callback function. This simplifies the action client code, as in the snippet below from `example_generic_cart_move_ac.cpp`.

```
//return to pre-defined pose:
ROS_INFO("back to waiting pose");
rtn_val=arm_motion_commander.plan_move_to_waiting_pose();
rtn_val=arm_motion_commander.execute_planned_path();

//get tool pose
rtn_val = arm_motion_commander.request_tool_pose();
tool_pose = arm_motion_commander.get_tool_pose_stamped();
ROS_INFO("tool pose is: ");
xformUtils.printPose(tool_pose);
//alter the tool pose:
tool_pose.pose.position.x += 0.2; // move 20cm, along x in torso frame
// send move plan request:
rtn_val=arm_motion_commander.plan_path_current_to_goal_gripper_pose(tool_pose);
//send command to execute planned motion
rtn_val=arm_motion_commander.execute_planned_path();
```

These lines of code invoke the behaviors of moving to a pre-defined waiting pose, then moving to a specified tool pose along a Cartesian path. At this level of abstraction, the client does not need to be aware of what type of robot it is controlling. Rather, the client can focus on moving a tool in task space while leaving details of inverse kinematics and Cartesian-space planning to the action server. In fact, the example generic action client works equally well with the Baxter robot or with the UR10 robot with no changes, in spite of the fact that these robots have very different kinematic properties, including numbers of joints.

One minor complication is that the relevant frame names (*e.g.* gripper frame and base-link frame) of different robots may have different names, which limits the re-usability of application code. A fix for this is to define additional frames on these robots that present generic names. For example, the Baxter robot's right arm has a gripper with a frame between the fingertips called `right_gripper`. The UR10 robot does not have such a frame name. The last link of the UR10 has a frame called `tool0`. Further, Baxter robot kinematics is computed with respect to the `torso` frame, whereas the UR10 robot has no torso frame. Instead, UR10 kinematics is computed with respect to its `base_link` frame. To have the Baxter robot present a common interface, `baxter_static_transforms.launch` from the `cartesian_planner` package is run. This launch file consists merely of:

```
<launch>
<node pkg="tf" type="static_transform_publisher" name="system_ref_frame" args="0 0 ↵
    -0.91 0 0 0 1 torso system_ref_frame 100" />
<node pkg="tf" type="static_transform_publisher" name="generic_gripper_frame" args="0 ↵
    0 0 0 0 0 1 right_gripper generic_gripper_frame 100" />
</launch>
```

By running this launch file, two additional frames are published on `tf`: `generic_gripper_frame` and `system_ref_frame`. The generic gripper frame in this case is prescribed to be identical to the `right_gripper` frame, since the argument list specifies $x, y, z = 0, 0, 0$, and quaternion $x, y, z, w = 0, 0, 0, 1$. With this publication, one can refer to `generic_gripper_frame` as a synonym for `right_gripper` to invoke planning and motion execution. The other frame published, `system_ref_frame`, relates a generic reference frame name to the relevant Baxter frame, `torso`. In this case, the torso is specified to be 0.91 m above the system reference frame (which is defined to be on the ground plane). Object poses can be defined with respect to the system reference frame, and the transform from system reference frame to torso frame can be calibrated to correspond to the mounting height of the robot.

Correspondingly for the UR10, the model frames to be referenced are `base_link` and `tool0` (unless a gripper is added to the model, in which case `tool0` should be substituted for an appropriate frame on the gripper). These frames are associated with `system_ref_frame` and `generic_gripper_frame`, respectively. This is done by launching `ur10_static_transforms.launch` within the `cartesian_planner` package. The contents of this launch file are:

```
<launch>
<node pkg="tf" type="static_transform_publisher" name="system_ref_frame" args="0 0 0 0↵
    0 0 1 world system_ref_frame 100" />
<node pkg="tf" type="static_transform_publisher" name="generic_gripper_frame" args="0 ↵
    0 0 0 0 0 1 tool0 generic_gripper_frame 100" />
</launch>
```

Launching this file results in `generic_gripper_frame` being synonymous with `tool0`, and `system_ref_frame` being synonymous with the `world` frame. This assumes that the robot is launched with a transform describing the robot's `base_link` with respect to the world frame. With these additional transforms, one can control the UR10 by referring to the `generic_gripper_frame` in the `system_ref_frame`.

The example generic Cartesian-move action client can be run on the UR10 robot as follows. First, start a UR10 simulation with:

```
roslaunch ur_gazebo ur10.launch
```

Then, start the static transform publications for gripper and reference frames with:

```
roslaunch cartesian_planner ur10_static_transforms.launch
```

Start a Cartesian-move action server for the UR10 robot by running:

```
rosrun cartesian_planner ur10_cart_move_as
```

At this point, the robot is ready to receive and act on Cartesian-motion planning and execution goals. The simple example action client can be run with:

```
rosrun cartesian_planner example_generic_cart_move_ac
```

This example code will prompt the user for input, including asking for a desired delta-z value for a Cartesian motion. The action client will run to conclusion, but the action server will persist, ready to accept new goals from new action clients.

To run the same, generic Cartesian-move client node with the Baxter robot, first start the Baxter robot simulation with:

```
roslaunch baxter_gazebo baxter_world.launch
```

Wait for the start-up to conclude (which may take 30 seconds, or so). Then enable the motors with the command:

```
rosrun baxter_tools enable_robot.py -e
```

Start a trajectory-streamer action server with:

```
rosrun baxter_trajectory_streamer rt_arm_as
```

Start the transforms that define the generic gripper frame and the system reference frame:

```
roslaunch cartesian_planner baxter_static_transforms.launch
```

Start the Cartesian-motion action server for Baxter's right arm:

```
rosrun   cartesian_planner baxter_rt_arm_cart_move_as
```

The Baxter robot is now ready to accept Cartesian-motion goals from an action client. The generic action client can be run, identical to the case of the UR10 robot, with:

```
rosrun cartesian_planner example_generic_cart_move_ac
```

Running this node results in Baxter's right-gripper frame executing Cartesian trajectories in exactly the same way as the UR10 moves its tool0 frame. This ability to refer to Cartesian-space motions of generic gripper frames offers the possibility of writing higher-level code that is independent of any specific robot arm design. By abstracting the programming problem in this way, the programmer can focus on thinking in task space while ignoring the details of how any one robot could carry out the desired task-space trajectories. This will be discussed further in the context of an object_grabber package, after introducing some further detail of the Baxter simulator.

13.4 WRAP-UP

Controlling a robot arm ultimately reduces to specifying a stream of coordinated joint-angle commands in time. Computing a desirable motion can be highly complex, and arm motion planning is a broad and open field. One example planning approach that is applicable to kinematically redundant robots (with low null-space dimensionality) was presented using dynamic programming. The result of the planning computation is a trajectory message, which can be communicated to a trajectory action server running on the robot. It was shown how Cartesian-motion planners can be embedded in action servers, and these action servers can refer to generic gripper and system reference frames. Such an approach supports the possibility of writing generic action client nodes that focus on task-space operations and are re-usable across multiple robot designs.

We next consider development of these concepts specifically with respect to the Baxter robot.

Arm Control with Baxter Simulator

CONTENTS

I NTRODUCTION
     In the context of learning about control of robot arms, the Baxter robot simulator, which runs in Gazebo, is helpful. The Baxter simulator, composed by the manufacturer ReThink Robotics, is a reasonably faithful model of the robot. The simulator and the physical robot have identical ROS interfaces, so programs developed under simulation can be ported quickly to the corresponding physical robot.

     The Baxter simulator can be found at: `http://sdk.rethinkrobotics.com/wiki/Baxter_Simulator`. It is downloaded and installed by default as part of the installation scripts used here (see `https://github.com/wsnewman/learning_ros_setup_scripts`), which installs ROS and all of the example code that accompanies this text.

     In this chapter, the Baxter model and simulator will be used to further explore forward and inverse kinematics, use of the joint-space planner, a Baxter-specific Cartesian planner, and a higher-level action server for planning and execution of object grasping.

14.1   RUNNING BAXTER SIMULATOR

To start the Baxter simulator, run:

```
roslaunch baxter_gazebo baxter_world.launch
```

This process launches Gazebo with a model of the Baxter robot. The start-up process can be slow (expect about 45 sec). The simulator is ready once the launch window displays `Gravity compensation was turned off`. The Gazebo display will appear as in Fig 14.1. Once the simulator launch is complete (*i.e.* once the gravity message is displayed), the joints

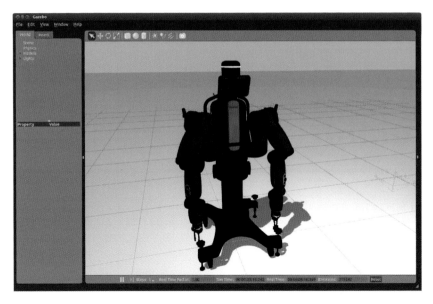

Figure 14.1: Gazebo view of Baxter simulator in empty world

can be enabled to respond to motion commands. This is done by issuing:

```
rosrun baxter_tools enable_robot.py -e
```

which will result in the response **Robot Enabled**.

Some example motions can be invoked with provided demos, including:

```
rosrun baxter_tools tuck_arms.py -t
```

which moves the robot's arms into a pose convenient for shipping, as shown in Fig 14.2. The complementary command:

```
rosrun baxter_tools tuck_arms.py -u
```

will untuck the arms.

Another interesting example is:

```
rosrun baxter_examples joint_velocity_wobbler.py
```

This program will pre-position the arms to a suitable initial pose then drive all of the arm joints through small sinusoidal motions. With this node running, one can confirm that the joint servos are active and visualize the mobility of the arms.

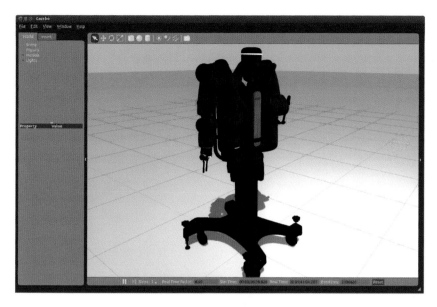

Figure 14.2: Gazebo view of Baxter in tuck pose

14.2 BAXTER JOINTS AND TOPICS

The Baxter robot has 15 servoed degrees of freedom, including seven joints of the right arm, seven joints of the left arm and a pan (left–right swivel) motion of the neck. The head (display) can also nod, although this is only a binary command (tilt of the neck is not servo controlled). The Baxter robot and simulator include three cameras: one on the display (head) and one each on the wrists. Other sensors include sonar sensors around the crown, a short-range infra-red distance sensor in each tool flange, joint angle sensors on each servoable joint, and joint torque sensors.

Launching `rviz` allows a user to visualize more information about the robot. Figure 14.3 shows the robot model with some frames displayed. The torso frame is a useful reference. Its z axis points up, x is forward, and y is to the robot's left.

Frames corresponding to arm joints are also displayed for the robot's right arm. For each of these seven frames, the blue axis corresponds to a joint rotation axis. (For forearm rotation, the blue axis is not visible, as it is buried within the forearm shell in this view.) It can be seen that the seven arm joints are organized has having a twist of 90 degrees for each sequential pair of joints (roll-pitch-roll-pitch-roll-pitch-roll).

Mathematically, to control six degrees of freedom of a desired gripper pose, one typically needs at least six independent joint degrees of freedom. However, once joint limits are imposed, it is typically difficult to satisfy all six constraints of a desired pose with six controllable joints. Having a seventh joint (analogous to a human arm) dramatically improves manipulability. At the same time, this presents additional challenges for inverse kinematics, given that the arm is kinematically redundant. Increasingly, industrial robots are offering seven controllable joints, and thus addressing kinematic redundancy is of growing importance.

An additional challenge for Baxter kinematics is that this robot does not have spherical wrists. In Chapter 12, the examples of the ABB IRB120 and the abstracted NASA satellite servicer arm both had spherical wrists, since the final three joint axes all intersect at a point. This kinematic property was exploited to help solve inverse kinematics analytically. For the

Baxter robot, the last two axes (blue axes of the last two frames displayed) do intersect, and we may refer to this intersection point as the wrist point. However, the forearm rotation axis (not shown) does not intersect this point; it misses intersecting the wrist point by a small offset. Since the offset is small, it can serve as a useful approximation to treat the wrist as spherical, which helps in the computation of fast, approximate inverse-kinematic solutions.

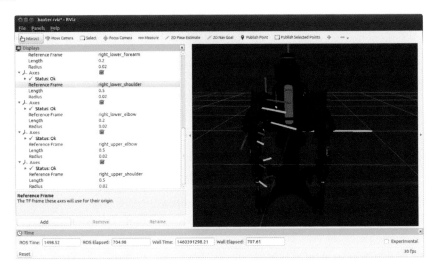

Figure 14.3: `rviz` view of Baxter simulator model illustrating right-arm frames

With the Baxter simulator (or physical Baxter) running, executing

```
rostopic list
```

reveals that approximately 300 topics are used. Nearly 40 of these topics appear under cameras, although Baxter has only three built-in cameras. The various camera topics include options of alternative image encoding and pre-processing. Having this many camera topics is not a bandwidth concern, since camera topics are published and subscribed using **image_transport**. One of the features of **image_transport** is that no data is published to these topics unless there is at least one subscriber.

There are also nearly 250 topics under the name space **robot**, including joint states, sensor values, robot I/O devices, controller properties and states, and display (**face**) topics. The **tf** and **tf_static** topics carry messages describing pair-wise coordinate-frame transforms. If running in simulation, there are also several Gazebo topics.

One of the most useful is the **/robot/joint_states** topic, which is described further below. Running: **rosnode list** shows that a **robot_state_publisher** is running, which is necessary for **rviz** to display the robot model with correctly positioned joints (whether running a Gazebo simulation or the actual robot).

Invoking **rosparam list** indicates that a **robot_description** has been loaded onto the parameter server (performed as part of the Baxter launch process).

The topic **/robot/joint_states** can be examined by running:

```
rostopic info /robot/joint_states
```

which shows that this topic carries messages of type **sensor_msgs/JointState**, which is a standard ROS means of publishing robot joint states. The `JointState` message type can be examined by running:

```
rosmsg show sensor_msgs/JointState
```

which shows that this message contains the following fields:

```
std_msgs/Header header
  uint32 seq
  time stamp
  string frame_id
string[] name
float64[] position
float64[] velocity
float64[] effort
```

Running:

```
rostopic hz /robot/joint_states
```

shows that messages on this topic are being updated at 50 Hz. To get a glimpse of the data published on this topic, run:

```
rostopic echo /robot/joint_states
```

A screenshot of this output is shown in Fig 14.4. This output shows position, velocity and

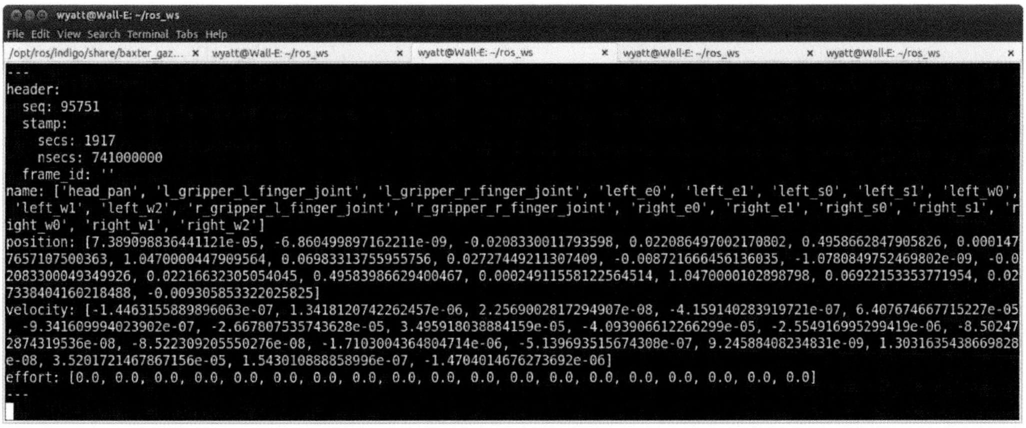

Figure 14.4: `rostopic echo` of `robot/joint_states` for Baxter

effort for each of 19 named joints. These joints include the 14 arm joints, the head pan joint and four gripper-finger joints. In the simulation, all of the reported efforts are zero. However, for the physical robot, effort values for the arm joints are measured and published.

The joint states are reported in the same order as the listing of joint names. The order used by Baxter is somewhat odd; conventionally, joints are reported in order of the chain they describe, from torso to gripper. ROS, however, is generally indifferent to the order of joints for reporting states and accepting commands; the order used in the **name** vector is the order ROS will use, and this may change from iteration to iteration.

To clarify Baxter's joint ordering and naming, the joints defined sequentially from torso to wrist for right arm are:

```
right_s0, right_s1, right_e0, right_e1, right_w0, right_w1, right_w2
```

For the left arm, the naming sequence is the same, but substitute left for right. All of `position[]`, `velocity[]` and `effort[]` are reported in the order listed by order of names (as seen from the `rostopic echo`).

The topic `/tf` publishes a large number of link-to-link transforms. All link-to-link relationships described by a connecting joint are updated rapidly. These can be viewed by running `rostopic echo tf` (although this means of viewing is inconvenient, since the display scrolls too quickly). To view a specific relationship, it is more convenient to use `tf_echo`. For example, to display the pose of the right hand (actually, the tool flange frame of the right arm) with respect to the robot's torso frame, run:

```
rosrun tf tf_echo torso right_hand
```

The position and orientation of the `right_hand` frame with respect to the `torso` frame will then be printed, updated once per second.

The camera topics and laserscan topics are useful for perceptual processing. If more sensors are added to Baxter, the sensor values should be published on topics as well, to be made available for interpretation.

14.3   BAXTER'S GRIPPERS

The grippers on the Baxter simulation model are electric, parallel-jaw grippers available from ReThink Robotics. One communicates with the grippers via ROS publish and subscribe. Gripper states are published on `/robot/end_effector/right_gripper/state` and `/robot/end_effector/left_gripper/state` using the message type `baxter_core_msgs/EndEffectorState`. This message definition defines multiple mnemonics and 16 fields. Notably, the gripper state includes values of `position` and `force`, which can be useful for detecting whether a part has been grasped successfully, or if the gripper is empty.

To command the grippers, one publishes to topics `/robot/end_effector/right_gripper/command` and `/robot/end_effector/left_gripper/command` with message type `baxter_core_msgs/EndEffectorCommand`. This message type includes 12 defined mnemonic strings for communicating with the gripper and five fields in the command message.

For the present purposes, it is sufficient to simply command the grippers in position mode, and to command positions of fully open (100) or fully closed (0), which corresponds to a throw of 44 mm. If a graspable object (neither too wide nor too narrow) is between the gripper fingers, commanding fully closed motion will result in the gripper fingers stalling before reaching the fully closed position. This can be detected by examining gripper state.

To simplify the gripper interface, a `simple_baxter_gripper_interface` library is defined in a package of the same name. This library defines a class, `BaxterGripper`, that sets up publishers and subscribers to the left and right grippers. Gripper command messages are defined and populated with:

```
    gripper_cmd_open.command ="go";
    gripper_cmd_open.args = "{'position': 100.0}'";
    gripper_cmd_open.sender = "gripper_publisher";

    gripper_cmd_close.command ="go";
    gripper_cmd_close.args = "{'position': 0.0}'";
    gripper_cmd_close.sender = "gripper_publisher";
```

Position feedback from the grippers can be noisy. If threshold testing for open and closed is sensitive, noise on the position signals can cause false triggers. Low-pass filters for position feedback are implemented in callbacks as:

```
void BaxterGripper::right_gripper_CB(const baxter_core_msgs::EndEffectorState& ↵
    gripper_state) {
    //low-pass filter the gripper position for more reliable threshold tests
    right_gripper_pos_ = (1.0- gripper_pos_filter_val_)*right_gripper_pos_ + ↵
        gripper_pos_filter_val_*gripper_state.position;
}
```

The value of `gripper_pos_filter_val_` is set to 0.1 in the `BaxterGripper` constructor. This value can range between 0 and 1.0. If it is increased to 1.0, the filter will merely return the current gripper position as reported by the gripper state publication. As the value is decreased toward 0, the `right_gripper_pos_` is more heavily filtered.

Note that the filtering occurs only when the callback functions are triggered, and thus use of these filters requires `ros::spin()` or `ros::spinOnce()` executions by the user program. (It would be preferable to have these callbacks run in their own thread, so user-program timing does not affect filter bandwidth.)

Member functions of the BaxterGripper class are simple, and thus they are defined in the header file `simple_baxter_gripper_interface.h`. The public functions defined in the header are:

```
void right_gripper_close(void) { gripper_publisher_right_.publish(gripper_cmd_close)↵
    ;};
void left_gripper_close(void) { gripper_publisher_left_.publish(gripper_cmd_close);};
void right_gripper_open(void) { gripper_publisher_right_.publish(gripper_cmd_open);};
void left_gripper_open(void) { gripper_publisher_left_.publish(gripper_cmd_open);};
double get_right_gripper_pos(void) { return right_gripper_pos_;};
double get_left_gripper_pos(void) { return left_gripper_pos_;};
```

Use of the simple Baxter gripper library is illustrated in the example main program `baxter_gripper_lib_test_main`, which appears in Listing 14.1.

Listing 14.1: `baxter_gripper_lib_test_main.cpp`: test main program using `BaxterGripper` library

```
1  // baxter_gripper_lib_test_main:
2  // illustrates use of library/class BaxterGripper for simplified Baxter gripper I/O
3
4  #include<ros/ros.h>
5  #include<simple_baxter_gripper_interface/simple_baxter_gripper_interface.h>
6  //using namespace std;
7
8  int main(int argc, char** argv) {
9      ros::init(argc, argv, "baxter_gripper_test_main"); // name this node
10     ros::NodeHandle nh; //standard ros node handle
11
12     //instantiate a BaxterGripper object to do gripper I/O
13     BaxterGripper baxterGripper(&nh);
14     //wait for filter warm-up on right-gripper position
15     while (baxterGripper.get_right_gripper_pos()<-0.5) {
16         ros::spinOnce();
17         ros::Duration(0.01).sleep();
18         ROS_INFO("waiting for right gripper position filter to settle; pos = %f", ↵
                baxterGripper.get_right_gripper_pos());
19     }
20
21     ROS_INFO("closing right gripper");
22     baxterGripper.right_gripper_close();
23     ros::Duration(1.0).sleep();
24     ROS_INFO("opening right gripper");
25     baxterGripper.right_gripper_open();
```

```
26      ros::spinOnce();
27      ROS_INFO("right gripper pos = %f; waiting for pos>95", baxterGripper.↩
           get_right_gripper_pos());
28      while (baxterGripper.get_right_gripper_pos() < 95.0) {
29          baxterGripper.right_gripper_open();
30          ros::spinOnce();
31          ROS_INFO("gripper pos = %f", baxterGripper.get_right_gripper_pos());
32          ros::Duration(0.01).sleep();
33      }
34
35      ROS_INFO("closing left gripper");
36      baxterGripper.left_gripper_close();
37      ros::Duration(1.0).sleep();
38      ROS_INFO("opening left gripper");
39      baxterGripper.left_gripper_open();
40
41      ROS_INFO("closing right gripper");
42      baxterGripper.right_gripper_close();
43      ros::spinOnce();
44      ROS_INFO("right gripper pos = %f; waiting for pos<90", baxterGripper.↩
           get_right_gripper_pos());
45      while (baxterGripper.get_right_gripper_pos() > 90.0) {
46          baxterGripper.right_gripper_close();
47          ros::spinOnce();
48          ROS_INFO("gripper pos = %f", baxterGripper.get_right_gripper_pos());
49          ros::Duration(0.01).sleep();
50      }
51
52      return 0;
53  }
```

In Listing 14.1, an object of type **BaxterGripper** is instantiated on line 14. This object initializes the gripper finger positions to impossible values of -1. If the gripper state is being published, the gripper position low-pass filters will eventually converge on the actual positions. In lines 16 through 20, the main program waits to observe that the right gripper position has increased to at least -0.5, which indicates that the gripper state publisher of the right gripper is running.

Once communication with the right gripper is established, gripper commands are sent. Line 23:

```
baxterGripper.right_gripper_close();
```

causes the right gripper to close. This simple function can be used conveniently in application programs. Similarly, line 26:

```
baxterGripper.right_gripper_open();
```

opens the right gripper. Lines 29 through 34 repeatedly test the filtered right-gripper position to confirm when the gripper has opened to a value of 95 or greater. (A value of 100 is fully open). Such a check can be important, since gripper opening can take some time, and one should confirm the gripper is fully open before attempting to approach a grasp position.

The test main program then closes and opens the left and right grippers, waiting 1 sec after sending gripper commands. Lines 46 through 51 test that the right gripper has closed to a (filtered) position of 90 or less.

If an object of known size is to be grasped, the finger positions should also be checked to confirm that the object is preventing full finger closure. If the gripper position goes to zero, the gripper has failed to grasp the object. The gripper finger position may also be examined to evaluate whether the grasped object is the expected size.

Alternatively, the **force** component of the gripper state may be examined to sense if an object has been grasped. Using force sensing would be more tolerant of object size uncertainty versus looking for an expected finger closure position for grasp.

An alternative interface for Baxter's right gripper is given in the `generic_gripper_` `services` package in the source code `rethink_rt_gripper_service.cpp`. This node presents a service called `generic_gripper_svc`, which communicates via messages defined in the `srv` directory of this package. The contents of the `srv` message are:

```
1  #generic gripper service interface message
2  uint8 TEST_PING = 0
3  uint8 GRASP = 1
4  uint8 RELEASE = 2
5  uint8 TEST_GRASP = 3
6  uint8 GRASP_W_PARAMS=4 #useful for Baxter gripper: provide optional param values
7                         #to test for successful grasp completion of a known object
8
9  uint8 cmd_code
10 float64 test_upper_val #may need these as parameters to check status
11 float64 test_lower_val #e.g., fingers opened/closed or object is grasped
12 ---
13 #response:
14 bool success
```

The intent of this service and message is to provide a generic interface, independent of Baxter-specific messages or commands. Additional gripper interfaces can be defined similarly, but with appropriate interfaces to specific hardware. For example, a vacuum gripper could still respond to generic commands `GRASP`, `RELEASE` and `TEST_GRASP`, although implementation would involve enabling and disabling suction and testing for vacuum status, instead of opening and closing fingers. This gripper abstraction will be useful in Chapter 15 in introducing the object-grabber package.

14.4   HEAD PAN CONTROL

A simple example of Baxter joint control is control of the pan (swivel) of the head (display). The following listing, `baxter_head_pan_zero.cpp` in package `baxter_head_pan`, shows how one can send angle commands to control the head pan.

Listing 14.2: Head pan control program to set pan angle to 0.0

```
1  //utility to send head pan angle to zero
2  #include <ros/ros.h>
3  #include <baxter_core_msgs/HeadPanCommand.h>
4  using namespace std;
5
6  int main(int argc, char **argv) {
7      ros::init(argc, argv, "baxter_head_pan_zero");
8      ros::NodeHandle n;
9      //create a publisher to send commands to Baxter's head pan
10     ros::Publisher head_pan_pub = n.advertise<baxter_core_msgs::HeadPanCommand>("/↵
           robot/head/command_head_pan", 1);
11
12     baxter_core_msgs::HeadPanCommand headPanCommand; //corresponding message type for ↵
           head-pan control
13     headPanCommand.target = 0.0; //set desired angle
14     headPanCommand.enable_pan_request=1;
15     ros::Rate timer(4);
16     for (int i=0;i<4;i++) { //send this command multiple times before quitting
17         //if node sends message and then quits immediately, message can get lost
18         head_pan_pub.publish(headPanCommand);
19         timer.sleep();
20     }
21     return 0;
22 }
```

Publishing commands to the head pan servo requires use of a message type defined in code associated with Baxter control (from Rethink Robotics). Specifically, line 3 includes the header file for `baxter_core_msgs/HeadPanCommand` and line 12 instantiates an object of this type. The components of the `headPanCommand` object are set to zero angle command and motion-control enable. A publisher that uses this message type and publishes to the topic `/robot/head/command_head_pan` is defined (line 10).

This node simply commands zero angle to the head. This can be useful given that the Baxter simulator can launch with an inconvenient head angle. This node simply publishes the motion command then completes. Note, though, that if a message is sent by a node that returns (dies) quickly, the message can get lost before the intended recipient receives it. For this reason, the command is sent several times with sleeps between publications.

A variant on this node in the same package is `baxter_head_pan.cpp`. This node prompts the user for an amplitude and frequency then continuously sends oscillating head pan angle commands at the specified amplitude and frequency.

To run either of these nodes, first start the Baxter simulator:

```
roslaunch baxter_gazebo baxter_world.launch
```

Wait for the simulation to finish launching, then enable the motors with:

```
rosrun baxter_tools enable_robot.py -e
```

Then run one of the head-pan control nodes, *e.g.*:

```
rosrun baxter_head_pan baxter_head_pan
```

Reply to the amplitude and frequency prompts, then observe the motion of the head pan in Gazebo.

This is the simplest example of commanding one of Baxter's joints. Joint commands for the right and left limbs are more involved, as discussed next.

14.5 COMMANDING BAXTER JOINTS

The Baxter simulator (and Baxter robot) subscribe to the topics `/robot/limb/right/joint_command` and `/robot/limb/left/joint_command` to accept joint position commands for the right and left arms, respectively. These topics carry messages of type `baxter_core_msgs/JointCommand`. We can examine this message type with:

```
rosmsg show baxter_core_msgs/JointCommand
```

```
int32 POSITION_MODE=1
int32 VELOCITY_MODE=2
int32 TORQUE_MODE=3
int32 RAW_POSITION_MODE=4
int32 mode
float64[] command
string[] names
```

The code snippet below illustrates how to command joint angles to joints of Baxter's right arm in position-control mode:

```
//Here is an instantiation of the proper message type:
baxter_core_msgs::JointCommand right_cmd;
//Assuming we have desired joint angles in the array vector qvec[],
//we can fill the command message with:
right_cmd.mode = 1; // position-command mode

//Define a right-arm publisher as:
joint_cmd_pub_right =
nh.advertise<baxter_core_msgs::JointCommand>("/robot/limb/right/joint_command", 1);

// Define the ordering of joints to be commanded as follows:
// define the joint angles 0-6 to be right arm, from shoulder to wrist;
// we only need to do this part once, and we can subsequently re-use this message,
// changing only the position-command data
right_cmd.names.push_back("right_s0");
right_cmd.names.push_back("right_s1");
right_cmd.names.push_back("right_e0");
right_cmd.names.push_back("right_e1");
right_cmd.names.push_back("right_w0");
right_cmd.names.push_back("right_w1");
right_cmd.names.push_back("right_w2");
//Note: in the above, we have created room for 7 joint names. The push_back() command
// should not be repeated, else the list of names will grow with every iteration.
// The joint-command object can be retained, and this ordering of joint names can be
// persistent, so this step can be treated as an initialization. Within a control loop
// only the desired joint values would need to be changed within this command message.

//Assuming a pose of interest is in qvec[], we can specify the right-arm joint-angle
//commands (in radians) with:
for (int i = 0; i < 7; i++) {
right_cmd.command[i] = qvec[i];
}
//and publish this command as:
joint_cmd_pub_right.publish(right_cmd);
```

To smoothly interpolate and execute trajectories, it is useful to have an action server, as described in Section 11.5 for the minimal two-link arm and in Section 11.6 for the seven-DOF arm. Corresponding joint trajectory interpolation action servers for Baxter's right and left arms are in package `baxter_trajectory_streamer`, nodes `rt_arm_as` and `left_arm_as`, respectively. These action servers utilize functions within the class `Baxter_traj_streamer`, which is compiled as a library from the source `baxter_trajectory_streamer.cpp` in the `baxter_trajectory_streamer` package.

The right and left arm trajectory interpolation action servers respond to action clients that send goals via the action messages in `baxter_trajectory_streamer/trajAction.h`, which is generated from the `traj.action` specification within the `baxter_trajectory_streamer` package. The `goal` part of this action message contains a component called `trajectory` that is of type `trajectory_msgs/JointTrajectory`.

The right and left arm action servers interpret received goals (trajectories), interpolate between joint-space sub-goals contained in these goal trajectories, and send joint commands at a fixed update rate of 50 Hz. (This frequency is specified in the `baxter_trajectory_streamer.h` header file with the parameter `const double dt_traj = 0.02;`.)

The trajectory interpolation action servers can be started with:

```
rosrun baxter_trajectory_streamer rt_arm_as
```

and

```
rosrun baxter_trajectory_streamer left_arm_as
```

These servers respond with `ready to receive/execute trajectories`, then wait for incoming goal requests from action clients.

An example client of the joint trajectory interpolation servers is in the same `baxter_trajectory_streamer` package, called `traj_action_client_pre_pose.cpp`. This client defines hard-coded goal poses for the right and left arms,

```
Eigen::VectorXd q_pre_pose_right,q_pre_pose_left;
q_pre_pose_right.resize(7);
q_pre_pose_left.resize(7);
q_pre_pose_right << -0.907528, -0.111813, 2.06622, 1.8737, -1.295, 2.00164, 0;
//corresponding values to mirror the left arm pose:
q_pre_pose_left  <<  0.907528, 0.111813, -2.06622, 1.8737, 1.295, 2.00164, -2.87179;
```

A minimal path in joint-space (for a specified arm) is defined in terms of two joint-space points. To avoid sudden jumps, the starting pose should be chosen to be the same as the current pose. In the example client, the starting points for the left and right arms are established by reading the current joint angles, with the help of a **Baxter_traj_streamer** object:

```
q_vec_right_arm = baxter_traj_streamer.get_q_vec_right_arm_Xd();
q_vec_left_arm = baxter_traj_streamer.get_q_vec_left_arm_Xd();
```

The start and end points are installed for the right and left arms in respective **path** objects:

```
std::vector<Eigen::VectorXd> des_path_right, des_path_left;
des_path_right.push_back(q_vec_right_arm); //start from current pose
des_path_right.push_back(q_pre_pose_right);

des_path_left.push_back(q_vec_left_arm);
des_path_left.push_back(q_pre_pose_left);
```

The minimal paths are converted to trajectory messages with the help of the **Baxter_traj_streamer** object. The **stuff_trajectory** functions use the path points to populate trajectory messages and assign reasonable arrival times. This is performed in the code as:

```
trajectory_msgs::JointTrajectory des_trajectory_right,des_trajectory_left; // empty ↩
    trajectories

baxter_traj_streamer.stuff_trajectory_right_arm(des_path_right, des_trajectory_right);
baxter_traj_streamer.stuff_trajectory_left_arm(des_path_left, des_trajectory_left);
```

and the resulting trajectory messages are copied into respective goal messages that are interpretable by the trajectory interpolation action servers:

```
baxter_trajectory_streamer::trajGoal goal_right,goal_left;
goal_right.trajectory = des_trajectory_right; // copy traj to goal:
goal_left.trajectory = des_trajectory_left;
```

Action clients of `rightArmTrajActionServer` and `leftArmTrajActionServer` are instantiated:

```
actionlib::SimpleActionClient<baxter_trajectory_streamer::trajAction> ↩
    right_arm_action_client("rightArmTrajActionServer", true);
actionlib::SimpleActionClient<baxter_trajectory_streamer::trajAction> ↩
    left_arm_action_client("leftArmTrajActionServer", true);
```

and these action clients are used to send the goal messages:

```
right_arm_action_client.sendGoal(goal_right, &rightArmDoneCb);
left_arm_action_client.sendGoal(goal_left, &leftArmDoneCb);
```

Assuming the Baxter simulator is running along with the left and right-arm trajectory interpolation action servers, the pre-pose node can be run with:

```
rosrun baxter_trajectory_streamer pre_pose
```

which results in the pose shown in Fig 14.5. Poses such as this can be useful in preparing

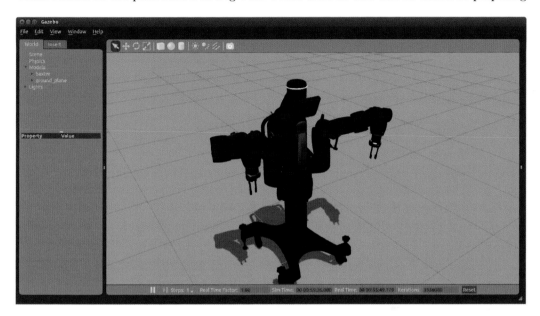

Figure 14.5: Result of running **pre_pose** action client of Baxter arm servers

for manipulation. If objects of interest are on a work surface within reach and within view of Baxter's sensors, this pre-pose can avoid blocking the view from a head camera while preparing the grippers for descent from above toward object grasp.

The example action client described here has limited usefulness, because it embeds hard-coded poses within the C++ code. More generally, desired poses should come from more flexible sources and should not require recompiling nodes. More general interfaces are considered next, starting with desired trajectories recorded in simple text files, and building up to a more general **object_grabber** action server.

14.6   USING ROS JOINT TRAJECTORY CONTROLLER

To this point, we have been using simplistic, custom-designed joint trajectory interpolation action servers. A more sophisticated package for joint interpolation is **joint_trajectory_controller**. (See http://wiki.ros.org/joint_trajectory_controller.) This package offers a variety of improvements, including cubic or quintic spline interpolation, and the ability to pre-empt and replace trajectories while the robot is running. Trajectory controllers must be configured at a minimum to specify the type of controller used (*e.g.* a position controller) and the topic names for commanding joints. For a specific robot, joint constraints and tolerances should be specified appropriately to override the default values.

For the Baxter robot, after starting and enabling the robot (in simulation or with a physical robot), the Baxter-specific trajectory controller should be started with:

```
rosrun baxter_interface joint_trajectory_action_server.py --mode position
```

This node offers action servers named `robot/limb/right/follow_joint_trajectory` and `robot/limb/left/follow_joint_trajectory`. These action servers use a standard ROS action message defined in the package `control_msgs`, including `control_msgs/FollowJointTrajectoryGoal`, as well as the corresponding `Feedback` and `Result` messages. The full message structure can be viewed with:

```
rosmsg show control_msgs/FollowJointTrajectoryAction
```

An action client of this service should include the corresponding ROS message header file as:

```
#include <control_msgs/FollowJointTrajectoryAction.h>
```

Two examples of action client nodes that use the joint trajectory action server are given in `baxter_jnt_traj_ctlr_client_home.cpp` and `baxter_jnt_traj_ctlr_client_pre_pose.cpp`. These nodes send Baxter's right arm to home (all zero joint angles) or to a hard-coded pre-pose position, respectively. With the joint trajectory action server running, a pre-pose example node can be run with:

```
rosrun baxter_jnt_traj_ctlr_client baxter_jnt_traj_ctlr_client_pre_pose
```

Baxter can be observed moving its right arm to the defined pose.

The other client node example, `baxter_jnt_traj_ctlr_client_home.cpp`, prompts the user for a move time. By experimentally trying different values, one can observe the behavior of the trajectory controller when the move time violates the velocity constraints, as well as observe the quality of motion of a legal, fast move.

As with the previous, simplistic joint interpolation and execution nodes, it is possible to send trajectory goals that are very sparse (*e.g.* only start and end goals). More generally, a candidate trajectory should be generated before execution, so it can be evaluated (*e.g.* for potential collisions) before it is allowed to be executed. Consequently, spline interpolation by the trajectory controller of a pre-computed, fully vetted and densely specified trajectory may not be useful, since the fully evaluated plan should not be altered by a lower-level (and less aware) controller.

14.7 JOINT-SPACE RECORD AND PLAYBACK NODES

One often convenient way of programming robots is teach and playback. If the robot's environment is structured such that workpieces are always presented at reproducible poses, useful programs can be run to perform blind manipulation.

Commonly, programming in this context is done with a teach pendant, with which the programmer drives the robot via key presses on a teach pendant device to send the robot to key poses. Each such pose is recorded, and sequences of these poses can be used to construct motion programs.

With the Baxter robot, a teach pendant is not necessary. One can simply grab either wrist (or both, simultaneously) to invoke a follower mode, then physically move the arm to points of interest. The Baxter code examples include a teach–playback interface. (See `http://sdk.rethinkrobotics.com/wiki/Joint_Trajectory_Playback_Example`.) However, these nodes are written in Python, and only the usage of these nodes is explained. Counterparts of these nodes have been written in C++ in the `learning_ros` repository, and these are explained here.

The C++ counterparts are in package `baxter_playfile_nodes`. The source code `get_and_save_jntvals.cpp`, with executable by the same name (`get_and_save_jntvals`) is useful for recording individual key points. With either the Baxter simulator or a physical Baxter robot running, this program can be started with:

```
rosrun baxter_playfile_nodes get_and_save_jntvals
```

The node will respond with: `enter 1 for a snapshot, 0 to finish:`. Then one or both of the arms can be moved to a new pose, and "1" can be entered to record the corresponding joint angles. The results are saved in files named `baxter_r_arm_angs.txt` and `baxter_l_arm_angs.txt`. As more points are recorded, they are added as lines to these files. When done recording key poses, responding to the prompt with "0" will close the files and terminate the program.

The files created by this process are simple ASCII text files. An example right-arm file content is:

```
-0.272241, 1.047, -0.010058, 0.49904, -0.0810748, 0.0224539, 0.0263518, 1
-0.9, -0.109987, 2.05995, 1.87001, -1.3, 2, 3.50788e-08, 2
```

This file simply contains the right-arm joint angles, listed in order of joints in Denavit–Hartenberg order (from the torso to the gripper). Each line of the file lists seven joint angles. A corresponding left-arm file contains the left-arm joint angles in the same sequence.

Working with the data recorded via the `get_and_save_jntvals` node requires some manual editing. The values could be hard-coded in action clients of the joint interpolation action servers as path points, then converted into trajectory messages (with the addition of arrival times) and sent as goals to the trajectory action servers.

Alternatively, one may record entire trajectories, including timing, with another node from the `baxter_playfile_nodes`: `baxter_record_trajectory.cpp` (with executable node name `baxter_recorder`). This program is run with:

```
rosrun baxter_playfile_nodes baxter_recorder
```

Upon start-up, the node prompts the user with: "enter 1 to start capturing, then move arms in desired trajectory; control-C when done recording." Once recording has started, joint angles of the right and left arms are stored in files `baxter_r_arm_traj.jsp` and `baxter_l_arm_traj.jsp`, respectively. (The `jsp` suffix is a mnemonic for joint space play-file.) Joint angles are sampled at 5 Hz, and each such sample is appended to the files. The files are simple ASCII text files. An example of the recording of a right-arm file (renamed `shy.jsp`) has the first few lines as follows:

```
-0.586748, 0.541879, 2.96327, 2.22964, 1.57195, -1.57386, 0.965257, 0.2
-0.585597, 0.552617, 2.95406, 2.23769, 1.57578, -1.57386, 0.965641, 0.4
-0.562204, 0.565272, 2.92377, 2.27221, 1.59764, -1.57348, 0.973694, 0.6
```

Each line has seven joint angles and an arrival time. The arrival times increment by 0.2 sec, since the sample rate of recording is 5 Hz.

The resulting recording can be used subsequently to replay the recorded motion. A package that illustrates this process is `baxter_playfile_nodes`. In simulation, one cannot physically grab Baxter's wrists to define lead-through trajectories. However, the recording process can be illustrated by recording any motion, *e.g.* as executed by an action client with hard-coded poses. Further, some example recorded joint-space trajectories from a physical Baxter robot are contained in the `baxter_playfile_nodes` package, including `shy.jsp`, `hug.jsp`, `wave.jsp`, `pre_pose_right.jsp`, `pre_pose_left.jsp`, `shake.jsp`, and `stick_em_up.jsp`.

To play any of the recorded trajectories, follow these steps:

To run Baxter in simulation, start a Gazebo world with:

```
roslaunch baxter_gazebo baxter_world.launch
```

(alternatively, start a physical Baxter robot instead of a simulated Baxter robot).

When the robot (real or simulated) is ready, enable the motors with:

```
rosrun baxter_tools enable_robot.py -e
```

Start the right and left arm trajectory-interpolation action servers with:

```
rosrun baxter_trajectory_streamer rt_arm_as
```

and (in a separate terminal):

```
rosrun baxter_trajectory_streamer left_arm_as
```

(Note: the `joint_trajectory_controller` could be used instead, with relatively minor edits to the `baxter_playback` node to utilize the action-message type and action server associated with the ROS `joint_trajectory_controller`.)

To run the `baxter_playback` node, open a terminal and navigate to the `baxter_playfile_nodes` package (since the example jsp files reside in this package):

```
roscd baxter_playfile_nodes
```

Then run the `baxter_playback` node with chosen file names for the right arm and (optionally) the left arm. For example, running:

```
rosrun baxter_playfile_nodes baxter_playback pre_pose_right.jsp pre_pose_left.jsp
```

moves the arms smoothly from the robot's initial pose to a defined pre-pose, resulting in the pose shown in Fig 14.6.

Optionally, one can specify a single playfile, in which case this is assumed to apply to the right arm. For example, running:

```
rosrun baxter_playfile_nodes baxter_playback shy.jsp
```

controls the right arm to move smoothly to a pose in which the right gripper is in front of Baxter's face, as shown in Fig 14.7 The source code `baxter_playfile_jointspace.cpp` is fairly long (nearly 380 lines) but is conceptually simple. A function, `read_traj_file(fname,trajectory)` opens and parses the file `fname`, which is assumed to be in `.jsp` format (seven joint angles and one arrival time per line). The files are checked for consistency (seven joint angles and one arrival time per line). With this information, the `jsp` file is interpreted and the data is used to populate a `trajectory_msgs::JointTrajectory` object. This is performed for both the right and (optionally) left arm files, populating trajectory objects `des_trajectory_right` and `des_trajectory_left` in the main program. It should

Figure 14.6: Result of running `baxter_playback` with `pre_pose_right.jsp` and `pre_pose_left.jsp` motion files

be recognized, however, that the first point in the recorded trajectory may be far away from the robot's current pose. This could result in wild transients if the recorded trajectory is played verbatim. To address this, an approach trajectory (for each arm) is computed to connect the current joint angles to the first point in the `jsp` file(s). A path object (comprised of a pair of, seven-DOF joint-space points) is converted to a trajectory with the help of member functions `stuff_trajectory_right_arm()` and `stuff_trajectory_left_arm` of the class `Baxter_traj_streamer` (an object of which is instantiated in the main program).

Action clients of the right and left arm joint-interpolation action servers are instantiated, and goal messages are populated with the approach trajectories and sent to the respective servers. When these motions are complete, the arms are at the starting points of the recorded jsp files, and the desired right and left arm trajectores are then commanded (by sending corresponding goal requests to the respective jointspace interpolation servers). The playback node waits for callbacks from the action servers, then terminates.

Another variation on `baxter_playback` is the node `baxter_multitraj_player` (with source code in `baxter_multitraj_player.cpp`) in the `baxter_playfile_nodes` package. This node is largely similar to `baxter_playback`, except that it can be invoked by other ROS nodes. Instead of taking a command line option of a file name, this node subscribes to the topic `playfile_codes` and listens for messages of type `std_msgs::UInt32`. When the corresponding callback function is awakened, incoming data is copied to the global variable `g_playfile_code`, and a flag `g_got_code_trigger` is set to `true`. In the main program, a timed loop is run continuously, and when the `g_got_code_trigger` is true, a switch–case statement selects a playfile based on the code in `g_playfile_code`. The corresponding jsp file is parsed and executed in the same manner as `baxter_playback`.

The node `baxter_multitraj_player` is also somewhat more convenient than `baxter_playback`, since it does not need to be started from the directory containing the jsp playfiles. To accomplish this, the `baxter_multitraj_player` builds a string containing the full path to the `baxter_playfile_nodes` package. This is accomplished by first asking the operating system for the environment variable `ROS_WORKSPACE` (the path to `ros_ws`), then concatenating the relative path to the package `baxter_playfile_nodes` using the following two lines of code:

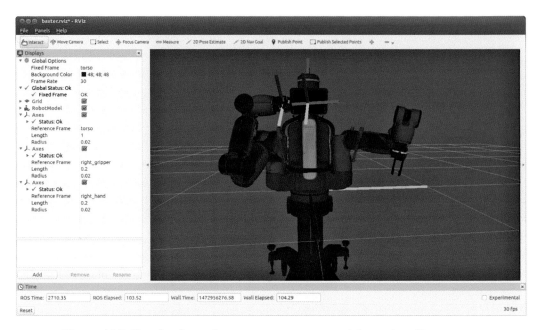

Figure 14.7: Result of running `baxter_playback` with motion file `shy.jsp`

```
std::string ros_ws_path = getenv("ROS_WORKSPACE"); //get the ros-workspace path
//append path to playfiles relative to ros_ws:
std::string path_to_playfiles= ros_ws_path+"/src/learning_ros/Part_5/baxter/↩
    baxter_playfile_nodes/";
```

Subsequently, each jsp file to be executed has its file name prepended with this path to `baxter_playfile_nodes`. This flexibility is particularly convenient in allowing `baxter_multitraj_player` to be started from a launch file.

To run `baxter_multitraj_player` (assuming physical or simulated Baxter is running, as well as the joint interpolation action servers) enter:

```
rosrun baxter_playfile_nodes baxter_multitraj_player
```

This file player is now ready to accept codes corresponding to recorded jsp files. These pre-recorded moves can be invoked by other programs by publishing codes to the topic `playfile_codes`. For testing, one can publish manually from a terminal. For example:

```
rostopic pub playfile_codes std_msgs/UInt32 0
```

results in sending playfile code "0", which invokes right and left arm motions corresponding to files `pre_pose_right.jsp` and `pre_pose_left.jsp`. In `baxter_multitraj_player`, codes 0 through 6 are defined for various jsp files. The code is easily extensible to include additional cases for additional jsp files.

The `baxter_multitraj_player` is a useful node that can run simultaneously with other motion control nodes. In controlling sequences of operations, it can be convenient to interleave pre-recorded motions with perceptually derived motions. By starting the `baxter_multitraj_player` node, a higher-level coordination node would be able to invoke any of the available playfile codes on demand, as needed.

To simplify start-up, `baxter_playfile_nodes.launch` is defined in package `baxter_launch_files`. With simulated or physical Baxter running (and ready to be enabled), run this launch file with:

```
roslaunch baxter_launch_files baxter_playfile_nodes.launch
```

This launch file will invoke `enable_robot.py` (which will enable the robot's motors), will start the right and left arm trajectory-interpolator action servers, start `rviz` (with a pre-defined configuration file), run `baxter_playback` with file-name arguments of `pre_pose_right.jsp` and `pre_pose_left.jsp`, and start the `baxter_multitraj_player` node.

The `baxter_playback` node with these file name arguments will invoke a start-up motion that will move the arms from the initial pose (arms hanging down) to the pose in Fig 14.6. The `baxter_playback` will run to conclusion, but the `baxter_multitraj_player` node will remain active to respond to playfile codes published on the `playfile_codes` topic.

A variation on `baxter_multitraj_player` is `baxter_playfile_service`. This node has logic essentially identical to `baxter_multitraj_player`. However, instead of using publish and subscribe, this node offers a service. A service message is defined in the `srv` sub-directory of `baxter_playfile_nodes`. This service message includes definitions of multiple mnemonics, such as `int32 PRE_POSE=0`, so clients can send mnemonic codes instead of numerical values to refer to `playfiles`.

The `baxter_playfile_service` node finds the path to `baxter_playfile_nodes`, instantiates a `Baxter_traj_streamer`, instantiates a pair of action clients corresponding to the `rightArmTrajActionServer` action server and the `leftArmTrajActionServer` action server, and instantiates `playfile_service`, which is associated with the `srv_callback()` callback function. Thereafter, the main program goes into a `spin()` while the service callback function does all the work. Within the callback function, requests are interpreted in terms of mnemonic codes that correspond to named `playfiles`.

An example client of the `playfile_service` is the node `example_baxter_playfile_client.cpp`, which appears in Listing 14.3.

Listing 14.3: `example_baxter_playfile_client.cpp`: example playfile client of Baxter playfile service

```
1   // example_baxter_playfile_client
2   #include<ros/ros.h>
3   #include<baxter_playfile_nodes/playfileSrv.h>
4
5   int main(int argc, char** argv) {
6       ros::init(argc, argv, "example_baxter_playfile_client"); // name this node
7       ros::NodeHandle nh;
8       //create a client of playfile_service
9       ros::ServiceClient client = nh.serviceClient<baxter_playfile_nodes::playfileSrv>("↵
            playfile_service");
10      baxter_playfile_nodes::playfileSrv playfile_srv_msg; //compatible service message
11      //set the request to PRE_POSE, per the mnemonic defined in the service message
12      playfile_srv_msg.request.playfile_code = baxter_playfile_nodes::playfileSrvRequest↵
            ::PRE_POSE;
13
14      ROS_INFO("sending pre-pose command to playfile service: ");
15      client.call(playfile_srv_msg);
16      //blocks here until service call completes...
17      ROS_INFO("service responded with code %d", playfile_srv_msg.response.return_code);
18      return 0;
19  }
```

This client includes the service message header defined for the `playfile` server (line 3). It instantiates a client of `playfile_service` with the corresponding service message type (line

9). It creates an instance of the `playfileSrv` message (line 10) and sets the `playfile_code` field of the request to `baxter_playfile_nodes::playfileSrvRequest::PRE_POSE`. Referring to codes in this fashion is long-winded, but is ultimately helpful in keeping definitions in the right places. In this instance, the definition of `PRE_POSE` is specific to the service message `playfileSrv` in the package `baxter_playfile_nodes`. If codes are later changed or augmented, placing them in the `playfileSrv` will propagate these changes to all nodes that use the `playfile` service. Note, also, that definitions can be defined in two places: in the `request` field, or in the `response` field. Definitions in the service message may be used by pre-pending the definition with the corresonding field name. For example, to refer to the constant `PRE_POSE` defined within the `request` field of the `playfile` service message, use: Correspondingly, these messages `playfileSrvRequest::PRE_POSE`.

The example client can be run by starting the Baxter simulator (or physical Baxter robot) and waiting for the simulator or robot to finish its start-up procedure. Then, launch

```
roslaunch baxter_launch_files baxter_playfile_service_nodes.launch
```

This file performs several tasks, including enabling the robot's motors, starting the left and right arm trajectory streamers, and starting the `playfile` service.

In another terminal, run the client with:

```
rosrun baxter_playfile_nodes baxter_playfile_client
```

There are advantages of using the `playfile` service over `baxter_multitraj_player`. With the service and client model, the client can be sure the service received its request. With publish and subscribe, if the node that subscribes to commands is not ready before the request is published, the message may be dropped. Second, a client of a service will block (suspend) until the request is concluded. Therefore clients of the service can know when a move is complete before requesting an additional move. Although client and service communications require more lines of code, the advantages of guaranteed peer-to-peer communications can be valuable in such instances.

While teach and playback programming has useful applications, greater autonomy is achieved with trajectories defined on the fly, ideally based on perception. To generate useful trajectories dynamically, Baxter requires inverse kinematics solutions and Cartesian-space planning. Baxter's kinematics is discussed next.

14.8 BAXTER KINEMATICS

Chapter 12 introduced robot arm kinematics using examples of a two-joint arm, a six-joint ABB IRB120 arm, and a seven-joint robot arm. As noted, forward kinematics is solved in general for open kinematic chains, and the same solution approach can be applied to Baxter's arms (including use of the KDL package). Computing inverse kinematics is more difficult, and IK algorithms are specialized for each robot type.

Baxter's arms are most similar to the `7DOFarm` example, since both of Baxter's arms are seven-DOF. However, the Baxter robot does not have spherical wrists. Consequently, the simplification invoked in both the ABB and seven-DOF arm cases cannot be invoked rigorously. Nonetheless, since Baxter's wrist offsets are relatively small, a useful first approximation is to assume a spherical wrist and invoke the same inverse-kinematic strategy as the seven-DOF arm example. This will yield IK solutions that are approximately correct. As necessary, numerical iterations with Jacobians can be invoked to compute more precise solutions. Given a good approximation of an IK solution as a seed value, the numerical iterations tend to converge stably and rapidly.

Implementation of forward and inverse kinematics functions for Baxter's right arm appears in package `baxter_fk_ik` as a library with source code `baxter_fk_ik.cpp`. Functions within this library compute the six-DOF pose of the robot's tool flange (deceptively called `right_hand`) with respect to the torso frame. Computing kinematics with respect to the tool flange is a useful approach, since different tools or grippers can be installed. Tool-flange kinematics can be re-used in these variations with only a tool transform to describe a fixed coordinate transform from the tool flange to a frame of interest on the gripper or tool. A fixed transform from the torso frame to the shoulder frame is also used, so as to conform with the Denavit–Hartenberg conventions for defining an initial frame (aligned with the first shoulder joint, in this case).

The `baxter_fk_ik.cpp` code is tedious, but some of the available functions, as declared in the header file of the `baxter_fk_ik` library, are worth highlighting. The source code defines classes `Baxter_fwd_solver` and `Baxter_IK_solver`.

The key function in the forward solver class is:

```
Eigen::Affine3d fwd_kin_flange_wrt_torso_solve(const Vectorq7x1& q_vec);
```

This function takes a vector of right-arm joint angles and returns an Eigen-style affine object describing the pose of the right gripper frame with respect to Baxter's torso frame. This function is fast, precise and well behaved.

A key function of the `Baxter_IK_solver` class is:

```
int ik_solve_approx_wrt_torso(Eigen::Affine3d const& desired_flange_pose,std::vector<↵
    Vectorq7x1> &q_solns);
```

This function invokes a spherical wrist approximation to compute a set of IK solutions. The solutions are indexed with respect to samples of joint $q0$. Candidate values of $q0$ are considered over the range of motion of $q0$ at sample intervals of `const double DQS0 = 0.05;`, as declared in the header file `baxter_fk_ik.h`. (This value may be increased or decreased, as desired by editing the header file's value and recompiling.) At each value of $q0$, there are up to eight IK solutions. The IK function returns the total number of viable solutions found, accounting for joint limits (but not considering possible collisions with the environment or with the robot's own body).

When it is adequate to have approximate Cartesian-space motion, the approximate IK function will be suitable. However, when approaching pick-up of an object, a more precise solution may be required. A function for this purpose is:

```
bool Baxter_IK_solver::improve_7dof_soln_wrt_torso(Eigen::Affine3d const& ↵
    desired_flange_pose_wrt_torso, Vectorq7x1 q_in, Vectorq7x1 &q_7dof_precise) {
```

which returns **true** a valid, improved solution is computed.

In performing solution improvement, it is assumed that the approximate solution provided has a desirable angle for $q0$, and this angle will be held fixed. The shoulder elevation, humerus twist and elbow angle will be perturbed to place the wrist point closer to the desired wrist point. Based on the approximate solution, a wrist-point error is computed with precise forward kinematics. The helper function `precise_soln_q123()` computes a 3×3 Jacobian, relating joints $q1$, $q2$ and $q3$ to wristpoint coordinates x, y, z. This 3×3 Jacobian is inverted and used in numerical iterations in a gradient search to find improved values for $(q1, q2, q3)$ to minimize the wrist-point position errors. The resulting solution will be a perturbation on the approximate solution, and the resulting Cartesian-space pose error of the tool flange will be small if the iterations succeed. An example using the approximate and precise IK algorithms is the source code `baxter_cartesian_planner.cpp` in package `cartesian_planner`.

14.9 BAXTER CARTESIAN MOVES

A Cartesian motion planner and corresponding action server for Baxter, in package `cartesian_planner`, were introduced in Section 13.3. Baxter's Cartesian motion planner is very similar to the `arm7dof_planner` introduced in Chapter 13, as both robots have seven joints (per arm), and thus in both cases one must address how to manage kinematic redundancy. In addition, the Baxter robot does not have an exact analytic inverse kinematics solution, so some additional functions must be invoked to address Baxter's non-spherical wrist.

The Baxter Cartesian planner source code is `baxter_cartesian_planner.cpp`, which is a library of planning functions. This library is illustrative, but it has numerous limitations. Joint range-of-motion limitations are taken into consideration. However, planned paths do not check for self collisions nor collisions with the environment. At a minimum, resulting candidate paths should be checked for collisions, and a more sophisticated planner would have to be invoked if a collision test fails. Nonetheless, this library is helpful in understanding the various steps involved in arm motion planning.

Two of the available planning functions are actually joint-space motion planners:

```
bool jspace_trivial_path_planner(Vectorq7x1 q_start, Vectorq7x1 q_end, std::vector<↵
    Eigen::VectorXd> &optimal_path);
bool jspace_path_planner_to_affine_goal(Vectorq7x1 q_start, Eigen::Affine3d ↵
    a_flange_end, std::vector<Eigen::VectorXd> &optimal_path);
```

The first function is merely for convenience; it does not perform any Cartesian-space path planning. This function accepts a start joint-space pose and and end joint-space pose, and it repackages these as an optimal path, consisting of only these two joint-space points. The format of the optimal path vector is suitable for conversion to a full trajectory message.

The second function also computes a joint-space trajectory. However, it considers alternative goal solutions that are IK solutions of a provided Cartesian goal pose. Each of the candidate goal solutions is evaluated in terms of preference for an efficient joint-space move.

Another four planning functions enforce the robot's end-effector move along a straight-line Cartesian path. These functions (from the header file `baxter_cartesian_planner.h`) include:

```
//these planners assume Affine args are right-arm flange w/rt torso
///specify start and end poses w/rt torso. Only orientation of end pose will be ↵
    considered; orientation of start pose is ignored
bool cartesian_path_planner(Eigen::Affine3d a_flange_start,Eigen::Affine3d ↵
    a_flange_end, std::vector<Eigen::VectorXd> &optimal_path);
/// alt version: specify start as a q_vec, and goal as a Cartesian pose (w/rt torso)
bool cartesian_path_planner(Vectorq7x1 q_start,Eigen::Affine3d a_flange_end, std::↵
    vector<Eigen::VectorXd> &optimal_path, double    dp_scalar = ↵
    CARTESIAN_PATH_SAMPLE_SPACING);
///this version uses a small Cartesian step size and refined IK
bool fine_cartesian_path_planner(Vectorq7x1 q_start,Eigen::Affine3d a_flange_end, std↵
    ::vector<Eigen::VectorXd> &optimal_path);
///alt version: compute path from current pose with cartesian move of delta_p with R ↵
    fixed
/// return "true" if successful
bool cartesian_path_planner_delta_p(Vectorq7x1 q_start, Eigen::Vector3d delta_p, std::↵
    vector<Eigen::VectorXd> &optimal_path);
```

It should be noted that all Cartesian poses expressed as `Eigen::Affine3d` objects are interpreted to mean the right arm tool flange frame with respect to the torso frame. Typically, one would care more about gripper frames with respect to some world or sensor frame. However, this code is more re-usable by referring to torso and tool-flange frames. For any given task, a gripper frame can be transformed to a corresponding tool-flange frame, and the `frame_id` of the desired gripper frame can be transformed into the torso frame. Thus, the present restriction on frame references is sufficiently general.

Of the above functions in `baxter_cartesian_planner`, the first one is the foundation. This function accepts arguments of two poses (expressed as Eigen-type affine objects) and it fills in a recommended sequence of joint-space poses to move the tool flange from the specified initial pose to the desired final pose. The interpolation invoked to define this move is expressed in the following lines (extracted from `baxter_cartesian_planner.cpp`, function `cartesian_path_planner(...)` with start and end affines as arguments):

```
Eigen::Affine3d a_flange_des;
Eigen::Vector3d dp_vec, del_p, p_start, p_end;
Eigen::Matrix3d R_des = a_flange_end.linear();
a_flange_des.linear() = R_des;
p_start = a_flange_start.translation();
p_end = a_flange_end.translation();
del_p = p_end - p_start;
dp_vec = del_p / nsteps;
for (int istep = 0; istep < nsteps; istep++) {
    a_flange_des.translation() = p_des;
    cartesian_affine_samples_.push_back(a_flange_des);
    p_des += dp_vec;
}
```

In this chosen interpolation, the robot is commanded to move its tool-flange origin in a straight line from specified start pose to desired end pose, where intermediate Cartesian positions along this line are sampled at resolution `CARTESIAN_PATH_SAMPLE_SPACING`, which is specified in this library's header file. This straight-line motion is intuitive. However, interpolation of orientation is less clear. If the start and end poses have the same tool-flange orientation, this orientation will be preserved throughout the move. This would be convenient, *e.g.* for carrying a tray of objects without spilling them (by keeping the tool-flange z axis vertical). If it is necessary to interpolate between different start and end orientations, this function simply assumes that the tool-flange orientation should change from start orientation to goal orientation as quickly as is feasible, then maintain the goal orientation for the remainder of the move.

Sampling in Cartesian space yields a sequence of desired affine poses. For each of these poses, there can be zero feasible IK solutions, or there can be infinitely many (since there is a one-DOF null-space due to redundant kinematics). For each Cartesian pose in the desired Cartesian path, a finite number of IK solutions is computed by invoking an approximate analytic IK function for joints 1 through 6, while joint 0 is indexed through samples of $q0$ (using the function `ik_solve_approx_wrt_torso(a_flange_des, q_solns)` within the `Baxter_IK_solver` class defined in the `baxter_fk_ik` library). If there are zero IK solutions for any desired pose along the Cartesian path, the requested Cartesian move is not feasible, and the function returns `false`. Note, though, that this may be an unnecessarily harsh test. In requesting a Cartesian move through free space, typically it is not necessary that the robot follow a precise Cartesian trajectory. In fact, a joint-space plan (using the function `jspace_path_planner_to_affine_goal()` may be acceptable, provided the end-effector arrives at the desired Cartesian pose. On the other hand, some Cartesian moves do more than merely pre-position the end effector, but are instead critically path dependent. Examples include laser cutting, sealant dispensing, seam welding and precision approach to a grasp or assembly pose.

A second `cartesian_path_planner(...)` function takes arguments of an intial joint-space pose and a destination Cartesian-space pose. As with the first flavor of Cartesian planner, the third argument is a reference to a `std::vector<Eigen::VectorXd>` object, which will be populated by this function with a computed joint-space trajectory. A fourth argument specifies the resolution of sampling along the Cartesian path (which defaults to `CARTESIAN_PATH_SAMPLE_SPACING` if this argument is not provided). This alternative planner first computes the affine Cartesian pose corresponding to the specified initial joint-space pose, using:

```
a_flange_start = baxter_fwd_solver_.fwd_kin_flange_wrt_torso_solve(q_start);
```

It then invokes the same logic as the first planner. Importantly, this second function specifies an unambiguous joint-space start pose, not a vector of start-pose options. This is important because robot trajectories need to start from the actual robot joint angles. To plan on the fly, the current robot joint angles should be obtained (from the `joint_states` topic), and these angles should be used as the starting point for planning Cartesian moves.

The third Cartesian planning function, `fine_cartesian_path_planner()`, also takes arguments of a starting joint-space pose and a destination Cartesian-space pose. However, this function refines the trajectory to obtain a more precise Cartesian-space motion than the solutions obtained by approximate IK solutions. To do so, this function first calls the Cartesian-space planner with starting joint angles and desired Cartesian goal, but it specifies a higher-resolution Cartesian-space sampling (using `CARTESIAN_PATH_FINE_SAMPLE_SPACING`). This results in a joint-space plan in which each joint-space solution is an approximate solution of the desired Cartesian path. This plan is refined using the function `refine_cartesian_path_plan(optimal_path)`, such that each joint-space solution is iterated numerically (with a 3×3 Jacobian inverse using only joints 1, 2 and 3) to obtain IK solutions that more precisely match each of the desired Cartesian subgoals.

An illustrative test `main()` program using the Cartesian-planning library is `example_baxter_cart_path_planner_main.cpp`. The following lines are extracted from this program:

Listing 14.4: Example use of Baxter Cartesian planner; lines extracted from `example_baxter_cart_path_planner_main.cpp`

```
1    for (x_des = 0.2; x_des < 1.5; x_des += 0.1) {
2        for (y_des = -1.5; y_des <= 1.5; y_des += 0.1) {
3            for (flange_theta = 0.0; flange_theta < 6.28; flange_theta += 0.2) {
4                flange_b_des << cos(flange_theta), sin(flange_theta), 0;
5                flange_t_des = flange_b_des.cross(flange_n_des);
6                R_flange_horiz.col(0) = flange_n_des;
7                R_flange_horiz.col(1) = flange_t_des;
8                R_flange_horiz.col(2) = flange_b_des;
9                a_toolflange_start.linear() = R_flange_horiz;
10               a_toolflange_end.linear() = R_flange_horiz;
11
12               flange_origin << x_des, y_des, z_des; //specify flange pose at grasp ↵
                     position
13               a_toolflange_start.translation() = flange_origin;
14               a_toolflange_end.translation() = flange_origin - L_depart*flange_b_des↵
                     ;
15
16               found_path = cartTrajPlanner.cartesian_path_planner(a_toolflange_start↵
                     , a_toolflange_end, optimal_path);
17
18               if (found_path) {
19                   ROS_INFO("found path; x= %f, y= %f, flange_theta = %f", x_des, ↵
                         y_des, flange_theta);
20                   outfile << x_des << ", " << y_des << ", " << flange_theta << endl;
21
```

```
22              } else {
23                  ROS_WARN("no path found; x= %f, y= %f, flange_theta = %f", x_des, ↵
                        y_des, flange_theta);
24              }
25          }
26      }
27  }
```

This program performs offline planning to consider the space of viable approach poses of a certain type. The assumed approach is a horizontal slide of the gripper (z axis of tool flange is horizontal) with the tool-flange x axis pointing up. This pose, with gripper fingers open, would be suitable for grabbing an upright cylinder in a power grasp by sliding horizontally to enclose the cylinder with the gripper fingers. The approach distance is proposed to be a slide along the tool-flange z axis of 0.25 m. This program considers candidate tool-flange origins in the range x (with respect to the torso, *i.e.* in front of Baxter) from 0.2 m to 1.4 m, y ranging from -1.5 m to $+1.5$ m, and $z = 0$ (the height of the torso, and slightly higher than the height of typical tables). The approach angle is tested at increments of 0.2 rad from 0 to 2π. This function tests whether each candidate grasp pose can be approached with a Cartesian move over 0.25 m. Roughly 13,000 approach paths are analyzed, which requires a few minutes of computation time. The results are stored in a file named **approachable_poses.dat**, which lists (line by line in ASCII text) the (x, y, θ) samples for which a 0.25 m horizontal Cartesian approach is viable. These results can be used to plan how to navigate to robot base poses to assure grasp plans can be executed.

In addition to offline Cartesian planning analyses, the Cartesian planning library can be used to plan motions online. For this purpose, the planner library specific to Baxter's right arm is incorporated within an action server, **baxter_rt_arm_cart_move_as.cpp** within the **cartesian_planner** library. This action server accepts goals via the general-purpose action message **cart_move.action**, defined in the **cartesian_planner** package. This action message refers to Cartesian-space move types, but it does not assume any specific robot design. Thus, action clients can be generic across robots, although the complementary action server must be specialized for each type of robot.

To support making robot agnostic action client nodes, motions may be requested with respect to general frames **generic_gripper_frame** and **system_ref_frame**. Note, though, that the IK functions for Baxter's right arm depend on referencing the right arm tool flange relative to the torso. To achieve the desired generality, incoming goals can specify motions of the generic gripper frame relative to the system reference frame, and these commands are transformed into robot-specific frames by the Cartesian motion action server. To do so, the constructor of the helper class **ArmMotionInterface** within the **baxter_rt_arm_cart_move_as** node finds the transform between the right tool-flange frame and the generic gripper frame (coincident with the fingertip frame of the right gripper). This is done using the following line of code:

```
tfListener_->lookupTransform("generic_gripper_frame","right_hand", ros::Time(0), ↵
    generic_toolflange_frame_wrt_gripper_frame_stf_);
```

Using the resulting transform, all references to the generic gripper frame are transformed into equivalent right arm tool-flange frame goals. Similarly, the same constructor uses the **tfListener** to find the transform between the torso frame and the system reference frame:

```
tfListener_->lookupTransform("system_ref_frame", "torso", ros::Time(0), ↵
    torso_wrt_system_ref_frame_stf_);
```

With knowledge of this transform, all incoming goal requests can be transformed appropriately into tool-flange motions with respect to the torso, which is required by the IK library.

With these transforms available, the right-arm Cartesian action server can act on motion requests defined in the `cart_move.action` file, including `PLAN_PATH_CURRENT_TO_WAITING_POSE`, `PLAN_JSPACE_PATH_CURRENT_TO_CART_GRIPPER_POSE`, `PLAN_PATH_CURRENT_TO_GOAL_DP_XYZ`, and `PLAN_PATH_CURRENT_TO_GOAL_GRIPPER_POSE|`. These goals refer to how a generic gripper frame is to be moved, and this is independent of the specific robot used.

The Cartesian motion action server responds with a result message that includes status codes. These status codes include `PATH_IS_VALID`, `PATH_NOT_VALID`, `COMMAND_CODE_NOT_RECOGNIZED`, and `SUCCESS`. Clients of the action server will receive a status message with one of the defined result messages in the field `return_code` of the returned `result` message.

The action server `baxter_cart_move_as.cpp` defines a class `ArmMotionInterface`. An object of this class is instantiated in the `main()` program, then the main program merely goes into a (timed) spin while callbacks of the `ArmMotionInterface` object do all of the work. The `executeCB()` function of `ArmMotionInterface` receives goal messages from action clients and processes them. This callback function first inspects the `command_code` element, then switches to a case to handle this command code. Depending on the command code, additional components of the goal message may need to be examined (*e.g.* a gripper goal pose must also be specified in the goal message if a Cartesian plan is requested).

If a plan is requested, the callback function will return a code indicating whether a viable motion plan was found. A successfully computed plan is retained in memory of the action server. An action client may then request execution of the plan by sending a command code of `EXECUTE_PLANNED_PATH`.

The example action client `example_generic_cart_move_ac.cpp` was introduced in Section 13.3. This action client illustrates use of the Cartesian-move action server, with the help of a class `ArmMotionCommander` defined in a library `cart_motion_commander` (with source code of the same name). The class `ArmMotionCommander` encapsulates some of the action server–action client interactions, which helps to make the Cartesian-move action client simpler.

Both planning and execution are invoked using member functions of `arm_motion_commander`. The example client illustrates use of eight defined action codes of the Cartesian-motion action server, including planning requests, pose requests, and execution requests. The example generic Cartesian-move action client can be run on the Baxter robot as presented in Section 13.3.

14.10   WRAP-UP

The Baxter simulator described in this chapter is a realistic and relevant robot model. Using this simulator, layers of motion control have been demonstrated, from low-level joint-space trajectory interpolators through Cartesian-space planners and action servers.

Considerable offline robot programming can be developed using this simulator. Subsequently running the developed code on a physical Baxter robot requires only specifying a remote ROS master. The human–machine interface used in development can be re-used as-is. One can even develop perceptual processing in simulation and subsequently run this software on the physical robot. With ROS's simulation tools, code thus developed will be largely ready to deploy to physical robots. Inevitably, there will be (at least) required tuning and calibration to take into account differences between the simulation model and the physical system. Further, lighting conditions, environment contact conditions, inadequately modeled sensor features, and insufficiently modeled virtual environments will lead to unexpected differences between simulation and physical reality. Nonetheless, offline development in ROS and Gazebo offers attractive productivity improvements in code development.

In the remainder of this text, the Baxter simulator will be used for illustrative examples.

An Object-Grabber Package

CONTENTS

I NTRODUCTION
Control of robot arms for manipulation incorporates considerable detail. It is desirable to construct solutions that decompose the challenge into separate layers that can encapsulate detail and allow progressively more abstract interaction. ROS capabilities support logical decomposition through mechanisms of the parameter server, publish and subscribe communications between independent nodes, client–service interactions and action servers. Exploiting these capabilities for constructing a general-purpose manipulation system is described in this chapter.

15.1   OBJECT-GRABBER CODE ORGANIZATION

Figure 15.1 shows a graphical interpretation of the code organization for the object-grabber development described here. At the top level, an object-grabber client interacts with the object-grabber action server. These nodes, which reside in the package `object_grabber`, communicate via an action message `object_grabber.action`, defined in the same package. At this level of abstraction, clients express goals in terms of a high-level action code, such as `GRAB_OBJECT`, together with an object identifier code and (stamped) pose of the object of interest. A grasp strategy can also be selected, if there are multiple known viable options to perform the grasp, or a default strategy can be assumed.

From the viewpoint of the client, it is not necessary to know the type of robot to be used, how this robot is mounted (or its current pose in the world), nor what type of gripper is on the robot. Rather, the client can focus on relocating parts from initial poses to goal poses.

The node `object_grabber_action_server` presents an action server named `object_ grabber_action_service`, which accepts goals via `object_grabber.action` messages. This action server must be aware of some more of the specific detail involved in carrying out a manipulation goal request. Specifically, this action server must understand what interactions are required between the specific end-of-arm tooling (*e.g.* gripper) that is present on the robot and the specified object to be manipulated. For a given gripper and object, there must be one or more strategies for successful manipulation if the goal is to be achieved. How to identify an appropriate grasp strategy (given a specific object and a specific gripper)

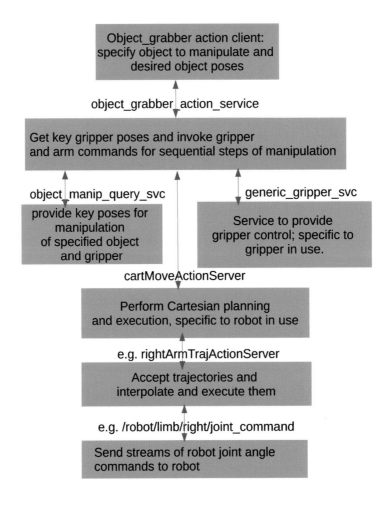

Figure 15.1: Code hierarchy for object-grabber system

---

is a difficult problem in general and an ongoing area of research. However, in the code organization, this problem can be encapsulated by creating an object manipulation query service. The intent is: given an object ID and a gripper ID, reply with key gripper poses for object grasp or object placement. This presumed oracle for answering queries regarding object manipulation may be replaced with future, more sophisticated instantiations. A primitive example is provided here. Replacing this example with a more capable query service can be performed without disruption to the rest of the system, provided the same service message interface is used.

In performing object acquisition and placement, the `object_grabber_action_server` must also have control over gripper actuation. A wide array of grippers is possible, *e.g.* a fingered hand, a parallel-jaw actuator, a vacuum gripper, or simple non-actuated tools such as a hook or spatula. For any given gripper type, the object-manipulation query service will have key poses to recommend, including a gripper-dependent pose suitable for initiating grasp. Generically, the object-grabber action server should interact with grippers by commanding whatever gripper action is suitable for grasp or release of an object of interest. A GRASP command may involve closing fingers (*e.g.* for the Baxter grippers), or enabling suction for a vacuum gripper, or doing nothing (for a hook or spatula). In each case, the gripper should perform the appropriate action. For this purpose, the object-grabber action server depends on a `generic_gripper_svc` service. Service requests can be sent as defined action codes, such as GRASP or RELEASE. When launching nodes for a particular system, the appropriate gripper service should be started, where this service incorporates the specifics of how to drive the gripper used.

The object-grabber action server also interacts with (sends goals to) a Cartesian-motion action server. As described in Section 13.3, a Cartesian-motion action client can be robot agnostic. If the intent is to move a specific gripper in space, the action client can request motion planning and motion execution to achieved the desired gripper motion. The Cartesian-motion action server, however, needs to be aware of the arm being used, since this node will need to invoke inverse kinematics calls that are robot-specific. Although the Cartesian-move action server must be robot-aware, it can nonetheless present a generic interface using a common server name (`cartMoveActionServer`) and a common action message (`object_grabber.action`).

To invoke planned trajectories, the Cartesian-planner action server bundles planned trajectories into messages of type `trajectory_msgs/JointTrajectory` and sends these to a lower-level interaction server to be executed. For the right arm of the Baxter robot, this can be either the provided `ros_controller` action service, `robot/limb/right/follow_joint_trajectory` or the simple, custom joint trajectory action server described herein, `rt_arm_as` (in the `baxter_trajectory_streamer` package). These action servers have direct control over the robot's joint commands, and thus they are also specific to a robot design and its ROS interface.

The action servers `cartMoveActionServer` and below were introduced previously. Descriptions of the `object_manip_query_svc`, `generic_gripper_svc`, `object_grabber_action_service` and an example object-grabber action client are offered in this chapter.

15.2   OBJECT MANIPULATION QUERY SERVICE

The code `object_manipulation_query_svc.cpp` in package `object_manipulation_properties` offers a service, `object_manip_query_svc`, that communicates via service message `objectManipulationQuery.srv`, defined in the same package. Use of this service is illustrated via an example client in the same package, `example_object_manip_query_client.cpp`. The intent of this service is to provide recommendations and options for how a specific gripper can acquire or place a specific object.

In this service, manipulation actions are abstracted in terms of key gripper poses associated with steps of acquiring and placing objects. Providing such advice can be quite complex, since there are myriad grasp options and strategies. For some simple robots, such as four-DOF SCARA[1] designs, manipulation is limited to approaching objects from above. If the gripper is a vacuum gripper, grasp options are constrained by supporting features on the object, *e.g.* horizontal, planar surfaces that are sufficiently wide, smooth and flat to provide a seal with a suction cup. Further, such features must be located close enough to the object's center of mass that the part will not fall due to excessive moments about the

---

[1]Selective Comp;iance Assembly Robot Arm

vacuum gripper. As a result, the options for successful grasp in such cases may be quite limited, corresponding to an (x, y, z) location on the part's surface relative to the part's reference frame. To recommend this grasp strategy, the object-manipulation query service would respond to an inquiry (which specifies gripper ID and object ID) by specifying a (stamped) pose of the object's frame relative to the gripper frame. (As noted earlier, the gripper frame may be referred to as `generic_gripper_frame`, provided a static transform publisher makes this frame known to `tf`).

In addition to specifying a relative grasp pose (object frame with respect to gripper frame), approach and depart strategies must also be specified. For the simple case of a vacuum gripper grasping from above, a typical strategy would be to move the gripper to a pose directly above the intended grasp pose; descend vertically to the desired grasp pose; enable the vacuum gripper to grasp the object; and depart from the work surface with a pure z translation upward. The strategy for grasping a part can thus be specified succinctly in terms of three key poses: the gripper pose (relative to the object frame) for approach, grasp, and depart. If Cartesian motions are executed from approach to grasp to depart, acquisition should be successful. The part would not be disturbed before grasp, and it would not be bumped out of grasp when the part is lifted. Similarly, for placement, three key gripper poses can be expressed, including the gripper pose required to achieve the desired part placement pose, and the corresponding approach and depart poses.

If multiple viable grasp sites are available, the object-manipulation query service should offer these as alternatives. A default (preferred) grasp option should be presented as the first option. But if this option is not possible (*e.g.* is out of reach of the robot or is occluded), alternative grasp options may be considered for successful manipulation.

For a six-DOF arm with a parallel-jaw gripper, more manipulation options are possible. For example, a cylinder sitting upright on a horizontal surface (*e.g.* a can or water bottle) could be grasped from above or from the side. If the intent is to pour from a bottle, grasp from the side would be preferred. If the object is a peg to be inserted in a hole, grasp from above may be necessary. Further, symmetry of the object in this case would leave a degree of freedom available for any one grasp strategy. For example, an upright cylinder could be approached sideways, with the gripper's z axis pointing radially toward the object's centerline. However, such radial approaches could be used at an arbitrary polar angle, providing a degree of freedom of approach and grasp options. Multiple options may need to be explored to find an approach that is kinematically reachable.

Even when the grasp pose is chosen, approach and depart poses have options. For example, an upright cylinder in a power grasp (with the gripper z axis horizontal) could be achieved by approaching the object along the gripper's z axis (*i.e.* by sliding the gripper parallel to the table surface). Alternatively, the same grasp pose could be achieved by orienting the gripper in the grasp orientation, but starting from a position above the cylinder and descending in the $-z$ direction. In either case, the desired strategy can be described in terms of 3 key poses: pose of the gripper frame relative to the object frame at the grasp pose, pose of the gripper frame relative to the object at the approach pose, and pose of the gripper frame relative to the (original pose of) the object for departure. Having established a grasp strategy, it is only necessary to specify these 3 key poses. (Note: in this implementation, key poses are expressed as pose of the object frame relative to the gripper frame).

Use of the object-manipulation query service requires reference to gripper ID and object ID. In the present implementation, these are specified in the `object_manipulation_properties` package in the `include` sub-directory in files `gripper_ID_codes.h` and `object_ID_codes.h`. The grippers and objects included are extremely sparse, at present, but these are placeholders for a future, more general knowledge database.

Use of the query service for grasping an object is illustrated in `object_grabber.cpp` in the `object_grabber` package. This code includes the function `get_default_grab_poses()`, which is displayed in part in Listing 15.2.

```
bool ObjectGrabber::get_default_grab_poses(int object_id,geometry_msgs::PoseStamped ←
    object_pose_stamped) {
    //fill in 3 necessary poses: approach, grasp, depart_w_object
    //find out what the default grasp strategy is for this gripper/object combination:
    manip_properties_srv_.request.gripper_ID = gripper_id_; //this is known from ←
        parameter server
    manip_properties_srv_.request.object_ID = object_id;
    manip_properties_srv_.request.query_code =
            object_manipulation_properties::objectManipulationQueryRequest::←
                GRASP_STRATEGY_OPTIONS_QUERY;
    manip_properties_client_.call(manip_properties_srv_);
    int n_grasp_strategy_options = manip_properties_srv_.response.←
        grasp_strategy_options.size();
    ROS_INFO("there are %d grasp options for this gripper/object combo; choosing 1st ←
        option (default)",n_grasp_strategy_options);
    if (n_grasp_strategy_options<1) return false;
    int grasp_option = manip_properties_srv_.response.grasp_strategy_options[0];
    ROS_INFO("chosen grasp strategy is code %d",grasp_option);
    //use this grasp strategy for finding corresponding grasp pose


    manip_properties_srv_.request.grasp_option = grasp_option; //default option for ←
        grasp strategy
    manip_properties_srv_.request.query_code = object_manipulation_properties::←
        objectManipulationQueryRequest::GET_GRASP_POSE_TRANSFORMS;
    manip_properties_client_.call(manip_properties_srv_);
    int n_grasp_pose_options = manip_properties_srv_.response.gripper_pose_options.←
        size();
    if (n_grasp_pose_options<1) {
                ROS_WARN("no pose options returned for gripper_ID %d and object_ID←
                    %d",gripper_id_,object_id);;
                return false;
            }

    grasp_object_pose_wrt_gripper_ = manip_properties_srv_.response.←
        gripper_pose_options[0];
```

This function populates a query message with a gripper ID code, an object ID code and the action code `GRASP_STRATEGY_OPTIONS_QUERY`. The client calls the query service with this request message and receives a reply. The function examines how many grasp strategies have been returned, and if there are no known grasp solutions, the function returns with failure. Otherwise, the function chooses the first available strategy, which (by design) is the default preferred grasp strategy.

The function `get_default_grab_poses()` then specifies use of this grasp option in the service request and specifies the action code `GET_GRASP_POSE_TRANSFORMS`. In the response, the field `geometry_msgs/Pose[] gripper_pose_options` will be populated with candidate grasp poses for the chosen grasp strategy. The first option in this function is selected by default. More generally, alternatives may be examined if the default pose is unreachable.

The `get_default_grab_poses()` repeats this process twice more to obtain approach poses and depart poses for the default grasp strategy.

The object-manipulation query service returns poses corresponding to the object frame with respect to the (generic) gripper frame. These poses must be converted into desired gripper poses with respect to some reference frame (*e.g.* sensor frame, torso, world frame or generic `system_ref_frame`). Lines of code from `get_default_grab_poses()` that perform this transformation for the grasp transform are:

```
    tf::StampedTransform object_stf =
            xformUtils.convert_poseStamped_to_stampedTransform(object_pose_stamped, "←
                object_frame");
    geometry_msgs::PoseStamped object_wrt_gripper_ps;
```

```
    object_wrt_gripper_ps.pose = grasp_object_pose_wrt_gripper_;
    object_wrt_gripper_ps.header.frame_id = "generic_gripper_frame";
    tf::StampedTransform object_wrt_gripper_stf =
            xformUtils.convert_poseStamped_to_stampedTransform(object_wrt_gripper_ps, ←
                "object_frame");
    //ROS_INFO("object w/rt gripper stf: ");
    //xformUtils.printStampedTf(object_wrt_gripper_stf);
    tf::StampedTransform gripper_wrt_object_stf = xformUtils.stamped_transform_inverse←
        (object_wrt_gripper_stf); //object_wrt_gripper_stf.inverse();
    //ROS_INFO("gripper w/rt object stf: ");
    //xformUtils.printStampedTf(gripper_wrt_object_stf);
    //now compute gripper pose w/rt whatever frame object was expressed in:
    tf::StampedTransform gripper_stf;
    if (!xformUtils.multiply_stamped_tfs(object_stf,gripper_wrt_object_stf,gripper_stf←
        )) {
        ROS_WARN("illegal stamped-transform multiply");
        return false;
    }
    //extract stamped pose from stf; this is the desired generic_gripper_frame w/rt a ←
        named frame_id
    // that corresponds to desired grasp transform for given object at a given pose w/←
        rt frame_id
    grasp_pose_ = xformUtils.get_pose_from_stamped_tf(gripper_stf);
```

In this process, representations are converted to `transform` objects for use in performing transformations, with the help of utility functions in the `XformUtils` package. First, the object's location in space is specified as a `poseStamped`. That is, a pose is expressed with respect to a named `frame_id`, but the pose does not specify a child frame. To convert the `poseStamped` to a `StampedTransform`, a `child_frame_id` is named `object_frame`. Note that the parent frame in which the object pose is expressed is arbitrary. It could be a sensor frame, world frame, or any other relevant frame. It will be referred to here as `object_parent_frame`.

The desired grasp pose, as recommended by the object-manipulation service, is provided merely as a `geometry_msgs/Pose`. This is converted to a `StampedTransform` by providing a parent `frame_id` (`generic_gripper_frame`) and a `child_frame_id` (`object_frame`). The resulting transform of object frame with respect to the gripper frame is inverted to obtain the gripper frame with respect to the object frame. This inversion is post-multiplied times the `StampedTransform` object to obtain the `StampedTransform` corresponding to the `generic_gripper_frame` with respect to the `object_parent_frame`. At this point, one of the key frames for object acquisition is known and expressed as a `StampedTransform` of the generic gripper frame with respect to some named frame (`object_parent_frame`). Later, this will be transformed to express the corresponding tool-flange pose with respect to the robot's base link, which will then be suitable for use of the inverse kinematics functions.

Corresponding transformations of the approach and depart poses are computed as well.

The function `get_default_dropoff_poses()` invokes the same process, yielding three key poses. However, these poses are specialized for part placement rather than part acquisition. They are transformed to express desired gripper poses with respect to a named `frame_id`.

Having obtained the key poses, the remaining step is to compute trajectories that will achieve these poses in sequence, and these trajectories should be Cartesian motions from approach to grasp to depart poses.

It is anticipated that this simple example object-manipulation query service will be replaced in the future with increasingly competent nodes. Notably, this oracle should be able to grow its knowledge base by multiple means, including manually visualized and coded grasp options, computed grasp options, and grasps that are discovered by online robotic experimentation and learning. The present implementation is intended merely for illustrative purposes in a simple context. Nonetheless, this implementation is already suitable for a variety of relatively simple factory automation tasks.

15.3   GENERIC GRIPPER SERVICES

It is intended that an object-grabber client be independent of any specific arm or grip-per. Rather, its focus should be on the task level, specifying desired object motions. To carry out such requests, the object-grabber action server must be aware of how to control the actual gripper being used. To keep the object-grabber action server general, the system refers to generic grippers. A generic gripper interface was described in Section 14.3, specifically for the ReThink gripper on the right arm of the Baxter simulator. The node `rethink_rt_gripper_service.cpp` in the package `generic_gripper_services` provides a wrapper for this gripper. Requests to this service use a generic service message, defined in `genericGripperInterface.srv`. Generic command codes, such as GRASP, can be sent using this service message to a service named `generic_gripper_svc`. Within the `generic_gripper_services` package are multiple nodes that host a service by this name and use the `genericGripperInterface` service message. Only one of these should be launched for use with a given robot system and should correspond to the actual gripper in use. Although the various generic gripper services accept common commands, any specific gripper service node should map these commands onto the appropriate actuations for a specific gripper.

The node `virtual_vacuum_gripper_service` in the `generic_gripper_services` package simulates a vacuum gripper. This service communicates with a custom Gazebo plug-in, called `sticky_fingers`, that behaves like a vacuum gripper (but lacking in pneu-matic or fluid-flow details). The sticky-fingers plug-in is defined in the `sticky_fingers` package within the complementary repository `learning_ros_external_packages`. This package creates a library from source code `sticky_fingers.cpp`, which compiles to `libsticky_fingers.so`. This library is used as a Gazebo plug-in. An illustration is in `ur10_launch` (a package within `Part_5` of the accompanying `learning_ros` repository), in the model file `ur10_on_pedestal_w_sticky_fingers.xacro`. This model file appears in Listing 15.1.

Within Listing 15.1, lines 6 and 7 import the UR10 model as installed from ROS (from package `ur_description`, referred to via included file `ur10_robot.urdf.xacro`). Line 10 brings in a simple model of a pedestal. Lines 14 through 19 specify a joint to attach the UR10 model to the pedestal, and lines 22 through 27 attach the pedestal to the world frame.

Lines 30 through 49 describe a simple, cylindrical link that will emulate the body of the virtual vacuum gripper. Importantly, within the description of the collision model, a name is assigned to the collision model: `vvg_collision`. This link is attached to the tool flange of the robot with a link, specified in lines 52 through 56.

Lines 59 through 64 bring in the sticky-fingers plug-in within a Gazebo tag. The sticky-fingers library limits the load capacity of the virtual gripper to a value set (in this case) to 10 kg. The sticky-fingers plug-in behaves as follows. The stickiness of the vvg link is enabled or disabled via a call to the service `/sticky_finger/wrist_3_link`, which merely sets the status to `true` or `false`. When status is set to `true`, the plug-in consults Gazebo to determine whether there is contact between the `vvg_collision` boundary and some object in the environment with mass less than 10 kg (a value set in the `capacity` tag). If so, a new, temporary joint is created to attach the identified object to a distal link of the robot model (the final wrist link). Thereafter, as the robot moves, it will also carry the contacted object and will experience the gravity, inertial and collision effects of carrying this object.

When the sticky-fingers status is set to `false`, the temporary link is removed. In this state, the virtual gripper will not attach to any objects and an object that was formerly held will be released.

Communication with the sticky-fingers state is via a corresponding service and service message defined within the `sticky_fingers` package. The node `virtual_vacuum_`

Listing 15.1: `ur10_on_pedestal_w_sticky_fingers.xacro`

```
1   <?xml version="1.0"?>
2   <robot
3     xmlns:xacro="http://www.ros.org/wiki/xacro" name="ur10_on_pedestal">
4
5   <!--ur10 description -->
6     <xacro:include filename="$(find ur10_launch)/ur10_robot.urdf.xacro">
7       <xacro:arg name="gazebo" value="${gazebo}"/>
8     </xacro:include>
9     <!--pedestal model-->
10    <xacro:include filename="$(find ur10_launch)/ur10_pedestal.xacro" />
11
12    <!-- attach robot base link to the pedestal -->
13      <link name="world"/>
14    <joint name="ur10_base_joint" type="fixed">
15      <parent link="pedestal_base_link" />
16      <!--ur10 base link is child-->
17      <child link="base_link" />
18      <origin rpy="0 0 0 " xyz="0 0 0.426"/>
19    </joint>
20
21    <!--attach the pedestal to the world-->
22    <joint name="glue_base_to_world" type="fixed">
23      <parent link="world" />
24      <child link="pedestal_base_link" />
25      <origin rpy="0 0 0 " xyz="0 0 0.426"/>
26      <origin rpy="0 0 0 " xyz="0 0 0"/>
27    </joint>
28
29    <!--define a link to emulate vacuum gripper-->
30    <link name= "vvg">
31        <origin rpy="0 0 0 " xyz="0 0 0"/>
32        <visual>
33          <geometry>
34            <cylinder length="0.02" radius="0.04"/>
35          </geometry>
36          <material name="blue">
37            <color rgba="0 0 .8 1"/>
38          </material>
39        </visual>
40        <collision name="vvg_collision">
41          <geometry>
42                <cylinder length="0.02" radius="0.04"/>
43          </geometry>
44        </collision>
45        <inertial>
46          <mass value="0.01"/>
47          <inertia ixx="1.0" ixy="0.0" ixz="0.0" iyy="1.0" iyz="0.0" izz="1.0"/>
48        </inertial>
49     </link>
50
51    <!--attach virtual vacuum gripper, vvg, to tool flange-->
52    <joint name="gripper_joint" type="fixed">
53      <parent link="tool0" />
54      <child link="vvg" />
55      <origin rpy="0 0 0" xyz="0 0 0.01"/>
56    </joint>
57
58    <!-- bring in the sticky-fingers plug-in to simulate a vacuum gripper-->
59      <gazebo>
60        <plugin name="virtual_vacuum_gripper_finger" filename="libsticky_fingers.so">
61          <capacity>10</capacity>
62          <link>ur10_on_pedestal::wrist_3_link</link>
63        </plugin>
64      </gazebo>
65
66  </robot>
```

`gripper_service` abstracts use of the vacuum gripper by creating another service with the generic service name `generic_gripper_svc`, which uses the generic service message `genericGripperInterface.srv` defined in the `generic_gripper_services` package. By running the node `virtual_vacuum_gripper_service`, one can send generic commands such as GRASP and RELEASE using the generic service name and generic service message. As a result, an object-grabber action client can command grasp and release without needing to know what type of gripper is used.

Use of the virtual vacuum gripper also requires that the corresponding static transform publisher specify a generic gripper frame at the desired location—the face of the virtual vacuum gripper link. The virtual vacuum gripper will be illustrated in the context of the object-grabber action server–action client operation.

15.4 OBJECT-GRABBER ACTION SERVER

The object-grabber action server (in package `object_grabber`) accepts manipulation goals from an action client (without reference to any specific arm or gripper) and generates plans and commands to achieve the desired manipulation goals.

Performing a successful grasp requires several steps. For example, consider grasping an object on a table surface from above using a parallel-jaw gripper. One must command the chosen arm to open the gripper; move the arm sufficiently high above the table surface (without bumping the table); orient the gripper for a suitable approach pose; perform a Cartesian move to approach the object, resulting in gripper fingers surrounding the object; close the gripper fingers to an appropriate finger separation; and perform a Cartesian move to depart from the table top (ideally, normal to the table surface).

These steps are accomplished with the help of two services and an action server. The generic gripper service translates GRASP and RELEASE commands into corresponding actuation commands appropriate for the target gripper (*e.g.* fingers versus vacuum gripper). The object-manipulation query service provides recommendations for key gripper poses, given an object ID and a gripper ID. Cartesian moves among the key poses (approach, grasp, depart) achieve appropriate approach and departure, such that gripper components do not interfere with the object before grasp, and the grasped object does not interfere with the support surface as it is withdrawn.

The object-grabber action service does not receive a gripper ID from its client. Rather, this service must be aware of what type of gripper is on the arm being controlled. This is accomplished using the parameter server. When launching the object-grabber nodes, one of the nodes should perform the action of placing the gripper ID on the parameter server. For example, the node `set_baxter_gripper_param.cpp` in the `object_grabber` package contains:

```
//example to show how to set gripper_ID programmatically
#include <ros/ros.h>
#include <object_manipulation_properties/gripper_ID_codes.h>

int main(int argc, char **argv) {
    ros::init(argc, argv, "gripper_ID_setter");
    ros::NodeHandle nh; // two lines to create a publisher object that can talk to ROS
    int gripper_id = GripperIdCodes::RETHINK_ELECTRIC_GRIPPER_RT;
    nh.setParam("gripper_ID", gripper_id);
}
```

Running this node results in setting the parameter `gripper_ID` to code `RETHINK_ELECTRIC_GRIPPER_RT` (which is defined in the header file `object_manipulation_properties/gripper_ID_codes.h`).

Upon start-up, the object-grabber node consults the parameter server to obtain the gripper ID that is used subsequently in sending queries to the object-manipulation query service to obtain key gripper poses. Also on start-up, the object-grabber node establishes connections with a generic gripper service, an object-manipulation query service, and a Cartesian motion action server. While the object-grabber action server does need to be aware of the gripper to be used for manipulation, it does not need to be aware of the type of robot arm being used. Rather, goal gripper poses are specified to a Cartesian motion action server, and this server is responsible for computing and executing coordinated joint motions that achieve the desired gripper motions.

The object-grabber code is long, but it has a style that is largely repetitive. The `executeCallback` function parses the goal message to extract the action code within the goal message. It uses this action code in a switch–case statement to direct computation appropriately. For the code `GRAB_OBJECT`, the corresponding case is as follows:

```
case object_grabber::object_grabberGoal::GRAB_OBJECT:
    ROS_INFO("GRAB_OBJECT: ");
    object_id = goal->object_id;
    grasp_option = goal->grasp_option;
    object_pose_stamped_ = goal->object_frame;
    //get grasp-plan details for this case:
    if (grasp_option != object_grabber::object_grabberGoal::←
        DEFAULT_GRASP_STRATEGY)
    {
        ROS_WARN("grasp strategy %d not implemented yet; using default ←
            strategy",grasp_option);
    }
    rtn_val = grab_object(object_id,object_pose_stamped_);
    ROS_INFO("grasp attempt concluded");
    grab_result_.return_code = rtn_val;
    object_grabber_as_.setSucceeded(grab_result_);
    break;
```

The key line in this case is `rtn_val = grab_object(object_id,object_pose_stamped_)`, which uses the `object_id` field of the goal message and the corresponding stamped pose within the goal message. These are used as arguments to the function `grab_object()`.

The `grab_object()` function is presented below. This function first invokes `get_default_grab_poses(object_id,object_pose_stamped)`, from which three key gripper poses are obtained: `approach_pose_`, `grasp_pose_`, and `depart_pose_`. The gripper is put in a state appropriate for object approach with the lines:

```
gripper_srv_.request.cmd_code = generic_gripper_services::←
    genericGripperInterfaceRequest::RELEASE;
gripper_client_.call(gripper_srv_);
```

In the case of a fingered gripper, the fingers are opened; for a vacuum gripper, the suction is disabled.

With assistance from the class `ArmMotionCommander`, which encapsulates some of the detail of communication with a Cartesian motion action server, an arm motion plan is computed for a joint-space motion from the current pose to the computed approach pose using:

```
rtn_val=arm_motion_commander_.plan_jspace_path_current_to_cart_gripper_pose(←
    approach_pose_);
```

The return value is inspected to see whether a motion plan was successfully computed. If so, the Cartesian motion action server is instructed to execute this motion plan through the function call:

```cpp
int  ObjectGrabber::grab_object(int object_id,geometry_msgs::PoseStamped ↩
    object_pose_stamped){
    //given gripper_ID, object_ID and object poseStamped,
    // and assuming default approach, grasp and depart strategies for this object/↩
        gripper combo,
    // compute the corresponding required gripper-frame poses w/rt a named frame_id
    // (which will be same frame_id as specified in object poseStamped)
    int rtn_val;
    bool success;
    if(!get_default_grab_poses(object_id,object_pose_stamped)) {
        ROS_WARN("no valid grasp strategy; giving up");
        return object_grabber::object_grabberResult::↩
            NO_KNOWN_GRASP_OPTIONS_THIS_GRIPPER_AND_OBJECT;
    }
    //invoke the sequence of moves to perform approach, grasp, depart:
    ROS_WARN("prepare gripper state to anticipate grasp...");
    gripper_srv_.request.cmd_code = generic_gripper_services::↩
        genericGripperInterfaceRequest::RELEASE;
    gripper_client_.call(gripper_srv_);
    success = gripper_srv_.response.success;
    if (success) { ROS_INFO("gripper responded w/ success"); }
    else {ROS_WARN("responded with failure"); }

    ROS_WARN("object-grabber as planning joint-space move to approach pose");
    //xformUtils.printPose(approach_pose_);

    rtn_val=arm_motion_commander_.plan_jspace_path_current_to_cart_gripper_pose(↩
        approach_pose_);
    if (rtn_val != cartesian_planner::cart_moveResult::SUCCESS) return rtn_val; //↩
        return error code

    //send command to execute planned motion
    ROS_INFO("executing plan: ");
    rtn_val=arm_motion_commander_.execute_planned_path();
    if (rtn_val != cartesian_planner::cart_moveResult::SUCCESS) return rtn_val; //↩
        return error code
    //ros::Duration(2.0).sleep();

    ROS_INFO("planning motion of gripper to grasp pose at: ");
    xformUtils.printPose(grasp_pose_);
    rtn_val=arm_motion_commander_.plan_path_current_to_goal_gripper_pose(grasp_pose_);
    if (rtn_val != cartesian_planner::cart_moveResult::SUCCESS) return rtn_val; //↩
        return error code
    ROS_INFO("executing plan: ");
    rtn_val=arm_motion_commander_.execute_planned_path();
    ROS_WARN("poised to grasp object; invoke gripper grasp action here ...");

    gripper_srv_.request.cmd_code = generic_gripper_services::↩
        genericGripperInterfaceRequest::GRASP;
    gripper_client_.call(gripper_srv_);
    success = gripper_srv_.response.success;
    ros::Duration(1.0).sleep(); //add some extra time to stabilize grasp...tune this
    if (success) { ROS_INFO("gripper responded w/ success"); }
    else {ROS_WARN("responded with failure"); }

    ROS_INFO("planning motion of gripper to depart pose at: ");
    xformUtils.printPose(depart_pose_);
    rtn_val=arm_motion_commander_.plan_path_current_to_goal_gripper_pose(depart_pose_)↩
        ;
    if (rtn_val != cartesian_planner::cart_moveResult::SUCCESS) return rtn_val; //↩
        return error code
    ROS_INFO("performing motion");
    rtn_val=arm_motion_commander_.execute_planned_path();


    return rtn_val;
}
```

```
rtn_val=arm_motion_commander_.execute_planned_path();
```

A Cartesian motion plan to move the arm from the approach pose to the grasp pose is then computed via the function call:

```
rtn_val=arm_motion_commander_.plan_path_current_to_goal_gripper_pose(grasp_pose_);
```

This function sends a corresponding planner action code in a goal message to the Cartesian-motion action server. Due to separation of the object-grabber action server node from the motion planner–action server node, the object-grabber node can be robot-agnostic, whereas the motion planner must be customized for a specific robot design.

The `grab_object()` function proceeds with additional calls to invoke grasp, then plans and executes a departure trajectory with the grasped object.

15.5   EXAMPLE OBJECT-GRABBER ACTION CLIENT

The intent of the object-grabber action service is to allow higher-level code to focus on the task level, independently of robot or gripper specifics. An example of this approach is described here with an action client of the object-grabber action service.

The node `example_object_grabber_action_client` contained in the `object_grabber` package appears in Listings 15.2 through 15.4.

In this program, the function `set_example_object_frames()`, lines 48 through 66, merely hard-code poses for pick-up and placement of an object. More generally, these poses would come from a database or a perceptual system.

The action client establishes connection with the `object_grabber_action_service` (lines 111 through 121). This action client is made available to external functions through the global pointer `g_object_grabber_ac_ptr`. The main program then uses the function `move_to_waiting_pose()` (defined in lines 68 through 74) to move the arm to a pre-defined pose above a presumptive table.

Next, the main program calls `grab_object(object_pickup_poseStamped)`, which is defined in lines 76 through 97. This function populates a goal message with an object ID, a pick-up pose, and a code to specify use of the default grasp strategy. When this goal is sent to the object-grabber service, the intended result is that the specified object is grasped by the robot's gripper from the specified pick-up location.

After the object is grasped, the main program calls `dropoff_object(object_dropoff_poseStamped)`, which is defined in lines 89 through 99. This function populates and sends a goal to perform the complementary operation of placing the grasped object at a defined location.

All coordinates specified refer to the object's frame, independent of any gripper or arm.

This example object-grabber action client can be run using different target robotic systems. It is only necessary that the named object actually be located at the prescribed pose. The action client is first illustrated with respect to the Baxter robot.

Figure 15.2 shows the frame named `right_gripper`, which is oriented with the z axis pointing out from the wrist and the y axis pointing from the right fingertip to the left fingertip. The origin is on the gripper-frame z axis, level with the tips of the fingers. Using the static-transform launch file `baxter_static_transforms.launch`, this frame is synonymous with a frame named `generic_gripper_frame`. Additionally, the `torso` frame used by Baxter's kinematics library is defined to be related to the `system_ref_frame`, which is located at the ground-plane level, 0.91 m directly beneath the torso frame.

For an object of interest, we will again consider the toy block used in Section 8.5. To

Listing 15.2: `example_object_grabber_action_client.cpp`: example use of object-grabber action server, preamble and functions

```cpp
1   // example_object_grabber_action_client: minimalist client
2   // use with object_grabber action server called "objectGrabberActionServer"
3   // in file object_grabber_as.cpp
4
5   //client gets gripper ID from param server
6   // gets grasp strategy options from manip_properties(gripper_ID,object_ID)
7   // two primary fncs:
8   //   **    object_grab(object_id, object_pickup_pose, grasp_strategy, approach_strategy ↩
                , depart_strategy)
9   //   **    object_dropoff(object_id, object_destination_pose, grasp_strategy, ↩
                dropoff_strategy, depart_strategy)
10  //      have default args for grasp_strategy, depart_strategy, ...
11  //      default is grab from above, approach and depart from above
12  //      grasp strategy implies a grasp transform--to be used by action service for ↩
                planning paths
13  //      all coords expressed as object frame w/rt named frame--which must have a
14  //      kinematic path to system_ref_frame (e.g. simply use system_ref_frame)
15
16
17  #include<ros/ros.h>
18  #include <actionlib/client/simple_action_client.h>
19  #include <actionlib/client/terminal_state.h>
20  #include <object_grabber/object_grabberAction.h>
21  #include <Eigen/Eigen>
22  #include <Eigen/Dense>
23  #include <Eigen/Geometry>
24  #include <xform_utils/xform_utils.h>
25  #include <object_manipulation_properties/object_ID_codes.h>
26  #include<generic_gripper_services/genericGripperInterface.h>
27
28  using namespace std;
29  XformUtils xformUtils; //type conversion utilities
30
31  int g_object_grabber_return_code;
32
33  void objectGrabberDoneCb(const actionlib::SimpleClientGoalState& state,
34          const object_grabber::object_grabberResultConstPtr& result) {
35      ROS_INFO(" objectGrabberDoneCb: server responded with state [%s]", state.toString ↩
                ().c_str());
36      g_object_grabber_return_code = result->return_code;
37      ROS_INFO("got result output = %d; ", g_object_grabber_return_code);
38  }
39
40
41  //test fnc to specify object pick-up and drop-off frames;
42  //should get pick-up frame from perception, and drop-off frame from perception or task
43
44  void set_example_object_frames(geometry_msgs::PoseStamped &object_poseStamped,
45          geometry_msgs::PoseStamped &object_dropoff_poseStamped) {
46      //hard code an object pose; later, this will come from perception
47      //specify reference frame in which this pose is expressed:
48      //will require that "system_ref_frame" is known to tf
49      object_poseStamped.header.frame_id = "system_ref_frame"; //set object pose; ref ↩
                frame must be connected via tf
50      object_poseStamped.pose.position.x = 0.5;
51      object_poseStamped.pose.position.y = -0.35;
52      object_poseStamped.pose.position.z = 0.7921; //-0.125; //pose w/rt world frame
53      object_poseStamped.pose.orientation.x = 0;
54      object_poseStamped.pose.orientation.y = 0;
55      object_poseStamped.pose.orientation.z = 0.842;
56      object_poseStamped.pose.orientation.w = 0.54;
57      object_poseStamped.header.stamp = ros::Time::now();
58
59      object_dropoff_poseStamped = object_poseStamped; //specify desired drop-off pose ↩
                of object
60      object_dropoff_poseStamped.pose.orientation.z = 1;
61      object_dropoff_poseStamped.pose.orientation.w = 0;
62  }
```

Listing 15.3: `example_object_grabber_action_client.cpp`: example use of object-grabber action server, manipulation functions

```
64  void move_to_waiting_pose() {
65      ROS_INFO("sending command to move to waiting pose");
66      g_got_callback=false; //reset callback-done flag
67      object_grabber::object_grabberGoal object_grabber_goal;
68      object_grabber_goal.action_code = object_grabber::object_grabberGoal::↵
            MOVE_TO_WAITING_POSE;
69      g_object_grabber_ac_ptr->sendGoal(object_grabber_goal, &objectGrabberDoneCb);
70  }
71
72  void grab_object(geometry_msgs::PoseStamped object_pickup_poseStamped) {
73      ROS_INFO("sending a grab-object command");
74      g_got_callback=false; //reset callback-done flag
75      object_grabber::object_grabberGoal object_grabber_goal;
76      object_grabber_goal.action_code = object_grabber::object_grabberGoal::GRAB_OBJECT;↵
            //specify the action to be performed
77      object_grabber_goal.object_id = ObjectIdCodes::TOY_BLOCK_ID; // specify the object↵
            to manipulate
78      object_grabber_goal.object_frame = object_pickup_poseStamped; //and the object's ↵
            current pose
79      object_grabber_goal.grasp_option = object_grabber::object_grabberGoal::↵
            DEFAULT_GRASP_STRATEGY; //from above
80      object_grabber_goal.speed_factor = 1.0;
81      ROS_INFO("sending goal to grab object: ");
82      g_object_grabber_ac_ptr->sendGoal(object_grabber_goal, &objectGrabberDoneCb);
83  }
84
85  void   dropoff_object(geometry_msgs::PoseStamped object_dropoff_poseStamped) {
86      ROS_INFO("sending a dropoff-object command");
87      object_grabber::object_grabberGoal object_grabber_goal;
88      object_grabber_goal.action_code = object_grabber::object_grabberGoal::↵
            DROPOFF_OBJECT; //specify the action to be performed
89      object_grabber_goal.object_id = ObjectIdCodes::TOY_BLOCK_ID; // specify the object↵
            to manipulate
90      object_grabber_goal.object_frame = object_dropoff_poseStamped; //and the object's ↵
            current pose
91      object_grabber_goal.grasp_option = object_grabber::object_grabberGoal::↵
            DEFAULT_GRASP_STRATEGY; //from above
92      object_grabber_goal.speed_factor = 1.0;
93      ROS_INFO("sending goal to dropoff object: ");
94      g_object_grabber_ac_ptr->sendGoal(object_grabber_goal, &objectGrabberDoneCb);
95  }
```

Listing 15.4: `example_object_grabber_action_client.cpp`: example use of object-grabber action server, main program

```
101  int main(int argc, char** argv) {
102      ros::init(argc, argv, "example_object_grabber_action_client");
103      ros::NodeHandle nh;
104      geometry_msgs::PoseStamped object_pickup_poseStamped;
105      geometry_msgs::PoseStamped object_dropoff_poseStamped;
106
107      //specify object pick-up and drop-off frames using simple test fnc
108      //more generally, pick-up comes from perception and drop-off comes from task
109      set_example_object_frames(object_pickup_poseStamped, object_dropoff_poseStamped);
110      //instantiate an action client of object_grabber_action_service:
111      actionlib::SimpleActionClient<object_grabber::object_grabberAction> ↩
                  object_grabber_ac("object_grabber_action_service", true);
112      g_object_grabber_ac_ptr = &object_grabber_ac; // make available to fncs
113      ROS_INFO("waiting for server: ");
114      bool server_exists = false;
115      while ((!server_exists)&&(ros::ok())) {
116          server_exists = object_grabber_ac.waitForServer(ros::Duration(0.5)); //
117          ros::spinOnce();
118          ros::Duration(0.5).sleep();
119          ROS_INFO("retrying...");
120      }
121      ROS_INFO("connected to object_grabber action server"); // if here, then we ↩
                  connected to the server;
122
123      //move to waiting pose
124      move_to_waiting_pose();
125      while(!g_got_callback) {
126          ROS_INFO("waiting on move...");
127          ros::Duration(0.5).sleep(); //could do something useful
128      }
129
130      grab_object(object_pickup_poseStamped);
131      while(!g_got_callback) {
132          ROS_INFO("waiting on grab...");
133          ros::Duration(0.5).sleep(); //could do something useful
134      }
135
136      dropoff_object(object_dropoff_poseStamped);
137      while(!g_got_callback) {
138          ROS_INFO("waiting on dropoff...");
139          ros::Duration(0.5).sleep(); //could do something useful
140      }
141      return 0;
142  }
```

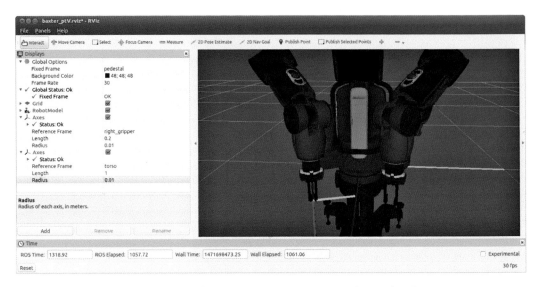

Figure 15.2: Torso frame and `right_gripper` frame for Baxter

introduce the table and block within reach of Baxter, first start Baxter in an empty world with:

```
roslaunch baxter_gazebo baxter_world.launch
```

Next, add a table and a toy block by launching:

```
roslaunch exmpl_models add_table_and_block.launch
```

In launch operations, there is no guarantee on timing. Thus, it can happen that the block model is spawned into Gazebo before the table model is spawned. In this case, the block will fall to the floor. There are two ways to reset the block. Within the Gazebo window, from the top menu, use the **edit** item and select **reset model poses**. The block will be reset to its originally defined pose. From the Gazebo window, one can see the result by opening the **models** item, selecting the toy-block model, and clicking on its **pose** item. As shown in Fig 15.3, the block's origin is at $(x, y, z) = (0.5, -0.35, 0.792)$. Note that these are the coordinates hard-coded in the example object-grabber client code.

An alternative way to reset the block pose is:

```
rosrun example_gazebo_set_state reset_block_state
```

This node includes code that interacts directly with Gazebo to set a named model to a specified pose (as introduced in Section 3.4).

After launching the Baxter simulator (or real robot), wait until the robot is ready to be enabled, then run:

```
roslaunch baxter_launch_files baxter_object_grabber_nodes.launch
```

This launch file starts 12 different nodes. Three of these nodes perform actions that run to completion and terminate. These nodes enable the robot's actuators, set the gripper ID on the parameter server, and run a **playfile** to move the arms to a defined waiting pose.

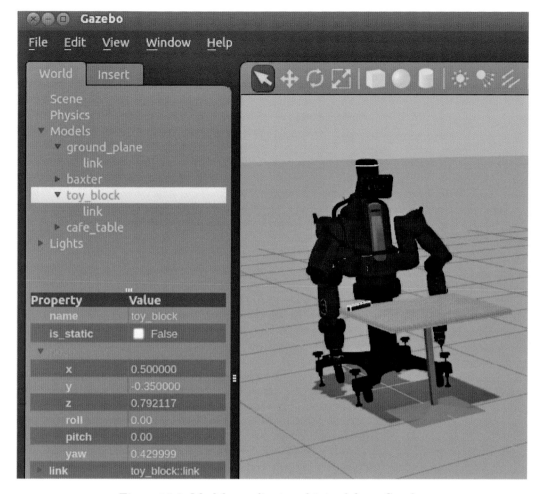

Figure 15.3: Model coordinates obtained from Gazebo

This launch file also starts up `rviz` with reference to a saved configuration file. It also includes (and runs) a launch file that starts two static-transform publishers that define the frames `generic_gripper_frame` and `system_ref_frame`.

Three services are started: a `playfile` service (which is not required for the object grabber, but can be useful); the object-manipulation properties service, and a generic gripper service that abstracts control of Baxter's right gripper.

Finally, four action servers are started. Right and left arm joint trajectory action servers are started; they can receive trajectories as goals and execute them with fine interpolation. (Alternatively, ReThink's joint trajectory action servers could be launched and used, although this would require modifying the Cartesian motion action server to use these.) The action server `baxter_rt_arm_cart_move_as` is launched, which provides Cartesian motion planning and execution capabilities specific to Baxter's right arm, but presents an interface to action clients that is robot-agnostic. The final action server launched is the `object_grabber_action_server`, which assumes use of a robot-independent Cartesian motion action server (which is `baxter_rt_arm_cart_move_as`, in this case). The `object_grabber_action_server` presents an action server interface to clients that enables clients to specify manipulation goals that are independent of specific robots or grippers.

The result of running this launch file is to command Baxter's arms to move to the defined waiting pose, then the object-grabber awaits goals. This initial state is shown in Fig 15.4. The example object-grabber action client program has a hard-coded object ID with

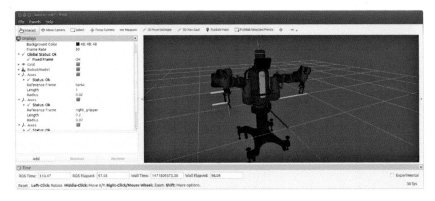

Figure 15.4: `rviz` view of Baxter simulator in pre-pose, illustrating right hand and right gripper frames

a hard-coded pose, consistent with the toy block on the cafe table spawned in Gazebo, as shown in Fig 15.3. The example client program can be run with:

```
rosrun object_grabber example_object_grabber_action_client
```

Running this node results in the robot picking up the block, rotating it, and setting it back down. The first step of this is sending a `GRAB_OBJECT` goal to the object-grabber action server. This invokes a sequence of actions enabling the robot to move its gripper to a recommended approach pose, open the gripper fingers, descend to a recommended grasp pose, close the fingers, and ascend to a recommended depart pose.

Execution of the default-strategy approach pose results in the scene in Fig 15.5, with the gripper 0.1 m directly above the block, and with gripper fingers open and aligned in preparation for descending to the grasp pose.

Figure 15.6 shows the result of moving to the grasp pose prescribed by the object-manipulation query service. The generic gripper frame, with origin defined between the fingertips, is positioned such that its origin is coincident with the block frame origin. The gripper frame orientation is specified to have the gripper z axis anti-parallel to the block frame z axis, and have the gripper frame x axis parallel to the major axis of the block (the block frame x axis). In this recommended grasp pose, the fingertips straddle the block near its center, preparing to grasp the block. After descending to the grasp pose and closing the gripper, the robot moves to its depart pose, presumably with the object successfully grasped, resulting in the pose shown in Fig 15.7. This concludes the `GRAB_OBJECT` action. The object-grabber action client next commands `DROPOFF_OBJECT`. In this sequence, the robot moves to an approach pose above the drop-off coordinates, descends to the destination pose, opens the gripper, then withdraws the gripper. Figure 15.8 shows the state of the robot just after the block has been placed at its destination coordinates and the gripper fingers have been opened, in preparation for withdrawing the gripper.

To illustrate the generality of the action client program, the same operation can be executed by a UR10 robot. First, start the UR10 robot on a pedestal with a virtual vacuum gripper added:

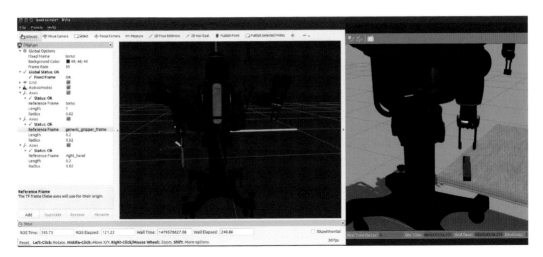

Figure 15.5: Approach pose from action client `object_grabber_action_client`, with pre-positioned block at known precise coordinates

```
roslaunch ur10_launch ur10_w_gripper.launch
```

Next, add a table and block and start the nodes needed for the object-grabber service:

```
roslaunch ur10_launch ur10_object_grabber_nodes.launch
```

This launch file is similar to the corresponding Baxter launch file. (One addition is that the UR10 launch file incorporates spawning the table and block.) It sets a gripper ID on the parameter server (in this case, a virtual vacuum gripper). It includes a launch file that starts static transform publishers for a generic gripper frame and a system reference frame relative to respective UR10 frames. It also starts `rviz` with reference to a pre-defined configuration file.

A gripper service is started that controls the virtual vacuum gripper, but it still presents a generic interface for defined GRASP and RELEASE command codes.

A Cartesian motion action server, `ur10_cart_move_as`, is started. This action server presents a generic interface to action clients, but its implementation is specific to the UR10 robot.

The object-manipulation query service is started. This service is identical to that used in the Baxter example. The object-grabber action service is started, identical to that used in the Baxter example.

The result of launching these nodes is as shown in Fig 15.9.

With the object-grabber action server (and supporting nodes) running, one can run an action client. The object-grabber action client demonstrated for the Baxter example can be run verbatim for the UR10 example:

```
rosrun object_grabber example_object_grabber_action_client
```

In performing the **GRAB_OBJECT** action, the UR10 is first sent to an approach pose, as shown in Fig 15.10. This approach pose is recommended by the object-manipulation query service as appropriate for the emulated vacuum gripper relative to the toy-block part.

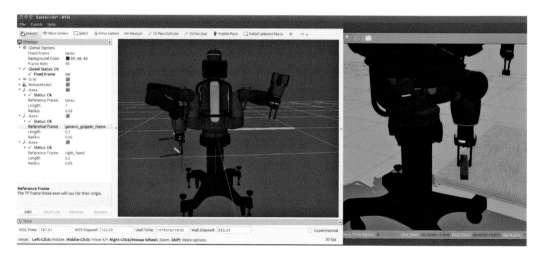

Figure 15.6: Grasp pose used by action client `object_grabber_action_client`

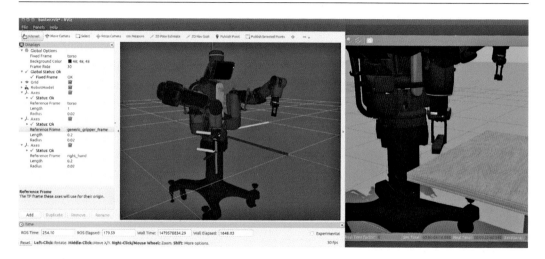

Figure 15.7: Depart pose used by action client `object_grabber_action_client`

Figure 15.11 shows the result of moving to the grasp pose prescribed by the object-manipulation query service specific for this gripper and object. The generic gripper frame, with origin defined at the center of the face of the vacuum gripper, is positioned such that its origin is coincident with the *top face* of the block. The gripper frame orientation is specified to have the gripper z axis anti-parallel to the block frame z axis. (The gripper frame x axis direction is arbitrary, but constrained to be perpendicular to the z axis.) In this recommended grasp pose, the vacuum gripper may be expected to form a seal with the top face of the toy-block object.

After descending to the grasp pose and commanding the gripper to GRASP, the robot moves to its depart pose, presumably with the object successfully grasped, resulting in the pose shown in Fig 15.12

This concludes the `GRAB_OBJECT` action. The object-grabber action client next commands `DROPOFF_OBJECT`. In this sequence, the robot moves to an approach pose above the drop-off coordinates, descends to the destination pose, invokes a RELEASE command to

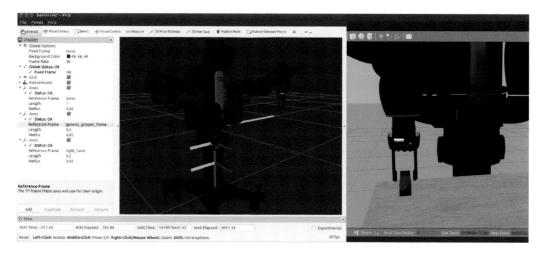

Figure 15.8: Drop-off pose used by action client `object_grabber_action_client`

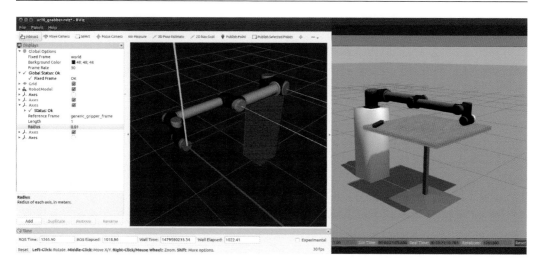

Figure 15.9: Initial state after launching UR10 object-grabber nodes

the gripper, then withdraws the gripper. Figure 15.13 shows the state of the robot just after the block has been placed at its destination coordinates but before the gripper has released the part.

The example object-grabber action client can successfully achieve the desired effect of moving a part from a specified initial pose to a desired destination pose. The same action client can be used with very different robot arms and grippers.

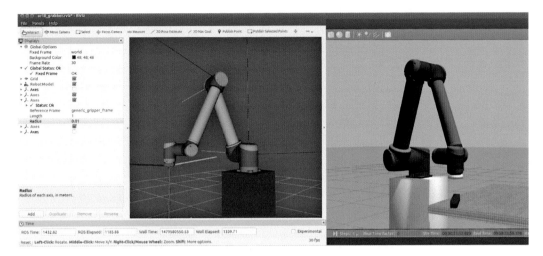

Figure 15.10: UR10 block approach pose

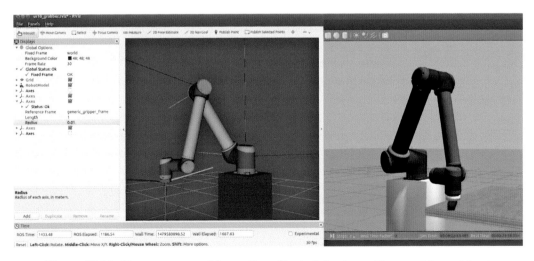

Figure 15.11: Grasp pose used by action client `object_grabber_action_client`

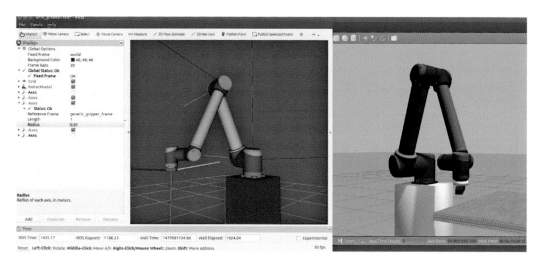

Figure 15.12: UR10 depart pose

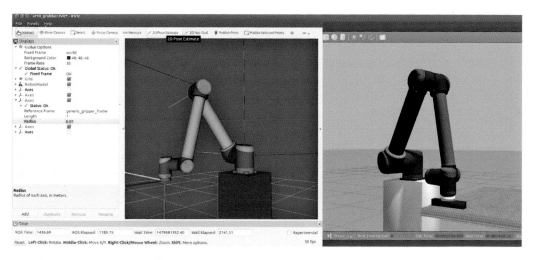

Figure 15.13: Drop-off pose used by action client `object_grabber_action_client`

15.6   WRAP-UP

The object-grabber action service described in this chapter is an example of how we can use ROS to make software more re-usable. The object-grabber action service is independent of any specific robot arm. For object-manipulation commands, the system must know what type of gripper is being used, and the approach taken here is to provide this information via the parameter server.

It was shown that an action client of the object-grabber action server can be composed in a manner that allows the programmer to focus on task properties rather than robot properties. In the simple example provided, an object was to be relocated from a specified initial pose to a specified destination pose. This was accomplished using two very different robot arms with very different grippers, yet the same action client program could be used with both systems to accomplish the task goals.

In the final section of this text, we consider issues of system integration incorporating perception, mobility and manipulation.

VI

System Integration and Higher Level Control

I NTRODUCTION
In this final part, focus is on system integration and higher level control. Developments from Section II (modeling), Section III (perception), Section IV (mobility) and Section V (manipulation) are combined to model and control a mobile manipulator. This part shows how these aspects can be combined to construct a fairly complex system that is able to execute goal-directed behaviors using perception, navigation planning and execution, and manipulation planning and execution. The resulting system embodies components of future robotic systems that will be able to reason about their environment to achieve specified goals, rather than merely repeat stereotyped mechanical motions.

Perception-Based Manipulation

CONTENTS

I NTRODUCTION
Before developing mobile manipulation, we first consider combining perception and manipulation. Combining perception and manipulation also requires hand–eye calibration, and thus extrinsic camera calibration is discussed first.

16.1 EXTRINSIC CAMERA CALIBRATION

Performing perception-based manipulation requires camera calibration. Intrinsic camera calibration was presented in Section 6.2. In addition, extrinsic camera calibration must be performed to establish the kinematic transform between a reference frame on the robot and the reference frame of the sensor. With good extrinsic calibration, one can perceive an object of interest, then transform the object's coordinates from the sensor frame to the robot's base frame. If this transformation is precise, a motion plan can be constructed and executed that enables the robot to successfully grasp the object.

Figure 16.1 illustrates a result of virtually perfect extrinsic camera calibration. This `rviz` view corresponds to data from a Kinect sensor mounted on the Baxter robot model. (This modified Baxter model will be described in the next section.) The modified Baxter model can be launched by first bringing up an empty world:

```
roslaunch gazebo_ros empty_world.launch
```

then launching:

```
roslaunch baxter_variations baxter_on_pedestal_w_kinect.launch
```

The screenshot of Fig 16.1 is taken looking over the robot's shoulder, from a perspective similar to that of the Kinect sensor. In the `rviz` display, point cloud points from the Kinect are colorized based on the z heights of the points. From this view, we can see many points on the surface of the model of the robot's right arm and hand. These points correspond to the Kinect seeing the robot's arm. The point-cloud points appear almost perfectly in agreement with the robot model, almost as if they were painted on the model. This agreement illustrates ideal hand–eye calibration. This level of precision is unrealistically good and is made possible only by self-consistency of the robot and Kinect models. In simulation, the Kinect sensor gets its points by ray tracing to surface points on the robot's arm. These surface points follow from the robot model (in the URDF) and from knowledge of the robot's state (joint angles). The transform of the Kinect with respect to the robot's torso frame is precise, because this transform is part of the robot model. As a result, there are no calibration errors between the robot model the Kinect perception of that model.

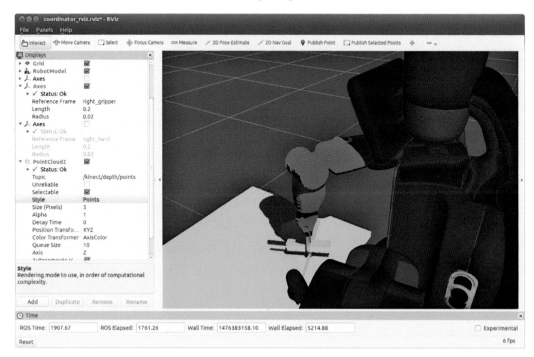

Figure 16.1: Point cloud including Kinect's partial view of robot's right arm

In practice, the CAD model of a physical robot exhibits significant differences between model and reality. Further, precise knowledge of the poses of the robot's links depends on the precision of the joint sensors, the quality of the robot's home-angle calibration, and the precision of the robot's kinematic model. In practice, the CAD models can be high quality, and the robot's kinematic model and home-angle calibrations can be quite precise, and thus this contribution to hand–eye miscalibration may be small.

The largest source of error in hand–eye calibration typically would derive from uncertainty of the transform between the sensor (Kinect) frame and the robot's torso (or base) frame. Finding this transform with sufficient precision is a necessity for successful perception-based manipulation. Finding this transform corresponds to *extrinsic* sensor calibration.

If the robot's URDF model (including the kinematic model and the visualization properties described within the link's CAD models) is sufficiently precise, visualization such as that in Fig 16.1 can be used to help identify the sensor transform experimentally.

A modified launch of the Baxter robot with a Kinect sensor can be started with:

```
roslaunch gazebo_ros empty_world.launch
```

```
roslaunch baxter_variations baxter_on_pedestal_w_kinect2.launch
```

The difference between the above launch file and the previous `baxter_on_pedestal_w_kinect.launch` launch file is removal of the line:

```
<node pkg="tf" type="static_transform_publisher" name="kinect_broadcaster2" args="0 ←
    0 0 -0.500 0.500 -0.500 0.500 kinect_link kinect_pc_frame 100" />
```

This line is inserted instead in a separate `kinect_xform.launch` file, the contents of which are:

```
<launch>
  <!--imperfect sensor transform, to illustrate extrinsic calibration -->
  <node pkg="tf" type="static_transform_publisher" name="kinect_broadcaster2" args="←
      0.1 0.1 0.1 -0.50 0.50 -0.50 0.5 kinect_link kinect_pc_frame 100" />
</launch>
```

The numerical values in this transform are deliberately incorrect (offset by 0.1 m in x, y and z directions). This transform publisher is started with:

```
roslaunch baxter_variations kinect_xform.launch
```

The control nodes and `rviz` can then be launched with:

```
roslaunch coordinator coord_vision_manip.launch
```

The robot's arm can be positioned conveniently within view of the Kinect by running (from the `baxter_playfile_nodes` package directory):

```
rosrun baxter_playfile_nodes baxter_playback can_grasp_pose.jsp
```

Under these conditions, the Kinect's point cloud appears as in Fig 16.2. Due to the imperfect Kinect frame transform, the point cloud no longer lies on the surface of the robot model's arm. The point cloud is recognizably related to the shape of the robot arm and gripper, but it is offset from the robot model. The `kinect_xform.launch` can be halted, edited, and re-launched (without needing to re-start any other nodes). By altering the transform values interactively, one can attempt to find values that result in point cloud alignment with the robot model, and this condition will indicate good hand–eye calibration.

(Note that the sensor frame in Gazebo simulation is called `kinect_pc_frame`, but when running the driver for a physical Kinect, this is called the `camera_link`, and thus different transform publications are needed for simulation versus physical equipment.)

As an alternative, a vision target may be attached to the arm with a known attachment transform, and the arm can be commanded to move the target to different poses in space. By associating Kinect snapshots with corresponding forward kinematics solutions, one can compute a best-fit explanation of the data in terms of a camera transform with respect to the robot's base frame.

Assuming extrinsic sensor calibration has been achieved to adequate precision, one can use sensory information to plan and execute manipulation. An example of this using the Baxter simulator is introduced next.

Figure 16.2: Offset perception of arm due to Kinect transform inaccuracy

16.2 INTEGRATED PERCEPTION AND MANIPULATION

In Section 8.5, an object-finder action server was described. Coordinates of an object of interest can be inferred from point-cloud images. In Section 15.4, an object-grabber action server was presented, with which an action client can specify manipulation goals, including an object ID and its pose. Putting these two capabilities together, a robot can be capable of perceiving objects of interest and manipulating them.

To combine these capabilities, a Kinect sensor is retrofit to the Baxter robot (model and/or physical robot). A model file that achieves this is `baxter_on_pedestal.xacro` in the package `baxter_variations`. This model file, below, merely combines individual model files.

```
<?xml version="1.0"?>
<robot
  xmlns:xacro="http://www.ros.org/wiki/xacro" name="baxter_on_pedestal">

  <!-- Baxter Base URDF -->
  <xacro:include filename="$(find baxter_description)/urdf/baxter_base/baxter_base.urdf.xacro">
    <xacro:arg name="gazebo" value="${gazebo}"/>
  </xacro:include>
  <!--grippers-->
  <xacro:include filename="$(find baxter_description)/urdf/left_end_effector.urdf.xacro" />
  <xacro:include filename="$(find baxter_description)/urdf/right_end_effector.urdf.xacro" />

  <!--retrofit with Kinect and pedestal-->
  <xacro:include filename="$(find baxter_variations)/kinect_link.urdf.xacro" />
  <xacro:include filename="$(find baxter_variations)/baxter_pedestal.xacro" />

  <!-- attach baxter torso to the pedestal -->
  <!-- results in torso 0.760 above ground plane-->
    <link name="world"/>
    <joint name="baxter_base_joint" type="fixed">
```

```
    <parent link="base_link" />
    <child link="base" />
    <origin rpy="0 0 0 " xyz="0 0 0.91"/>
  </joint>
  <joint name="glue_base_to_world" type="fixed">
    <parent link="world" />
    <child link="base_link" />
    <origin rpy="0 0 0 " xyz="0 0 0"/>
  </joint>

</robot>
```

In the above, three files (the base and two grippers) are included from the package `baxter_description`, which is a model of the Baxter robot provided by Rethink Robotics. Two additional included files are from the package `baxter_variations` (within the accompanying repository) that are a simple pedestal (a rectangular prism) and a Kinect model. The file `kinect_link.urdf.xacro` is essentially identical to `simple_kinect_model2.xacro` used in the object finder of Section 8.5, except this model is not attached to the world. (It is only attached to a link called `base_link`.) The `baxter_pedestal.xacro` model constitutes the `base_link`. The two joints declared in `baxter_on_pedestal.xacro` rigidly attach the Baxter base to the pedestal and the pedestal to the world. This combination of models integrates the Kinect sensor (which is used in the object-finder package) with the robot and grippers (which are used in the object-grabber package).

This model can be launched by first bringing up an empty world:

```
roslaunch gazebo_ros empty_world.launch
```

then launching:

```
roslaunch baxter_variations baxter_on_pedestal_w_kinect.launch
```

This will bring up the Baxter model, mounted on a pedestal and augmented with a Kinect sensor and located in front of a table holding a block, as shown in Fig 16.3. (Note: if the block has fallen off the table, reset its position via the Gazebo GUI, under **edit**, choose "Reset Model Poses.")

Once the Baxter simulator is ready, control nodes can be launched with:

```
roslaunch coordinator coord_vision_manip.launch
```

This launch file starts 18 nodes, including six action server nodes, four service server nodes, a marker-display listener node, two static-transform publishers, `rviz`, and four simple nodes that run to a quick conclusion. Most of these nodes were introduced earlier.

The four nodes that run to conclusion perform actions to enable Baxter's motors, set the gripper ID on the parameter server, command the head-pan to zero angle (in case it starts up rotated, occluding Kinect vision), and run a playfile to move the arms to an initial pose (with via points that avoid hitting the cafe table).

The two static-transform publisher nodes are brought in by including a launch file from the `cartesian_planner` package. These publishers establish frames `generic_gripper_frame` and `system_ref_frame`.

An `rviz` display is started with reference to a pre-defined configuration file. The `triad_display` node from the `example_rviz_marker` package is started up, which assists in displaying object-finder results within `rviz`. The four services started include the Baxter

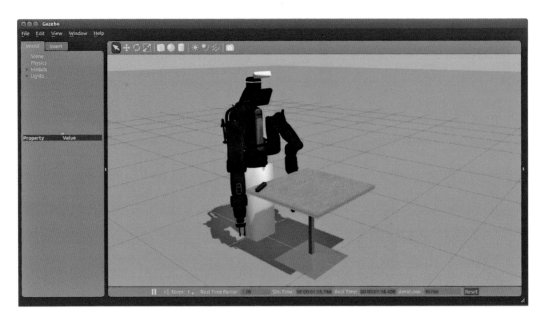

Figure 16.3: Result of launching `baxter_on_pedestal_w_kinect.launch`

`playfile` service (which is not necessary, but potentially useful); the object-manipulation query service introduced in Section 15.2; the generic gripper service that controls Baxter's right gripper, as introduced in Section 15.3; and a `set_block_state` service, from the `example_gazebo_set_state` package. This latter service is not necessary, but it can be useful for running manipulation tests. Each time this service is called, it re-sets the toy-block model to a random pose, constrained to lie on the table and within reach of the robot.

With assistance from the above nodes and services, the primary work of the vision and manipulation system is performed by six action servers. Four servers support object manipulation, as illustrated in Fig 15.1 in Section 15.1. These include trajectory streamers for both the left and right arms (as described in Section 14.5), a Cartesian move action server (as described in Section 14.9) and an object-grabber action server (Section 15.4). The object-grabber action server awaits goals to perform manipulations.

To combine perception and manipulation, an object-finder action server (from Section 8.5) is also launched. The server awaits goals to find object poses.

The sixth action server, introduced here, is a `command_bundler` from the `coordinator` package.

With these nodes launched, the Baxter robot will be in the pose shown in Fig 16.4.

In this pose, the arms are pre-positioned to prepare to descend on objects from above, without hitting the table. Further, the arms are positioned to avoid occluding the view of the Kinect sensor. The Kinect point cloud is displayed in `rviz`, from which a block on the table is apparent.

To invoke object perception and manipulation, a client of the coordinator can be started with:

```
rosrun coordinator acquire_block_client
```

This client asks the coordinator to: find the table surface, perceive a block, plan and execute grasp of the block, and move the arms (with grasped block) to the pre-pose position. Figure 16.5 shows the result of a request to locate a toy block on the table. The object finder

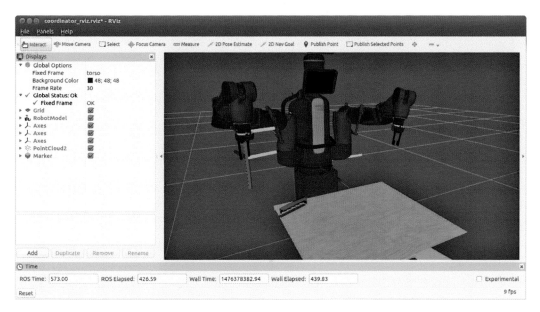

Figure 16.4: Result of launching `coord_vision_manip.launch`

computes the location of the block from point-cloud data, and it uses the `triad_display` node to place a marker in `rviz` to illustrate the result of this computation. An example result of this step is shown in Fig 16.5. This behavior is consistent with the object finder described in Section 8.5. The only difference is that the Kinect is mounted to the Baxter robot, and the frame of the Kinect sensor is knowable with respect to Baxter's torso frame (since they are both connected to a defined `base_link` associated with the pedestal).

Given the object frame, the next objective is to grab the object. This action is invoked similar to the example `example_object_grabber_action_client` described in Section 15.5. Code from this example was incorporated in the `command_bundler` action server to invoke the GRASP goal for the object-grabber action server. The robot's arm moves to approach from above (based on the perceived object coordinates and recommendation from the object-manipulation query service), opens the gripper fingers (using the generic gripper service interface), descends to the grasp pose, closes the gripper, and departs from the table with the part grasped.

Figure 16.6 shows an intermediate state of this process, in which the gripper is positioned at the grasp pose. The axes displayed in `rviz` show that the gripper and object frames have the desired grasp transform relationship.

Once the object is grasped, the robot performs a depart move, corresponding to lifting the object in the z direction. The result of this move is shown in Fig 16.7. A Cartesian move is appropriate for this step so that the robot does not bump the object against the table and lose grasp.

After a Cartesian-move departure from the table, the robot performs a joint-space move to a defined pre-pose (still grasping the block).

A second action client can be invoked to drop off the block, with:

```
rosrun coordinator dropoff_block_client
```

This client specifies a hard-coded drop-off pose for the object. This move relies on the object-grabber action server to invoke a joint-space motion to the approach pose, Cartesian

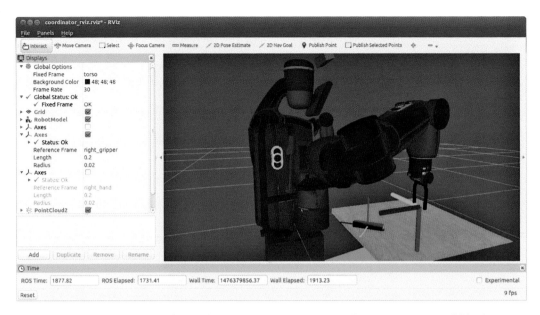

Figure 16.5: Result of coordinator action service invoking perception of block

descent to the drop-off pose, opening of the gripper, and Cartesian withdrawal from the drop-off pose.

Figure 16.8 shows the robot at the drop-off pose, just before release.

After opening the gripper and performing a Cartesian-move departure in the vertical direction, the robot is commanded to return to its pre-pose position.

Clients of the command bundler action server receive results back from the action server. If any errors are reported, the client has the opportunity to examine these codes and attempt error recovery.

A third client node, `coordinator_action_client_tester`, combines object pick-up and drop-off in a loop. This action client can be run with:

```
rosrun coordinator coordinator_action_client_tester
```

This client node runs a continuous loop sending goals to the coordinator and logging results. It repeatedly asks the coordinator to perceive a block, plan and execute grasp of the block, and place the block at specified coordinates. After each result from a goal request, the client evaluates the result, logs data regarding failures, then invokes the `set_block_state` service to reposition the block (at a randomized, but reachable pose) on the table to prepare for another iteration.

Having introduced the components of integrated perception and manipulation, we now examine some of the details within the code.

The launch file `coord_vision_manip.launch` in the `coordinator` package consolidates a variety of behaviors useful for perception and manipulation. The node `command_bundler.cpp` has action clients of both an object-grabber action server, as described in Section 15.4, and an object-finder action server, as described in Section 8.5. The command bundler node provides an action server, `manip_task_action_service`, that communicates with action clients via the `ManipTask.action` message. This action message defines goal components as follows:

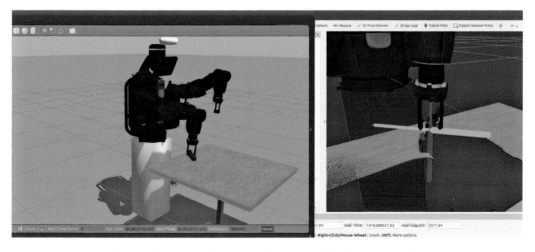

Figure 16.6: Result of coordinator action service invoking move to grasp pose

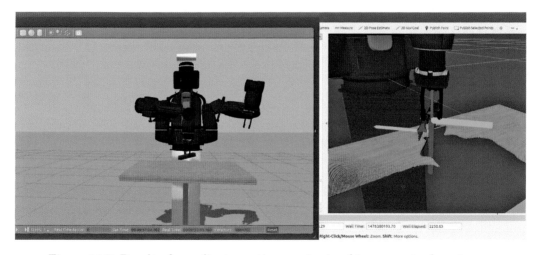

Figure 16.7: Result of coordinator action service invoking move to depart pose

```
#goal specification:
int32 action_code #what should we do with the named object?
int32 object_code #refer to a-priori known object types by object-ID codes
geometry_msgs/PoseStamped pickup_frame #specify object coords for pickup
geometry_msgs/PoseStamped dropoff_frame #specify desired drop-off
                                        #coords of object's frame
int32 perception_source  #e.g. name a camera source
```

The `action_code` field selects one of the capabilities of the action server within node `command_bundler.cpp`. The action codes, defined within the goal field of the ManipTask.action message, include: `FIND_TABLE_SURFACE`, `GET_PICKUP_POSE`, `GRAB_OBJECT`, `DROPOFF_OBJECT`, and `MOVE_TO_PRE_POSE`. The simplest of these action codes is `MOVE_TO_PRE_POSE`, which merely commands the robot to raise its arms to a pose that does not interfere with the Kinect sensor's view.

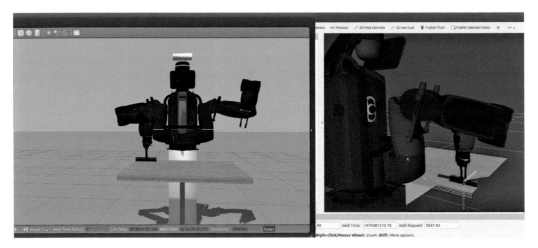

Figure 16.8: Result of coordinator action server invoking object drop-off

The action code `FIND_TABLE_SURFACE` invokes functionality of the object-finder action server that identifies the height of a horizontal surface. Identifying this surface makes it simpler to find objects sitting on the surface.

Most actions are performed with respect to a specific object. An object is referred to by a unique identifier code in the `object_code` field of the goal message. These codes are defined in the `object_manipulation_properties/object_ID_codes.h` header file.

The action code `GET_PICKUP_POSE` requires additional specifications. The object to be grasped must be specified via an object ID in the `object_code` field. Further, the field `perception_source` must be set to a perception source ID (which is currently defined only as `PCL_VISION`). If `PCL_VISION` is specified, the object finder is invoked to find the specified object on the table surface. In this case, if the object finder successfully locates the specified object, the object pose is returned in the `result` message in the field `object_pose`.

The action code `GRAB_OBJECT` invokes motion planning and execution that includes a joint-space move to an object-approach pose, a Cartesian motion with open gripper to straddle the object with the gripper fingers, grasp of the object, and Cartesian departure from the grasp pose. The `GRAB_OBJECT` action requires specification of the object ID, which is necessary for the motion planner to infer the desired gripper pose based on the object pose and a desired grasp transform.

These actions are illustrated in the client node `acquire_block_client.cpp` in the `coordinator` package. This node starts by sending the arms to a hard-coded pre-pose, which has the arms out of the way of the Kinect's view, and has the grippers positioned to prepare for manipulation. The following lines invoke this behavior:

```
ROS_INFO("sending a goal: move to pre-pose");
g_goal_done = false;
goal.action_code = coordinator::ManipTaskGoal::MOVE_TO_PRE_POSE;
action_client.sendGoal(goal, &doneCb, &activeCb, &feedbackCb);
while (!g_goal_done) {
    ros::Duration(0.1).sleep();
}
if (g_callback_status != coordinator::ManipTaskResult::MANIP_SUCCESS)
{
    ROS_ERROR("failed to move quitting");
    return 0;
}
```

The action code MOVE_TO_PRE_POSE requires no other goal fields to be populated. After sending the goal, the main program inspects the global variable g_goal_done, which will be set by the action client's callback function when the action server returns a result message. The action client's callback function will receive a result message from the action server. The result field manip_return_code should return MANIP_SUCCESS if the goal was successfully accomplished. This value is inspected, and the client program quits if the goal was not successful.

Next, the acquire_block_client node invokes the find-table behavior with the lines:

```
//send vision request to find table top:
ROS_INFO("sending a goal: seeking table top");
g_goal_done = false;
goal.action_code = coordinator::ManipTaskGoal::FIND_TABLE_SURFACE;

action_client.sendGoal(goal, &doneCb, &activeCb, &feedbackCb);
while (!g_goal_done) {
    ros::Duration(0.1).sleep();
}
```

For finding the table top, only the action code is required in the goal message. As with the MOVE_TO_PRE_POSE goal, after the goal is sent, the main program inspects g_goal_done, which will be set by the action client's callback function when the action server returns a result message.

Next, the action client sends a goal to find the pose of a toy block object on the table surface, using the following lines:

```
//send vision goal to find block:
ROS_INFO("sending a goal: find block");
g_goal_done = false;
goal.action_code = coordinator::ManipTaskGoal::GET_PICKUP_POSE;
goal.object_code= ObjectIdCodes::TOY_BLOCK_ID;
goal.perception_source = coordinator::ManipTaskGoal::PCL_VISION;
action_client.sendGoal(goal, &doneCb, &activeCb, &feedbackCb);
while (!g_goal_done) {
    ros::Duration(0.1).sleep();
}
if (g_callback_status!= coordinator::ManipTaskResult::MANIP_SUCCESS)
{
    ROS_ERROR("failed to find block quitting");
    return 0;
}
g_object_pose = g_result.object_pose;
```

In this case, use of point-cloud processing of the Kinect data (perception-source PCL_VISION) is specified within the goal message. If finding the object is successful, the result message returned to the action client callback function will contain the object's pose. In the current example, failure to find the object causes the action client node to quit. More generally, a recovery behavior could be invoked, e.g. to look elsewhere for the object.

Assuming the object was found, the robot is next commanded to grasp the object, using the following lines:

```
//send command to acquire block:
ROS_INFO("sending a goal: grab block");
g_goal_done = false;
goal.action_code = coordinator::ManipTaskGoal::GRAB_OBJECT;
goal.pickup_frame = g_result.object_pose;
goal.object_code= ObjectIdCodes::TOY_BLOCK_ID;
action_client.sendGoal(goal, &doneCb, &activeCb, &feedbackCb);
while (!g_goal_done) {
    ros::Duration(0.1).sleep();
}
```

```
        if (g_callback_status!= coordinator::ManipTaskResult::MANIP_SUCCESS)
    {
        ROS_ERROR("failed to grab block; quitting");
        return 0;
    }
```

In the above, the goal action code is set to **GRAB_OBJECT**, the object ID is set and its pose is specified (per the result returned by the prior vision call). As in the previous instances, the main program examines the **g_goal_done** flag to determine when the action server has finished. If the return code does not indicate success, the program quits. Alternatively, error recovery could be attempted (*e.g.* to look for the part again and re-attempt grasp).

If the object is successfully grasped, the **MOVE_TO_PRE_POSE** behavior is again invoked, and this action client then concludes.

A second illustrative action client is **dropoff_block_client.cpp**. This node hard-codes a drop-off pose, then uses it within the **DROPOFF_OBJECT** behavior with the following lines:

```
    g_goal_done = false;
    goal.action_code = coordinator::ManipTaskGoal::DROPOFF_OBJECT;
    goal.dropoff_frame = dropoff_pose; //pre-defined pose
    goal.object_code= ObjectIdCodes::TOY_BLOCK_ID; //assumes robot is holding ↵
        TOY_BLOCK object
    action_client.sendGoal(goal, &doneCb, &activeCb, &feedbackCb);
    while (!g_goal_done) {
        ros::Duration(0.1).sleep();
    }
        if (g_callback_status!= coordinator::ManipTaskResult::MANIP_SUCCESS)
    {
        ROS_ERROR("failed to drop off block; quitting");
        return 0;
    }
```

In the above, the goal field **dropoff_frame** is set, as well as the object ID. The drop-off frame refers to the desired pose of the grasped object, and thus the object ID is required to deduce the corresponding tool-flange pose from the associated grasp transform. As before, the **g_goal_done** flag is polled, then the result code is inspected to evaluate whether the drop-off was successful.

The node **coordinator_action_client_tester** illustrates combining object acquisition and object drop-off. The goal **dropoff_frame** is set with hard-coded values. Perception, grasp and drop-off are called repeatedly, using lines of code identical to those in the pick-up and drop-off action client examples. After each attempt, the block pose is randomized, and the perception, grasp and placement actions are repeated.

16.3   WRAP-UP

In this chapter, perception and manipulation were combined to enable vision-based manipulation. The previously introduced object-finder action server and object-grabber action server were used together, coordinated by a command bundler. For vision-based manipulation to work, it is important that the imaging source is calibrated to the robot. It was shown that **rviz** can be used to visualize the quality of extrinsic camera calibration, and it can be used to perform calibration interactively.

The example command bundler action server and the corresponding action client nodes presented here illustrate how to integrate perception and manipulation. Even greater flexibility can be realized if the manipulator is mobile, as discussed next.

Mobile Manipulation

CONTENTS

I NTRODUCTION: Mobile manipulation can be illustrated by combining the Baxter model with our previous mobot mobile platform. The mobile manipulator can then take advantage of all of the developments presented in sections III (perception), IV (mobility) and V (manipulation).

17.1   MOBILE MANIPULATOR MODEL

The mobile-manipulator model is contained in the file `baxter_on_mobot.xacro` in the package `baxter_variations`. The contents of this model file are:

```
<?xml version="1.0"?>
<robot
  xmlns:xacro="http://www.ros.org/wiki/xacro" name="baxter_on_mobot">
  <xacro:include filename="$(find baxter_variations)/mobot_base.xacro" />

  <xacro:include filename=
    "$(find baxter_variations)/baxter_base.urdf.xacro">
    <xacro:arg name="gazebo" value="${gazebo}"/>
  </xacro:include>
  <xacro:include filename=
    "$(find baxter_description)/urdf/left_end_effector.urdf.xacro" />
  <xacro:include filename=
    "$(find baxter_description)/urdf/right_end_effector.urdf.xacro" />
  <xacro:include filename=
    "$(find baxter_variations)/kinect_link.urdf.xacro" />

  <!-- attach baxter torso to the mobile robot -->
  <joint name="baxter_base_joint" type="fixed">
    <parent link="mobot_top" />
    <child link="base" />
    <origin rpy="0 0 0 " xyz="0.1 0 0.06"/>
  </joint>
</robot>
```

This model brings together a mobile base model (`mobot_base.xacro`), the Baxter model, left and right end effectors, and a Kinect model. The Baxter model is a modified version of the original with the pedestal removed (as well as multiple sensors disabled). The mobile base is nearly identical to the mobot-with-Lidar model in package `mobot_urdf` (described in Section II). However, the height of the mobot is elevated approximately to the same height as Baxter's pedestal.

The Baxter model `base` frame is attached to the `mobot_top` frame of the mobile base using a fixed `baxter_base_joint`, defined in the `baxter_on_mobot.xacro` model file.

A Kinect sensor is added to the robot model by including the Kinect model file `kinect_link.urdf.xacro` from package `baxter_variations`. This model file is the same file that was included in `baxter_on_pedestal.xacro` (also in the `baxter_variations`) in Section 16.2. The Kinect model file contains a joint that attaches the Kinect link to Baxter's torso frame.

The combined mobile base and Baxter robot appear as in Fig 17.1.

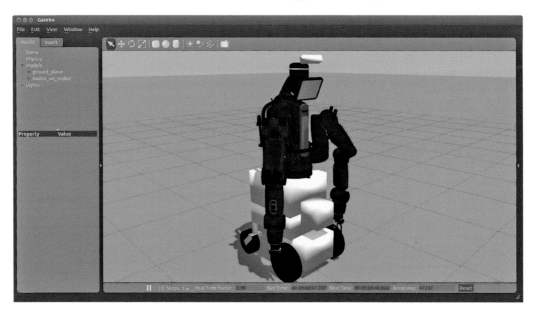

Figure 17.1: Gazebo view of a mobile manipulator model

This model includes the functionalities of the Baxter robot, the Kinect sensor and a mobile base with a LIDAR sensor. The associated sensors and controls support map making, localization, and navigation, as well as 3-D point cloud perception and robotic manipulation.

17.2 MOBILE MANIPULATION

Mobile manipulation is illustrated using the following procedure. First, launch Gazebo and the `baxter_on_mobot` model with:

```
roslaunch baxter_variations baxter_on_mobot.launch
```

(Note: mobot uses a LIDAR Gazebo plug-in, which requires a GPU or suitable emulation.) Next, start the manipulator controls. Wait for simulator to stabilize. Then, launch a variety of nodes and services, including the command bundler with:

```
roslaunch coordinator command_bundler.launch
```

This launch file is nearly identical to that used in Section 16.2, except that it includes nodes for mobility. The file `baxter_variations/mobot_startup_navstack.launch` is included. This launch file starts the AMCL localizer, loads the costmap and `move_base` parameters, and starts the `move_base` node. Two utility services are also started: `open_loop_nav_service` and `open_loop_yaw_service`, which enable simple open-loop control over the base (*e.g.* for backing up a defined distance).

The mobile-manipulation process can be commanded in increments by running a sequence of nodes. Commanding:

```
rosrun coordinator acquire_block_client
```

invokes object perception and grasp, as in Section 16.2. Once the block has been grasped, the mobile base can be commanded to move backwards 1 m with the manual command:

```
rosservice call open_loop_nav_service -- -1
```

then it can be commanded to rotate counter-clockwise 1 radian with the service call:

```
rosservice call open_loop_yaw_service 1
```

These commands are used here to simplify the planning process within `move_base`. When the robot is close to the table, the `move_base` planner may consider the robot to be in a fatal pose from which the planner cannot recover. Further, the planner has some difficulty invoking reverse motion, which is essential for the robot to depart from its close approach to the table. By inserting manual motion commands to clear the robot from the table and point toward the exit, subsequent automated planning is simpler and more robust.

At this point, automated planning and driving can be initiated with:

```
rosrun example_move_base_client example_move_base_client
```

This client requests that the mobile base move from its current pose to a goal pose near and facing a second table located outside the pen.

Once the robot has approached the second table, an open-loop command to approach the table closer can be invoked with:

```
rosservice call open_loop_nav_service 0.7
```

This command assumes that the `move_base` process successfully achieved a pose facing the second table, offset by approximately 0.7 m (to avoid perception of entering a fatal region). (This open-loop presumption will be relaxed later.)

With the robot having approached the second table, a block drop-off command can be invoked, using a hard-coded destination with respect to the robot's torso:

```
rosrun coordinator dropoff_block_client
```

The above incremental process illustrates the steps involved in mobile manipulation. A weakness, however, is that the pose of the robot with respect to the second table may be insufficiently precise for successful block drop-off. More robustly, one would invoke perception to fine-tune the approach pose and to perceive a drop-off location. Such improvements are illustrated in the node `fetch_and_stack_client`, which combines all of the above steps in addition to perception-based approach and drop-off. This node can be run with:

```
rosrun coordinator fetch_and_stack_client
```

The fetch-and-stack node includes action clients of both the command bundler (for perception and manipulation) and `move_base` (for navigation planning and execution). It also has clients of the open-loop base translation and yaw services. This node incorporates the separate commands introduced above, including block perception and grasp, open-loop base back-up and re-orientation, and invocation of navigation. In this node, the navigator is assisted by first commanding motion to a via point near the starting pen exit, then commanding approach to the second table. Once the `move_base` approach to the second table has concluded, the fetch-and-stack node consults tf to find the robot's base relative to the map frame (as computed by AMCL). From this pose, offset from the known map coordinates of the second table is computed, and the open-loop approach service is invoked for close approach to the table.

The initial and final operations of fetch-and-stack are illustrated in Figs 17.2 and 17.3. Figure 17.2 shows the mobile robot in both Gazebo and `rviz` views, positioned initially in front of a table with a block. The `rviz` view shows the robot's pose within a global cost map. The scatter of uncertainty of robot pose is initially fairly large, since the robot has not yet moved, and thus there is no corroborating evidence of its pose beyond the LIDAR view of its start-up pose. Nonetheless, grasp of the object can be accurate, since its pose is based on the robot's perception from Kinect data.

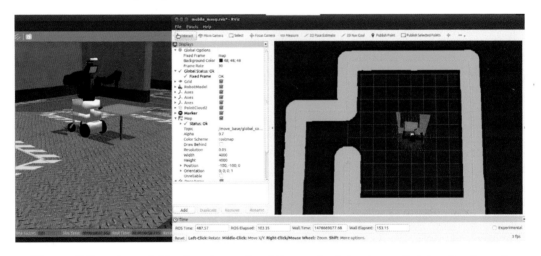

Figure 17.2: `rviz` and Gazebo views of mobile manipulator immediately after launch

Figure 17.3 shows the conclusion of the fetch-and-stack client, after the robot has followed its navigation plan, approached the second table, found the block on the second table, and stacked the grasped block on top. The `rviz` view in Fig 17.3 shows the computed path (thin blue line) and LIDAR pings (red spheres) aligned with the wall to the robot's left, which allows localization.

Lines 275 through 291 of `fetch_and_stack_client.cpp`, shown below, perform final approach to the second table. Desired coordinates of approach are known, and the tfListener is consulted to find the pose of the robot at the end of the `move_base` command (which is deliberately short of the final destination). The open-loop yaw service is invoked to correct

Figure 17.3: `rviz` and Gazebo views of mobile manipulator stacking fetched block

for any orientation misalignment, and the open-loop translation service is used to advance the robot to a computed distance from the table.

```
tfListener.lookupTransform("map","base_link", ros::Time(0), tfBaseLinkWrtMap);
current_pose = xform_utils.get_pose_from_stamped_tf(tfBaseLinkWrtMap);
xform_utils.printStampedPose(current_pose);
yaw = xform_utils.convertPlanarQuat2Phi(current_pose.pose.orientation);
ROS_INFO("yaw = %f",yaw);
//desired yaw is -90 deg = -1.57 rad
openLoopNavSvcMsg.request.move_distance= -1.57-yaw;
nav_yaw_client.call(openLoopNavSvcMsg);
//and move forward as well:
 tfListener.lookupTransform("map","base_link", ros::Time(0), tfBaseLinkWrtMap);
current_pose = xform_utils.get_pose_from_stamped_tf(tfBaseLinkWrtMap);
xform_utils.printStampedPose(current_pose);

ROS_INFO("approaching table 2");
//should be pointing in -y direction w/rt map; compute how much to creep up on ↩
    table:
openLoopNavSvcMsg.request.move_distance= -(table2_y_des  -current_pose.pose.↩
    position.y);
nav_move_client.call(openLoopNavSvcMsg);
```

To illustrate perception-based drop-off, previous block perception code is re-used. (The launch file has placed a block on this table.) The fetch-and-stack node requests perception of a block on the second table (lines 301 through 304):

```
goal.action_code = coordinator::ManipTaskGoal::GET_PICKUP_POSE;
goal.object_code= TOY_BLOCK_ID;
goal.perception_source = coordinator::ManipTaskGoal::PCL_VISION;
action_client.sendGoal(goal, &doneCb, &activeCb, &feedbackCb);
```

The fetch-and-stack client then computes the coordinates corresponding to stacking the grasped block on top of the existing block, then commands drop-off at these coordinates. This computation is performed on lines 325 through 330:

```
goal.action_code = coordinator::ManipTaskGoal::DROPOFF_OBJECT;
goal.dropoff_frame = g_object_pose; //frame per PCL perception
goal.dropoff_frame.pose.position.z+=0.035; //set height to one block thickness ↩
    higher
```

```
                                      // so new block will stack on prior  ↵
                                   block
    goal.object_code= ObjectIdCodes::TOY_BLOCK_ID;
    action_client.sendGoal(goal, &doneCb, &activeCb, &feedbackCb);
```

The result is that the robot successfully places the grasped block on top of the perceived block. More generally, one might perceive a tray or specific packaging and deduce placement coordinates from this view.

17.3   WRAP-UP

The example described in this section integrated elements described throughout this text. All four forms of ROS communications are used: parameter server, publish and subscribe, services and action servers. Robot simulation, sensor simulation, robot modeling and visualization are used to combine a mobile platform, a dual-arm robot and a Kinect sensor. Perceptual processing is used to locate objects of interest. Arm motion plans are computed and executed to manipulate objects. Navigation is performed with respect to a map using sensor-based localization, global motion planning, local motion planning and vehicle driving. Collectively, these components enable a mobile manipulator that uses its sensors to plan and act to accomplish specified goals.

The example covered here is intentionally minimalist for illustrative purposes. Many improvements can and should be made. The perceptual system has been demonstrated only to recognize a single, isolated object on a horizontal surface of approximately known height. The object-manipulation query service is a mere skeleton of a more capable future system populated with many object and gripper combinations. Execution of navigation plans is performed with poor driving precision, which could be improved with incorporation of precision steering algorithms. Most importantly, the example code does not incorporate error detection and correction. A practical system would be an order of magnitude larger to provide suitable error testing and contingency planning.

In spite of the many simplifications and limitations, the examples show the potential for ROS to simplify building large, complex robotic systems. ROS's use of packages, nodes, and messaging options encourages modularization, code re-use, collaboration, and ease of extensibility and testability. With these virtues, ROS promises hope for a foundation on which future, highly complex and highly capable robotic systems can be built.

## CHAPTER 18

Conclusion

Although the first industrial robot was installed over 50 years ago, progress in the field has been slow. In large part, this is because isolated efforts were built as unique systems, and little of this work was reusable in subsequent systems. Given the difficulty of building a complex, intelligent system, individual efforts struggled to surpass the achievements of earlier projects. With the advent of ROS and its acceptance by roboticists, current robotic systems are developed much more rapidly than in the past, enabling more attention on pushing back frontiers of robot competence. With ROS's communications infrastructure, separate but integrated nodes can be run concurrently, and these nodes can be distributed easily across multiple computers. These nodes can be contributions from collaborators world-wide. Specific algorithms shown to have world-leading performance can be adopted and incorporated in ROS-compatible systems easily and rapidly. Further, ROS leverages capabilities of independent open-source projects, including OpenCV, the Point Cloud Library, the Eigen library, and Gazebo (with underlying open-source physics engines).

This text has reviewed ROS in a structured presentation, starting with its foundations in communications. The concept of communication among nodes was covered in Section I, including the paradigms of publish and subscribe, services and clients, action servers and action clients and the parameter server. Tools to assist development and debugging include `rosrun`, `roslaunch`, `rosbag`, `rqt_plot`, `rqt_reconfigure` and `rqt_console`. These tools, together with the underlying communications, help developers create robotic systems faster than ever before.

Section II introduced simulation and visualization in ROS. A unified robot description format supports ease of robot modeling, including kinematics, dynamics, visualization and physical interaction (contacts and collisions). A key component of ROS's simulation capabilities is the ability to simulate the physics of sensors, including torque sensors, accelerometers, LIDAR, cameras, stereo cameras and depth cameras. Further, the open source of Gazebo simulation and `rviz` visualization supports additional plug-ins to extend these capabilities. `rviz` supports robot system development by providing visualization of robot models together with their sensory values (*e.g.* point-cloud displays) and visualization of plans (*e.g.* navigation plans within cost maps). In addition to providing sensory display, `rviz` can be used as a human interface. An operator can interact directly with displayed data, providing the robotic system with focus of attention or context. User-definable markers can help display the robot's "thoughts," which can help with debugging or validation of plans in a supervisory-controlled system. Interactive markers can also be constructed, allowing an operator to specify 6-D poses of interest.

Section III surveyed sensory interpretation, including camera calibration, use of OpenCV, stereo imaging, 3-D LIDAR sensing and depth cameras. Use of sensors is necessary for mapping, navigation, collision avoidance, object sensing and localization, manipulation and error detection and recovery. Perceptual processing is a large field, and the present introduction is not intended to teach machine vision or point-cloud processing in general. However, it is an asset of ROS that it integrates smoothly with perception, including bridges to OpenCV and PCL. It is also key that ROS provides extensive support for coordinate transformations. Coordinate transforms are essential for calibrating sensors for navigation, collision avoidance and hand–eye coordination. ROS's `tf` package and `tfListener` are used to integrate data for display in `rviz`, as well as support navigation and robot arm kinematics. An object-finder action server is described, in which specific, modeled objects can be recognized and localized. This example package is limited in its capabilities, but it illustrates functionality for more general perception packages.

Section IV reviewed ROS's support of mobile robots. The navigation stack is one of the popular successes of ROS. It integrates map-making, localization, global planning, local planning and steering. Nav-stack modules that work together include some of the best algorithms from researchers around the world. At the same time, improvements can be incorporated, for the field as a whole or for targeted, specific systems.

Section V presented ROS in the context of robot arm planning and control. At the joint-space level, the concept of the trajectory message unifies robot interfacing across a wide variety of common and novel robots, and the parallel ROS Industrial effort extends this to a growing base of industrial robots. The joint-level interface supports teach and playback–a common industrial programming approach. Perception-based manipulation, however, requires online kinematic planning. While the fields of robot kinematics (forward and reverse) and kinematic planning are too broad to be covered in this text, it is shown how kinematic libraries and kinematic planning can be implemented in ROS. Specific examples are provided using a realistic robot simulation of the Baxter robot. An object-grabber action server was introduced, which performs kinematic planning and execution in the context of manipulation of objects with known, desirable grasp transforms. It was shown how such functionality can be constructed to abstract manipulation programming at the task level. With such abstraction, manipulation programs can be re-usable across different types of robots with different types of tooling or grippers.

In Section VI, the focus was on system integration. Robot and sensor modeling from Section II was applied to integrate a dual-arm robot on a mobile base, together with LIDAR sensing for navigation and depth sensing for manipulation. The object finder of Section III, the nav-stack implementation from Section IV and the object-grabber package of Section V were integrated within a fetch-and-stack example. Using these capabilities, the robot was able to perceive an object, plan and execute grasp of the object, navigate to a drop-off destination, perceive a drop-off location, and place the grasped object accordingly. This demonstration is indicative of emerging applications in automated storage and retrieval, industrial kitting operations, logistics for filling customer orders, and future applications in domestic robotics.

While this presentation has surveyed wide-ranging aspects of robotics, it has barely scratched the surface of possibilities. ROS has thousands of open-source packages, allowing developers to build novel systems rapidly and incorporate specific expertise without requiring the developer to be an expert in each aspect. Additional capabilities and upgrades continue to emerge from contributors world-wide. Higher-level controls, including state machines and decision trees for more abstract planning, have not been addressed here. Sophisticated object recognition, grasp planning, environment modeling and kinematic planning in 3-D are available in ROS. The MoveIt! environment, which consolidates many of these components, has not been described. Nonetheless, it is the author's hope that this text will prepare the reader to be a more effective learner. The online tutorials on ROS offer extensive details not covered here, but this presentation should enable the reader to learn more effectively from these tutorials. Further, by building and dissecting the code accompanying this text, it is hoped that the reader will be more capable of adopting existing ROS packages, using and modifying them as necessary and contributing new packages and tools that will further help advance the field of robotics.

Bibliography

[1] C. E. Agero, N. Koenig, I. Chen, H. Boyer, S. Peters, J. Hsu, B. Gerkey, S. Paepcke, J. L. Rivero, J. Manzo, E. Krotkov, and G. Pratt. Inside the virtual robotics challenge: Simulating real-time robotic disaster response. *IEEE Transactions on Automation Science and Engineering*, 12(2):494–506, April 2015.

[2] H. H. An, W. I. Clement, and B. Reed. Analytical inverse kinematic solution with self-motion constraint for the 7-dof restore robot arm. In *2014 IEEE/ASME International Conference on Advanced Intelligent Mechatronics*, pages 1325–1330, July 2014.

[3] Haruhiko Asada and Jean-Jacques E. Slotine. *Robot analysis and control.* J. Wiley and Sons, New York, 1986. A Wiley Interscience publication.

[4] Gary Bradski and Adrian Kaehler. *Learning OpenCV.* O'Reilly Media Inc., 2008.

[5] DARPA Urban Challenge. `http://archive.darpa.mil/grandchallenge/`.

[6] Boston Dynamics. `http://www.bostondynamics.com/robot_Atlas.html`.

[7] J.F. Engelberger, D. Lock, and K. Willis. *Robotics in Practice: Management and Applications of Industrial Robots.* AMACOM, 1980.

[8] Open Dynamics Engine. `http://www.ode.org/`.

[9] C. Fitzgerald. Developing baxter. In *2013 IEEE Conference on Technologies for Practical Robot Applications (TePRA)*, pages 1–6, April 2013.

[10] Open Source Robotics Foundation. `http://www.osrfoundation.org/`.

[11] J. Funda and R. P. Paul. A comparison of transforms and quaternions in robotics. In *Proceedings. 1988 IEEE International Conference on Robotics and Automation*, pages 886–891 vol.2, Apr 1988.

[12] Willow Garage. `http://www.willowgarage.com/`.

[13] Patrick Goebel. *ROS By Example.* Lulu, April 2013.

[14] L. Gomes. When will google's self-driving car really be ready? it depends on where you live and what you mean by "ready" [news]. *IEEE Spectrum*, 53(5):13–14, May 2016.

[15] V. Hayward and R. Paul. Introduction to rccl: A robot control amp;c amp; library. In *Proceedings. 1984 IEEE International Conference on Robotics and Automation*, volume 1, pages 293–297, Mar 1984.

[16] P. Kazanzides, Z. Chen, A. Deguet, G. S. Fischer, R. H. Taylor, and S. P. DiMaio. An open-source research kit for the da vinci; surgical system. In *2014 IEEE International Conference on Robotics and Automation (ICRA)*, pages 6434–6439, May 2014.

[17] Desmond King-Hele. Erasmus darwin's improved design for steering carriages–and cars. *Notes and Records of the Royal Society of London*, 56(1):41–62, 2002.

[18] DaVinci Research Kit. `https://github.com/jhu-dvrk/dvrk-ros`.

[19] Nathan Koenig and Andrew Howard. Design and use paradigms for gazebo, an open-source multi-robot simulator. In *IEEE/RSJ International Conference on Intelligent Robots and Systems*, pages 2149–2154, Sendai, Japan, Sep 2004.

[20] Eigen Library. `http://eigen.tuxfamily.org/`.

[21] PointCloud Library. `http://pointclouds.org/`.

[22] Hokuyo LIDAR. `https://www.hokuyo-aut.jp/02sensor/07scanner/urg_04lx_ug01.html`.

[23] Sick LIDAR. `www.sick.com/us/en/product-portfolio/detection-and-ranging-solutions/3d-laser-scanners/c/g282752ation`.

[24] Labview Robotics Module. `http://www.ni.com/white-paper/11564/en/`.

[25] Richard M. Murray, Zexiang Li, and S. Shankar Sastry. *A mathematical introduction to robotic manipulation*. CRC Press, Boca Raton, 1994.

[26] Andrew Y. Ng, Stephen Gould, Morgan Quigley, Ashutosh Saxena, and Eric Berger. Stair: Hardware and software architecture. In *AAAI 2007 Robotics Workshop*, 2007.

[27] Jason M. O'Kane. *A Gentle Introduction to ROS*. Independently published, October 2013. Available at `http://www.cse.sc.edu/~jokane/agitr/`.

[28] OpenCV. `http://opencv.org/`.

[29] M. W. Park, S. W. Lee, and W. Y. Han. Development of lateral control system for autonomous vehicle based on adaptive pure pursuit algorithm. In *2014 14th International Conference on Control, Automation and Systems (ICCAS 2014)*, pages 1443–1447, Oct 2014.

[30] G. Pratt and J. Manzo. The darpa robotics challenge [competitions]. *IEEE Robotics Automation Magazine*, 20(2):10–12, June 2013.

[31] Yale Openhand Project. `https://www.eng.yale.edu/grablab/openhand/`.

[32] Selected Patch Publisher. `https://github.com/xpharry/publish_selected_patch`.

[33] Selected Points Publisher. `https://github.com/tu-rbo/turbo-ros-pkg`.

[34] Morgan Quigley, Brian Gerkey, and William D. Smart. *Programming Robots with ROS*. O'Reilly Media, 12 2015.

[35] Carnegie Robotics. `http://carnegierobotics.com/multisense-sl`.

[36] Bruno Siciliano, Lorenzo Sciavicco, and Luigi Villani. *Robotics : modelling, planning and control*. Advanced Textbooks in Control and Signal Processing. Springer, London, 2009. 013-81159.

[37] Baxter simulation model. `http://sdk.rethinkrobotics.com/wiki/Baxter_Simulator`.

[38] J. Solaro. The kinect digital out-of-box experience. *Computer*, 44(6):97–99, June 2011.

[39] IEEE Spectrum. `http://spectrum.ieee.org/automaton/robotics/robotics-software/microsoft-shuts-down-its-robotics-group`.

[40] Microsoft Robotics Studio. `http://www.techspot.com/downloads/3690-microsoft-robotics-studio.html`.

[41] Intuitive Surgical. `http://www.intuitivesurgical.com/`.

[42] Velodyne. `http://velodynelidar.com/`.

Index